Discrete Linear Control Systems

T0332577

Mathematics and Its Applications (*Soviet Series*)

Managing Editor:

M. HAZEWINKEL
Centre for Mathematics and Computer Science, Amsterdam, The Netherlands

Editorial Board:

A. A. KIRILLOV, *MGU, Moscow, U.S.S.R.*
Yu. I. MANIN, *Steklov Institute of Mathematics, Moscow, U.S.S.R.*
N. N. MOISEEV, *Computing Centre, Academy of Sciences, Moscow, U.S.S.R.*
S. P. NOVIKOV, *Landau Institute of Theoretical Physics, Moscow, U.S.S.R.*
M. C. POLYVANOV, *Steklov Institute of Mathematics, Moscow, U.S.S.R.*
Yu. A. ROZANOV, *Steklov Institute of Mathematics, Moscow, U.S.S.R.*

Volume 67

Discrete Linear Control Systems

by

V. N. Fomin
Leningrad University, U.S.S.R.

KLUWER ACADEMIC PUBLISHERS
DORDRECHT / BOSTON / LONDON

Library of Congress Cataloging-in-Publication Data

Fomin, V. N. (Vladimir Nikolaevich), 1937-
[Metody upravleniia lineĭnymi diskretnymi ob"ektami. English]
Discrete linear control systems / by V.N. Fomin.
p. cm. -- (Mathematics and its applications (Soviet series) ; v. 67)
Translation of: Metody upravleniia lineĭnymi diskretnymi ob"ektami.
Includes bibliographical references.
ISBN 0-7923-1248-1 (alk. paper)
1. Control theory. I. Title. II. Series: Mathematics and its applications (Kluwer Academic Publishers). Soviet series ; 67.
QA402.3.F638 1991
629.8'312--dc20 91-12990

ISBN 0-7923-1248-1

Published by Kluwer Academic Publishers,
P.O. Box 17, 3300 AA Dordrecht, The Netherlands.

Kluwer Academic Publishers incorporates
the publishing programmes of
D. Reidel, Martinus Nijhoff, Dr W. Junk and MTP Press.

Sold and distributed in the U.S.A. and Canada
by Kluwer Academic Publishers,
101 Philip Drive, Norwell, MA 02061, U.S.A.

In all other countries, sold and distributed
by Kluwer Academic Publishers Group,
P.O. Box 322, 3300 AH Dordrecht, The Netherlands.

Printed on acid-free paper

This is the translation of the original work
МЕТОДЫ УПРАВЛЕНИЯ ЛИНЕЙНЫМИ
ДИСКРЕТНЫМИ ОБЪЕКТАМИ
Published by Leningrad University Press, © 1985.

All Rights Reserved
© 1991 Kluwer Academic Publishers
No part of the material protected by this copyright notice may be reproduced or utilized in any form or by any means, electronic or mechanical, including photocopying, recording or by any information storage and retrieval system, without written permission from the copyright owner.

Printed in the Netherlands

SERIES EDITOR'S PREFACE

'Et moi, ..., si j'avait su comment en revenir, je n'y serais point allé.'

Jules Verne

The series is divergent; therefore we may be able to do something with it.

O. Heaviside

One service mathematics has rendered the human race. It has put common sense back where it belongs, on the topmost shelf next to the dusty canister labelled 'discarded non-sense'.

Eric T. Bell

Mathematics is a tool for thought. A highly necessary tool in a world where both feedback and non-linearities abound. Similarly, all kinds of parts of mathematics serve as tools for other parts and for other sciences.

Applying a simple rewriting rule to the quote on the right above one finds such statements as: 'One service topology has rendered mathematical physics ...'; 'One service logic has rendered computer science ...'; 'One service category theory has rendered mathematics ...'. All arguably true. And all statements obtainable this way form part of the raison d'être of this series.

This series, *Mathematics and Its Applications*, started in 1977. Now that over one hundred volumes have appeared it seems opportune to reexamine its scope. At the time I wrote

> "Growing specialization and diversification have brought a host of monographs and textbooks on increasingly specialized topics. However, the 'tree' of knowledge of mathematics and related fields does not grow only by putting forth new branches. It also happens, quite often in fact, that branches which were thought to be completely disparate are suddenly seen to be related. Further, the kind and level of sophistication of mathematics applied in various sciences has changed drastically in recent years: measure theory is used (non-trivially) in regional and theoretical economics; algebraic geometry interacts with physics; the Minkowsky lemma, coding theory and the structure of water meet one another in packing and covering theory; quantum fields, crystal defects and mathematical programming profit from homotopy theory; Lie algebras are relevant to filtering; and prediction and electrical engineering can use Stein spaces. And in addition to this there are such new emerging subdisciplines as 'experimental mathematics', 'CFD', 'completely integrable systems', 'chaos, synergetics and large-scale order', which are almost impossible to fit into the existing classification schemes. They draw upon widely different sections of mathematics."

By and large, all this still applies today. It is still true that at first sight mathematics seems rather fragmented and that to find, see, and exploit the deeper underlying interrelations more effort is needed and so are books that can help mathematicians and scientists do so. Accordingly MIA will continue to try to make such books available.

If anything, the description I gave in 1977 is now an understatement. To the examples of interaction areas one should add string theory where Riemann surfaces, algebraic geometry, modular functions, knots, quantum field theory, Kac-Moody algebras, monstrous moonshine (and more) all come together. And to the examples of things which can be usefully applied let me add the topic 'finite geometry'; a combination of words which sounds like it might not even exist, let alone be applicable. And yet it is being applied: to statistics via designs, to radar/sonar detection arrays (via finite projective planes), and to bus connections of VLSI chips (via difference sets). There seems to be no part of (so-called pure) mathematics that is not in immediate danger of being applied. And, accordingly, the applied mathematician needs to be aware of much more. Besides analysis and numerics, the traditional workhorses, he may need all kinds of combinatorics, algebra, probability, and so on.

In addition, the applied scientist needs to cope increasingly with the nonlinear world and the extra mathematical sophistication that this requires. For that is where the rewards are. Linear models are honest and a bit sad and depressing: proportional efforts and results. It is in the non-linear world that infinitesimal inputs may result in macroscopic outputs (or vice versa). To appreciate what I am hinting at: if electronics were linear we would have no fun with transistors and computers; we would have no TV; in fact you would not be reading these lines.

There is also no safety in ignoring such outlandish things as nonstandard analysis, superspace and anticommuting integration, p-adic and ultrametric space. All three have applications in both electrical engineering and physics. Once, complex numbers were equally outlandish, but they frequently proved the shortest path between 'real' results. Similarly, the first two topics named have already provided a number of 'wormhole' paths. There is no telling where all this is leading - fortunately.

Thus the original scope of the series, which for various (sound) reasons now comprises five sub-series: white (Japan), yellow (China), red (USSR), blue (Eastern Europe), and green (everything else), still applies. It has been enlarged a bit to include books treating of the tools from one subdiscipline which are used in others. Thus the series still aims at books dealing with:

- a central concept which plays an important role in several different mathematical and/or scientific specialization areas;
- new applications of the results and ideas from one area of scientific endeavour into another;
- influences which the results, problems and concepts of one field of enquiry have, and have had, on the development of another.

Mathematically, a control system (in discrete time) is something like a dynamical system with inputs or controls and observations. For instance, the set of equations $x_{t+1}=F(x_t,u_t)$, $y_t=h(x_t)$, $x_t\in\mathbf{R}^n$, $u_t\in\mathbf{R}^m$, $y_t\in\mathbf{R}^p$; though much more complicated situations are often considered and are relevant for applications. Here, u_t are the inputs or controls and y_t the observations. There may well be extra complications such as additional external noise; the equations may be stochastic equations; etc. Central problems in control theory involve finding feedback laws, $u_t=K(y_{t-1})$, such that the resulting closed loop system has some sort of prescribed behaviour or is stable, or .... The task of the control mathematician is to design such feedback laws and to do this also when there are extra difficulties such as in the case that some of the parameters of the system are uncertain or unknown (adaptive control).

Control theory as a specialism came out of the engineering world and much of the literature has an engineering flavour, particularly the monographic literature. That has made the field less known and less accessible to mathematicians. That is a great pity because the topic abounds with unsolved problems and mathematical challenges. In particular, there are all sorts of adaptive schemes that appear to work without any solid mathematical proof or understanding of them. Nor do we have the foggiest idea whether these schemes are in any way more or less optimal or not.

The present volume may do much to help bridge the gap. Written by a well known and powerful contributor to the topic, it gives an elegant and concise treatment of the central questions of control theory. The book is written by a mathematician for mathematicians in mathematical style and takes the reader efficiently to the point where he can start working on some of the many important open problems of the control world.

The shortest path between two truths in the real domain passes through the complex domain.

J. Hadamard

La physique ne nous donne pas seulement l'occasion de résoudre des problèmes ... elle nous fait pressentir la solution.

H. Poincaré

Never lend books, for no one ever returns them; the only books I have in my library are books that other folk have lent me.

Anatole France

The function of an expert is not to be more right than other people, but to be wrong for more sophisticated reasons.

David Butler

Amsterdam, July 1991

Michiel Hazewinkel

PREFACE (TO THE RUSSIAN EDITION)

The book is dedicated to the statement of a few topics of modern control theory chosen as per interest and capability of the author. The models of the plants in the discrete time domain have been examined. Such a restriction is explained not only by growing importance of discrete analysis meant for large scale use of computer control, but also by the author's desire to avoid unnecessary introduction of mathematics into the theory, so that its technicality does not become detrimental to the contents. Majority of the results obtained for discrete processes can also be applied to continuous systems.

Various aspects of control theory have been discussed including the conventional ones of finite period control. Control with or without noise, with complete or partial information about the states of the plant, and various other cases have also been discussed.

Based on above choice it was necessary to study adaptive control problems when the plant parameters are not defined. Synthesis of adaptive strategies is often realized through identification technique with simultaneous conduction of optimisation and identification of the processes. Optimal control of sufficiently known plants and their operation has also received considerable attention. Similarly identification of dynamic systems from output observability has been studied.

The complexity in identification problem is determined by the plant noise and the structure of the plant. Linear objects with stochastic noise have been mainly studied in the book. Quite a few new results have been obtained in case of bounded noise not having useful stochastic characteristics.

Major attention has been paid to the foundation of adaptive control techniques. Mathematical results for effectiveness are carried out through well known recursive adaptive algorithms (such as Kalman - Bucy filter for stochastic approximation) and their new modifications.

The book is written for specialists in the area of control theory, for senior class students and research workers specializing in theoretical cybernetics : study of their operations. The contents of the book are based on two-semester courses delivered by the author to the university students specializing in 'Applied Mathematics'. For convenience all the proofs are presented separately. The model examples presented to illustrate major theoretical results must help better understanding of the main material. Nevertheless a complete understanding of the materials presented in the book demands sufficient mathematical background of the reader, such as completion of first three years in a higher academic institution.

Major portion of the book is based on new results obtained by the colleagues of the Leningrad University. The author derived pleasure in working several years with them. The origin of the book owes to the creative atmosphere prevailing in this group and the author takes the courage of assertaining that the present work is the collective view of the group regarding adaptive control problems. The author takes the opportunity of expressing gratitude to his colleagues – to start with to S.A. Agafonov, G.S. Axenov, A.E. Barabanov, T.P. Krasulina, N.A. Sankina, S.G. Semenov and B.M. Sokolov. The author deeply acknowledges the useful help rendered by innumerable communications and discussions on the problems presented in the book, with V.A. Iakubovich, U.F. Kazarinov, O.U. Kul'chitskii, V.G. Sragovich and Ia. Z. Tsypkin. The manuscript was examined with interest By V. Ia. Katkovnik and V.A. Yakubovich. Their comments and suggestions have been taken care of. Great help was rendered by T.V. Pliako, A.N. Kirillova and I.S. Galkina in organizing the format of the book. The author is deeply grateful to all of them.

CONTENTS

1

Basic Concepts and Statement of Problems in Control Theory

§ 1.1 Initial Premises

Control, in a broad sense, is the action responsible for evolution (with time) of a process so as to attain given specifications. The process may be related to any of the phenomena from the surrounding universe or from the sphere of human activities, such as, pollution of international seas, formation of hailstorm free clouds, socio-economic life of people and their government, diplomacy and war affairs, technology and science, control of varieties of machines and mechanisms and activity of a particular organism, etc.

The control action is realised by the controller which can be a human being, a real or an artificial organ (a device) etc. It is important to note that selection or interpretation of objectives of control (desired specifications of the controlled process) is a prerogative of man or of a group of people.

Any control problem demands definite answers, to the following : 'what to control' and 'why control'. Fulfilment of the above conditions is followed by the premises to the solution of the control problem, viz. 'how to control'. Control theory is mainly based on finding the answer to the last question.

First two questions appear to be superficial when examined in the context of mathematical theory of control. The mathematical model of the process and the necessary conditions of its states at some instant are usually assumed. This means that the question 'what to control' is clear. Moreover, the objective of control is formulated at the initial stage and hence 'why control' is also understood. It is true that there exist in control theory many types of models and control objectives. But the possibility of their satisfactory application to real systems and selection of unconventional control objectives will not be discussed here. Such approaches require wide mathematical treatment. Practical significance of the results, thus obtained, will depend on the nature of the adopted models of the process and the objective of control.

Considerable progress has been made in modelling of processes through dynamics of simple physical and mechanical systems. Mathematical methods of control serve as the guide to modelling of surrounding systems like economic systems, sociological systems and other complex systems.

§ 1.2 Basic Concepts of Control Theory

The basis of the control theory rests on three concepts : Object (or process or plant), objective of control (or control cost) and control strategy (or control algorithm or control law). The plant is characterized by its inputs (control variables and noise signals), outputs (variables to be controlled or outputs of the process) and the relationship between the input and output variables. The plant will be defined by an operator representing a few variables at its input and output stages. The represented form many be a complex functional, but it satisfies the causal conditions, *i.e.* the magnitude of the output variable at any instant does not depend on the value of the input variable at a future instant. Any variation of the input variable leads to a variation of the output variable. The control objective should be such that the process output possessed given characteristics. Attainment of control objective is possible by selection of control variables. The latter is formulated by a few rules (algorithms) and is known as the *control strategy.* The control strategy must be an admissible one, *i.e.*, it will make use of only those which are related to the process and accessible at the moment under consideration. The control problem consists of determining the class of admissible strategies. Its solution rests in choice of the class of strategy that ensures the objective of control.

After the above discussion in initial concepts of control theory, their formal expressions will be taken up. The process is considered in discrete intervals of time t and can be a member (unless otherwise mentioned) of

$$T_{t_*} = \{0,\ 1,\ 2,\ \ldots,\ t_*\}, t_* \leq \infty.$$

The inequality $t_* < \infty$ corresponds to control with finite time period. Infinite time period of control T_∞ will be the main consideration of the present treatment.

1.2.1 *The Control Object*

Let us introduce the abstract sets $\mathrm{Y} = \{y\}, \mathrm{U} = \{u\}, \mathrm{V} = \{v\}$, where the elements of the sets y, u and v are output variables, control variables and disturbance (or noise) variables, respectively. By inputs, disturbances and outputs of the plant we shall mean the following sequences of the variables* :

$$\text{control} : u_0^{t_*-1} \quad = (u_0,\ \ldots,\ u_{t_*-1})$$

$$\text{disturbance} : v_0^{t_*} = (v_0,\ \ldots,\ v_{t_*}) \text{ output} : y_0^{t_*} = (y_0,\ \ldots,\ y_{t_*})$$

Let the sets $\mathbf{U}^{t_*-1}$, $\mathbf{V}^{t_*}$ and $\mathbf{Y}^{t_*}$ be separated and the defined sequence of functions be $Y^{t_*}(\cdot) = (Y_0(\cdot),\ \ldots,\ Y_{t_*}(\cdot))$. The functions assume values from the set $\mathbf{Y}$ and satisfy the

---

*If the least value of the index in the sequence $u_s^t = (u_s, u_{s-1}, \ldots, u_t)$, $t \geq s$, is evident or insignificant then u^t along with u_s^t is used.

condition of unpredictability *i.e.*, for any three $(y^{t_*}, u^{t_*-1}, v^{t_*}) \in \mathbf{Y}^{t_*} \otimes \mathbf{U}^{t_*-1} \otimes \mathbf{V}^{t_*}$ the following relationship holds good

$$Y_t\big(y^{t_*}, u^{t_*-1}, v^{t_*}\big) = Y_t(y^{t-1}, u^{t-k}, v^t), \quad t = 1, 2, \; \ldots, \; t_*,$$

where $u^{t-k} = (u_0, \ldots, u_{t-k})$, $v^t = (v_0, \ldots, v_t)$, $y^{t-1} = (y_0, \ldots, y_{t-1})$, k = natural number, $1 \le k \le t_*$. It is assumed that

$$Y_t(y^{t-1}, u^{t-k}, v^t) = Y_t(y^{t-1}, v^t) \text{ for } 1 \le t \le k, \; Y_0(y^{t_*}, u^{t_*-1}, v^{t_*}) = Y_0(v_0).$$

The sequence $Y_0^t(\cdot)$ of such functions is identified with the plant if y^{t_*} given by

$$y_t = Y_t(y^{t-1}, u^{t-k}, v^t), \quad t = 0, 1, \ldots, t_*, \tag{2.1}$$

belongs to the set $\mathbf{Y}^{t_*}$ for any $u^{t_*-1} \in \mathbf{U}^{t_*-1}$ and $v^{t_*} \in \mathbf{V}^{t_*}$. The natural number k is known as the control delay*.

It is not difficult to prove that the expression (2.1) actually represents the input set $\mathbf{U}^{t_*-1} \otimes \mathbf{V}^{t_*}$ in the output set $\mathbf{Y}^{t_*}$. As a consequence, the unpredictability of functions $\mathbf{Y}_t(\cdot)$ ensures output causality of the plant as referred before.

1.2.2 *Control Algorithm*

Let us examine the possible methods of control. It may be remembered that at the instant t, the output variables $y^{t-s} = (y_0, \ldots, y_{t-s})$ (s = non-negative integer = measurement delay) are known. Consequently this information may be updated when the control signal u_t is formulated**. Assuming $s < t_*$, the set of expanded output variables is expressed as $y_{-s+1}^{t_*-s} = (y_{-s+1}, \ldots, y_0, y_1, \ldots, y_{t_*-s})$.

It is supplemented by output variables $y_{-s+1}, \ldots, y_0$ to facilitate the expressions to follow. The supplementing values are independent of inputs $u_0^{t_*-1}$, $v_0^{t_*}$ and are assumed to be chosen constant quantities. The set of expanded output variables is designated as $\mathbf{Y}^{t_*-s}$.

(*a*) *Control strategy for deterministic systems.* Let the sequence of functions $U_0^{t_*-1}(\cdot) = (U_0(\cdot), \ldots, U_{t_*-1}(\cdot))$, be defined in the set $\mathbf{Y}^{t_*-s} \otimes \mathbf{U}^{t_*-1}$.

---

* More formal expression for the plant is given by the combination of symbols $(\mathbf{U}, \mathbf{V}, \mathbf{Y}, k, t_*, \mathbf{U}^{t_*-1}, \mathbf{V}^{t_*}, \mathbf{Y}_0^{t_*}(\cdot))$. The expressions in (2.1) describe the plant in the time domain. If the sets of inputs and outputs are evident or insignificant then linking of the equation (2.1) with the plant (more precisely with its mathematical model) is permitted.

** Thus, the output variables of the plant are identified with the values observed (measured) from the plant. Collection of all the measuring devices is sometimes known as the sensor. In technical systems it is the collection of transducers, in biological systems these are the sensing organs for sight, hearing, feel etc. In general treatment of control theory, the term *sensor* will be associated with output variables. For particular systems it may have some physical meaning. In this respect observed or measured quantities, such as, the internal disturbance measuring component, may appear unrealistic.

The functions can assume values from the set U and possess the property of unpredictability

$$U_0\left(u_0^{t_*-1},\ y_{-s+1}^{t_0}\right)=U_0(y_{-s+1}),\ U_t\left(u_0^{t_*-1},y_{-s+1}^{t_*-1}\right)=$$

$$=U_t(u_0^{t-1},\ y_{-s+1}^{t-s}).$$

Each sequence of functions will be called as admissible regular control strategy (*i.e.*, control strategy for deterministic systems), if the control $u_1^{t_*-1}$ belongs to the set U^{t_*-1} for all $y^{t_*-s}\in y^{t_*-s^*}$*, where

$$u_0=U_0(y_{-s+1}),\ \ u_t=U_t(u^{t-1},\ y^{t-s}),\ t=1,\ 2,\ \ldots,\ t_*-1, \tag{2.2}$$

Equation (2.2) represents the rules for realising control law and describes the function of a control system or a *controller***. The plant and the controller together are known as the *control system*. Figure 1 depicts the general structure of a control system.

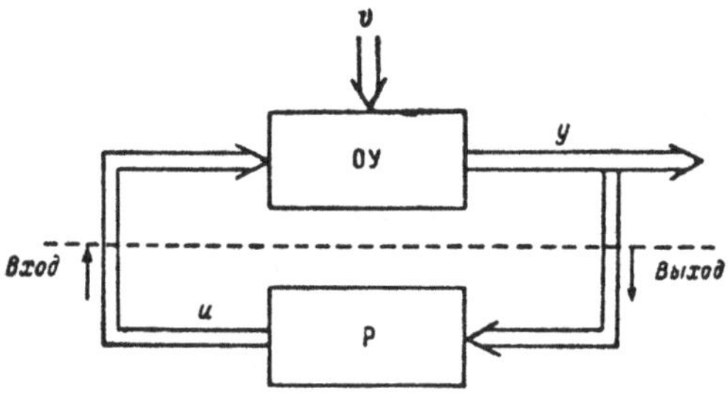

Fig. 1. General structure of a control system.
P = plant, C = controller.

Let us consider a set $\mathbf{U}^a=\left\{U_0^{t_*-1}(\cdot)\right\}$ from the admissible control strategies. Corresponding to different control laws there will be different outputs of the plant. It means that there is a possibility of selecting the control law from the set $\mathbf{U}^a$. The correspondence

---

*The admissible control consists of only accessible information. Such plant data can be divided into structural and current information. Reduction of the plant structure and its parameters, reduction of noise signals, etc. represent the structural data. The current data are given by the sensors. This type of division appears to be helpful because the current information may also present structural information such as, information received by identification methods. The expression (2.2) essentially represents control strategy based on current information. Structural information may be considered during selection of the set of admissible control. Inclusion of some structural data into the sensor group may often prove to be advantageous.

** The formal expression for control system, as explained before, is given by $(\mathbf{Y},\mathbf{U},s,t_*,\mathbf{Y}^{t_*-s},\mathbf{U}^{t_*-1},U^{t_*-1}(\cdot))$. If the set $\mathbf{Y}^{t_*-s},U^{t_*-1}$ is evident or insignificant, then the equation (2.2) can be identified as the control law.

The symmetry between the plant and the control law may be noted.

between the output of the plant and the control strategy is usually not simple. The chosen strategy determines the set of output variables depending on the dependence of the plant output on noise (equations 2.1 and 2.2).

If the functions $U_t(.)$ in eqn. (2.2) are independent of output variables, then u^{t_*-1} is known as the *programmed control.* The control algorithm is formulated without considering the noise and is determined a priori (before the control action starts). Considerable progress has been made in determination of admissible control strategies using feedback, where the control strategy is framed taking the output variable into account. The feedback principle is the basis for various mechanisms of control in technical and biological systems.

(*b*) *Random Control Strategy.* In the control strategy for deterministic systems, the control variable u^t as shown, is determined simply by the sequences u_0^{t-1}, y_{-s+1}^{t-s} (eqn. (2.2)). The control strategy loses its simplicity in some cases (as discussed later) and the random control strategy is to be adopted.

Let $\mathbf{W} = \{w\}$ be the abstract set of w known as noise variables (with respect to control). Let $\mathbf{W}^{t_*-1} = \{w_0^{t_*-1}\}$ be the separated set of sequences $w^{t_*-1} = (w_0, \ldots, w_{t_*-1})$, $w_t \in \mathbf{W}$.

The functions $U_t(.)$ are defined in the set $\mathbf{U}^{t_*-1} \otimes \mathbf{Y}^{t_*-s} \otimes \mathbf{W}^{t_*-1}$ with values in $\mathbf{U}$. The unpredictable sequences $U_0^{t_*-1}(\cdot)$ of such functions will be called as the control strategies. The control signal u_t can be computed from

$$u_0 = U_0(y_{-s+1}, w_0),\ u_t = U_t(u_0^{t-1},\ y_{-s+1}^{t-s},\ w_0^t),$$

$$t = 1, \ldots, t_* - 1, \tag{2.3}$$

where $u_0^{t^*-1} \in \mathbf{U}^{t^*-1}$ are chosen from admissible strategies $\mathbf{U}^a$. Equation 2.3 shows that for a priori fixed t_0^{t-1}, y_{-s+1}^{t-s}, the control signal u_t can not be determined simply as it also depends on the noise w^{t^*-1}.

The noise signal w^{t_*-1} is of interest depending on the specific problem. The external noise signal w_t may play its role with respect to the test signal* generated by the controller whose values at later instants may not be known. The current values of the noise signal or its values during the whole control interval may be known. The noise signal may have its values not known or only its general characteristics may be known.

1.2.3 *Control Objective*

After the discussion on the plant and its possible control strategies, let us formalise the objective of control. In this context an important concept 'ensemble operation' will be introduced. Let

---

*Test signals are usually applied when the control problem is associated with identification of the plant.

$$\mathbf{Q} = \{q(\cdot)\} \tag{2.4}$$

be the set of functions $q = q(v^{t_*}, w^{t_*-1})$, defined in the set $\mathbf{E} = \mathbf{V}^{t_*} \otimes \mathbf{W}^{t_*-1}$. Its real component $q(\cdot)$ is $\mathbf{E} \to \mathbf{R}^1$.*. Any functional $\tilde{\mathbf{M}} : \mathbf{Q} \to \mathbf{R}^1$ is known as ensemble operation of the set **E**.**

It is further assumed that any ensemble operation $\tilde{\mathbf{M}}$ of ensemble **E** is defined on all functions of the set (2.4) Let

$$q = q\left(y^{t_*},\ u^{t_*-1}\right), \tag{2.5}$$

be the non-negative objective function in the set $\mathbf{Y}^{t_*} \otimes \mathbf{U}^{t_*-1}$. The variables y_t, u_t in the function (2.5) can be ruled out because of the relationship given by expressions (2.3) and (2.1). Consequently the function can be seen as a functional defined in the set $\mathbf{E} \otimes \mathbf{U}^a$:

$$q = \tilde{q}\left(v^{t_*}, w^{t_*-1};\ U^{t_*-1}(\cdot)\right). \tag{2.6}$$

It is assumed that inclusion of $\tilde{q}(\cdot;\ \cdot;\ U^{t_*-1}(\cdot)) \in \mathbf{Q}$ is justified for all control strategy $U^{t_*-1}(\cdot)) \in \mathbf{U}^a$. Then the function in $\mathbf{U}^a$ can be determined as

$$J\left[U^{t_*-1}(\cdot)\right] = \tilde{\mathbf{M}}\tilde{q}\left(v^{t_*},\ w^{t_*-1};\ U^{t_*-1}(\cdot)\right). \tag{2.7}$$

This can be termed as control objective functional (or cost functional).

Let the control objective be to ensure the inequality

$$J\left[U^{t_*-1}(\cdot)\right] \le r^{-1}, \tag{2.8}$$

where r = a positive number – the index of control quality.

Thus, the control objective is to select the control strategy $U^{t_*-1}(\cdot)$ from the admissible set $\mathbf{U}^a$. The quality functional in this case does not cross the limit (inverse proportional level of quality of control).

The common form for formula (2.7) is

$$J = \tilde{\mathbf{M}}q\left(y^{t_*},\ u^{t_*-1}\right), \tag{2.9}$$

where the set on which ensemble operation is carried out is implicit. In many cases, this type of expression does not require explanation, if the context of the control strategy is known and the transition from function (2.5) to function (2.6) is understood. If such clarity does not exist then the quality functional (2.9) may demand a complex interpretation or may have no meaning at all.

There can be control objectives that require existence of a few inequalities. In such cases the control objective (2.5) is chosen as a vector and if each component of the

---

*For generalisation's sake it is assumed that the noise is present. In case it is absent then the functions $q = q(v^{t_*})$ is defined in the set $\mathbf{E} = \mathbf{V}^{t_*}$.

Typical applications of ensemble operation can be found in control theory for example in maximization operation $\tilde{\mathbf{M}}q(\cdot) = \sup_{e \in \mathbf{E}} q(e)$ and in averaging operation $\tilde{\mathbf{M}}q(\cdot) = \int_{\mathbf{E}} q\ (e)\, dF$, where F represents probabilistic measure defined in σ – algebra of subsets of set **E. Other operations related to computation of characteristics of functions q(.) of the ensemble **E** are possible. It is implied that ensemble operation $\tilde{\mathbf{M}}$ is monotonous : $q(\cdot) \ge \tilde{q}(\cdot) \Rightarrow \tilde{\mathbf{M}}q \ge \tilde{\mathbf{M}}\tilde{q}$.

function (2.5) belongs to the set Q[*], then the function $J[U^{t_*-1}(\cdot)]$ (eqn. (1.7)) is a vector function. In the control objectives, it may be necessary to ensure inclusion of

$$J\left[U^{t_*-1}(\cdot)\right] \subseteq R, \tag{2.10}$$

where R is a given set in the value space of the function $J(.)$. The control objective of eqn. (2.10) is the natural generalisation of the expression in (2.8).

§ 1.3 Modelling of Control Objects and their General Characteristics

Along with the input-output relationship of the plant as given by eqn.(2.1), the states of the object are also made use of.

An important class of the plant is the linear system where input-output relationship is linear. The coefficients of the equation, *i.e.*, the values of the input and output variables in linear systems, are completely described by mathematical models. If these coefficients are time invariant then stationary models of objects are obtained, which may be of special interest from application point of view.

A detailed discussion of various methods of describing a plant (basically linear systems) is carried out in the paragraphs to follow :

1.3.1 *State Equations of Discrete Processes*

Recursive equations of the object require the knowledge of initial values *i.e.*, collection of values that can determine output of the process when the input is known. The problem of minimizing collection of such values is solved naturally in terms of the plant state.

Let x be the state variables or phase variables of the plant and elements of the abstract set $\mathbf{X} = \{x\}$. The set of sequences of the state variables, viz., $x^{t_*} = (x_o, \ldots, x_{t_*})$ will be expressed through $\mathbf{X}^{t_*} = \{x^{t_*}\}$. As before $\mathbf{U}^{t_*-1} \otimes \mathbf{V}^{t_*}$ will separate the sets of control and noise. It is assumed that the sequence of functions $\mathbf{X}^{t_*}(\cdot)$ is defined on the set $\mathbf{X}^{t_*} \otimes \mathbf{U}^{t_*-1} \otimes \mathbf{V}^{t_*}$ with its values in $\mathbf{X}$ and having the property of unpredictability

$$X_0\left(x^{t_*}, u^{t_*-1}, v^{t_*}\right) = X_0(v_0),$$

$$X_t\left(x^{t_*}, u^{t_*-1}, v^{t_*}\right) = X_t(x^{t-1}, u^{t-k}, v^t), \quad t = 1, 2, \ldots,$$

where the natural number k is known as the control delay. The relationship

$$x_t = X_t(x^{t-1}, u^{t-k}, v^t), \quad t = 0, 1, \ldots, t_*, \tag{3.1}$$

will be known as state equation. It is assumed that the sequence x^{t_*} in this relationship belongs to set $\mathbf{X}^{t_*}$. The state equation (3.1) is supplemented by the plant output equations

$$y_t = \bar{Y}_t(x^t, v^t), \quad t = 0, 1, \ldots, t_*, \tag{3.2}$$

where $\bar{Y}_o^{t_*}(\cdot)$ is the sequence of unpredictable functions defined in the set $\mathbf{X}^{t_*} \otimes \mathbf{V}^{t_*}$ and having its values from the set $\mathbf{Y}$. If the state variables of eqn. (3.1) are included in

---

[*]It is possible to have a situation where different components of the vector function (2.5) will have different ensemble operations.

eqn. (3.2), then the type of eqn.(2.1) based on input-output variables is obtained. The inverse problem is more interesting *i.e.*, is it possible to express the input-output equation in the state equation form ? It appears that it is possible for natural assumptions of functions $Y_t(.)$ and that too by methods different from each other. If **Y**, **X**, **U** are considered as the vector spaces then the problem of determination of state equations with minimum measurement of state spaces will be also included and it will be solved quite easily.

1.3.2 *Observability and Controllability*

Analysis of the plant in the context of equations (3.1) and (3.2) gives rise to the problem of simple correspondence between outputs and states. Or more precisely : is it possible to establish sequence of states x^{t^*} when y^{t^*} is known ? For noise free cases it is possible and then it is said to satisfy the condition of observability.

Let x' and x'' be the states of the plant for which admissible control strategies (as in Sec. 1.2) U^a exist where x'' is value at $t'' < t^*$ obtained from the value x' at $t' = 1$, then it is said to satisfy condition of controllability. For a wide class of linear stationary plants, the conditions of observability and controllability may be obtained quite effectively.

In presence of unpredictable noise it is usually not possible to establish plant states from its observed output independent of the duration of control T_{t*}. In such cases attempts may be made to obtain estimates of the plant states, optimal in some sense. Theory of state estimation for the process from its observed output or the theory of filtration has made substantial progress for control of random stationary processes.

1.3.3 *Linear Process*

(*a*) *Standard form of the equation.* Let us assume that sets **X**, **Y**, **U**, **V** are the subsets of finite dimensional euclidean space, generally speaking, of different dimensions. If the equations (3.1) and (3.2) are expressed as

$$x_{t+1} = A_t x_t + B_t u_{t-k+1} + v'_{t+1}, \quad y_t = C_t x_t + v_t'', \tag{3.3}$$

then the plant is said to be linear, when A_t, B_t, C_t are the linear operators for any t, $A_t:\mathbf{X} \to \mathbf{X}, B_t:\mathbf{U} \to \mathbf{X}, C_t:\mathbf{X} \to \mathbf{Y}, v_t = (v'_t, v''_t)$, $\mathbf{V} = \mathbf{X} \otimes \mathbf{Y}$. Here v'_t, v''_t are the cumulative signals ($\{v'_t\}$ is the noise in the plant and $\{v''_t\}$ is the noise in the observer). The relationships (3.3) are also known as equations of the control system in the standard form. Choosing proper bases for the sets **X**, **Y**, **U**, it may be possible to represent the variables x, u, v', v'', y as column vectors and the operators A_t, B_t, C_t as orthogonal matrices corresponding to the measurements. A_t is a square matrix. For time invariant matrices $A_t \equiv A, B_t \equiv B, C_t \equiv C$ and the linear system of equn. (1.13) can be expressed as

$$x_t = Ax_{t-1} + Bu_{t-k} + v'_t, \; y_t = Cx_t + v''_t, \; t = 1, 2, \ldots t_*, \tag{3.4}$$

and is known as a stationary system. Exclusion of state variables in equations (3.4) will also lead to linear stationary equation in terms of input-output variables as follows :

$$y_t + a_1 y_t + \quad \ldots \quad + a_p y_{t-p} = b_k u_{t-k} + b_{k+1} u_{t-k-1} + \quad \ldots \quad +$$

$$+ b_r u_{t-r} + c_o v_t' + \quad \ldots \quad + c_q v'_{t-q} + v''_t + a_1 v''_{t-1} + \quad \ldots \quad + a_p v''_{t-p}, \tag{3.5}$$

where p, q, r are the natural numbers, coefficients a_j, b_j, c_j are the matrices of dimensions corresponding to elements of matrices A, B, C of eqn. (3.4). Introducing operation ∇ one step backward, *e.g.* $\nabla y_t = y_{t-1}$, $\nabla u_t = u_{t-1}$, $\nabla v_t = v_{t-1}$ etc. the equation (3.5) may be written in the following compact form

$$a(\nabla)y_t = b(\nabla)u_t + c(\nabla)v'_t + a(\nabla)v''_t, \tag{3.6}$$

$$\text{where } a(\lambda) = I + \lambda a_1 + \quad \dots \quad + \lambda^p a_p,$$

$$b(\lambda) = \lambda^k b_k + \quad \dots \quad + \lambda^r b_r,$$

$$c(\lambda) = c_o + \lambda c_1 + \quad \dots \quad + \lambda^q c_q.$$

(*b*) *Transition from standard form to equation involving variables input-output.* Let us introduce euristic method of obtaining the equation (3.6) using varying input-output, from the equation (3.4). Rewriting eqn. (3.4) we get

$$(I_n - A\nabla)x_t = B\nabla^k u_t + v'_t, \quad y_t = Cx_t + v''_t,$$

$x_t \in \mathbf{R}^n$, I_n = identity matrix of $n \times n$ dimensions. The first equation for x_t will be

$$x_t = \nabla^k (I_n - A\nabla)^{-1} Bu_t + (I_n - A\nabla)^{-1} v'_t.$$

Then from the second equation

$$y_t = \nabla^k C(I_n - A\nabla)^{-1} Bu_t + C(I_n - A\nabla)^{-1} v'_t + v''_t. \tag{3.7}$$

Matrix $(I_n - A\lambda)^{-1}$ is fractionally rational. Let

$$a(\lambda) = \det(I_n - \lambda A)^{-1}. \tag{3.8}$$

Then $a(\lambda)(I_n - \lambda A)^{-1}$ is a polynomial matrix. Operation $a(\nabla)$ on eqn. (3.7) and considering eq. (3.6), we have polynomial matrices

$$b(\lambda) = a(\lambda)\lambda^k C(I_n - \lambda A)^{-1} B,$$

$$c(\lambda) = a(\lambda) C(I_n - \lambda A)^{-1} \tag{3.9}$$

Thus, the transition from linear stationary state equation to linear stationary input-output equation has been achieved. Moreover, the formulae (3.8) and (3.9) relating the coefficients of these equations have been obtained. Precise basis for the correspondence is not difficult to obtain using discrete Fourier Transforms.

(*c*) *Controllability and Observability of Linear Stationary Systems.* The conditions for controllability and observability have been explained in (3.4) where for $k = 1$ it has algebraic characteristics. For noisefree ($v'_t \equiv 0$, $v''_t \equiv 0$) systems, when the dimension of x is not less than that of y, u, then the system controllability (3,4) is equivalent to the condition.

$$\det DD^* > 0, \tag{3.10}$$

Where D = controllability matrix

$$D = \| B, \quad AB, \quad \dots, \quad A^{n-1}B \|, \tag{3.11}$$

n = dimension of the state vector x,

det = determinant of respective matrices,

and D^* = conjugate of matrix D.

Similarly the observability condition is given by

$$\det EE^* > 0, \tag{3.12}$$

where E = observability matrix

$$E = \| C^*, \quad A^*C^*, \quad \dots, \quad (A^*)^{n-1}C^* \|. \tag{3.13}$$

Since the controllability and observability of linear systems (3.4) are completely determined by matrices of its coefficients, then the pair (A, B) can be termed as controllable if the inequality (3.10) is satisfied and the corresponding pair (A, C) will be termed as observable, if the inequality (3.12) is satisfied. It is well known that for the control pair (A, B) there exists a matrix K, such that the matrix $A + BK$ is stable *i.e.*, all its eigenvalues are arranged in open unit disk. This property of matrix pairs is of importance in control theory and is termed as stability. Precisely the matrix pair (A, B) is known as stabilising, if there exists a matrix K of proper dimensions such that the matrix $A+BK$ is stable. Duality of stability is detectibility - matrix pair (A,B) is detectable if the pair (A^*, B^*) is stable. Stability follows controllability and detectibility follows observability.

If simultaneously controllability and observability of a plant are present, then the standard expression (3.4) is known as minimal.

Many standard forms (3.4) exist for plants expressed through input-output variables. The plants may or may not possess may properties of controllability and observability, but the minimal form will be always present. All minimal forms have identical dimensions and can be transformed in a similar way.

(*d*) *Output stability, control stability and noise stability.* A few useful properties of linear systems, based on infinite time control ($t^* = \infty$) will be introduced here.

The linear system is *output stable* if there exist time invariant constants C_u, C_v such that the following inequalities are satisfied

$$\overline{\lim_{t\to\infty}} \; |y_t| \le C_u \sup_{t\ge 0} \; |u_t| + C_v \sup_{t\ge 0} \; |v_t|,$$

where y_t, u_t, v_t are the arbitrary values of output, control and noise variables related to the plant. $|\,|$ designates euclidean norms for respective variables. (It may be recalled that the variables may assume values in euclidean spaces of different dimensions.)

The linear system will be called *control stable* (or *noise stable*) if there exist constants C_y, C_v (or C_y, C_u) such that the following inequalities are satisfied.

$$\overline{\lim_{t\to\infty}} \; |u_t| \le C_y \sup_{t\ge 0} |y_t| + C_v \sup_{t\ge 0} |v_t|$$

$$\text{or} \left(\overline{\lim_{t-\infty}} \; |v_t| \le C_y \sup_{t\ge 0} |y_t| + C_u \sup_{t\ge 0} |u_t|\right).$$

For stationary systems (3.4) output stability Y_t is equivalent to absence of the roots (for $|\lambda| \le 1$. in the characteristic polynomial (3.8). Polynomials like $a(\lambda)$ will be henceforth known as stable. Thus, the plant in (3.6) is stable if the polynomial $a(\lambda)$ in the equation is stable. Stability w.r.t. states x_t (3.4) is equivalent to stability of matrix A.

Control stability of scalar systems (3.6) ($\mathbf{Y} \subseteq \mathbf{R}^1$, $\mathbf{U} \subseteq \mathbf{R}^1$, $\mathbf{V} \subseteq \mathbf{R}^1$) is known as minimum-phase systems. Minimal phase is equivalent to absence of the roots (for $0 <| \lambda | \le 1$.) in the polynomial $b(\lambda)$.

For scalar objects (3.6) noise stability is equivalent to stability of polynomial $c(\lambda)$ (for $v''_t = 0$).

§ 1.4 Precising the Statement of the Control Problem

1.4.1 *Classification of Control Objectives*

The control objectives can be classified based on index of performance.

(*a*) *Dissipatibility (w.r.t. functional J)*. The inequality (2.8) must be always satisfied and eqn. (2.8) can be then rewritten as

$$J\left[U_0^{t_*-1}(\cdot)\right] < \infty. \tag{4.1}$$

(*b*) *Stabilisation (w.r.t. functional J)*. The inequality (2.8) must be ensured for the specific index of performance.

(*c*) *Optimisation (w.r.t. functional J)*. The inequality (2.8) must be satisfied for the index of performance.

$$r^{-1} = r_{opt}^{-1} = \inf_{U_0^{t_*-1} \in \mathbf{U}^a} J\left[U_0^{t_*-1}(\cdot)\right], \tag{4.2}$$

where $\mathbf{U}^a$ is the set of admissible control strategies.

(*d*) *Suboptimisation (w.r.t. functional J)*. The index of performance is chosen from the condition

$$r = \rho^{-1} r_{opt},$$

where ρ = degree of suboptimisation, $\rho > 1$.

Amongst all the categories dissipatibility appears to be the weakest. Fairly complex control objectives are related to optimisation or suboptimisation. If the function $J\,[U_0^{\infty}(\cdot)]$ is a vector function then there is a choice for hybrid control objectives *i.e.*, different components of this function may be chosen from different indices of performance, such as, dissipatibility, optimisation, etc.

1.4.2 *Optimisation of Control*

Let the admissible control strategy $U_0^{t_*-1}(\cdot)$, satisfying (4.2) be termed as $(J, \mathbf{U}^a)$ - optimal. The optimisation problem may be modified when determination of $(J, \mathbf{U}^a)$ optimal strategy (optimal in any sense) is not advisable. One such case may arise when synthesis of $(J, \mathbf{U}^a)$ - optimal strategy becomes complex from practical point of view. Hence, it is suggested that the admissible control strategy is to be determined from another class $\mathbf{U}^r$ to be known as the class of realisable control strategies. The control cost in this case requires ensuring of the following inequality

$$J\left[U_0^{t_*-1}(\cdot)\right] \le r_{opt}^{-1}, \quad U_0^{t_*-1}(\cdot) \in \mathbf{U}^r. \tag{4.3}$$

As before r_{opt} is determined by relationship (4.2). Thus, the admissible control strategy $\mathbf{U}^a$ determines index of performance, the latter in its turn demands the control strategy to belong to realisable class $\mathbf{U}^r$.

Such problems will be called optimising and the strategy $U^{t_*-1}(\cdot)$ ensuring the control objective (4.3) will be called $(J, \mathbf{U}^r, \mathbf{U}^a)$ – optimal control strategy. For $\mathbf{U}^r = \mathbf{U}^a$, the earlier optimisation problem will be attained. In practical situations the sets $\mathbf{U}^r$ and $\mathbf{U}^a$ may not be subsets of each other or even they may not intersect each other.

1.4.3 *Observations on selection of Control Strategies*

Selection of admissible (in some class) control strategy has been discussed in connection with formulation of control objective. If the sets $\mathbf{U}^a$ and $\mathbf{U}^r$, from which the admissible control is to be chosen, are bounded then the above selection carries a clear sense. Usually the sets $\mathbf{U}^a$ and $\mathbf{U}^r$ are characterised by very general properties and contain innumerable different strategies, out of which the required strategy is chosen through, precise statement. In practical problems the class of $\mathbf{U}^r$, the realisable strategies, is given precisely upto finite selection of parameters. These are determined either from the beginning by analysing the problem provided the data available are enough or they may be 'self tuned' from the observed output of the controlled process. Self tuning itself usually presents a few algorithms (from the class of probable algorithms) and hence here also the question of admissible strategies of tuning or about its class arises, etc.

Application of tuning strategies for solution of control problems in the finite period of control ($t_* < \infty$) appears to be quite limited. If $t_* = \infty$ then there exists a broad class of control problem, in which such a tuning is advisable. This class is defined as quality functions (2.7), in which the function (2.5) depends on plant output when control period $t \to \infty$. For example, either

$$q = q(y_0^\infty, u_0^\infty) = \overline{\lim_{t \to \infty}} \ \Phi(y_t, u_t) \tag{4.4}$$

$$\text{or } q = q(y_0^\infty, u_0^\infty) = \overline{\lim_{t \to \infty}} \ \frac{1}{t} \sum_{s=0}^{t-1} \Phi(y_s, u_s), \tag{4.5}$$

where $\Phi(y, u)$ is the given function of output and control variables. The control problems, in which quality functional depends only on bounded values of input and output, permit different methods of estimation and identification. It is only important that these methods are able to realise the actual process. Optimal control of such quality functional is naturally termed as limiting optimal control. In the solution of *limited optimal control* problems the transient response of the system is neglected and this may appear to be the important defect in the problem formulation. Moreover, the optimisation problem may appear to be a degenerated one. This may permit construction (synthesis) of simple (realisable) control strategies which will be also limited optimal*.

---

*Analogous to this we have analytical solution of equation $f(x) = 0$. When the equation appears to be a complex problem, it is usual to adopt simple successive methods to determine roots of the equation.

2

Finite Time Period Control

The problem related to synthesis of optimal control is investigated in this and the subsequent chapters under assumption that all the required data for optimal control are known. A few sections of adaptive control systems *i.e.*, the problem of synthesis of control under the conditions of insufficiently described control object and inputs, are devoted in chapters 4 and 6.

For optimal control synthesis problems, the most complete studies have been made in the area of control on a finite time interval. Here one finds a powerful method like Dynamic Programming and its stochastic analog viz. Bayesian method of optimization. However, from practical point of view, these methods solve the general optimization problem only in principle because of large computational difficulties. By avoiding unnecessary generalisation of problems (*i.e.*, by narrowing the class of control objects or by limiting the noise, etc.), simplification of the solution to the optimization problems has been made possible and considerable progress has been made in this direction.

In the following paragraphs, the general method of optimization is carried out simultaneously for deterministic and stochastic linear control systems operating in discrete time.

§ 2.1 Dynamic Programming

2.1.1 *Statement of the Optimization Problem*

Let the noise free nonlinear control object be described through state variables as

$$x_t = X_t(x_{t-1}, u_{t-1}), \quad t = 1, 2, \ldots, t_*, \tag{1.1}$$

where, x = state variable, $x \in \mathbf{X} \subseteq \mathbf{R}^n$; u = control variable, $u \in \mathbf{U} \subseteq \mathbf{R}^m$. The initial state $x_o \in \mathbf{X}$ of the control object is assumed to be known.

The control objective consists of selection of the control $u_0^{t_*-1} \in \mathbf{U}^{t_*-1}$ from the condition for minimization of performance criterion

$$J\left(u^{t_*-1}\right)=\sum_{t=1}^{t_*} q_t(x_t, u_{t-1}), \tag{1.2}$$

where $q_t(.)$ are the given non-negative functions of state and control variables*.

Theoretically, solution to the problem formulated is simple: in view of eqn. (1.1), the state variables $x_1^{t_*}$ may be expressed as

$$x_t=\tilde{X}_t(x_0, u_0^{t-1}). \tag{1.3}$$

The optimal control $u_0^{t_*-1}$ is then determined by the formula

$$u_0^{t_*-1} = \underset{\tilde{u}_0^{t_*-1} \in U^{t_*-1}}{\operatorname{argmin}} \sum_{t=1}^{t_*} q_t[\tilde{X}_t(x_0, \tilde{u}_0^{t-1}), \tilde{u}_{t-1}], \tag{1.4}$$

i.e., up to the determination of an algorithm for minimization of a known function, the optimal control problem is solved accurately.

As a matter of fact the problem of minimization of multivariable functions is rather complex and to obtain its effective computational algorithm, some speciality of the minimizing function is needed to be considered. In Dynamic Programming, which is discussed below, a specific character of the function (1.2) is taken into account.

2.1.2 *Description of the Dynamic Programming Methods*

We assume the control variables $u_0, u_1, ..., u_{t-2}$ to be known by some means. Then using eqn. (1.3) concrete values of the state variables $x_1^{t_*-1}$ can be determined. The functional (1.2) can be written as

$$J=\sum_{t=1}^{t_*-1} q_t(x_t, u_{t-1})+q^{(1)}\left(x_{t_*-1}, u_{t_*-1}\right), \tag{1.5}$$

where $$q^{(1)}(x,u)=q_{t_*}\left[X_{t_*}(x,u);u\right], \tag{1.6}$$

The functional can be considered as a function of control signal u_{t_*-1}. Since the last term in eqn. (1.5) depends only on u_{t_*-1} then for optimal control, u_{t_*-1} must be determined from the conditions for minimization of function $q^{(1)}(x_{t_*-1}, u)$:

$$u_{t_*-1}=U_{t_*-1}(x_{t_*-1})=\underset{u \in U}{\operatorname{argmin}}\, q^{(1)}(x_{t_*-10}, u). \tag{1.7}$$

It is assumed that eqn. (1.7) gives an effective formula in the sense that there exists sufficiently simple and known algorithm for minimization of $q^{(1)}(x,u)$ for all $x \in \mathbf{X}$.

In view of eqn. (1.7) the variable u_{t_*-1} can be eliminated from eqn. (1.5) to determine the optimal value u_{t_*-2} for fixed values of $u_0, \ldots, u_{t_*-3}$. This problem can be solved in

---

*The performance criterion (1.2) does not contain ensemble operations. It is a simple function of control $u_0^{t_*-1}$ (and initial state x_0). In view of eqn. (1.1), the state variables x_t may be eliminated.

an analogous manner discussed above. The k-th step of the optimization algorithm ($k = 2, 3, \ldots$) is

$$j = \sum_{t=1}^{t_*-k} q_t(x_t, u_{t-1} + q^{(k)}(x_{t_*-k}, u_{t_*-k}), \tag{1.8}$$

$$q^{(k)}(x,u) = q_{t_*-k+1}\left[\mathbf{X}^1_{t_*-k+1}(x,u), u\right]$$

$$+q^{(k-1)}\left[\mathbf{X}_{t_*-k+1}(x,u), U_{t_*-k+1}\left(X_{t_*-k+1}(x,u)\right)\right], \tag{1.9}$$

$$U_{t_*-k+1}(x) = \underset{u \in \mathbf{U}}{\operatorname{argmin}}\, q^{(k-1)}(x,u). \tag{1.10}$$

Then u_{t_*-k} is determined from

$$u_{t_*-k} = U_{t_*-k}\left(x_{t_*-k}\right) = \underset{u \in \mathbf{U}}{\operatorname{argmin}}\, q^{(k)}\left(x_{t_*-k}, u\right). \tag{1.11}$$

The method for determination of sequence of functions $q^{(k)}(x,u)$ and for minimizing the functions $U_{t-k}(x)$ will give the optimal control $u_0^{t_*-1}$ in the form of recursive relations

$$u_0 = U_0(x_0), \quad u_1 = U_1[\tilde{X}_1(x_0, u_0)], \ldots, u_{t_*-1}$$

$$= U_{t_*-1}\left[\tilde{X}_{t_*-1}\left(x_0, u_0^{t_*-2}\right)\right], \tag{1.12}$$

where $\tilde{X}_t(\cdot)$ are the functions from eqn. (1.3). Formulae (1.12) represent programmed control that can be computed *a priori* upto the start of the control action. In addition to that, if the variables x_t are 'observable' at the respective instants, then it is not necessary to eliminate them because of eqn. (1.3) and the optimal control can be presented in the form of 'feedbacks':

$$u_0 = U_0(x_0), \qquad u_1 = U_1(x_1), \ldots, u_{t_*-1} = U_{t_*-1}\left(x_{t_*-1}\right). \tag{1.13}$$

The sequence of functions $U_0^{t_*-1}(\cdot)$ gives the optimal control strategy*.

2.1.3 *Bellman's Equation*

The above optimization method can be presented in a more elegant manner. Let

$$V_t(x) = \min_{u \in \mathbf{U}} q^{(t_*-t)}(x,u). \tag{1.14}$$

Then from the recursive relationship of eqn. (1.10) we get

$$V_t(x) = \min_{u \in \mathbf{U}} \{V_{t+1}[X_{t+1}(x,u)] + q_{t+1}[X_{t+1}(x,u), u]\}, \tag{1.15}$$

$$t = 0, 1, \ldots, t_* - 1,$$

*For absence of unobservable variables, any closedloop strategy results in some programmed control. In general, it is not so.

$$V_{t_*-1}(x) = \min_{u \in \mathbf{U}} q^{(1)}(x, u). \tag{1.16}$$

The functional equation of (1.15) is known as *Bellman's equation*. The relationship (1.16) plays the role of the initial condition in Bellman's equation. If $V_t(x)$, the solution to eqn. (1.15), is found out, then the optimal control is determined as follows :

$$u_t = U_t(x_t) = \operatorname*{argmin}_{u \in \mathbf{U}}\{V_{t+1}[X_{t+1}(x_t, u)]$$

$$+ q_{t+1}[X_{t+1}(x_t, u), u]\}, \tag{1.17}$$

where x_t is the state of the control object due to control signal u_0^{t-1}. Evidently, the value $V_0(x_0)$ coincides with the minimum of the performance criterion. In the above solution, the positive function $V_t(x)$ satisfies the following inequality (because of eqn. (1.16)) :

$$V_t(x_t) \geq V_{t+1}[X_{t+1}(x_t, u_t)],$$

i.e., it is the Liapunov function for the control system described by equations (1.1) and (1.17). Owing to this the solution of eqn. (1.15) is known as *Bellman-Liapunov function*. Thus, the synthesis of optimal control appears to be related to the problem of construction of special type Liapunov function.

There are various generalized forms of Dynamic Programming (vide, for example, paragraph 2.1.5 and § 2.3)

2.1.4 Example: Linear - quadratic deterministic system

Let the control object be described by the linear equation

$$x_{t+1} = A_t x_t + B_t u_t, \quad y_t = x_t, \quad t = 0, 1, \quad \ldots, t_* - 1, \tag{1.18}$$

where $x_{t+1} \in \mathbf{X} = \mathbf{R}^n$, $u_t \in \mathbf{U} = \mathbf{R}^m$, A_t, B_t are the time variant matrices of proper dimensions. It is assumed that

$$\det B_t^* B_t \neq 0. \tag{1.19}$$

The control objective is to minimize the performance criterion

$$J\left(u_0^{t_*-1}\right) = \sum_{t=1}^{t_*} (x_t^* Q_{t-1} x_t + u_{t-1}^* R_{t-1} u_{t-1}), \tag{1.20}$$

where Q_t, R_t are symmetric matrices such that

$$R_t > 0, \quad Q_t \geq 0, \quad t = 0, 1, \ldots, t_* - 1, \tag{1.21}$$

the inequality (1.21) is understood in the quadratic sense*.

Theorem 2.1.1 *Optimal control $u_0^{t_*-1}$ in the linear quadratic problem of (1.18) and (1.20) is determined by the relations*

$$\bar{u}_t = K_t x_t, \quad t = 0, 1, \ldots, t_* - 1, \tag{1.22}$$

$$K_t = -L_{t+1} B_t^* P_{t_*-t} A_t, \tag{1.23}$$

---

* $Q_t > 0 \Leftrightarrow x^* Q_t x > 0$ for any $x \in \mathbf{R}^n, x \neq 0$.

$$L_{t+1}^{-1} = R_t + B_t^* P_{t_*-1} B_t, \tag{1.24}$$

where symmetric matrices P_t are determined by

$$P_{t+1} = A_{t_*-t}^* P_t A_{t_*-t}$$

$$-A_{t_*-t}^* P_t B_{t-t_*} L_{t_*-t+1} B_{t_*-t}^* p_t A_{t_*-t} + Q_{t_*-t-1} \tag{1.25}$$

and by the initial condition

$$P_1 = Q_{t_*-1}. \tag{1.26}$$

Then

$$J\left(\bar{u}_0^{t_*-1}\right) = \min J\left(u_0^{t_*-1}\right) = x_0^* P_{t_*} x_0. \tag{1.27}$$

Theorem 2.1.1 presents a distinct solution to optimization problem by means of linear feedback concept (1.22), in which the gain matrix coefficient (1.23) is determined by recursive relations (1.24) to (1.26).

Matrix equation (1.25) is known as (*discrete*) *Riccati equation*, which appears in investigation of various problems of optimization, identification, filtration and adaptive.

2.1.5 *Generalisation of Bellman's Equation for Infinite Time Control Problems*

The form of Bellman's equation (1.15) does not depend on finiteness of t_*. The finiteness of the control interval is affected only with selection of initial condition for $V_{t_*-1}(x)$ given at the terminal instants of control. It is natural to expect that if the performance criterion has the form

$$J(u_0^\infty) = \sum_{t=0}^{\infty} q_t(x_t, u_t), \tag{1.28}$$

and the assumed control strategies are set to satisfy convergence of series in (1.28), then the optimal control must be also related to the solution of Bellman's equation (1.15) satisfying the 'initial' condition

$$\lim_{t \to \infty} V_t(x) = 0. \tag{1.29}$$

It appears that the above discussion can be proved.

Theorem 2.1.2 *Let the class of admissible control strategies* $\mathbf{U}^a$ *consisting of all possible functions* $U_0^\infty(\cdot)$ *ensuring finiteness of function (1.28) independent of choice of initial state of control object (1.1) be not permitted. If there exists a solution for the equation*

$$V_t(x) = \min_{u \in \mathbf{U}} \{V_{t+1}[X_{t+1}(x,u)] + q_t(x,u)\},$$

$$t = 0, 1, \ldots, \tag{1.30}$$

that satisfies the condition (1.29), then the control strategy given by the formulae

$$U_t(x) = \underset{u \in \mathbf{U}}{\operatorname{argmin}} (V_{t+1}[X_{t+1}(x,u)] + q_t(x,u), \tag{1.31}$$

will be optimal and

$$\min_{U_0^\infty(\cdot) \in \mathbf{U}^a} J[U_0^\infty(\cdot)] = V_0(x_0). \tag{1.32}$$

The eqn. (1.30) differs from the eqn. (1.15) in the sense that in the former, Bellman-Liapunov function is determined in the infinite control interval $[0, \infty]$ and not in the finite control interval $[0, t_*-1]$. Also in the equation (though not so important) $q_{t+1}[X_{t+1}(x,u), u]$ finds place along with $q_t(x, u)$ conditioned by the performance criterion of (1.28).

The theorem 2.1.2 presents the sufficient conditions for existence of optimal control, which are, however, close to the necessary conditions. As a matter of fact, if there exists an optimal control strategy $U_0^\infty(\cdot)$ which is not dependent on initial state x_0 and the objective functions $q_t(x, u)$ are positive (*i.e.* $q_t(x, u) > 0$ for $|x| + |u| \neq 0$) then, it can be easily proved that the function

$$V_t(x) = \sum_{s=t}^{\infty} q_s[x_s, U_s(x_s)] \mid x_t = x, \tag{1.33}$$

satisfies the relations (1.30) and (1.29), *i.e.*, it becomes Bellman-Liapunov function.

From the proof of the theorem 2.1.2 it follows that by imposing additional equality condition, viz. $\lim_{t \to \infty} V_t(x_t) = 0$ for any control strategy $U_0^\infty(\cdot) \in \mathbf{U}^a$ (or even for optimal strategy only, if $V_t(x)$ is a non-negative function) along with the condition in (1.29), the latter can be relaxed.

§ 2.2 Stochastic Control Systems

2.2.1 *Statement of the Problem*

The control systems are known as *stochastic*, if their description requires the help of random processes. Usually randomness is apparent in external noises (which are cumulative disturbances) or else there may be random changes in control object parameters with time (parametric noises).

Let the control object be described through state variables as

$$x_0 = X_0(v_0), \quad x_t = X_t(x_0^{t-1}, u_o^{t-1}, v_o^t),$$
$$t = 1, 2, \ldots, t_*, \tag{2.1}$$

and let the output be determined by relations

$$y_t = Y_t(x_0^t, v_0^t), \quad t = 0, 1, \ldots, t_*. \tag{2.2}$$

The state variables x, output variables y and the noise variables v assume the values from the sets $\mathbf{X} \subseteq \mathbf{R}^n$, $\mathbf{Y} \subseteq \mathbf{R}^l$, $\mathbf{V} \subseteq \mathbf{R}^m$, respectively.

The noise $v_0^{t_*}$ will be assumed to be the realisation of the random process. From eqn. (2.1) it follows that the initial state of the control object can also be random in nature.

The control objective is assumed to ensure the inequality

$$J\left[x_0, U_0^{t_*-1}(\cdot)\right] \leq r^{-1} \tag{2.3}$$

where r is the equality level and J the cost functional :

$$J\left[x_0, U_0^{t_*-1}(\cdot)\right] = \mathbf{M}q\left(x^{t_*}, u^{t_*-1}\right), \tag{2.4}$$

Here, $q(\cdot)$ is the non-negative (objective) functions of state variables and control variables, M is the symbol for mathematical expectation, $U_0^{t_*-1}(\cdot)$ is the control strategy that realises the control u^{t_*-1} (vide para 1.2.2).

Let us assume that the class of admissible control strategies $\mathbf{U}^a$ are separated out. Study of the optimization problem will be limited to those having the control quality level r in eqn.(2.3) as

$$r^{-1} = r_{\text{opt}}^{-1} = \inf_{U_0^{t_*-1}(\cdot) \in \mathbf{U}^a} J\left[x_0, U_0^{t_*-1}(\cdot)\right]$$

i.e., it is the highest. Thus, the optimization of discrete stochastic control problems in (2.1), (2.2) consists of determination (choice, construction, synthesis) of strategy $U^{t-1}(\cdot) \in \mathbf{U}^a$, that satisfies the highest level of control quality.

2.2.2 *Dependence of the Optimal Solution on the Choice of the Admissible Control Strategies*

The possibility of solving the optimization problem (as presented in para 2.2.1) and determination of optimal synthesis algorithm by choosing the class of admissible control strategies $\mathbf{U}^a$, have been established. As a simple example, dependence of three important classes of admissible strategies will be illustrated here. Let the sealer process with parametric disturbance $\{v_t\}$ be expressed as

$$x_{t+1} = x_t + u_t v_{t+1}, \quad y_t = x_t. \tag{2.5}$$

The set $\mathbf{V}$ of noise variables is assumed to possess two values *i.e.*, $\mathbf{V} = \{-1, 1\}$. Moreover, the independent random variables v_t have identical distribution

$$\mathbf{P}\{v_t = 1\} = \frac{3}{4}, \quad \mathbf{P}\{v_t = -1\} = \frac{1}{4}. \tag{2.6}$$

The control variable set $\mathbf{U}$ will also be chosen to have two values, $\mathbf{U} = \{-1, 1\}$. Let the initial state of the control object be a null quantity, $x_0 = 0$. The control problem will be studied in the first two intervals of time ($t = 0,1$). The control objective is minimization of the functional

$$J = \mathbf{M} \sum_{t=0}^{1} q(x_t, X_{t+1}) \tag{2.7}$$

where q(.) is a non-negative objective function given for a pair of states of the control object. Because of the imposed conditions, the set of state variables X will have five values, viz. $X = \{-2, -1, 0, 1, 2\}$. The plant dynamics can be conveniently illustrated in the form of a graph in the $\{t, x\}$ plane (Fig.2) where the transition from one state to another is shown by arrows and along the path are shown 'cost of transition' $q(x_s, x_t)$.

(*a*) *Program Control.* The class of program control strategy $\mathbf{U}^a$ consists of four elements

$$\mathbf{U}^a = \{(u_0, u_1)\} = \{(1,1), \quad (1,-1), \quad (-1,1), \quad (-1,-1)\}.$$

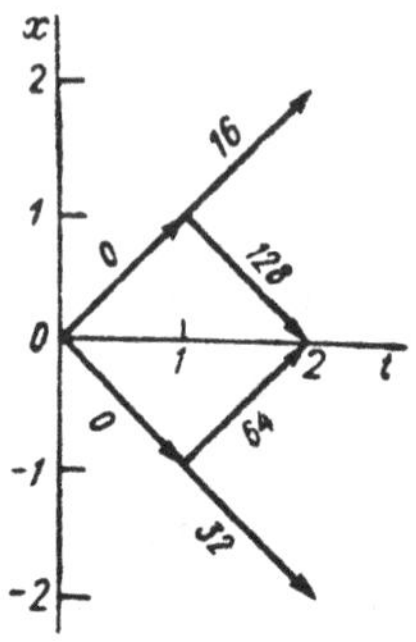

Figure 2 : Graphical Representation of Transition Dynamics of the State for the Example under Consideration

It may be recalled that the programmed strategy realises the control law under the conditions that when the output of the control object (here it is $y_t = x_{t)}$ is not being observed and hence the signal $u_t = \pm 1$ is performed independent of what states the control object attains at $t = 1$.

(b) Closedloop Control (Feedback). Possibility of observing the control objects output may improve the quality of control. The admissible strategies of the closed loop control give rise to pairs U_0, $U_1(u_0, x_1)$, each of which assumes a value in the set $\mathbf{U} = \{-1,1\}$. In this case, optimal strategy can be easily determined using dynamic programming.

(c) Program - closedloop Control. Another class of control strategy, which is between the program control and the closedloop control will be presented. This class is known as *program - closedloop control.*

The admissible strategies from this class are determined as follows: program control is formed at the initial stage. As a result of the control action u_0 of the program control, the control object attains a state x_1. At $t = 1$ new program control is synthesised assuming that the state x_1 is known, etc.

The control algorithm partially makes use of the feedback signals. Only the current state of the control object (output variable) is assumed to be known. But the possible availability of the future state of the control object output is not assumed.

(d) Summary of Computed Results of Optimal Control for Various Classes of Admissible Strategies.

The computed results for various control strategies applied to the given example, are shown in a tabular form

Program Control	*Program-closedloop Control*	*Closedloop Control*
$u_0 = 1$	$u_0 = 1$	$u_0 = 1$
$u_1 = 1$	$u_1 = \begin{cases} 1, & \text{for } x_1 = 1 \\ -1, & \text{for } x_1 = -1 \end{cases}$	$u_1 = \begin{cases} 1, & \text{for } x_1 = 1 \\ -1, & \text{for } x_1 = -1 \end{cases}$
$J_{opt} = 47$	$J_{opt} = 43$	$J_{opt} = 41$

The example chosen is such that optimal control law for various classes of strategies and the corresponding minimum values of the functional did not become indentical. In general, optimal program-closedloop control gives no worse quality of control than that by optimal program-control method. But it does not give a better result than that by optimal closedloop control method. However, synthesis of optimal program-closedloop control is easier than the optimal closedloop control. Moreover, in many practical cases the quality of control by the latter two methods appears to be identical.

§ 2.3 Stochastic Dynamic Programming

2.3.1 *Description of the Method*

The generalised approach to synthesis of optimal feedbacks in problems (2.1), (2.2) and (2.4) will be now described. Let the class of admissible control strategies

$$\mathbf{U}^a = \mathbf{U}_0^a \otimes \mathbf{U}_1^a \otimes \ldots \otimes \mathbf{U}_{t_*-1}^a$$

which consist of sequences of functions $U_0^{t_*-1}(\cdot)$, $U_t(\cdot) \in \mathbf{U}_t^a$, be fixed. The latter functions have values from the set U and the control signals are constructed as per following rules

$$u_0 = U_0(y_0, w_0), \quad u_t = U_t(u^{t-1}, y^t, w^t),$$

$$t = 1, 2, \ldots, t_* - 1. \tag{3.1}$$

Here measurement delay is not present ($s = 0$), but the strategy may be randomised $w \in \mathbf{W}$.

For the above method it is essential that the control interval is finite ($t_* < \infty$). Minimization of functional (2.4) is carried out 'backward in the time domain' starting at the instant t_*-1.

Let any strategy $U^{t_*-1}(\cdot)$ be chosen. The variables x_t, y_t, u_t then become random variables with probability distribution determined by the functions $X_t(.)$, $Y_t(.)$, $U_t(.)$ in equations (2.1), (2.2), (3.1) and by probability distributions of random values v^{t_*}, w^{t_*}. Based on this the value

$$q^{(1)}\left(z^{t_*-1}, u^{t_*-2}; U^{t_*-1}\right) = \mathbf{M}\left\{q\left(x^{t_*}, u^{t_*-1}\right) \mid z^{t_*-1}, u^{t_*-2}\right\} \tag{3.2}$$

is determined. Here $z = (y, w) \in \mathbf{Z} = \mathbf{Y} \otimes \mathbf{W}$ and $\mathbf{M}\{q \mid z^{t_*-1}, u^{t_*-2}\}$ is the conditional mathematical expectation of the random variable q. Formula (3.2) determines random variables $q(\cdot, \cdot; U^{t_*-1})$ as a deterministic function of random variables z^{t+1}, u^{t-1} for all values taken from the set $\mathbf{Z}^{t_*-1}$, $\mathbf{U}^{t_*-2}$. The form of the function $q^{(1)}$ (.) is assumed to depend on the chosen strategy U^{t_*-1} (.) and is reflected on it. Taking (3.2) into account the functional (2.4) can be expressed as

$$J\left[U^{t_*-1}(\cdot)\right] = \mathbf{M}q^{(1)}\left(z^{t_*-1}, u^{t_*-2}; U^{t_*-1}\right), \tag{3.3}$$

where the mathematical expectation is for random variables z^{t_*-1}, u^{t_*-2}. It follows from (3.3) that for optimal control the function $U_{t_*-1}(\cdot)$ must minimize the function $q^{(1)}(.)$ and it can be conditionally expressed as

$$\tilde{U}_{t_*-1}\left(y^{t_*-1},u^{t_*-2};\omega^{t_*-1}\right)=\underset{U_{t_*-1}(\cdot)\in \mathbf{U}^a_{t_*-1}}{\operatorname{argmin}}\ q^{(1)}\left(\cdot,\cdot;U^{t_*-1}\right). \tag{3.4}$$

The possibility of dependence of functions $\tilde{U}_{t_*-1}(\cdot)$ on w^{t_*-1} means that the control signal $u_{t_*-1}=\tilde{U}_{t_*-1}\left(u_{t_*-2}y^{t_*-1}\omega^{t_*-1}\right)$, for fixation of variables u^{t_*-1}, y^{t_*-1}, is determined as a random variable, whose properties are given by the form of the function $\tilde{U}t_{t_*-1}(\cdot)$ and by the distribution of random variable ω^{t_*-1}. The problem of actual determination of function (3.4) will not be considered now but we shall assume that the function (3.4) can be obtained. Keeping the strategy $\left(U^{t_*-2}(\cdot),\tilde{U}_{t_*-1}(\cdot)\right)$ fixed, determination of the optimal function $\tilde{U}_{t_*-2}(\cdot)$ can be continued. Thus, at the k-th stage of the process, we get

$$q^{(k)}\left(z^{t_*-k},u^{t_*-k-1}.U^{t_*-k},\tilde{U}^{t_*-1}_{t_*-k+1}\right)$$

$$=\mathbf{M}\left\{q^{(k-1)}\left(z^{t_*-k+1},u^{t_*-k};U^{t_*-k},\tilde{U}^{t_*-1}_{t_*-k+1}\right)\mid z^{t_*-k};u^{t_*-k-1}\right\} \tag{3.5}$$

Minimization of (3.5) w.r.t. U_{t_*-k} is the optimum :

$$\tilde{U}_{t_*-k}\left(y^{t_*-k},u^{t_*-k-1},w^{t_*-k}\right)$$

$$=\underset{U_{t_*-k}(\cdot)\in \mathbf{U}^a_{t_*-k}}{\operatorname{argmin}}\ q^{(k)}\left(\cdot,\cdot;U^{t_*-k},\tilde{U}^{t_*-1}_{t_*-k+1}\right). \tag{3.6}$$

On completion of the procedure we get the optimal strategy $U^{t_*-1}(\cdot)$ as in (3.1).

2.3.2 *Bellman's equation for stochastic control systems*

Dynamic programming as given above is closely related to Bellman's functional equation. We introduce the functions $V_t(\cdot):\mathbf{Z}^{t_*-1}\otimes\mathbf{U}^{t_*-1}\otimes\mathbf{U}^a\to\mathbf{R}^1$, given by the formulae

$$V_t\left(z^{t_*-t},u^{t_*-t-1};U^{t_*-t-1}\right)=\min_{U\in\mathbf{U}^a_{t_*-t}} q^{(t)}\left(\cdot,\cdot;U^{t_*-t-1},U,\tilde{U}^{t_*-1}_{t_*-t+1}\right), \tag{3.7}$$

where functions $q^{(t)}(\cdot)$ are determined by relations (3.5). Then

$$V_t\left(z^{t_*-t},u^{t_*-t-1};U^{t_*-t-1}\right)$$

$$=\min_{U\in\mathbf{U}_{t_*-t}}\mathbf{M}\left\{V_{t-1}\left(z^{t_*-t+1},u^{t_*-t};U^{t_*-t-1},U\right)\mid z^{t_*-t},\ u^{t_*-t-1}\right\} \tag{3.8}$$

i.e., we resort to stochastic version of Bellman's equation (1.15). The equation is to be supplemented by initial condition

$$V_1\left(z^{t_*-1},u^{t_*-2};U^{t_*-2}\right)=\min_{U\in\mathbf{U}^a_{t_*-1}} q^{(1)}\left(z^{t_*-1},u^{t_*-2};U^{t_*-2},U\right), \tag{3.9}$$

that follows eqns. (3.2), (3.7). The structure of eqn. (3.8) is considerably more complex than that of eqn. (1.15), since in the former Bellman's function $V_t(\cdot\,,\cdot\,;\quad U^{t_*-t-1)}$ at any instant t is not only the function of the variables z^{t_*-t}, u^{t_*-t-1} but also of the preceding control strategy $U^{t_*-t-1}(\cdot)$. Moreover, in (3.8) there is averaging by ensemble operation.

2.3.3 *Example: Linear quadratic problem with randomly varying coefficients and observable states of the control object.*

The control object equation given in (1.18) will be examined again but this time the coefficient matrices A_t, B_t of the equation will be taken as random quantities.

The cost functional then takes the form

$$J\left[U_0^{t_*-1}(\cdot)\right] = \mathbf{M}\sum_{t=1}^{t_*}(x_t^* Q_{t-1} x_t + u_{t-1}^* R_{t-1} u_{t-1}), \tag{3.10}$$

where the non-negative deterministic matrices Q_t, R_t possess the properties (1.21). The admissible strategies class $\mathbf{U}^a$ consists of sequences of functions $U_0^{t_*-1}(\cdot), U_t(\cdot)\quad : \mathbf{U}^{t-1}\otimes \mathbf{X}^t \otimes \mathbf{W}^t \to \mathbf{U} = \mathbf{R}^m$. The control $u_0^{t_*-1}(\cdot)$ is determined when the strategy $U_0^{t_*-1}(\cdot)$, given by the following relation, is fixed :

$$u_0 = U_0(w_0), u_t = U_t(u^{t-1}, x^t, w^t),$$

$$t = 1, 2, \ldots\ldots, t_* - 1. \tag{3.11}$$

Regarding the stochastic properties of the matrices $\{A_t, B_t\}$, the following assumptions are made.

1. The pair of matrices $\{A_t, B_t\}$ are (stochastically) independent for all values of t.
2. Matrices $\{A_t, B_t\}$ are in totality independent with noise ω^{t_*-1} randomising the admissible control strategies.
3. Elements of the random matrices $A_t^* A_t$, $B_t^* B_t$ have finite mathematical expectations, and

$$\mathbf{M}B_t^* B_t > 0, \qquad t = 1, \ldots\ldots, t_* - 1. \tag{3.12}$$

Theorem 2.3.1 *Optimal control u^{t_*-1} in linear quadratic problem (1.18), (3.10) with independently varying coefficients that satisfy the assumptions 1 to 3, and with noise free observable states of the object, can be determined by the formula*

$$u_t = K_t x_t, \tag{3.13}$$

where matrix coefficient K_t is given by the relations

$$K_t = -L_{t+1} T_t, \quad L_{t+1}^{-1} = R_t + N_t,$$

$$T_t = \mathbf{M}\left(B_t^* P_{t_*-t} A_t\right),$$

$$N_t = \mathbf{M}\left(B_t^* P_{t_*-t} B_t\right) \tag{3.14}$$

The non-negative matrices P_t are determined from the matrix equation

$$P_{t+1} = M_{t_*-t} - T^*_{t_*-t} L_{t_*-t+1} T_{t_*-t} + Q_{t_*-t-1}, \tag{3.15}$$

$$M_t = \mathbf{M}\left(A_t^* P_{t_*-t} A_t\right), \tag{3.16}$$

with the initial condition

$$P_1 = Q_{t_*-1}. \tag{3.17}$$

Hence

$$\min_{U_0^{t_*-1}(\cdot) \in \mathbf{U}^a} J\left[U_0^{t_*-1}(\cdot)\right] = x_0^* P_{t_*} x_0.$$

Matrix equation of (3.15) is analog of the Riccati equation (1.25). If A_t, B_t are deterministic matrices then formulae (3.13) to (3.16) become identical to (1.22) to (1.25). Matrices P_t are non-negative for all t.

2.3.4 *Example: Linear stationary object with control delay*

When control delay is present, synthesis of optimal feedback control, as given in para 2.3.1, requires some modification. The procedure is illustrated with the control object described by scalar 'input-output' variables

$$a(\nabla) y_t = b(\nabla) u_t + v_t, \quad t = 0, 1, \ldots, t_*. \tag{3.18}$$

Here, ∇ is the operation in time backward (vide para 1.3.3.),

$$a(\lambda) = 1 + \lambda a_1 + \ldots + \lambda^n a_n,$$

$$b(\lambda) = \lambda^k b_k + \ldots + \lambda^n b_n, \tag{3.19}$$

$v_0^{t_*}$ is the random noise with independent values and properties

$$\mathbf{M} v_t = 0, \quad \mathbf{M} v_t^2 = \sigma_v^2. \tag{3.20}$$

Equation (3.18) is to be supplemented with initial conditions of the variables for $t < 0$. Let us assume, for simplicity's sake, $y_t = u_t = 0$ for $t < 0$. The control delay is given by the natural number k, $0 < k < t_*$.

Let the control objective be minimization of performance criterion (cost functional)

$$J\left[U_0^{t_*-k}(\cdot)\right] = \mathbf{M}\, | y_{t_*} |^2. \tag{3.21}$$

Here, the problem is that of *terminal point control.* It is evident that the control process ends at the moment $t_* - k$ and the later control actions can not change y_{t_*}. In the given case control strategy appears to be linear. With the establishment of this fact, it will be easier to make use of the following algebraic assertion.

Lemma 2.3.1 *Let the polynomials* $a(\lambda)$, $b(\lambda)$ and $g(\lambda)$ *be given. For existence of polynomials* $\alpha(\lambda)$ and $\beta(\lambda)$ *that satisfy the relationship*

$$a(\lambda)\alpha(\lambda) - b(\lambda)\beta(\lambda) = g(\lambda), \tag{3.22}$$

for any complex λ, *the necessary and sufficient condition is that the polynomial* $g(\lambda)$ *is completely divisible by the greatest common divisor (G.C.D) of* $a(\lambda)$, $b(\lambda)$, *i.e., by* $\nu(\lambda) = \langle a(\lambda) ; b(\lambda) \rangle$.

The polynomials $\alpha(\lambda)$, $\beta(\lambda)$ *are determined uniquely by any of the following conditions, when the relationship* (3.22) *is solvable:*

$$\deg \alpha(\lambda) < \deg [b(\lambda)\nu^{-1}(\lambda)],$$

$$\deg \beta(\lambda) < \deg [a(\lambda)\nu^{-1}(\lambda)], \tag{3.23}$$

where the symbol 'deg' indicates the degree (index) of the corresponding polynomial. If the inequality $\deg g(\lambda) < \deg a(\lambda) + \deg b(\lambda) - \deg \gamma(\lambda)$ *is satisfied, then each of the conditions in* (3.23) *leads to one and same pair of polynomials* $(\alpha(\lambda), \beta(\lambda))$.

For actual determination of the polynomials $\alpha(\lambda)$, $\beta(\lambda)$ from the relationship (3.22) one can make use of either euclidean algorithm or equation (3.22) for linear algebraic systems related to coefficients of polynomials under consideration.

Returning to the problem of optimal control, the relationship

$$a(\lambda)F(\lambda) - \lambda^k G(\lambda) = 1 \tag{3.24}$$

for polynomials $F(\lambda)$, $G(\lambda)$, is examined. Since $a(0) = 1$, then the polynomials $a(\lambda)$ and λ^k are mutually unreducible and because of lemma 2.3.1 the eqn. (3.24) can be determined uniquely if the condition $\deg F(\lambda) < k$ is enforced.

Considering the eqn. (3.18) for $t = t_*$, and using the relationship in (3.24), the operation $F(\nabla)$ can be determined as

$$y_{t_*} = -G(\nabla)y_{t_*-k} + F(\nabla)b(\nabla)u_{t_*} + F(\nabla)v_{t_*}.$$

Since

$$F(\lambda) = F_0 + \lambda F_1 + \ldots + \lambda^{k-1}F_{k-1}, \quad F_0 = 1, \tag{3.25}$$

then random values $F(\nabla)v_{t_*} = v_{t_*} + F_1 v_{t_*-1} + \ldots + F_{k-1}v_{t_*-k+1}$ are centered and are independent of random values $G(\nabla)y_{t_*-k} - F(\nabla)b(\nabla)u_{t_*}$ and hence, because of (3.20), we have

$$\mathbf{M}\left\{| y_{t_*} |^2 | y^{t_*-k}, u^{t_*-k-1}\right\} = | -G(\nabla)y_{t_*-k} + b_k\overline{U}_{t_*-k}$$

$$+ [F(\nabla)b(\nabla) - b_k\nabla^k]U_{t_*} |^2 + \sum_{s=0}^{k-1} F_s^2\sigma_v^2$$

$$+ b_k^2 \mathbf{M}\left\{| \overline{U}_{t_*-k} - U_{t_*-k} |^2 | y^{t_*-k}, u^{t_*-k-1}\right\}, \tag{3.26}$$

where $\overline{U}_{t_*-k} = \mathbf{M}\left\{U_{t_*-k}\left(y^{t_*-k}, u^{t_*-k-1}, w^{t_*-k}\right) | y^{t_*-k}, u^{t_*-k-1}\right\}$.

Using (3.26) the following result may be obtained.

Theorem 2.3.2 *Optimal control of problem* (3.18) *to* (3.21) *is determined by the relationship*

$$F(\nabla)b(\nabla)u_{t_*} = G(\nabla)y_{t_*-k} \tag{3.27}$$

where $F(\lambda)$, $G(\lambda)$ are the polynomials that can be determined uniquely by relationship (3.24), *deg $F(\lambda) < k$. Thus,*

$$\min_{U^{t_*-k}(\cdot)\in U^a} J\left[U_0^{t_*-k}(\cdot)\right] = \sum_{s=0}^{k-1} F_s^2\sigma_v^2. \tag{3.28}$$

Thus, in the given problem optimal control is realised in the form of linear nonrandom feedback loop.

From the relationship of (3.27) the control signal u_{t_*-k} is determined. Choice of previous control signals is in no way regulated. The noteworthy character of feedback loop in (3.27) is the independence of its loop coefficients from time t_*, which requires minimization of output variable dispersion. If the control signal for all values of t can be formulated as per following law

$$F(\nabla)b(\nabla)u_t = G(\nabla)y_{t-k}, \tag{3.29}$$

then the output variable dispersion will be the least (for fixed initial conditions) w.r.t. the unpredicted arbitrary strategies for all $t > k$. It the delay in control is minimum (k = 1), then from (3.24) it follows that $F(\lambda) = 1$ and

$$G(\lambda) = \frac{(a(\lambda)-1)}{\lambda} = a_1 + \lambda a_2 + \ldots + \lambda^{n-1}a_n.$$

Equation (3.29) is then re-written as

$$b(\nabla)u_t = [a(\nabla)-1]y_t \tag{3.30}$$

or when (3.19) is taken into account, it is expressed as

$$b_1u_t + \ldots + b_nu_{t-n+1} = a_1y_t + a_2y_{t-1} + \ldots + a_ny_{t-n+1}.$$

This result is a natural consequence of (3.18). With control (3.30) the equation (3.18) takes the form $y_t = v_t$, and because of the independence of the random variables $\{v_t\}$ it provides the least output variable dispersion for all $t > 1$.

§ 2.4 Bayesian Control Strategy

Application of dynamic programming to stochastic control objects necessitates computation of conditional mathematical expectation of some random values when others are fixed. Since the control strategy is fixed, all the random values can be obviously expressed through random signals and this, in principle, permits computation of necessary conditional average. Such a technique of excluding the intermediate variables for computation of conditional average is not always advisable. It is usually convenient to carry out evaluation of their a posteriori probabilistic distribution (instead of elimination of intermediate variables) and then carry out conditional averaging with the help of this a posteriori distribution. This method of computation of conditional averaging (functions (3.5) and (3.8)), as required in dynamic programming, is related to application of Bayesian formulae for multiplication of probabilities. It is known as *Bayesian method of optimization* and the optimal control strategy determined by the method as *Bayesian* strategy. The basis of the Bayesian method of synthesis of optimal strategy consists of formulae for recursive computation of *a posteriori* distributions as given by R.L. Stratovich in his studies on conditional Markov processes.

The Bayesian method for control problem is given below in a form simpler than that in § 2.3.

2.4.1 *Bayesian approach to the optimization problems*

Let the control object be described by equations

$$x_0 = X_0(v'_0), \quad x_t = X_t(x_{t-1}, u_{t-1}, v'_t),$$

$$t = 1, 2, \ldots, t_*, \tag{4.1}$$

and the output by

$$y_t = Y_t(x_t, v''_t), \quad t = 0, 1, \ldots, t_*. \tag{4.2}$$

Here, $v_t = \text{col}\,(v'_t, \ \ v''_t)$ is the noise signal (at instant *t)* in the control object and in the measurement channel. Let the noise v^{t_*} be random and its probability distribution $\mu_v(\cdot)$ be known. From (4.1) it follows that the initial state x_0 of the control object may be random quantities.

The control objective is minimization of the performance criterion

$$J = \mathbf{M} \sum_{t=1}^{t_*} q_t(x_t, u_{t-1}), \tag{4.3}$$

where $q_1^{t_*(\cdot)}$ is the collection of non-negative functions of state and control variables.

Let the admissible control strategies class $\mathbf{U}^a$ consists of sequences $U^{t_*-1}(\cdot)$ of unpredictable random functions. Fixation of a sequence like this results in control u^{t_*} given by formulae

$$u_0 = U_0(y_0, w_0), \quad u_t = U_t(u^{t-1}, y^t, w^t), \quad t = 1, 2, \ldots, t_*. \tag{4.4}$$

Here, w^{t_*} is the noise in the control randomizing the control strategy (vide para 1.2.2b) when the strategy $U^{t_*-1}(\cdot)$ is fixed, the functional (4.3) can be written as

$$J\left[U^{t_*-1}(\cdot)\right] = \mathbf{M} \sum_{t=1}^{t_*-1} q_t(x_t, \ \ u_{t-1}) + \mathbf{M}q^{(1)}\left(y^{t_*-1}, u^{t_*-2}\right), \tag{4.5}$$

where

$$q^{(1)}\left(y^{t_*-1}, u^{t_*-2}\right) = \mathbf{M}\left\{q_{t_*}\left(x_{t_*}, U_{t_*-1}\right) \mid y^{t_*-1}, u^{t_*-2}\right\}. \tag{4.6}$$

In the formula (4.6) all the variables $x_{t_*}, y^{t_*-1}, u^{t_*-2}$ are essential random variables (according to relations (4.4), (4.1) and (4.2)), which are single valued functions of v^{t_*}, w^{t_*-1}. The random quantity $q^{(1)}(\cdot)$ is given by relationship (4.6) as a deterministic function of all random variables y^{t_*-1}, u^{t_*-2}. (The function assumes the form depending on chosen strategy $U_0^{t_*-1}(\cdot)$.) It will be further assumed that the function $q^{(1)}(\cdot)$ is determined for all values of $y^{t_*-1} \in \mathbf{Y}^{t_*-1}, u^{t_*-2} \in \mathbf{U}^{t_*-2}$. For this type of formulation of random variables y^{t_*-1}, u^{t_*-2}, in the form of their arguments (as determined by relations (4.1), (4.2) and (4.4)), will lead to random variables which satisfy the relationship (4.6) with probability 1.

Since only the last term in the right hand side of the formula (4.5) depends on choice of $U_{t_*-1}(\cdot)$, then, according to dynamic programming, synthesis of optimal control begins with the choice of $U_{t_*-1}(\cdot)$ from the minimal condition of function $q^{(1)}(\cdot)$.

For precise realization of the choice, collection of conditional distributions will be introduced, and the relationship between them will be expressed through Bayesian formulae.

2.4.2 *A Posteriori Distribution and Bayesian Formula*

The conditional probability distribution for synthesis of optimal control can be related to random values x_t, u_t, y_t when control strategy is fixed. To express these distributions, $\mathbf{A}_x$, $\mathbf{A}_u$, $\mathbf{A}_y$, $\mathbf{A}_{xu}$ will designate the borel sets corresponding to sets $\mathbf{X}$, $\mathbf{U}$, $\mathbf{Y}$, $\mathbf{X} \otimes \mathbf{U}$, etc. Again to indicate, for example, that $\mathbf{A}_x$ is an arbitrary borel set then the symbol $\forall\, \mathbf{A}_x$ is used. $I_A(\cdot)$ is the indicator of the set $\mathbf{A}$*.

Because of the properties of conditional mathematical expectations we get from the equality $I_{A_x}(x_{t+1})I_{A_u}(u_t) = I_{A_x \otimes A_u}(x_{t+1}, u_t)$ the following expression

$$\mathbf{M}\{I_{A_x \otimes A_u}(x_{t+1}, u_t) \mid y^t, u^{t-1}\}$$

$$= \mathbf{M}\left[I_{A_u}(u_t)\mathbf{M}\{I_{A_x}(x_{t+1}) \mid y^t, u^t\} \mid y^t, u^{t-1}\right]. \tag{4.7}$$

Let

$$\mu_{x_{t+1},u_t}(\mathbf{A}_x \otimes \mathbf{A}_u \mid y^t, u^{t-1}) = \mathbf{M}\{I_{A_x \otimes A_u}(x_{t+1}, u_t) \mid y^t, u^{t-1}\}, \tag{4.8}$$

$$\mu_{x_{t+1}}(\mathbf{A}_x \mid y^t, u^t) = \mathbf{M}\{I_{A_x}(x_{t+1}) \mid y^t, u^t\}, \tag{4.9}$$

$$\mu_{u_t}(\mathbf{A}_u \mid y^t, u^{t-1}) = \mathbf{M}\{I_{A_u}(u_t) \mid y^t, u^{t-1}\}. \tag{4.10}$$

Formulae (4.8) to (4.10) determine the conditional probability distributions $\mu_{x_{t+1},u_t}(\cdot \mid y^t, u^{t-1})$, $\mu_{x_{t+1}}(\cdot \mid y^t, u^t)$, $\mu_{u_t}(\cdot \mid y^t, u^{t-1}$ as measurable functions of random variables corresponding to $(y^t,\ u^{t-1})$, $(y^t,\ u^t)$, $(y^t,\ u^{t-1})$. Obviously,

$$\mu_{x_{t+1},u_t}(\mathbf{X} \otimes \mathbf{A}_u \mid y^t, u^{t-1}) = \mu_{u_t}(\mathbf{A}_u \mid y^t, u^{t-1}) \tag{4.11}$$

for all $\mathbf{A}_u \subseteq \mathbf{U}$ and for almost all $(y^t, u^{t-1}) \in \mathbf{Y}^t \otimes \mathbf{U}^{t-1}$.

If $\phi(u_t)$ is an arbitrary function summed up over distribution $\mu_{u_t}(\cdot \mid y^t, u^{t-1})$ then for random variables u^t, y^t, u^{t-1} with probability 1, we get the equality

*Thus, $I_{A_x}(x) = 1$, if $x \in \mathbf{A}_x$; otherwise $I_{A_x}(x) = 0$.

$$\mathbf{M}\{\phi(u_t)I_{A_u}(u_t) \mid y^t, u^{t-1}\} = \int_{A_u} \phi(u)\mu_{u_t}(du \mid y^t, u^{t-1}), \tag{4.12}$$

which follows from the properties of the Lebeg-Stiltes integral. Partially when $\phi(u_t) = \mu_{x_{t+1}}(\mathbf{A}_x \mid y^t, u^t)$we get

$$\mathbf{M}\{\mu_{x_t+1}(\mathbf{A}_x \mid y^t, u^t)I_{A_u}(u_t) \mid y^t, u^{t-1}\}$$

$$= \int_{A_u} \mu_{x_{t+1}}(\mathbf{A}_x \mid y^t, u^t)\mu_{u_t}(du_t \mid y^t, u^{t-1}).$$

This formula along with (4.12) permits rewriting of expression (4.7) as

$$\mu_{x_{t+1},u_t}(\mathbf{A}_x \otimes \mathbf{A}_u \mid y^t, u^{t-1})$$

$$= \int_{A_u} \mu_{x_{t+1}}(\mathbf{A}_x \mid y^t, u^t) \cdot \mu_{u_t}(du_t \mid y^t, u^{t-1}). \tag{4.13}$$

The above equality is the *Bayesian formula* for multiplication of probabilities. It is more known in the differential form

$$\mu_{x_{t+1},u_t}(dx_{t+1}du_t \mid y^t, u^{t-1})$$

$$=\mu_{x_{t+1}}(dx_{t+1} \mid y^t, u^t)\mu_{u_t}(du_t \mid y^t, u^{t-1}). \tag{4.14}$$

Assuming $\mathbf{A}_u = \mathbf{U}$, we get from (4.13)

$$\mu_{x_{t+1}}(\mathbf{A}_x \mid y^t, u^{t-1} = \int_{\mathbf{U}} \mu_{x_{t+1}}(\mathbf{A}_x \mid y^t, u^t)\mu_{u_t}(du_t \mid y^t, u^{t-1}) \tag{4.15}$$

the convolution distribution along u_t. The conditional distribution $\mu_{x_{t+1}}(\bullet \mid y^t, u^{t-1})\,(\mu_{u_t}(\bullet \mid y^t, u^{t-1}))$ is known as *a posteriori distribution* of random variables x_{t+1} (corresponding to u_t) during observation of y^t, u^{t-1} and conditioned by randomisation of control strategy*.

With a determined strategy, the distribution $\mu_{u_t}(\cdot)$ is concentrated on some point $u = u(y^t, u^{t-1})$, *i.e.*, for every $\mathbf{A}_u \subseteq \mathbf{U}$ it's fulfilled

$$\mu_{u_t}(\mathbf{A}_u \mid y^t, u^{t-1}) = \begin{cases} 1, & \text{if } u(y^t, u^{t-1}) \subseteq \mathbf{A}_u, \\ 0 & \text{otherwise.} \end{cases}$$

---

*Random control is often expressed in the form of a family of conditional distributions $\mu_{u_t}(\cdot \mid y^t, u^{t-1})$ along with relations (4.4). The latter determines the computational algorithm for control signal u_t for realising value of noise w_t (when the quantities y^t, u^{t-1} are fixed). Only the mechanism of generation of random variables w^t, representing the noise (of elementary event), is not defined. For the problem of conditional distributions $\mu_{u_t}(\cdot \mid y^t, u^{t-1})$ algorithm to formulate control signals is not shown, and it is not required for assuming the control objective. It is so because the performance criterion does not change its value for any method of generation of u_t. So the conditional distribution $\mu_{u_t}(\cdot \mid y^t, u^{t-1})$ remains unchanged.

The following simple assertion is necessary for the latter statement.

Lemma 2.4.1 *Let us assume that*

$$\mu_{v_{t+1},w_t}(\cdot \mid x^t, y^{t-1}, u^{t-1}, v^t, w^{t-1})$$

$$=\mu_{v_{t+1},w_t}(\cdot \mid y^{t-1}, u^{t-1}, v^t, w^{t-1}) \tag{4.16}$$

for every strategy $U^{t_*-1}(\cdot) \in \mathbf{U}^a$, i.e., a posteriori distribution of random variables v_{t+1} and w_t does not depend on x_t^*.

Then the a posteriori distribution $\mu_{x_{t+1}}(\cdot \mid y^t, u^t)$ does not depend on the choice of control strategy $U^{t_*-1}(\cdot) \in \mathbf{U}^a$.

2.4.3 *Regularity in Bayesian Control Strategy*

The independence of the form of functions $\mu_{x_{t+1}}(\cdot) \mid y^t, u^t)$ from the choice of control strategy permits us to establish that nonrandomised (regular) control strategy can be chosen as the optimal one.

Theorem 2.4.1 *Let us assume that $\mathbf{U}^a$ consists of all possible sequences of functions $U_t(\cdot):\mathbf{U}^{t_*-1} \otimes \mathbf{Y}^{t_*} \otimes \mathbf{W}^t \to \mathbf{U}$. Then a regular optimal control strategy exists in the class $\mathbf{U}^a$ if the conditions in lemma 2.4.1 are satisfied.*

Let us explain the method of determining existence of regular strategy and its synthesis procedure. Thus, the function $U_{t_*-1}(\cdot)$ is to be determined from the condition for minimum function (4.5) for fixed values of y^{t_*-1}, u^{t_*-2} and for functions $U^{t_*-2}(\cdot)$. Using a posteriori distributions $\mu_{x_{t+1},\mu_t}(\cdot), \mu_{u_t}(\cdot), \mu_{x_{t+1}}(\cdot)$ and Bayesian formula (4.14) eqn. (4.6) can be rewritten as

$$q^{(1)}\left(y^{t_*-1}, u^{t_*-2}\right) = \int_{\mathbf{X}\otimes\mathbf{U}} q_{t_*}\left(x_{t_*}, y_{t_*-1}\right)$$

$$\times \mu_{x_{t_*}, u_{t_*-1}}\left(dx_{t_*} du_{t_*-1} \mid y^{t_*-1}, u^{t_*-2}\right)$$

$$= \int_{\mathbf{U}} \phi_1\left(y^{t_*-1}, u^{t_*-1}\right) \mu_{u_{t_*-1}}\left(du_{t_*-1} \mid y^{t_*-1}, u^{t_*-2}\right) \tag{4.17}$$

where

$$\phi_1\left(y^{t_*-1}, u^{t_*-2}, u_{t_*-1}\right)$$

$$= \int_{\mathbf{X}} q_{t_*}\left(x_{t_*}, u_{t_*-1}\right) \mu_{x_{t_*}}\left(dx_{t_*} \mid y^{t_*-1}, u^{t_*-1}\right). \tag{4.18}$$

---

* The relationship of (4.16) is fulfilled if $\{v_t, w_t\}$ are independent in the aggregate of random values.

In the conditions laid down in lemma 2.4.1 the function $\phi_1\left(y^{t_*-1}, u^{t_*-2}, u_{t_*-1}\right)$ does not depend on choice of $U_{t_*-1}(\cdot)$ in admissible control strategy $U^{t_*-1}(\cdot)$. Designating the point, where the minimum value of function $\phi_1(\cdot)$ w.r.t. u_{t_*-1} is attained, through $\tilde{U}_{t_*-1}$, we have

$$\tilde{U}_{t_*-1} = \tilde{U}_{t_*-1}\left(y^{t_*-1}, u^{t_*-2}\right) = \underset{u \in \mathbf{U}}{\operatorname{argmin}}\ \phi_1\left(y^{t_*-1}.u^{t_*-2}, u\right). \tag{4.19}$$

It follows from the formula (4.17) that the minimum value of the function $q^{(1)}(.)$ is attained at the function $U_{t_*-1}(\cdot)$ which meets the *a posteriori* distribution $\mu_{u_{t_*-1}}(\cdot \mid y^{t_*-1}, u^{t_*-2})$ concentrated at points where minimum of function $q^{(1)}(\cdot)$ occurs. In particular the function (4.19) is such a function. Then,

$$q^{(1)}\left(y^{t_*-1}, u^{t_*-2}\right) = \phi_1(y^{t_*-1}, u^{t_*-2}, \tilde{U}_{t_*-1}\left(y^{t_*-1}, u^{t_*-2}\right). \tag{4.20}$$

It also gives the possibility of choice for determining the function $U_{t_*-1}(\cdot)$ in an optimal control strategy.

Continuing the above reasoning it is possible to prove the theorem.

Procedure for synthesis of optimal strategy which follows the argument in theorem 2.4.1 will be described. Let $q^{(k)}(\cdot), \phi_k(\cdot), \tilde{U}_{t_*-k}(\cdot)$ be the functions determinable by recursive process

$$\phi_k\left(y^{t_*-k}, u^{t_*-k}\right) = \int_{\mathbf{X}} \left[q_{t_*-k+1}\left(x, u_{t_*-k}\right)\right]$$

$$+ q^{(k-1)}\left[y^{t_*-k}, \overline{Y}_{t_*-k+1}(x), u^{t_*-k}\right] \mu_{x_{t_*-k+1}}\left(dx \mid y^{t_*-k}, u^{t_*-k}\right), \tag{4.21}$$

$$\tilde{U}_{t_*-k} = U_{t_*-k}\left(y^{t_*-k}, u^{t_*-k-1}\right) = \underset{u \in \mathbf{U}}{\operatorname{argmin}}\ \phi_k\left(y^{t_*-k}, u^{t_*-k-1}, u\right), \tag{4.22}$$

$$q^{(k)}\left(y^{t_*-k}, u^{t_*-k-1}\right) = \phi_k\left[y^{t_*-k}, u^{t_*-k-1}, \tilde{U}_{t_*-k}\left(y^{t_*-k}, u^{t_*-k-1}\right)\right], \tag{4.23}$$

$$Y_t(x) = \mathbf{M}\{Y_t(x_t, v'_t) \mid x_t = x\}, \tag{4.24}$$

and the functions $\phi_1(\cdot), q^{(1)}, \tilde{U}_{t_*-1}(\cdot)$ are determined by formulae (4.17) to (4.20). In (4.24) averaging w.r.t. the disturbance (for fixed state of the control object) is carried out. The functions $\tilde{U}_t(\cdot)$ determined by this method give Bayesian control strategy.

2.4.4 *Recursive Formulae for Computations of a Posteriori Distributions*

During realisation of procedure in eqns. (4.17) to (4.24) it is necessary to compute *a posteriori distribution* $\mu_{x_{t+1}}(\cdot \mid y^t, u^{t-1})$, of random values x_{t+1} when y^t, u^{t-1} are observed. As per conditions in lemma 2.4.1 the admissible strategies may be considered regular, *i.e.*

$$\mu_{x_t+1}(\cdot \mid y^t, u^{t-1}) = \mu_{x_t+1}(\cdot \mid y^t), \tag{4.25}$$

as the random values y^t (for fixed strategy) uniquely determine the random values u^{t-1}. Hence, a shorter form will be used now onwards, but it is to be remembered that the expression for distribution $\mu_{x_t+1}(\cdot \mid y^t)$ depends on chosen control strategy.

Theorem 2.4.2 *For conditions laid down in lemma 2.4.1 the conditional distributions* $\mu_{x_t}(\cdot \mid y^{t-1}), \mu_{x_t+1}(\cdot \mid y^t)$ *for any* $\mathbf{A}_x \subseteq \mathbf{X}, \mathbf{A}_y \subseteq \mathbf{Y}$ *satisfy the following relationship*

$$\int_{\mathbf{A}_y} \mu_{x_{t+1}}(\mathbf{A}_x \mid y^t)\mu_{y_t}(dy_t \mid y^{t-1}) = \int_{\mathbf{X}} \mu_{x_{t+1}}, y_t(\mathbf{A}_x \times \mathbf{A}_y \mid x_t, y^{t-1})$$

$$\times \mu_{x_t}(dx_t \mid y^{t-1}), \quad t = 1, 2, \ldots\ldots, t_* - 1, \tag{4.26}$$

$$\mu_{x_1}(\mathbf{A}_x) = \int_{\mathbf{A}_x} X_1(v_1)\mu_{v_1}(dv_1). \tag{4.27}$$

Here the a posteriori distributions $\mu_{y_t}(\cdot \mid y^{t-1}), \mu_{x_{t+1}, y_t}(\cdot \mid x_t, y^{t-1})$ *are determined by relations* $(\forall \mathbf{A}_y, \mathbf{A}_x)$

$$\mu_{y_t}(\mathbf{A}_y \mid y^{t-1}) = \mathbf{M}\{I_{A_y}(y_t) \mid y^{t-1}\},$$

$$\mu_{x_{t+1}, y_t}(\mathbf{A}_x \otimes \mathbf{A}_y \mid x_t, y^{t-1}) =$$

$$= \mathbf{M}\{I_{A_x} \otimes_{A_y} (x_{t+1}, y_t) \mid x_t, y^{t-1}\}. \tag{4.28}$$

Corollary I Let us assume that the following limits exist

$$\lim_{\varepsilon \to 0} \frac{\mu_{x_{t+1}}(D_\varepsilon(x_{t+1}) \mid y^t)}{V_\varepsilon(x_{t+1})} = p(x_{t+1} \mid y^t),$$

$$\lim_{\varepsilon \to 0} \frac{\mu_{x_{t+1}, y_t}(D_\varepsilon(x_{t+1}) \otimes D_\varepsilon(y_t) \mid x_t, y^{t-1})}{V_\varepsilon(x_{t+1}) V_\varepsilon(y_t)}$$

$$= p(x_{t+1}, y_t \mid x_t, y^{t-1}),$$

where $V_\varepsilon(x_{t+1}), V_\varepsilon(y_t)$ are the volumes of the spheres $D_\varepsilon(x_{t+1}), D_\varepsilon(y_t)$ with radius ε and centres at points x_{t+1}, y_t and x_t, y^{t-1} the arbitrary variables in the sets $\mathbf{X}$, $\mathbf{Y}^{t-1}$. Then the *a posteriori* density distributions $p(x_{t+1} \mid y^t)$ satisfy the recursive equations

$$p(x_{t+1} \mid y^t) = \frac{\int p(x_{t+1}, y_t \mid x_t, y^{t-1}) p(x_t \mid y^{t-1}) dx_t}{p(y_t \mid y^{t-1})}. \tag{4.29}$$

The expression $p(y_t \mid y^{t-1})$ in the denominator of the right hand side of the eqn. (4.29) is a function of variables $y^t = (y_0, \ldots, y_t)$ and is determined from the normalisation condition $\int p(x_{t+1} \mid y^t) dx_{t+1} = 1$, *i.e.*

$$p(y_t \mid y^{t-1}) = \int_{\mathbf{X}} \int_{\mathbf{X}} p(x_{t+1}, y_t \mid x_t, y^{t-1}) p(x_t \mid y^{t-1}) dx_t dx_{t+1}. \tag{4.30}$$

When the noise signals v'_t, v^*_{t-1} are independent, formula (4.29) takes a simpler form

$$p(x_{t+1} \mid y^t) = \frac{\int_{\mathbf{X}} p(x_{t+1} \mid x_t, u_t) p(y_t \mid x_t) p(x_t \mid y^{t-1}) dx_t}{\int_{\mathbf{X}} p(y_t \mid x_t) p(x_t \mid y^{t-1}) dx_t}. \tag{4.31}$$

The probability densities $p(x_{t+1} \mid x_t, u_t)$ and $p(y_t \mid x_t)$ are determined by equations (4.1) and (4.2). If the noise signal v_t in equations (4.1) and (4.2) is additive, *i.e.*,

$$X_t(x_t-1,\ u_t-1,\ v'_t) = X_t(x_{t-1},\ u_{t-1}) + v'_t,\quad Y_t(x_t,\ v''_t) = Y_t(x_t) + v''_t,$$

then $p(x_{t+1} \mid x_t,\ u_t) = p_{v'}[x_{t+1} - X_{t+1}(x_t,\ u_t)]$,

$$p(y_t \mid x_t) = p_{v''}[y_t - Y_t(x_t)],$$

where $p_{v'}(\cdot)$, $p_{v''}(\cdot)$ are the density functions of the random values v_t' and v_t''.

§ 2.5 Linear Quadratic Gaussian Problem

In general, realisation of Bayesian control strategy hardly meets any surmountable computational difficulties in relation to necessity for finding out *a posteriori* distributions and minimization of complex functions. The situation is considerably simpler in case of linear quadratic problem with additive Gaussian noise. Here the optimal control can be realised quite effectively. The following paragraph relates to this problem.

2.5.1 *Statement of the Problem*

Let the control object be described by the linear equation

$$x_0 = v'_0,\quad x_{t+1} = A_t x_t + B_t u_t + v'_{t+1},\ t = 0,\ 1,\ \ldots,\ t_* - 1, \tag{5.1}$$

with time variant matrices A_t, B_t which are of $n \times n$ and $n \times m$ dimensions respectively. Moreover, the noise $[v']_0^{t_*}$ is additive. The output of the control object is given by

$$y_t = C_t x_t + v_t'',\quad t = 0,\ 1,\ \ldots,\ t_*, \tag{5.2}$$

where C_t are the time variant rectangular matrices of dimensions $m \times l$,* and $[v'']_0^{t_*}$ are the measurement noise:

It is assumed that the noise signals $v_t =$ Col $(v_t',\ v''_{t-1})$ are independent Gaussian vector quantities.**

$$v_0 \sim N\left(\begin{pmatrix} \bar{x}_0 \\ 0 \end{pmatrix},\ R_v(0)\right),\quad v_t \sim N(0,\ R_v(t)). \tag{5.3}$$

---

* For scalar output C_t is a single row matrix (row vector).

** $v \sim N(\bar{v},\ R)$ means $v =$ Gaussian noise with average $\bar{v}$ and covariance matrix R, i.e., v possesses normal distribution with density function $p(v) \sim N(\bar{v},\ R)$ or (for det $R \neq 0$)

$$p(v) = (2\pi det\ R)^{-n/2} \exp\{-(1/2)(v - \bar{v})^* R^{-1} (v - \bar{v})\},$$

where $n =$ dimension of the space **V** of random values v.

The control objective is to minimize the quadratic functional

$$J\left[U_0^{t_*-1}(\cdot)\right] = \sum_{t=1}^{t_*} \mathbf{M}(x_t^* Q_{t-1} x_t + u_{t-1}^* R_{t-1} u_{t-1}) \tag{5.4}$$

with non-negative matrices Q_t, R_t. In case of $l = n$, $C_t = I_n$ and $v_t'' = 0$, *i.e.* when the states of the control object are observed without any noise, the problem of optimal control synthesis is studied as a noise free problem in para 2.1.4. Presence of noise in the measurement channel sharply complicates the control problem and makes it considerably interesting from the point of view of mathematics.

As shown in § 2.4 Bayesian control strategy is a regular one. Hence, at the beginning itself we choose functions $U_t(.)$ from the set of admissible control strategies $\mathbf{U}^a$ which are deterministic functions of control and output variables.

2.5.2 *Conditional Gaussism of the States and Sufficient Statistics*

Let an admissible control strategy be fixed. Due to non-linear dependence of controls on output, the state variables do not remain Gaussian random values. However, the important characteristic of the problem is the conditional Gaussism of the states and output variables. The conditional Gaussism of random values x_{t+1} signifies that their corresponding *a posteriori* distribution $\mu_{x_{t+1}}(\cdot \mid y^t)$ possesses normal density $p(x_{t+1} \mid y^t)$. Normal density of a posteriori distribution is characterized completely by its own parameters by mean value $\hat{x}_{t+1}$ and covariance distribution matrix :

$$p(x_{t+1} \mid y^t) \sim N(\hat{x}_{t+1}, P_{t+1}). \tag{5.5}$$

These parameters, which are functions of previous observations $\hat{x}_{t+1} = \hat{x}_{t+1}(y^t)$, $P_{t+1}(y^t)$ are known as *sufficient statistics*. Knowledge of sufficient statistics gives the same information about random values as is given by knowledge of *a posteriori* distribution. In conditional Gaussian cases the recursive formulae for computation of *a posteriori* densities may be replaced by recursive formulae for determination of sufficient statistics and it will make the optimal control synthesis procedure effective. The recursive relations for the above statistics will be shown without its proof. The conclusion of these relations is extensive in nature but sufficiently simple.

Theorem 2.5.1 *For independent Gaussian noises* $v_t = \text{col}\,(v_t', v''_{t-1})$ *and arbitrary admissible control strategy* $U^\infty(\cdot)$,

$$u_0 = U_0(y_0),\ u_t = U_t(u^{t-1}, y^t),\ t = 1, \ldots, t_* - 1, \tag{5.6}$$

the sufficient statistics $\hat{x}_t$, P_t (vide (5.5)) *satisfy the recursive relations*

$$\hat{x}_{t+1} = A_t \hat{x}_t + B_t u_t + K_t (y_t - C_t \hat{x}_t), \tag{5.7}$$

$$K_t = [r_{12}(t) + A_t P_t C_t^*]\,[r_{22}(t) + C_t P_t C_t^*]^{-1}, \tag{5.8}$$

$$P_{t+1} = r_{11}(t) + A_t P_t A_t^* - [r_{12}(t) + A_t P_t C_t^*]$$

$$\times [r_{22}(t) + C_t P_t C_t^*]^{-1}\,[r_{12}(t) + A_t P_t C_t^*]^*. \tag{5.9}$$

Here $t = 0, 1, \ldots, t_* - 1$, $r_{ij}(t) =$ *submatrices of matrix* $R_v(t)$:

$$R_v(t) = \begin{Vmatrix} r_{11}(t) & r_{12}(t) \\ r_{12}^*(t) & r_{22}(t) \end{Vmatrix}, r_{22} > 0, \tag{5.10}$$

u_t *are the control signals determined by formulae* (5.6). *The initial conditions in procedure* (5.7) *to* (5.9) *are (vide* (5.3))

$$\hat{x}_0 = \bar{x}_0, \quad P_0 = \mathrm{M}(v_0' - \bar{x}_0)(v'_0 - \bar{x}_0)^*. \tag{5.11}$$

2.5.3 *Bayesian Control Strategy*

It appears that the optimal strategy of the present problem can be simply expressed through sufficient statistics $\hat{x}_t, P_t$ and it will present the Bayesian strategy in the final form. The corresponding well-known and important result in control theory will be formulated below without any proof.

Theorem 2.5.2 *For conditions laid down in theorem 2.5.1 the optimal control (for positive matrices* R_t*) is determined by formulae*

$$u_t = -L_t B_t^* \tilde{Q}_t z_t, \; t = 0, 1, \ldots, t_* - 1. \tag{5.12}$$

Here $z_t = A_t \hat{x}_t + K_t (y_t - C_t \hat{x}_t)$,

$$L_t = (R_t + B_t^* \tilde{Q}_t B_t)^{-1}, \tag{5.13}$$

$$\tilde{Q}_t = Q_t + A_{t+1}^* \tilde{P}_{t_*-t-1} A_{t+1}, \tag{5.14}$$

$$\tilde{P}_{t_*-t} = \tilde{Q}_t - \tilde{Q}_t B_t L_t B_t^* \tilde{Q}_t, \quad \tilde{Q}_{t_*-1} = Q_{t_*-1}, \tag{5.15}$$

vectors $\hat{u}_t$ *and matrices* K_t *are determined in the manner shown in* (5.7) to (5.11).

For optimal control (5.12) *the performance criterion* (5.4) *will have the value*

$$\min J = \bar{x}_0^* A_0^* \tilde{P}_{t_*} A_0 \bar{x}_0 + \sum_{s=1}^{t_*} Sp Q_{s-1}^{1/2} P_s Q_{s-1}^{1/2}$$

$$+ \sum_{s=1}^{t_*-1} Sp \tilde{P}_{t_*-s}^{1/2} K_s [C_s P_s C_s^* + r_{22}(s)] K_s^* \tilde{P}_{t_*-s}^{1/2}. \tag{5.16}$$

We may note that for $v_{12}(t) \equiv 0$ feedback (5.12) takes from

$$u_t = L_t B_t^* \tilde{Q}_t A_t \bar{x}_t, \tag{5.17}$$

$$\bar{x}_t = \hat{x}_t + P_t C_t^* [v_{22}(t) + C_t P_t C_t^*]^{-1} (y_t - C_t \hat{x}_t). \tag{5.18}$$

For observed vector x_t, it is not difficult to show that the optimal feedback has the form $u_t = -L_t B_t^* \tilde{Q}_t A_t x_t$. We observe that the vector (5.18) is the best estimation in the mean square sense of the vector $x_t (\bar{x}_t$ is the mean value of the *a posteriori* density $p(x_t \mid y^t))$ and come to the well-known division theorem for optimal control with incomplete observable state vectors. Here the optimal control has the same form as with observable states. But in addition it uses their best estimates at any instant t. Thus, the control and identification appear to be separable.

§ 2.A Appendix

In this section general forms and basic assertions of the probability theory, as used in the book, will be briefly enumerated. This will be done following axiomatic of A.N. Kolmogorov [80], and can be found in [105].

2.A.1 *General forms of probability theory*

The *probability space* consists of the trio (Ω, **A**, **P**), where Ω is the abstract set of *elementary events w*, **A** is the chosen σ-algebra of subsets of set Ω (*i.e.*, subsets **A** contain Ω, null set ϕ and are closed with respect to operations of union, intersection and addition used in accounting number), and **P** is the *measure of probability* defined in the sets of **A** *i.e.*, $\mathbf{P}(A) \geq 0, A \in \mathbf{A}$, $\mathbf{P}(\Omega) = 1$. Sets are known as *events*. Magnitude **P**(A) is known as the *probability of event A*.

(*a*) *Random variables*. Reflection of sets Ω on some euclidean space $\mathbf{R}^N$ of dimension N will be considered.

The natural σ-algebra **B** of borel sets (the least σ-algebra containing all open sets in $\mathbf{R}^N$ is contained in $\mathbf{R}^N$) is contained in $\mathbf{R}^N$. The reflection

$$\xi : \Omega \to \mathbf{R}^N \tag{A.1}$$

is now as *(vector) random variable* when the prototype of each of the open sets in $\mathbf{R}^N$, with reflection as in (A.1), is an event. (This means, if $B \in \mathbf{B}$, then $\{\omega : \xi(\omega) \in B\} \subseteq \mathbf{A}$.) The random variables ξ, η are equivalent if $\mathbf{P}\{\xi \neq \eta\} = 0$.*

(*b*) *Probability distribution and mean value of random variables*. Every random variable $\xi = \xi(\omega)$ results in some measures defined in sets of **B** in conformity with formula

$$\mu_\xi(B) = \mathbf{P}\{\xi(\omega) \in B\}. \tag{A.2}$$

Measure $\mu_\xi(\cdot)$ is known as probability distribution of random variable ξ. Let $y \in \mathbf{R}^N, \varepsilon > 0$ and $D_\varepsilon(y) = \{\bar{y} : |y - \bar{y}| < \varepsilon\}, V_\varepsilon = \int_{D_\varepsilon(y)} d\bar{y}$. Let there exist

$$\lim_{\varepsilon \to 0} V_\varepsilon^{-1} \mu_\xi(D_\varepsilon(y)) = p\xi(y). \tag{A.3}$$

Quantity $p_\xi(y)$ is known as *density function* $\mu_\xi(\cdot)$ at the point y.

If the random variable ξ can be summed up in respect of probability **P**, then $\mathbf{M}\xi = \int \xi d\mathbf{P} = \int y \mu_\xi(dy)$ is known as *mathematical expectation (mean value) of the random variable* ξ. If the density $p_\xi(y)$ exists for all $y \in \mathbf{R}^N$ then $\mathbf{M}\xi = \int y p_\xi(y) dy$. A random variable ξ is *central*, if $\mathbf{M}\xi = 0$. Matrix

$$R_\xi = \operatorname{cov} \xi = \mathbf{M}(\xi - \mathbf{M}\xi)(\xi - \mathbf{M}\xi)^* \tag{A.4}$$

is known as the *covariance* of the random variable ξ and $\sigma_\xi^2 = \operatorname{Sp} R_\xi$ is known as the *dispersion of the random variable* ξ.

---

*The equality $\xi = \eta$ signifies that the random values ξ and η are equivalent.

(c) *Independence of random variables.* Random variable $\xi(\omega)$ is known as $\mathbf{A}_1$*-measurable*
if σ-algebra $\mathbf{A}_1$ contains all prototypes of open sets. The least of the σ-algebras $\mathbf{A}_\xi$, for which random variable ξ is measureable, is known as the *σ-algebra of the random variable* ξ.

The events A_1 and A_2 are *independent* if $\mathbf{P}(A_1 \cap A_2) = \mathbf{P}(A_1) \cdot \mathbf{P}(A_2)$. The two σ-algebras $\mathbf{A}_1$, $\mathbf{A}_2$ are independent, if the arbitrary events $\mathbf{A}_1$ and $\mathbf{A}_2$, $A_1 \in \mathbf{A}_1$, $A_2 \in \mathbf{A}_2$, are independent.

The random variables ξ and η are *independent,* if their σ-algebras $\mathbf{A}_\xi$, $\mathbf{A}_\eta$ are independent. For independent random variable ξ, η the equality $\mathbf{M}\xi\eta^* = \mathbf{M}\xi(\mathbf{M}\eta)^*$ is true.

(d) *Conditional mathematical expectation of random variables.* Let $\mathbf{A}_1$ be an arbitrary σ-algebra and ξ be the random variable with finite $\mathbf{M}\xi$. *Conditional mathematical expectation* of a random variable ξ will be known as $\mathbf{A}_1$ - measurable random variable $\mathbf{M}(\xi \mid \mathbf{A}_1)$ for $\mathbf{A}_1$ which satisfies (for any $A \in \mathbf{A}_1$) the relationship

$$\int_A \mathbf{M}(\xi \mid \mathbf{A}_1) d\mathbf{P} = \int_A \xi d\mathbf{P}. \tag{A.5}$$

According to Radon-Nikodim theorem [105] these conditions of random variable $\mathbf{M}(\xi \mid \mathbf{A}_1)$ is determined uniquely (with precision of equivalence).

If $\mathbf{A}_1 = \mathbf{A}_\eta$, then $\mathbf{M}(\xi \mid \mathbf{A}_\eta) = \mathbf{M}(\xi \mid \eta)$ is a determining function of random variable η, *i.e.*, $\mathbf{M}(\xi \mid \eta) = f(\eta)$. This property is useful in interpretation of different relations in random variables. Operation $\mathbf{M}(\cdot \mid \mathbf{A}_1)$ is linear and singular. Thus, the basic properties of condtional mean are: (1) $\mathbf{M}(\xi \mid \mathbf{A}_2) = \mathbf{M}\{\mathbf{M}(\xi \mid \mathbf{A}_1) \mid \mathbf{A}_2\}$, if $\mathbf{A}_1 \supseteq \mathbf{A}_2$; (2) $\mathbf{M}(\xi \mid \mathbf{A}_\eta) = \mathbf{M}\xi$, if ξ, η are independent random variables ; (3) $\mathbf{M}(\xi_1\xi^* \mid \mathbf{A}_1) = \xi_1 \mathbf{M}(\xi^* \mid \mathbf{A}_t)$, if ξ_1 is $\mathbf{A}_1$ measurable.

2 A.2 Convergence of Random variables

The sequence $\{\xi_n\}$ of random variables converges to random value ξ *in probability sense* for $n \to \infty$ if $\mathbf{P}(|\xi_n - \xi| > \varepsilon) \to 0$ for any $\varepsilon > 0$; *with probability* 1 (*almost surely*), if $\mathbf{P}(\lim \xi_n = \xi) = 1$; and mean squarely, if $\mathbf{M} \mid \xi_n - \xi \mid^2 \to 0$.

The sequence ξ_1^∞ of scalar random variables is known as *supermartingale* with respect to nondecaying σ-algebras $\{\mathbf{A}_t\}$, $\mathbf{A}_t \subseteq \mathbf{A}_{t+1}$, if there exists $\mathbf{M}\xi_1$ and $\mathbf{M}(\xi_{t+1} \mid \mathbf{A}_t) \le \xi_t$ with probability 1. The fundamental result of convergence of supermartingales is due to *J.* Doob [54, 105 p. 414]. The corollary from Doob's theorem, as given in the following assertation, will be usually made use of in various cases.

Theorem 2A.1 *If for random variables* ξ_1^∞ *the following relationship is satisfied*

$$\mathbf{M}(\xi_{t+1} \mid \xi^t) \le (1 + \mu_t)\xi_t + v_t, \quad \mathbf{M}\xi_1 < \infty, \tag{A.6}$$

where μ_t *is a non-negative and* $v_t = v_t(\xi_1^t)$ *are such non-negative functions that satisfy* $\sum^\infty(\mu_t + \mathbf{M}v_t) < \infty$, *then* $\xi_t \to \xi$, $t \to \infty$, *with probability* = 1 , *and* $\mathbf{M} \mid \xi \mid < \infty$.

Proof of the theroem 2.A.1 is well-known and may be referred, for example, to [14, 170].

From the theorem 2A.1 the law of large numbers for independent random variables follows.

Theorem 2.A.2 *Let* ξ_1^∞ *be the sequence of independent centralised random variables with properties* $\mathbf{M}\,|\,\xi_t\,|^2 \le C_\xi < \infty$ and $\nu_1^\infty(\cdot)$ *be the sequence of functions* $\nu_t = \nu_t(\xi_1^{t-1})$. *For the latter, the inequality* $\sum^{\infty} t^{2(\varepsilon-1)}\mathbf{M}\,|\,\nu_t\,| < \infty$ *holds good for any* $\varepsilon \in [0, 1/2]$.

Then for random variables $\eta_t = t^{\varepsilon-1}\sum_{s=1}^{t-1}\nu_s\xi_s$ *with probability 1 and in the mean square sense we have* $\eta_t \to 0$ *as* $t\to\infty$.

Proof of the theorem 2.A.2 is based on assertion of the sequence of random variables η_t^2, which is close to supermartingality.

The important (for the appendix) form of law of large numbers is due to Cramer and Lidbetter [81].

Theorem 2.A.3 *Let the sequence* ξ_1^∞ *be composed of centralised scalar random variables, for which the function* $R_{t,s} = \mathbf{M}\xi_1\xi_s$ *satisfies the following inequality, for all* $t \ge 1,\quad s \ge 1$ *and for some* $C > 0,\quad p, q, 0 \le 2p < q < 1$,

$$|\,R_{t,s}\,| \le C\frac{t^p + s^p}{1 + |\,t - s\,|^q}.$$

Then for the random variables $n_t = t^{-1}\sum_{s=1}^{t}\xi_s$ *tend to zero with probability 1, and as* $t\to\infty$.

2.P Proofs of Lemmas and Theorems

2.P.1 *Proof of the Theorem 2.1.1*

The formula (1.20) will be written as

$$J\left[u^{t_*-2}, \tilde{u}_{t_*-1}\right] = \sum_{t=1}^{t_*-2}\left(x_t^*Q_{t-1}x_t + u_{t-1}^*R_{t-1}u_{t-1}\right)$$
$$+x_{t_*-1}^*P_2x_{t_*-1} + u_{t_*-2}^*R_{t_*-2}u_{t_*-2}. \tag{P.1}$$

The control variable $\tilde{u}_{t_*-2}$ in accordance with eqn. (1.10) is to be determined from the condition for minimization of the quadratic form

$$q^{(2)}(x,u) = u_{t_*-2}^*R_{t_*-2}u_{t_*-2} + \left(A_{t_*-2}x + B_{t_*-2}u\right)$$
$$\times\tilde{Q}_{t_*-2}\left(A_{t_*-2}x + B_{t_*-2}u\right) \tag{P.2}$$

where $\tilde{Q}_{t_*-2} = P_2$, Minimum of (P.2) is achieved for $\tilde{u}_{t_*-2} = -L_{t_*-1}\,B_{t_*-2}^*P_2B_{t_*-2}x_{t_*-2}$,

$L_{t_*-1}=\left(R_{t_*-2}+B^*_{t_*-2}P_2B_{t_*-2}\right)^{-1}$. For this $q^{(2)}(x_{t_*-2}, \tilde{u}_{t_*-2})=x^*_{t_*-2}\tilde{P}_3x_{t_*-2}$, $\tilde{P}_3= A^*_{t_*-2}P_2A_{t_*-2}-A^*_{t_*-2}P_2B_{t_*-2}L_{t_*-1}B^*_{t_*-2}P_2A_{t_*-2}$.

The functional (P.1) is then expressed as

$$J\left[u_t^{t_*-3}, \tilde{u}_{t_*-2}, \tilde{u}_{t_*-1}\right] = \sum_{t=1}^{t_*-2}(x_t^*Q_{t-1}x_t+u^*_{t-1}R_{t-1}u_{t-1})+$$

$$+x^*_{t_*-2}\tilde{P}_3x_{t_*-2}.$$

Continuing for $t=t_*-3, t_*-4, \ldots, 0$, the assertion in the theorem can be established.

2.P.2 *Proof of the Theorem 2.1.2*

The relations in (1.31) turn out to be identical to those in (1.30). Adding the latter and taking into account (1.29) we get (1.32). Let $\tilde{U}(\cdot)\in \mathbf{U}^a$ be an arbitrary strategy. Then in conformity with (1.30) we get

$$V_{t+1}(x_{t+1})-V_t(x_t)+q_t(x_t, \tilde{U}(x_t))\geq 0,$$

and thus

$$\Sigma_{t=0}^{\infty}[V_{t+1}(x_{t+1})(x_{t+1})-V_t(x_t)]+\Sigma_{t=0}^{\infty}q_t(x_t, \tilde{U}_t(x_t))\geq 0,$$

or because of (1.29), (1.32)

$$\Sigma_{t=0}^{\infty}q_t(x_t, \tilde{U}_t)\geq V_0(x_0)=\Sigma_{t=0}^{\infty}q_t(x_t, U_t),$$

this proves the optimality of the control strategy (1.31).

2.P.3 *Proof of the Theorem 2.3.1*

According to para 2.3.1 for function $q^{(1)}(.)$ ((3.2), (3.10)) we have

$$q^{(1)}\left(x^{t_*-1}, u^{t_*-2}; U^{t_*-1}\right)=\Sigma_{t=1}^{t_*-1}(x_t^*Q_{t-1}x_t+u^*_{t-1}R_{t-1}u_{t-1})$$

$$+\mathbf{M}\left\{x^*_{t_*}Q_{t_*-1}x_{t_*}+u^*_{t_*-1}R_{t_*-1}u_{t_*-1}\mid x^{t_*-1}, u^{t_*-2}\right\}. \tag{P.3}$$

Using (1.18) x_{t_*} is excluded, and taking into account independence of noise w^{t_*-1} and matrices $\{A_t, B_t\}$, the last term of the right hand side of (P.3) can be transformed into

$$\mathbf{M}\left\{x^*_{t_*}Q_{t_*-1}x_{t_*}+u^*_{t_*-1}R_{t_*-1}u_{t_*-1}\mid x^{t_*-1}, u^{t_*-2}\right\}$$

$$=x^*_{t_*-1}\left(M_{t_*-1}-T^*_{t_*-1}L_{t_*}T_{t_*-1}\right)x_{t_*-1}$$

$$+\left(\overline{U}_{t_*-1}+L_{t_*}T_{t_*-1}x_{t_*-1}\right)^*L^{-1}_{t_*}\left(\overline{U}_{t_*-1}+L_{t_*-1}T_{t_*-1}x_{t_*}\right)$$

$$+\mathbf{M}\left\{\left(U_{t_*-1}-\overline{U}_{t_*-1}\right)^*L^{-1}_{t_*}\left(U_{t_*-1}-\overline{U}_{t_*-1}\right)\mid x^{t_*-1}, u^{t_*-2}\right\}, \tag{P.4}$$

here, the expressions in (3.14) and (3.16) have been used alongwith

$$\overline{U}_{t_*-1}=\mathbf{M}\left\{U_{t_*-1}\left(x^{t_*-1},u^{t_*-2},w^{t_*-1}\right)\mid x^{t_*-1},u^{t_*-2}\right\}. \tag{P.5}$$

Matrices M_{t_*-1}, T_{t_*-1} exist due to conditions in (3), matrix L_{t_*} exists due to (3.12) and positiveness of matrix Q_{t_*-1}. Mean is taken in (P.5) w.r.t. random variables w^{t_*-1} (for fixed values of x^{t_*-1}, u^{t_*-2}).

From (P.3) it follows that for optimal control, choice of $U_{t_*-1}(\cdot)$ must be carried out from the condition of minimum of the last term in the right hand side. The expression (P.4) shows that for optimal control the function $U_{t_*-1}(\cdot)$ must be determined according to formula

$$\tilde{U}_{t_*-1}\left(u^{t_*-2},x^{t_*-1},w^{t_*-1}\right)=-L_{t_*}T_{t_*-1}x_{t_*-1}, \tag{P.6}$$

i.e., this function is the determined function of given observations and it varies (linearly) with last observation x_{t_*-1}. The optimality criterion (3.10) of this, following (P.6), can be expressed as

$$J\left[U^{t_*-2}(\cdot),\ \tilde{U}_{t_*-1}(\cdot)\right]=\mathbf{M}\sum_{t=1}^{t_*-1}(x_t^*Q_{t-1}x_t+u_{t-1}^*R_{t-1}u_{t-1})$$

$$+\mathbf{M}x_{t_*}^*\left(M_{t_*-1}-T_{t_*-1}^*L_{t_*}T_{t_*-1}\right)x_{t_*}. \tag{P.7}$$

Similarly, optimization w.r.t. u_{t_*-2}.... etc. can be carried out in (P.7) resulting in theorem 2.3.1.

2.P.4 ***Proof of the Lemma 2.3.1***

Solution of eqn. (3.22) can be established using Euclidean algorithm (For example, [177, pp. 184-185].

2.P.5 ***Proof of Theorem 2.3.2***

From (3.26) it follows that for optimal control, the function $\tilde{U}_{t_*-k}(\cdot)$ must be determinable $U_{t_*-k}=\overline{U}_{t_*-k}(\cdot)$, $\overline{U}_{t_*-k}(\cdot)$ is determined as follows

$$b_k\overline{U}_{t_*-k}-G(\nabla)y_{t_*-k}+[F(\nabla)b(\nabla)-b_k\nabla^k]u_{t_k}=0, \tag{P.8}$$

and it leads to (3.27). The equality (3.26) can be then expressed as

$$\mathbf{M}\left\{\mid y_{t_*}\mid^2\mid y^{t_*-k},\ u^{t_*-k-1}\right\}=\Sigma_{s=0}^{k-1}F_s^2\sigma_s^2. \tag{P.9}$$

Thus, expression in (3.28) and assertion in theorem 2.3.2 are proved.

2.P.6 ***Proof of the Lemma 2.4.1***

From equations (2.1), (2.2) and (3.1) the following properties of *a posteriori* distributions are obtained

$$\mu_{x_t+1}(\cdot\mid x^t,\ y^t,\ u^t,\ v^{t+1},\ w^t)=\mu_{x_{t+1}}(\cdot\mid x^t,\ u^t,\ v^{t+1}),$$

$$\mu_{y_t}(\cdot \mid x^t, y^{t-1}, v^{t-1}, w^t) = \mu_{y_t}(\cdot \mid x^t, v^t), \tag{P.10}$$

$$\mu_{u_t}(\cdot \mid x^t, y^{t-1}, u^{t-1}, v^{t+1}, w^t) = \mu_{u_t}(\cdot \mid y^{t-1}, w^t).$$

Using the formulae of (P.10), (4.16) and (4.14) the following recursive relationship is obtained

$$\mu_{t+1} = \mu(dv_{t+1}, dw_t \mid y^{t-1}, u^{t-1}, v^t, w^{t-1})\mu_t \times \frac{\mu(dx_{t+1} \mid x^t, y^t, v^{t+1})\mu(dy_t \mid x^t, v^t)\mu(du_t \mid y^{t-1}, w^{t-1})}{\mu(dy_t, du_t, dv_{t+1}, dw_t \mid y^{t-1}, u^{t-1}, v^t, w^{t-1})} \tag{P.11}$$

for *a posteriori* distribution

$$\mu_{t+1} = \mu_{x^{t+1}}(\cdot \mid y^t, u^t, u^{t+1}, w^t). \tag{P.12}$$

In expression (P.11) indexed variables have been ommitted for making it shorter. The variables have their corresponding distributions. Value $\mu(\cdot \mid y^{t-1}, u^{t-1}, v^t, w^{t-1})$is determined from the conditions for normalisation

$$\int \mu_{x^{t+1}}(dx^{t+1} \mid y^t, u^t, v^{t+1}, w^t)dx^{t+1} = 1. \tag{P.13}$$

Expression (P.11) is converted as (taking into account (P.13)),

$$\mu_{t+1} = \frac{\mu(dx_{t+1} \mid x^t, y^t, v^{t+1})\mu(dy_t \mid x^t, v^t)\mu_t}{\int \mu(dy_t \mid x^t, v^t)\mu_t dx^t}. \tag{P.14}$$

The recursive relations of (P.14) determine the *a posteriori* distribution of (P.12) as a function of y^t, u^t, v^{t+1}, w^t. This function, in accordance with the condition of the lemma, does not depend on the expression for strategy $U^{t_*-1}(\cdot) \in \mathbf{U}^a$. Hence the *a posteriori* distribution

$$\mu_{x_{t+1}}(dx_{t+1} \mid y^t, u^t) = \int_{A_{x^t} \otimes A_{v^{t+1}} \otimes A_{w^t}} \mu_{x^{t+1}}(dx^{t+1} \mid y^t, u^t, v^{t+1}, w^t) \times \mu(dv^{t+1}, dw^t \mid y^t, u^t) \quad \text{...(P.15)}$$

and thus does not depend on choice of strategy $U^{t_*-1}(\cdot) \in \mathbf{U}^a$. Hence the proof of the lemma.

2.P.7 *Proof of the Theorem 2.4.1*

The induced transition in formulae (4.22) to (4.24) is now to be proved. Let the functions $\tilde{U}^{t_*-1}_{t_*-k}(\cdot)$, $\phi_k(\cdot)$, $q^{(k)}(\cdot)$ corresponding to these formulae be determined. The functional (4.5) is expressed as

$$J\left[U^{t_*-k-1}(\cdot), \tilde{U}^{t_*-1}_{t_*-k}(\cdot), \right] = \mathbf{M} \sum_{t=1}^{t_*-k-1} q_t(x_t, u_{t-1}) + \mathbf{M}\left[q_{t_*-k}\left(x_{t_*-k}, u_{t_*-k-1}\right) + q^{(k)}\left(y^{t_*-k}, u^{t_*-k-1}\right)\right].$$

For determination of $U_{t_*-k-1}(\cdot)$ it is necessary to minimize w.r.t. u_{t_*-k-1} the following expression

$$\mathbf{M}\left[q_{t_*-k},\ u_{t_*-k-1}\right)+q^{(k)}\left(y^{t_*-k},\ u^{t_*-k-1}\right)\mid y^{t_*-k-1}\ u^{t_*-k-1}\right]$$
$$=\int_{\mathbf{X}} q_{t_*-k}\left(x_{t_*-k},\ u_{t_*-k-1}\right)\mu_{x_{t_*-k}}\left(dx_{t_*-k}\mid y^{t_*-k-1},\ u^{t_*-k-1}\right)$$
$$+\mathbf{M}\left\{q^{(k)}\left[y^{t_*-k-1},\ \overline{Y}_{t_*-k}(x_{t_*}-k),\ u^{t_*-k-1}\right]\mid y^{t_*-k-1},u^{t_*-k-1}\right\}$$
$$=\int_{\mathbf{X}}\{q_{t_*-k}\left(x_{t_*-k-1},\ u_{t_*-k-1}\right)+q^{(k)}[y^{t_*-k-1},$$
$$\overline{Y}_{t_*-k}\left(x_{t_*-k}\right),\ u^{t_*-k-1}]\}\mu_{x_{t_*-k}}\left(dx_{t_*-k}\mid y^{t_*-k-1},\ u^{t_*-k-1}\right)$$
$$=\phi_{k+1}\left(y^{t_*-k-1},\ u^{t_*-k-1}\right).$$

The induced transition concludes with the formula

$$\tilde{U}_{t_*-k-1}=\tilde{U}_{t_*-k-1}\left(y^{t_*-k-1},\ u^{t_*-k-2}\right)$$
$$=\underset{u\in\mathbf{U}}{\operatorname{argmin}}\,\phi_{k+1}\left(y^{t_*-k-1},\ u^{t_*-k-2},\ u\right).$$

The inductions are verified following the statement of the theorem. Thus the theorem 2.4.1 is proved.

2.P.8 ***Proof of the Theorem 2.4.2***

Using determination of *a posteriori* distributions and the known properties of conditional mean, we get

$$\int_{\mathbf{A}_y}\mu_{x_{t+1}}(\mathbf{A}_x\mid y^t)\mu_{y_t}(dy_t\mid y^{t-1})=\mathbf{M}\left\{I_{\mathbf{A}_x\otimes\mathbf{A}_y}(x_{t+1},\ y_t)\mid y^{t-1}\right\}$$
$$=\mathbf{M}\left(\mathbf{M}\left\{I_{\mathbf{A}_x\otimes\mathbf{A}_y}(x_{t+1},\ y_t)\mid x_t,\ y^{t-1}\right\}\mid y^{t-1}\right)$$
$$=\mathbf{M}\left[\mu_{x_{t+1},\ y_t}(\mathbf{A}_x\otimes\mathbf{A}_y\mid x_t,\ y^{t-1})I_{\mathbf{A}_x}(x_t)\mid y^{t-1}\right]$$
$$=\int_{\mathbf{X}}\mu_{x_{t+1},\ y_t}(\mathbf{A}_x\otimes\mathbf{A}_y\mid x_t,\ y^{t-1})\mu_{x_t}(dx_t\mid y^{t-1}].$$

Thus the theorem is proved.

3

Infinite Time Period Control

Dynamic programming or its analogs can not be applied when the process control period is not definite. Other universal methods for control optimization for such problems are so far unknown. Moreover, the unconstrained control interval leads to new essential restrictions on control and output variables (and may be on states) of the control object : they must be in some sense or other, uniformly time bounded. The control object is known as *stabilizable** if it provides the above property and the corresponding control strategy is known as *stabilizing*. Construction of stabilized control strategy often presents an independent and sufficiently interesting mathematical problem and may achieve the control objective. It will be further shown that not all the control strategies with infinite control interval, which appear to be true generalization of the optimal control strategies with finite control interval, possess stablizing properties. Infinite period of control makes the problem complex and does not permit generalised synthesis methods as available in finite period control problems. Hence the optimisation problems related to linear systems will be mainly studied and a few nonlinear systems, for which synthesis of stablizing control is possible, will be mentioned.

§ 3.1 Stabilization of Dynamic Systems using Liapunov's Method

The control object is described by the state equation

$$x_{t+1} = X_t(x_t,\ u_t,\ v_t), \qquad t = 0,\ \ldots, \tag{1.1}$$

and by the output equation (observable)

$$y_t = Y_t(x_t,\ v_t), \qquad t = 0,\ 1,\ \ldots. \tag{1.2}$$

Here $x \in \mathbf{R}^n$ is the state variable, $u \in \mathbf{R}^m$ is the control variable, $y \in \mathbf{R}^l$ is the output variable and $v \in \mathbf{V} \subseteq \mathbf{R}^q$ is the noise variable. It is assumed that $\mathbf{V}^\infty = \{v^\infty\}$ is the set of noise signals given by

$$\mathbf{V}^\infty = \{v^\infty : v_t \in \mathbf{V}\}; \tag{1.3}$$

*To be precise : stabilizable in the class of admissible control strategies $\mathbf{U}^a$.

the noise variable set **V** is assumed to be closed and constrained :

$$\sup_{v \in \mathbf{V}} | v | = C_v < \infty. \tag{1.4}$$

The class of admissible control strategies $\mathbf{U}^a$ is given by

$$\mathbf{U}^a = \overset{\infty}{\underset{t=0}{\pi}} \otimes \mathbf{U}_t^a, \tag{1.5}$$

where $\mathbf{U}_t^a$ is the admissible regular control laws* at time t.

The control objective is provided by the inequality

$$J[U^\infty(\cdot)] = \sup_{v^\infty \in \mathbf{V}^\infty} \overline{\lim_{T \to \infty}} \frac{1}{T} \sum_{t=0}^{T} q_t(y_t, u_t) \le r^{-1}, \tag{1.6}$$

where $q_t(\cdot)$ is the nonnegative objective function of control and output variables; r is the given level of control quality.

The stabilization problem with respect to the functional $J(\cdot)$ (as expressed by the left hand side of the inequality (1.6)) is thus examined.

3.1.1 *Description of Liapunov's Method*

We introduce the general method for synthesis of stabilizing control which is based on the condition of providing not too fast a growth of some positive functions of the control system trajectories. Such functions (Liapunov functions) are used in well-known Liapunov's direct method for studying stability of differential and difference equations.

Unfortunately, it is due to the absence of any generalised method for construction of Liapunov functions, that their application is restricted. Moreover, the construction is relatively simpler only for $y_t = x_t$. The generalised assertion for such cases will be formulated.

Theorem 3.1.1 *Let us assume existence of a nonnegative function*

$$V_t = V_t(x), \quad x \in \mathbf{R}^n, \quad t = 0, 1, \ldots,$$

with the following properties :

(1) *the inequality is valid for the finite function* $\psi_t(x, u)$ *for all* x, u

$$\psi_t(x, u) \ge \sup_{v \in \mathbf{V}} V_{t+1}[X_{t+1}(x, u, v)]; \tag{1.7}$$

(2) *the following inequality is true for all* x

$$\phi_t(x) = \inf_{v \in \mathbf{V}} [\psi_t(x, u) q_t(x, u)] - V_t(x) \le r^{-1} \tag{1.8}$$

(3) *for each* t *there is a solution,* $u = U_t(x) \in \mathbf{U}_t^a$, *of the equation*

$$\phi_t(x) = \psi_t(x, u) + q_t(x, u) - V_t(x). \tag{1.9}$$

Then the control strategy U_0^∞ *that can be determined from the above conditions will ensure the control objective (1.6).*

---

* *i.e.*, $\mathbf{U}_t^a$ is the separated set of functions $U_t(\cdot)$ of the form $U_t = U_t(u^{t-1}, y^t)$. It is used to form control signal u_t.

Stabilization problems having the following objective inequalities can be solved within the structure of Liapunov's method :

$$\sup_{v^\infty \in \mathbf{V}^\infty} \overline{\lim_{t\to\infty}}\, q_t(y_t,\ u_t) \le r^{-1}.$$

When it is noise free we can choose $r = \infty$ and then the control objective will be to ensure the limiting equality

$$\lim_{t\to\infty} q_t(y_t,\ u_t) = 0.$$

3.1.2 ***Stabilization of Linear Systems with Observable States***

A linear control object will be taken as an illustration for Liapunov's method. Its states are obtained from the equation

$$x_{t+1} = Ax_t + Bu_t + v_t, \tag{1.10}$$

where x, u, v have the connotations as before.

(*a*) *System whose states are not being observed at all instants of time.* Let the output of the control object be given by relations

$$y_t = x_{kT} \quad \text{for} \quad kT \le t < (k+1)T,\ k = 0,\ 1,\ \dots \tag{1.11}$$

Here T is the interval of time, at which states of the control object are accessible for observation.

The control objective is to ensure the inequality (1.6) for

$$q_t(x_t,\ u_t) = |\, y_t\,|^2. \tag{1.12}$$

The admissible control strategies $\mathbf{U}^a$ have the form (1.5), where $\mathbf{U}_t^a$ is the set of arbitrary reflections

$$U_t(\cdot) : \mathbf{U}^{t-1} \otimes \mathbf{Y}^t \to \mathbf{U}.$$

It is convenient to bring equation (1.10) in terms of the variables $y_t = \tilde{x}_t$:

$$\tilde{x}_{t+1} = A^T \tilde{x}_t + \tilde{B}\tilde{U}_t + w_t, \tag{1.13}$$

where the symbols used are as follows

$$\tilde{x}_t = x_{tT},\ \tilde{U}_t = \text{ col } (u_{tT},\ u_{tT+1},\ \dots,\ u_{(t+1)T-1}), \tag{1.14}$$

$$\tilde{B} = ||B,\ AB,\ \dots.,\ A^{T-1}B||,$$

$$w_t = \sum_{k=0}^{T-1} A^k v_{tT-k}. \tag{1.15}$$

The states $\tilde{x}_t$ of the system (1.13) are observable, for all t, by virtue of (1.11) and (1.14).

Let the Liapunov functions be chosen as

$$V_t(\tilde{x}) = |\,\tilde{x}\,|^2. \tag{1.16}$$

To explain the conditions for existence of stabilizing control strategies in conformity with the theorem 3.1.1, it is necessary to compute (not difficult in this case) functions $\psi_t(x,\ \tilde{u})$ and $\phi_t(x)$ using formulae (1.7) and (1.8). For this a few characteristics of noise in equation (1.13) will be introduced. The set W of possible values of function (1.15) is the *geometrical sum* of sets $\mathbf{V}_k = A^k\mathbf{V}$, $k = 0, 1, \ldots, T-1$.* The deviation $\rho_W(\tilde{w})$ of the set W (related to point $\tilde{w}$) will be determined for any point $\tilde{w}$, as

$$\rho_W(\tilde{w}) = \sup_{w \in W} |w - \tilde{w}|. \tag{1.17}$$

The point w_W corresponding to minimum deviation $\rho_W(\tilde{w})$ is known as centre of the set W :**

$$w_W = \operatorname{argmin} \rho_W(\tilde{w}). \tag{1.18}$$

ω_W and $\rho_W(\omega_W)$ can be computed for a given set **V** using formula (1.15).

Theorem 3.1.2 *Let the following conditions be satisfied :*

(1) pair $\{A,B\}$ in equation (1.10) *is controllable;* ***

(2) the observation interval T (vide (1.11)) satisfies condition $T \geq n$, where n is the measure of state space for the control object (1.10);

(3) the quality level r in (1.6) *and* (1.12) *satisfies the condition*

$$r\rho_W(w_W) \leq 1, \tag{1.19}$$

where w_W is the centre of set **W**, *which can be determined along with the set (1.3) using formula* (1.15).

Then in the class $\mathbf{U}^a$ *there exists a stabilizing strategy which ensures the control objective* (1.6) *and* (1.12) *and can be determined using relations*

$$\begin{aligned} &\operatorname{col}(u_{Tk}, u_{Tk+1}, \ldots, u_{Tk+(T-1)}) = \tilde{B}^*(\tilde{B}\tilde{B}^*)^{-1} \\ &\times (w_W - A^T y_{Tk}),\ k = 0, \quad 1, \ \ldots . \end{aligned} \tag{1.20}$$

We observe that by virtue of the first two conditions of the theorem, matrix $\tilde{B}\tilde{B}^*$ is not degenerate :

$$\begin{aligned} &\tilde{B}\tilde{B}^* = BB^* + BAA^*B^* + \ldots + BA^{T-1}(A^*)^{T-1}B^* > \\ &\geq BB^* + \ldots + BA^{n-1}(A^*)^{n-1}B^* > 0. \end{aligned}$$

The following observations are made on theorem 3.1.2.

1. The system of equations (1.10) and (1.20) is asymptotically stable when noise is absent ($v_t \equiv 0$) :

$$\lim_{t\to\infty} x_t = 0, \quad \lim_{t\to\infty} u_t = 0$$

and the limits are attained for the control period not exceeding T.

2. The control strategy determined from relations (1.20) is optimal for the functional

$$J[U_0^\infty(\cdot)] = \sup_{v^\infty \in V^\infty} \overline{\lim_{T\to\infty}} \frac{1}{T} \sum_{t=0}^{T} |y_t|^2 \tag{1.21}$$

---

*Geometrical sum **W** of sets $\{\mathbf{V}_k\}$ are the set

$$\mathbf{W} = \left\{ w : w = \sum_{k=0}^{T-1} v_k,\ v_k \in \mathbf{V}_k \right\}.$$

**If W is a sphere, w_W is its centre; if W is a parallele piped, w_W is intersection of its diagonals, etc.

***Vide section 1.3.3.c.

in the class of admissible control strategy $\mathbf{U}^a$. In fact for control (1.20), the equation (1.13) takes the form $\tilde{x}_{t+1} = w_W + w_t$ and so

$$J[U_0^\infty(\cdot)] = \rho_W^2(w_W). \tag{1.22}$$

This is the least value of the performance criterion in the class of arbitrary control strategies, where no additional information about noise is required.

3. The solution of the optimization problem becomes complicated, if the observable variables appear as linear combination of states. A few results, with additional restrictions on the class of admissible control strategies $\mathbf{U}^a$, will be presented in section 3.5.

4. The control strategy, optimal w.r.t. functional (1.21), is not unique when $T > n$: side by side with (1.20) the control determined by following relations will be optimal

$$u_{t*} = u_t + \hat{u}_t,$$

where $\hat{u}_t$ is arbitrary solution of the equation

$$B\hat{u}_t + AB\hat{u}_{t+1} + \ldots + A^{T-1}B\hat{u}_{t+T-1} = 0. \tag{1.23}$$

The equation (1.20) determines all the solutions from the set, for which magnitude $\sum_{s=t+kT}^{t+k(T+1)} | u_s |^2$ is the least for every $k = 0, 1, \ldots$. Obviously, the equation (1.23) has a null solution $\hat{u}_t = \hat{u}_{t+1} = \ldots = \hat{u}_{t+T+1} = 0$ for $T = n$.

(*b*) *Linear quadratic problem with scalar control.* Let the control object be again represented by (1.10) but now the control signal u is scalar and

$$y_t = x_t. \tag{1.24}$$

The control objective is to ensure inequality (1.6) for

$$q_t(x, u) = x^* Q x \tag{1.25}$$

with positive matrix Q. Let us assume

$$V_t(x) = x^* H x \tag{1.26}$$

where H is the positive matrix. The conditions for existence of stabilizing control will be explained. We assume that for this the noise signals satisfy the condition

$$| v_t | \le C_v \tag{1.27}$$

(constant $C_v > 0$) and for the rest they are arbitrary. In the present case we choose the functions $\psi(x, u)$ as

$$\psi_t(x, u) = | Ax + Bu |_H^2 \left(1 + \frac{C_v \Lambda_H^{1/2}}{| Ax + Bu |_H} \right)^2, \tag{1.28}$$

where Λ_H is the greatest eigen value of matrix H and expressed in the short form as

$$| Ax + Bu |_H^2 = (Ax + Bu)^* H (Ax + Bu).$$

The function (1.28) attains minimum w.r.t. u when

$$u = -\frac{B^* H A x}{B^* H B}, \tag{1.29}$$

and the inequality (1.8) takes the form

$$x_t^*\left[A^*HA-\frac{A^*HBB^*HA}{B^*HB}-H+Q\right]x_t$$

$$+2C_v\sqrt{x_t^*\left[A^*HA-\frac{A^*HBB^*HA}{B^*HB}\right]x_t\Lambda_H}+C_v^2\Lambda_H\leq r^{-1}. \tag{1.30}$$

If the inequality (1.30) is ensured for any x_t then by virtue of theorem 3.1.1, the equation determined by the relations (vide (1.29)), *i.e.*

$$u_t=-\frac{B^*HA}{B^*HB}x_t, \tag{1.31}$$

will be stabilizing. One of the possible results based on analysis of inequality (1.30) will be presented below.

Theorem 3.1.3. *Let us assume that the following conditions are fulfilled :*

(1) *there exist a vector-line K and a number* $\rho>0$, *such that all the eigenvalues of the matrix*

$$\tilde{A}=\sqrt{1+\rho}(A+BK) \tag{1.32}$$

are arranged in the region $|\lambda|<1$;

(2) *matrix H satisfies the matrix inequality*

$$\tilde{A}^*H\tilde{A}-H\leq -Q; \tag{1.33}$$

(3) *the numbers* ρ *from (1.32), r from (1.5) and* C_v *from (1.27) are connected by the inequality*

$$\Lambda_H r(1+\rho^{-1})C_v^2\leq 1. \tag{1.34}$$

Then the control u_0^∞ *determined from relations* (1.31) *will be stabilyzing. The inequality* (1.6) *with objective function* (1.25) *is fulfilled independent of choice of initial condition in the equation* (1.10).

A few observations are made on the theorem.

1. In general the synthesized control strategy will not be optimal, *i.e.*, it does not minimize the performance criterion

$$J[U_0^\infty(\cdot)]=\sup_{v^\infty\in V^\infty}\overline{\lim_{T\to\infty}}\frac{1}{T}\sum_{t=0}^{T-1}x_t^*Qx_t.$$

2. It is known that for the controllable pair (A,B) there exists a row matrix K such that the significant eigenvalues of the matrix $A+BK$ coincide with the roots of a given arbitrarily advanced polynomial of degree n with real coefficients. Design of stabilizing control based on assigned location of roots of the characteristic polynomial det $(A+BK-\lambda In)$ is known as *Modal Control*. In particular, the row vector K can be chosen for the controllable pair (A,B) in such a manner that the whole of the spectrum of matrix $A+BK$ will be concentrated at zero. Then for any $\rho>0$ the matrix (1.32) will satisfy the first condition of the theorem.

3. For the matrix $\tilde{A}$ with its spectrum within a unit circle the inequality of (1.33) is solved with respect to H with any positive matrix Q. Then the matrix $H=\sum_{k=0}^{\infty}(\tilde{A}^*)^kQ\tilde{A}^k$ is symmetric positive and satisfies the Liapunov equation $\tilde{A}^*H\tilde{A}-H=-Q$.

4. In the absence of any noise, the control system of (1.10) and (1.31) is asymptotically stable : $\lim_{t\to\infty} x_t = 0$. Then it justifies the summation estimation

$$\sum_{t=0}^{\infty} x_t^* Q x_t \le x_0^* H x_0, \tag{1.35}$$

characterizing the transient process. The inequality (1.35) is obvious from the established inequality $x_{t+1}^* H x_{t+1} - x_t^* H x_t + x_t^* Q x_t \le 0$. Putting $r = \infty$ in (1.8), we get the equation $\phi_t(x) = 0$, which can be considered to be the condition for choice of matrix H. In the present case, the equation has the form

$$x_t^* \left[A^* HA - \frac{A^* HBB^* HA}{B^* HB} - H + Q \right] x_t = 0.$$

The following matrix equation involving H will be obtained if the last relationship is to be satisfied for all $x_t \in \mathbf{R}^n$.

$$H = A^* HA - \frac{A^* HBB^* HA}{B^* HB} + Q. \tag{1.36}$$

The existence of matrix $H > 0$, satisfying equation (1.36) is well-investigated and the solution to this may be formulated in frequency terms. This problem will be discussed again (vide theorem 3.2.2).

3.1.3 *Stabilization of Linear Systems with Unobservable States*

Let the control object and the observation channel be described as

$$x_{t+1} = Ax_t + Bu_t + v'_t, \quad y_t = Cx_t + v_t''. \tag{1.37}$$

The control signal is given by the relations

$$u_t = K\hat{x}_t, \tag{1.38}$$

$$\hat{x}_{t+1} = A\hat{x}_t + Bu_t - L(y_t - C\hat{x}_t), \tag{1.39}$$

where K and L are matrices that determine the control strategy.

The control objective is given in the inequality form

$$\sup_{v^\infty \in \mathbf{V}^\infty} \overline{\lim_{t\to\infty}} |x_t| \le r^{-1}, \tag{1.40}$$

where $\mathbf{V}^\infty$ is the set of possible noises v^∞, $v_t = (v_t', v_t'')$, and r is a given constant.

Liapunov's method makes it possible to pick out the non-null set of matrices $\{K,L\}$, the control strategies of which (as determined by relations (1.38) and (1.39)) satisfy the control objective. Application of Liapunov function to the given problem will now be explained.

The relationship of (1.39) can be interpreted as the algorithm for prediction of state x_{t+1} from observation of y^t, u^t. The relationship (1.38) with $\hat{x}_t = x_t$ has the form of a linear stationary feedback loop that stabilizes the control object (1.37)*, when the matrix

$$A_K = A + BK \tag{1.41}$$

---

* *i.e.* provides a limit $|x_t|$ uniformily along t.

is stable. Hence assuming the matrix (1.41) to be stable (*i.e.* it does not have roots for $|\lambda| \geq 1$), the matrix L can be chosen from (1.39) so that the forecasting error

$$\varepsilon_t = x_t - \hat{x}_t \tag{1.42}$$

is small for sufficiently large t. Thus, the choice of K and L must be aimed at making $|x_t|$, $|\varepsilon_t|$ as small as possible. It is related to introduction of pair of Liapunov functions

$$V_1(x) = (x^* H_1 x)^{1/2}, \; V_2(\varepsilon) = (\varepsilon^* H_2 \varepsilon)^{1/2}, \tag{1.43}$$

where H_1, H_2 are positive matrices setting the metric for determination of smallness of values $|x_t|$, $|\varepsilon_t|$.

The equations (1.37) and (1.39) can be re-written as

$$\varepsilon_{t+1} = (A + LC)\varepsilon_t + v_t' + L v_t'', \tag{1.44}$$

$$x_{t+1} = (A + BK)x_t - BK_{\varepsilon_t} + v_t'. \tag{1.45}$$

From (1.44) it follows that the smallness of $|\varepsilon_t|$ can be achieved by choice of matrix L from the stability condition of matrix

$$A_L = A + LC, \tag{1.46}$$

and the smallness of $|x_t|$ can be achieved, in a similar manner, from the choice of the stable matrix (1.41). Thus, the pair of matrices (A,B) must be stabilizable and the pair of matrices (A,C) - detectable (vide section 1.3.3.c). To obtain quantitative estimates of smallness of $|x_t|$, $|\varepsilon_t|$ (for large t), it is necessary to coordinate choice of matrices K,L with matrices H_1, H_2. This coordination may be performed from analysis of the behaviour of the functions (1.43) on the system trajectories (1.44) and (1.45). One such possible analysis will be laid down.

Theorem 3.1.4 *Let us assume that the matrices H_1, H_2, K, L satisfy the inequalities*

$$(A + BK)^* H_1 (A + BK) \leq \alpha H_1, \tag{1.47}$$

$$(A + LC)^* H_2 (A + LC) \leq \beta H_2, \tag{1.48}$$

$$K^* B^* H_1 B K \leq \gamma^2 H_2 \tag{1.49}$$

with positive values α, β, γ where $\alpha < 1$, $\beta < 1$. Then for all t the following inequalities are justified:

$$V_1(x_t) \leq \rho_t', \; V_2(\varepsilon_t) \leq \rho_t'', \tag{1.50}$$

where ρ_t', ρ_t'' are determined by equations

$$\rho'_{t+1} = \alpha^{1/2}\rho_t' + \gamma\rho_t'' + C_v |H_1|^{1/2}, \tag{1.51}$$

$$\rho''_{t+1} = \beta^{1/2}\rho_t'' + C_v\left[|H_2|^{1/2} + (L^* H_2 L)^{1/2}\right], \tag{1.52}$$

$|H|$ = norm of the matrix H and

$$C_v = \sup_{v \in \mathbf{V}} |v|, \tag{1.53}$$

$\mathbf{V}$ = set of possible values of noise variables

$$v_t = \text{col}\,(v'_t, v_t''),$$

$$\rho'_0 = (x_0^* H_1 x_0)^{1/2}, \quad \rho_0'' = (\varepsilon_0^* H_2 \varepsilon_0)^{1/2}. \tag{1.54}$$

Corollary Since $0 < \alpha, \quad \beta < 1$, the following inequalities of limits are obtained from (1.51) and (1.52) :

$$\overline{\lim_{t\to\infty}}\, \rho_t'' \le C_v \frac{|H_2|^{1/2} + (L^* H_2 L)^{1/2}}{1-\sqrt{\beta}},$$

$$\overline{\lim_{t\to\infty}}\, \rho'_t \le C_v \frac{(1-\sqrt{\beta})\,|H_1|^{1/2} + \gamma[\,|H_2|^{1/2} + (L^* H_2 L)^{1/2}]}{(1-\sqrt{\alpha})(1-\sqrt{\beta})}. \tag{1.55}$$

If λ_{H_1} is the least eigen value of matrix H_1, $\lambda_{H_1} > 0$, then from (1.50), (1.55) and (1.43), we get

$$\overline{\lim_{t\to\infty}}\, |x_t| \le \frac{C_v}{\sqrt{\lambda_{H_1}}} \frac{(1-\sqrt{\beta})\,|H_1|^{1/2} + \gamma[\,|H_2|^{1/2} + (L^* H_2 L)^{1/2}]}{(1-\sqrt{\alpha})(1-\sqrt{\beta})}. \tag{1.56}$$

If the right hand side of the inequality (1.56) does not exceed r^{-1} then the special purpose inequality of (1.40) is achieved.

In the absence of noise ($v_t \equiv 0$) we get from (1.50) and (1.43)

$$\lim_{t\to\infty} x_t = 0, \quad \lim_{t\to 0}\, (x_t - \hat{x}_t) = 0.$$

This means that asymptotic stability of control system (1.37) to (1.39) is attained and the asymptotic realization of precise prediction of the states of control object is made possible.

The problem of existence of the matrix inequalities (1.47) to (1.49) with respect to positive matrices H_1, H_2 and orthogonal matrices L,K of proper dimensions requires additional investigation.

§ 3.2 Discrete Form for Analytical Design of Regulators

It has been already mentioned that dynamic programming can not be directly applied to optimization problems on infinite time period of control. At the same time it has been mentioned in section 2.1.5, that Bellman's equation can be used for synthesis of this type of optimal control problems. A few such cases are discussed below. Difficulty in solving Bellman's equation compels us to confine its application to linear quadratic control problems.

In the following sections (including the present one but excluding section 3.2.4) the control problem is studied with the prepositions that noise is absent. Synthesis of optimal control in the presence of statistical noise is realised in section 3.4 and the same in the presence of arbitrarily constrained noise is realized in section 3.5.

3.2.1 *Statement of the Problem*

Let the control object be described by the linear equation

$$x_{t+1} = A_t x_t + B_t u_t, \quad t = 0, 1, \quad \ldots, \tag{2.1}$$

with known $n \times n$ square matrices A_t and $n \times m$ rectangular matrices B_t. The states x_t of control object are assumed to be known. The performance criterion has the form

$$J[U_0^\infty(\cdot)] = \sum_{t=0}^{\infty} q_t(x_t, u_t), \tag{2.2}$$

where

$$q_t(x, u) = \begin{pmatrix} x \\ u \end{pmatrix}^* N_t \begin{pmatrix} x \\ u \end{pmatrix} \tag{2.3}$$

and

$$N_t = \begin{Vmatrix} Q_t & S_t \\ S_t^* & R_t \end{Vmatrix} \tag{2.4}$$

– symmetric nonnegative matrix.

The class of admissible control strategies $\mathbf{U}^a$ consists of all possible functions $U^\infty(\cdot)$ constituting the control u_0^∞ as per formula

$$u_t = U_t(u_0^{t-1}, x_0^t) \tag{2.5}$$

and in conformity with the inequality

$$\sum_{t=0}^{\infty} (|u_t|^2 + |x_t|^2) < \infty \tag{2.6}$$

independent of the choice of initial state x_0 of the control object. The inequality of (2.6) guarantees finiteness of the functional (2.2) if the matrices (2.4) are uniformly bounded along t.

It is required to formulate the control strategy from the class $\mathbf{U}^a$ that minimizes the functional (2.2).

In the stationary variant

$$A_t \equiv A, \quad B_t \equiv B, \quad Q_t \equiv Q, \quad S_t \equiv S, \quad R_t \equiv R \tag{2.7}$$

the formulated problem is known as analytical design of regulators.

3.2.2 *Reduction of the Optimization Problem to the Solvability of the Matrix Riccati Equation*

The theorem 2.1.2 will be used to design optimal control. Bellman's equation (1.30) in Chapter 2 can be written as

$$V_t(x) = \min_{u \in \mathbf{U}} \{V_{t+1}(A_t x + B_t u) + q_t(x, u)\}, \tag{2.8}$$

where $q_t(x, u)$ is the quadratic form of (2.3). Bellman - Liapunov function will also be searched in the quadratic form

$$V_t(x) = x^* H_t x. \tag{2.9}$$

The positive matrix H_t must be chosen in such a way that the equation (2.8) is satisfied. The ' limiting' condition

$$\lim_{t \to \infty} V_t(x_t) = 0 \tag{2.10}$$

will be satisfied if the matrices H_t are uniformly bounded along t. It follows from the inequality (2.6) which ensures admissible control strategy*.

Minimization of (2.8) with respect to u easily realizes and leads to the formula

$$u = u_t = -(R_t + B_t^* H_{t+1} B_t)^{-1} (B_t^* H_{t+1} A_t + S_t^*) x. \tag{2.11}$$

It is assumed that the matrix

$$L_t^{-1} = R_t + B_t^* H_{t+1} B_t \tag{2.12}$$

is non-degenerate.

Putting (2.11) in (2.8) and equating the quadratic form with respect to x it can be found that the function (2.9) satisfies the equation (2.8) if the matrix H_{t+1} satisfies the Riccati equation

$$H_t = A_t^* H_{t+1} A_t - (A_t^* H_{t+1} B_t + S_t) L_t (B_t^* H_{t+1} A_t + S_t^*) + Q_t, \tag{2.13}$$

where the matrix L_t is determined by the formula (2.12).

Thus, if there exists a uniformly bounded (along t) positive solution H_t for the equation (2.13) and the equation

$$x_{t+1} = [A_t - B_t L_t (B_t^* H_{t+1} A_t + S_t^*)] x_t \tag{2.14}$$

is asymptotically stable ($x_t \to 0$ for $t \to \infty$ independent of choice of initial condition x_0), then the control

$$u_t = -L_t (B_t^* H_{t+1} A_t + S_t^*) x_t \tag{2.15}$$

is optimal. In this case the functional (2.2) assumes the least value equal to

$$\min_{U^\infty(\cdot) \in U^a} J[U^\infty(\cdot)] = x_0^* H_0 x_0. \tag{2.16}$$

The linearity of optimal control strategy follows from (2.15), (2.13) and (2.12). The asymptotic stability of equation (2.14) and boundedness of functions H_t ensure admissability of the control strategy (2.15).

Unfortunately, nothing is known about any of the conditions satisfying existence of above mentioned solutions of Riccati equation. The sationary versions of the problem (2.1), (2.2) and (2.7) are investigated more or less in details that may be significant for their applications.

3.2.3 *Lur'e Equation and a Few of its Properties*

With the assumption (2.7), the Riccati equation takes the form of an algebraic equation in respect of matrix $H > 0$, *i.e.*

$$H = A^*HA - (A^*HB + S)L(B^*HA + S^*) + Q. \tag{2.17}$$

Here $$L = (B^*HB + R)^{-1}. \tag{2.18}$$

---

*It may appear that such a class of strategies is void and then optimization problem does not have a solution. Synthesis of optimal regulator implies not only proof of existence of optimal control strategy but also indications to its design.

(*a*) *Frequency condition for solvability of the algebraic Riccati equation.* If the matrix

$$N = \begin{Vmatrix} Q & S \\ S^* & R \end{Vmatrix} \tag{2.19}$$

is positive then the result of solvability of equation (2.17), in the class of positive matrices, is well-known. The result based on frequency theorem can be obtained applying the steps presented below.

In the assumption for stabilizability of the pair (A,B) and positiveness of the matrix (2.19) (in conformity with the theorem 3.A.1), there exist matrices $H = H^*$, $\Gamma > 0$ and K so that the following identity (along $x \in \mathbf{R}^n$ and $u \in \mathbf{R}^m$) is satisfied

$$\begin{pmatrix} x \\ u \end{pmatrix}^* N \begin{pmatrix} x \\ u \end{pmatrix} + (Ax + Bu)^* H (Ax + Bu)$$

$$= (u - Kx)^* \Gamma (u - Kx) + x^* H x, \tag{2.20}$$

when the matrix

$$A_k = A + BK \tag{2.21}$$

is stable, *i.e.*, it does not have eigenvalues for $|\lambda| \geq 1$*. From (2.20) it follows that

$$\min_u [x^*Qx + 2x^*Su + u^*Ru + (Ax + Bu)^* H (Ax + Bu)] = x^*Hx, \tag{2.22}$$

it also happens to be the Bellman's equation in respect of the function

$$V(x) = x^*Hx. \tag{2.23}$$

The left hand side of (2.22) attains the minimum with respect to u when

$$u = -(R + B^*HB)^{-1}(B^*HA + S^*)x. \tag{2.24}$$

Putting (2.24) in (2.22) the equation (2.17) is obtained. It proves its solvability in the class of symmetric matrices $H = H^*$. Moreover, from (2.20) and (2.24), it follows that

$$K = -(B^*HB + R)^{-1}(B^*HA + S^*) \tag{2.25}$$

and hence (2.20) can be re-written as

$$x^*(A_K H A_K - H)x = -\begin{pmatrix} x \\ Kx \end{pmatrix}^* N \begin{pmatrix} x \\ Kx \end{pmatrix} \equiv -x^*Gx. \tag{2.26}$$

Here the notation of (2.21) has been used. Because of positiveness of matrix (2.19), the matrix G is positive, and consequently H satisfies Liapunov equation

$$A_K^* H A_K - H = -G.$$

Since the matrix (2.21) is stable the solution of the above equation can be presented in the form

$$H = \sum_{S=0}^{\infty} (A_K^*)^S G A_K^S.$$

It proves positiveness of matrix H. It also establishes the solvability of equation (2.17) in a class of positive matrices H. Moreover equation (2.14), in the present case, has a form

$$x_{t+1} = A_K x_t. \tag{2.27}$$

---

*The expression (2.20) in the presence of stable matrix (2.21) is justified. It is so for nonnegative matrix (3.19), if frequency condition (D.2) is satisfied.

The control strategy given by $u_t = Kx_t$ is admissible since the matrix (2.21) is stable, *i.e.*, the solvability of optimization problem is established.

It may be noted that under the assumptions made so far, the equation (2.17) may possess a few nonnegative solutions but only one of them corresponds to the stable matrix (2.21), (2.25).

One of the possible methods of optimal control synthesis related to solution of equation (2.17) and permitting relaxation of positiveness of matrix N^* will be stated in section 3.3.

(*b*) *Example : Scalar object.* Let the control object be described as

$$x_{t+1} = ax_t + bu_t, \tag{2.28}$$

where a and b are scalars, $b \neq 0$. The matrix N has the form as in (2.19), where $S = 0$ and Q, R are scalars. The expression in (2.20) will be re-written as

$$Qx^2 + Ru^2 = [x^2 - (ax + bu)^2]H + (u - Kx)^2\Gamma.$$

Since x and u are arbitrary it is equivalent to relations involving Γ, H, K

$$Q = K^2\Gamma + (1 - a^2)H, \tag{2.29}$$

$$K\Gamma + abH = 0, \quad R = \Gamma - b^2H. \tag{2.30}$$

Because of (2.30), Γ and K can be eliminated as

$$\Gamma = R + b^2H, \qquad K = -\frac{abH}{R + b^2H}. \tag{2.31}$$

Putting these values in (2.29), algebraic scalar equation is obtained :

$$H = a^2H - \frac{a^2b^2H^2}{R + b^2H} + Q. \tag{2.32}$$

When $Q \neq 0$, this equation has a unique nonnegative solution

$$H = (2b^2)^{-1}[(a^2 - 1)R + b^2Q + \sqrt{((a^2 - 1)R + b^2Q)^2 + 4QRb^2}]. \tag{2.33}$$

For $Q = 0$, $\quad H = 0$, the solution is unique, if $a^2 \leq 1$.

Optimal control is determined by the feedback

$$u_t = -\frac{abH}{R + b^2H}x_t, \tag{2.34}$$

where H is given by the formula (2.33).

Permissibility of control strategies given by (2.34), can be shown when

$$||a| - 1| + Q \neq 0. \tag{2.35}$$

In the present case, asymptotic stability of equation (2.27) is equivalent to inequality $(a - 1)R < b^2H$. The latter can be re-written (taking into account of (2.33)) as

$$R(a - 1)^2 + b^2Q + \sqrt{(a^2R + b^2Q - R)^2 + 4QRb} > 0. \tag{2.36}$$

---

*In the assumptions the following cases of N, for which matrix (2.19) is non positive, are important.

$$N = \begin{vmatrix} Q & 0 \\ 0 & 0 \end{vmatrix} \text{ or } N = \begin{vmatrix} 0 & 0 \\ 0 & R \end{vmatrix}.$$

Since $b \neq 0$, the inequality (2.36) is satisfied when the condition (2.35) is valid.

The feedback (2.34) with the condition (2.35) ensures the performance criterion

$$J[U^{\infty}(\cdot)] = \sum_{t=0}^{\infty}(Qx_t^2 + Ru_t^2).$$

The least value is $\min_{U^{\infty}(\cdot) \in \mathbf{U}^{a}} J = Hx_0^2$ where x_0 is the initial state of the control object (2.28).

The feedback (2.34) takes a simpler form $u_t = -ab^{-1}x_t$ when $R = 0$ and $Q > 0$. Then $x_1 = x_2 = \ldots = 0$ and $\min J = Qx_0^2$.

The optimization problem is meaningful when $Q = 0$ and $R > 0$, even when the control object (2.28) is unstable ($a^2 > 1$). The optimal feedback is then $u_t = (ab)^{-1}(1 - a^2)x_t$ and it ensures the performance criterion $\min J = Rb^{-2}(a^2 - 1)x_0^2$. For $Q = 0$ and $a = 1$ the optimization problem does not have a solution since for the unique solution $H = 0$ of equation (2.32) the inequality (2.36) is not satisfied.

(c) *Solving equations of Lur'e.* The equality (2.20) establishes the link between the coefficient of the quadratic form and the matrices K, Γ, H:

$$q(x, u) = \begin{pmatrix} x \\ u \end{pmatrix}^{*} N \begin{pmatrix} x \\ u \end{pmatrix}$$

$$Q = K^{*}\Gamma K + H - A^{*}HA,$$

$$S = -K^{*}\Gamma - A^{*}HB, \quad R = \Gamma - B^{*}HB. \tag{2.37}$$

Equation (2.37) is the discrete analog of solving equations of Lure used in absolute stability theory. Solvability of these equations with respect to matrices K, Γ, H is directly linked with solvability of equation (2.17). If matrix $H > 0$ and the matrices K and Γ are determined from (2.25) and (2.37) respectively, then the first two equalities of (2.37) will be satisfied when H satisfies the equation (2.17)*. Thus, the equation (2.17) turns out to be a condition for solvability of relations (2.37). Due to this reason, the equation (2.17) will be further known as *Lur'e equation.*

(d) *Properties of Lur'e equation.* Let *f*(*H*) represent the matrix function

$$f(H) = A^{*}HA - (A^{*}HB + S)(B^{*}HB + R)^{-1}(B^{*}HA + S^{*}) + Q. \tag{2.38}$$

Solutions to Lur'e equation (2.17) will obviously be the stationary points of the reflection of $f(H)$ (for $f(H) = H$).

Theorem 3.2.1 *Let $R > 0$. Then the following implications are true :*

$$H \geq 0 \Rightarrow f(H) \geq Q - SR^{-1}S^{*}, \tag{2.39}$$

---

*The above has been worked out for a scalar equation (2.28) (vide (2.31) and (2.32)).

$$0 \le H_1 \le H_2 \Rightarrow f(H_1) \le f(H_2), \tag{2.40}$$

$$0 \le H_1 \le H_2 \Rightarrow f(H_2) - f(H_1) \le \Phi^*(H_1)(H_2 - H_1)\Phi(H_1). \tag{2.41}$$

$$\text{Here} \quad \Phi(H) = A + BK(H), \tag{2.42}$$

$$K(H) = -(B^*HB + R)^{-1}(B^*HA + S^*). \tag{2.43}$$

The monotony of matrix function $f(H)$ is useful in studying asymptotic properties of stationary Riccati equation (vide Lemma 3.4.3). It will also be used in verifying the following assertion.

Theorem 3.2.2 *Let us assume* $N \ge 0$, $R > 0$ *in (2.19) and the pair (A,B) be stabilizable.*

Then the assertions given below are true :

1. *Equation* (2.17) *is solvable in the class of non-negative matrices H.*
2. *The set of all nonnegative solutions of Lur'e equation contains the greatest and the smallest elements, i.e., there exist matrices* $0 \le H_{\min} \le H_{\max}$ *that satisfy the equation* (2.17) *and for any nonnegative solution H of this equation the following inequality will be satisfied*

$$H_{\min} \le H \le H_{\max}. \tag{2.44}$$

3. *The matrix of* (2.42), (2.43) *that corresponds to nonnegative solution H of Lur'e equation, may be stable only for the greatest element* $H_{\max}$ *(vide* (2.44)).
4. *If the matrix* $A - BR^{-1}S^*$ *does not have any unity modular eigenvalue then the stable matrix of* (2.42), (2.43)* *corresponds to each of the solutions* ($H > 0$) *of Lur'e equation.*
5. *If the pair of matrices* $(A - BR^{-1}S^*, \ [Q - SR^{-1}S^*]^{1/2})$ *is detectable (vide section* 1.3.3)**, *then the nonnegative solution of Lur'e equation is unique and the stable matrix of* (2.42), (2.43) *corresponds to this solution.*
6. *The sequence of matrices determined by the Riccati equation*

$$H_{t+1} = A^*H_tA - (A^*H_tB + S)(B^*H_tB + R)^{-1}(B^*H_tA + S^*) + Q \tag{2.45}$$

possesses the finite limit for the given initial condition $H_0 = 0$, *i.e.,*

$$H_\infty = \lim_{t \to \infty} H_t \tag{2.46}$$

which coincides with nonnegative solution $H_{\min}$ *of Lur'e equation.*

Thus, under the conditions of theorem 3.2.2 and assertion 5:

$$x_{t+1} = Ax_t + Bu_t, \quad t = 0, 1, \ldots, \tag{2.47}$$

$$J[U_0^\infty(\cdot)] = \sum_{t=0}^{\infty} q(x_t, u_t), \quad q(x,u) = \begin{pmatrix} x \\ u \end{pmatrix}^* N \begin{pmatrix} x \\ u \end{pmatrix} \tag{2.48}$$

---

*In accordance with the assertion 3, if the solution $H > 0$ exists and coincides with solution $H_{\max}$ (vide (2.44)).

**We note that the inequality $N \ge 0$, $R > 0$ involves the inequality $Q - SR^{-1}S^* \ge 0$. If matrix $A - BR^{-1}S^*$ is stable for any $N > 0$ then the above mentioned property of detectability is apparently satisfied.

the optimization problem is ensured a solvable optimal control strategy in the admissible class $\mathbf{U}^a$ that satisfies the inequality of (2.6). The strategy is determined by the linear feedback law

$$u_t = K(H)x_t, \tag{2.49}$$

where the coefficient gain matrix $K(H)$ has the form (2.43). With the control in (2.49), the functional (2.48) takes the least value $x_0^* H x_0$.

From the assertions 5 and 6, it follows that Riccati equation (2.45) may serve as the iterative procedure for finding out the required solution of Lur'e equation (analogous to the Lemma 3.4.3).

As shown by the above example, detectability of the pair $(A, Q^{1/2})$ is the sufficient condition for solvability of the optimization problem. When $a^2 > 1$ the pair $(a, 0)$ is not detectable but the optimal problem possesses a solution. In the present example, the necessary and sufficient condition for solvability of the optimization problem is fulfilment of the frequency inequality (vide theorem 3.A.l). This condition can be expressed as

$$Q\,|1-\lambda a|^{-2} + R \geq \varepsilon(1+|1-\lambda a|^{-2}). \tag{2.50}$$

It may be easily shown that when (2.35) is satisfied, the inequality (2.50) is also satisfied for any $|\lambda| = 1$ and $\varepsilon > 0$. The frequency condition for equivalent solubility of the optimization problem in a class of linear stationary strategies follows from the results in section 3.3.

3.2.4 *Analytical design of Regulators in the Presence of Additive Noise*

(*a*) *Optimal control in the presence of decreasing noise*. Analytical design of regulators is generalized for the case when additive noise is present in equation (2.1) and it decreases as $t \to \infty$. Let the control object be described by

$$x_{t+1} = Ax_t + Bu_t + v_{t+1}, \quad t = 0, 1, \ldots, \tag{2.51}$$

where v^∞ is the noise given by the sequence of random values with the following properties

$$\mathbf{M}v_t = 0, \quad \mathbf{M}v_t v_t^* = R_v(t), \quad \sum_{t=0}^{\infty} |R_v(t)| < \infty. \tag{2.52}$$

The performance criterion is in the form

$$J[U^\infty(\cdot)] = \sum_{t=0}^{\infty} \mathbf{M}q(x_t, u_t), \tag{2.53}$$

$$q(x, u) = \begin{pmatrix} x \\ u \end{pmatrix}^* N \begin{pmatrix} x \\ u \end{pmatrix}, \tag{2.54}$$

where the nonnegative matrix N is determined by the formula (2.19). The problem is that of minimization of the functional (2.53) in the class of admissible control strategies $\mathbf{U}^a$. Here, admissible strategies mean the sequence $U^\infty(.)$ of functions $U_t(\cdot)$ that constitute the control in accordance with relations (2.5) and also ensure fulfilment of inequality

$$\sup_t \mathbf{M}(|x_t|^2 + |u_t|^2) < \infty. \tag{2.55}$$

Solution through Bellman's equation will be made use of in synthesizing the optimal control strategy in the class $\mathbf{U}^a$.

The Bellman - Liapunov function will now be formulated in the form of (2.23) and with nonnegative matrix H. To do this, the following magnitude on the control system trajectories of (2.51) and (2.5) will be examined

$$W(x_t, u_t) = \mathbf{M}\{V(x_{t+1}) + q(x_t, u_t) \mid x^t, u^t\}.$$

Considering (2.23), (2.51) and (2.52), we get

$$W(x_t, u_t) = (Ax_t + Bu_t)^* H(Ax_t + Bu_t) + q)(x_t, u_t)$$

$$+\mathrm{Sp}\, H^{1/2} R_v(t+1) H^{1/2}. \tag{2.56}$$

Consequently, the Bellman's equation has the form

$$V(x) = \min_u W(x,u) - \mathrm{Sp}\, H^{1/2} R_v(t+1) H^{1/2}. \tag{2.57}$$

Minimization of the quadratic form $W(x,u)$ with respect to u has been already found out and its minimum is realized when $u = -(R + B^*HB)^{-1}(B^*HA + S^*)x$, (vide (2.11)). Putting this in (2.57), it will be observed that the function (2.9) will satisfy equation (2.56), if the matrix H turns out to be the solution of Lur'e equation (2.17) and (2.18). If the stable matrix of (2.42) (with K in the form of (2.43)) answers to matrix H then, as has been examined in section 3.2.3, the minimization problem of (2.53) for the equation (2.51) is in this case solvable. The optimal control is determined, in the same manner, by the feedback (2.49) (as it has been shown for the noise free case), but the functional has the value

$$\min_{U^\infty(\cdot) \in \mathbf{U}^a} J[U^\infty(\cdot)] = x_0^* H x_0 + \sum_{t=1}^{\infty} \mathrm{Sp}\, H^{1/2} R_v(t) H^{1/2}. \tag{2.58}$$

In fact, because of (2.57), for any strategy from $\mathbf{U}^a$ the following is true

$$\mathbf{M}V(x_t) \le \mathbf{M}[V(x_{t+1}) + q(x_t, u_t)] - \mathrm{Sp}\, H^{1/2} R_v(t+1) H^{1/2}.$$

Adding these inequalities, we obtain

$$\sum_{t=0}^{T} \mathbf{M}q(x_t, u_t) \ge V(x_0) - \mathbf{M}V(x_{T+1}) + \sum_{t=1}^{T+1} \mathrm{Sp}\, H^{1/2} R_v(t) H^{1/2}.$$

Turning to the limit $T \to \infty$ and taking into account $V(x) \ge 0$, we find out

$$\sum_{t=0}^{\infty} \mathbf{M}q(x_t, u_t) \ge x_0^* H x_0 + \sum_{t=0}^{\infty} \mathrm{Sp}\, H^{1/2} R_v(t) H^{1/2}. \tag{2.59}$$

With the feedback (2.49), the inequality of (2.59) is converted into an equality. It proves the optimality of the corresponding control strategy in $\mathbf{U}^a$.

(*b*) *Optimal control with discounted performance criterion.* The above problem of optimization will be studied with nondecaying noise and discounted performance criterion. Let the control object be described as before, by equation (2.5.1). But the noise in it v^∞ consists of independent random values having the following properties

$$\mathbf{M}v_t = 0, \quad \mathbf{M}v_t v_t^* = R_v. \tag{2.60}$$

Here covariance matrix R_v is independent of t.

Let the performance criterion be

$$J[U^{\infty}(\cdot),\gamma] = (1-\gamma^2)\sum_{t=0}^{\infty}\gamma^{2t}\mathbf{M}q(x_t,u_t), \tag{2.61}$$

where γ is a positive number, $\gamma<1$, and $q(x,u)$ is determined by the formula (2.54). Parameter γ is the weighting factor for $q(x,u)$ that is taken into account at some moment or other. It is, therefore, known as the *discounting parameter (forgetting factor).* The weightage of the objective function decays exponentially. The functional (2.61) will be refered as *discounted.* The problem is now to minimize it in the class of admissible strategies $\mathbf{U}^a$ (vide (2.5), (2.55)).

The notations given below will be used with respect to control problem of (2.51) and (2.53). The variables and the coefficients in it will be changed accordingly. Thus, if

$$\tilde{x}_t = \gamma^t x_t, \quad \tilde{u}_t = \gamma^t u_t, \quad \tilde{A} = \gamma A, \quad \tilde{B} = \gamma B, \quad \tilde{v}_t = \gamma^t v_t, \tag{2.62}$$

then $R_{\tilde{v}}(t) = \gamma^{2t} R_v$. (2.63)

The optimal control in the resulting variables will be determined by the feedback

$$u_t = -(B^*HB+\gamma^{-2}R)^{-1}(B^*HA+\gamma^{-2}S^*)x_t, \tag{2.64}$$

where the nonnegative matrix H satisfies Lur'e equation

$$\gamma^{-2}H = A^*HA+\gamma^{-2}Q-(A^*HB+\gamma^{-2}S)(B^*HB+\gamma^{-2}R)^{-1}$$

$$\times(B^*HA+\gamma^{-2}S^*). \tag{2.65}$$

The feedback (2.64) determines control strategy $\mathbf{U}^a$ if the matrix $A-B(B^*HB+\gamma^{-2}R)^{-1}(B^*HA+\gamma^{-2}S^*)$ is stable. The condition for existence of such a solution to Lur'e equation (2.65) is given by the theorem 3.2.2*.

The least value of the functional (2.61) (because of (2.58) and (2.63)) is equal to

$$\min_{U^{\infty}(\cdot)\in\mathbf{U}^a} J[U^{\infty}(\cdot),\gamma] = (1-\gamma^2)x_0^*Hx_0+\gamma^2\,\mathrm{Sp}\,H^{1/2}R_vH^{1/2}. \tag{2.66}$$

Formula (2.66) brings out clearly the correspondence between the contributes of level of control with initial condition and with noise dependent on the forgetting factor γ. The first term in the right hand side of the equation (2.66) is determined by the transient response of the control system and the second term is conditioned by the noise acting on the control object. For $\gamma\to 1$ the term depending on transient response becomes negligible. Hence it will be quite justified to re-write the functional (2.61) as

$$J[U^{\infty}(\cdot),\gamma] = \lim_{t\to\infty}\left(\sum_{s=0}^{t}\gamma^{2s}\right)^{-1}\sum_{s=0}^{t}\gamma^{2s}\mathbf{M}q(x_s,u_s),$$

then for $\gamma=1$, the functional will be

$$J[U^{\infty}(\cdot)] = \lim_{t\to\infty}\frac{1}{t}\sum_{s=0}^{t}\mathbf{M}q(x_s,u_s) \tag{2.67}$$

which is independent of system transient response. Minimization of functional (2.67) is studied in details in section 3.4. The results of the studies show that the minimizing

---

*Use of the notations A, B (vide 2.62) in equation (2.65) will lead to (2.17).

functional of (2.61) (for $\gamma \to 1$) is the same for that of (2.67).

For $\gamma \to 0$, the feedback (2.64) determines $u_0 = -R^{-1}S^* x_0$, equation (2.65) determines the matrix $H = Q - S^* \times R^{-1}S$ and the right hand side of (2.66) assumes the expected value of

$$x_0^* H x_0 = \begin{pmatrix} x_0 \\ u_0 \end{pmatrix}^* N \begin{pmatrix} x_0 \\ u_0 \end{pmatrix} = x_0^*(Q - SR^{-1}S^*)x_0.$$

§ 3.3 Transfer Function Method in Linear Optimization Problem

3.3.1 *Statement of the Linear Optimization Problem*

Optimal control of the object

$$x_{t+1} = Ax_t + Bu_t, \quad t = 0, 1, \ldots, \tag{3.1}$$

with performance criterion

$$J[U_0^\infty(\cdot)] = \sum_{t=0}^{\infty} \begin{pmatrix} x_t \\ u_t \end{pmatrix}^* N \begin{pmatrix} x_t \\ u_t \end{pmatrix}, \tag{3.2}$$

(where N has the form given in (2.19)) can be effectively realized in a class of linear stationary control strategies given by the feedback

$$\alpha(\nabla)u_t = \beta(\nabla)x_t, \quad t = 0, 1, \quad \ldots . \tag{3.3}$$

Here

$$\alpha(\lambda) = I_m + \lambda\alpha_1 + \quad \ldots \quad + \lambda^p \alpha_p,$$

$$\beta(\lambda) = \beta_0 + \lambda\beta_1 + \quad \ldots \quad + \lambda^p \beta_p \tag{3.4}$$

are the polynomials with matrix coefficients and I_m = unit $m \times m$ matrix , α_j = quadratic $m \times m$ matrices, and β = rectangular $m \times m$ matrices. The natural number p and the set of matrices $\{\alpha_j, \beta_j\}$ determine definite control strategy. We shall include, in the admissible control strategies $\mathbf{U}^a$, all similar types of strategies which differ with various p and the set of matrices α_j, β_j. The latter (strategies) ensure inequality (2.6) independent of choice of initial conditions of the control system (3.1) and (3.3)*. It can be easily shown that the necessary and sufficient condition for this is stability of the characteristic polynomial

$$g(\gamma) = \det \begin{Vmatrix} I_n - \lambda A & -\lambda B \\ \beta(\lambda) & -\alpha(\lambda) \end{Vmatrix}, \tag{3.5}$$

i.e., it should have no roots for $|\lambda| \leq 1$. The regulator (3.3) will be known as a *stabilizing* one for control object (3.1) when its polynomial (3.5) is stable.

---

*The initial conditions of the control system are the set of state and control variables which ensure uniqueness of the trajectory x^∞, u^∞. For the system of (3.1) and (3.3), these initial conditions consist of initial state x_0 of the control object and values of $x_{-1}, x_{-2}, x_{-p}, u_{-1}, \quad \ldots, \quad u_{-p}$ (3.3).

The class of admissible control strategies U^a is quite broad based and it includes the control strategies that result from feedback and may be optimal in the class of arbitrary feedback, *i.e.*

$$u_t = Kx_t. \tag{3.6}$$

According to (3.5) the regulator of (3.6) is stabilizing, if matrix (2.21) is stable.

3.3.2 *Transfer Functions of Control Systems and their Properties*

The matrix functions $W_1(\lambda)$, $W_2(\lambda)$ of complex argument λ are related to equations (3.1) and (3.3) as

$$W_1(\lambda) = [I_n - \lambda A - \lambda B\alpha^{-1}(\lambda)\beta(\lambda)]^{-1},$$

$$W_2(\lambda) = [\alpha(\lambda) - \beta(\lambda)(I_n - \lambda A)^{-1}\lambda B]^{-1}\beta(\lambda)(I_n - \lambda A)^{-1}, \tag{3.7}$$

combining them we get an (aggregate) *transfer function* of the control system of (3.1) and (3.3)

$$W(\lambda) = \begin{Vmatrix} W_1(\lambda) \\ W_2(\lambda) \end{Vmatrix} = \begin{Vmatrix} I_n - \lambda A & -\lambda B \\ \beta(\lambda) & -\alpha(\lambda) \end{Vmatrix}^{-1} \begin{Vmatrix} I_n \\ 0 \end{Vmatrix}. \tag{3.8}$$

The matrix elements of the transfer function (3.8) are *fractionally rational functions* (F.R.F's). Important properties of transfer function will now be given.

Theorem 3.3.1 *1. For existence of the stabilizing regulator* (3.3) *for the control object* (3.1), *it is necessary that the equation*

$$\| I_n - \lambda A, \ \ -\lambda B \| \, W(\lambda) = I_n \tag{3.9}$$

was solvable in the class of matrix F.R.F.'s. which are analytical in the unit circle and have a dimension equal to* $(n+m)\times n$.

2. Let $W(\lambda)$ *be such a solution and the pair* $\{A, B\}$ *be stabilizable (vide section* 1.33 *c). Then there exists a stablizing regulator of the form* (3.1) *for which the transfer function* $\alpha^{-1}(\lambda)\beta(\lambda)$ *is determined by the formula*

$$\alpha^{-1}(\lambda)\beta(\lambda) = W_2(\lambda)W_1^{-1}(\lambda). \tag{3.10}$$

here $W_1(\lambda), W_2(\lambda)$ *are the components of* $W(\lambda)$ (vide (3.8)).

The following procedure be followed to determine matrix polynomials $\alpha(\lambda), \beta(\lambda)$ *from the relation (3.10). Let K be a certain matrix for which the matrix of (2.21) is stable. We shall determine analytically in the unit disk F.R.F.*

$$\psi(\lambda) = W_2(\lambda) - KW_1(\lambda). \tag{3.11}$$

Let $\psi_1(\lambda), \psi_2(\lambda)$ *be the matrix polynomials of dimensions* $n\times m$ *and* $n\times m$ *respectively. Let the function* $\psi_1^{-1}(\lambda)$ *be analytical in the unit circle and the polynomials* $\psi_1(\lambda)$, $\psi_2(\lambda)$ *be related as*

$$\Psi(\lambda) = \psi_1^{-1}(\lambda)\psi_2(\lambda). \tag{3.12}$$

---

*A matrix function is analytical in the unit circle (disk), if its matrix elements are analytical in the region $|\lambda| < 1+\varepsilon$ for all $\varepsilon > 0$.

Then

$$\alpha(\lambda) = \psi_1(\lambda) + \psi_2(\lambda)\lambda B \ , \beta(\lambda) = \psi_1(\lambda)K + \psi_2(\lambda) - \lambda\psi_2(\lambda)A. \tag{3.13}$$

The matrix F.R.F. will be known as *stable** F.R.F if it is analytical in the unit circle.

In this way the set of stable F.R.F.'s $W(\lambda)$ satisfying equation (3.9) is adequate for the set of linear stabilizing regulators (3.3).

The functional (3.2) can be expressed through transfer function $W(\lambda)$ of the control system (3.1) and (3.3).

Lemma 3.3.1 *When the initial conditions of a stabilizing regulator (3.3) are*

$$x_{-1} = x_{-2} = \ldots = x_{-p} = 0,$$

$$u_{-1} = u_{-2} = \ldots = u_{-p} = 0 \tag{3.14}$$

then the performance criterion of (3.2) takes the form

$$J[W(\cdot)] = x_0^{*}\frac{1}{2\pi i}\oint W^{\nabla}(\lambda)NW(\Lambda)\frac{d\lambda}{\lambda}x_0$$

$$= x_0^{*}\frac{1}{2\pi i}\oint \tilde{G}[W_2(\lambda), \lambda]\frac{d\lambda}{\lambda}x_0. \tag{3.15}$$

Here

$$\tilde{G}[W_2, \lambda] = W_2^{\nabla}(\lambda)G(\lambda)W_2(\lambda) + W_2^{\nabla}(\lambda)L^{\nabla}(\lambda)$$

$$+L(\lambda)W_2(\lambda) + M(\lambda), \tag{3.16}$$

$$G(\lambda) = \left\| \begin{matrix} a^{-1}(\lambda)b(\lambda) \\ I_m \end{matrix} \right\|^{\nabla} N \left\| \begin{matrix} a^{-1}(\lambda)b(\lambda) \\ I_m \end{matrix} \right\|, \tag{3.17}$$

$$L(\lambda) = \left\| \begin{matrix} a^{-1}(\lambda) \\ 0 \end{matrix} \right\|^{\nabla} N \left\| \begin{matrix} a^{-1}(\lambda)b(\lambda) \\ I_m \end{matrix} \right\|, \tag{3.18}$$

$$M(\lambda) = \left\| \begin{matrix} a^{-1}(\lambda) \\ 0 \end{matrix} \right\|^{\nabla} N \left\| \begin{matrix} a^{-1}(\lambda) \\ 0 \end{matrix} \right\|, \tag{3.19}$$

$a(\lambda) = I_n - \lambda A$, $b(\lambda) = \lambda B$, $\oint d\lambda$ *is integral along the unit circle in the complex plane*** *and* $A^{\nabla}(\lambda) = A^T(\lambda^{-1})$.***

Thus, the performance criterion can be expressed through the second component of the transfer function $W(\lambda)$ (vide (3.8)).

---

*It may be noted that any polynomial is a stable F.R.F. in this sense, though the polynomial itself may not be stable.

**The direction of integration has been chosen such that $\oint \frac{d\lambda}{\lambda} = 2\pi i, \quad i = \sqrt{-1}$.

***The symbol $\| \quad \|^T$ indicates a matrix transposed from matrix (without the complex conjugate).

3.3.3 *Geometrical Interpretation of the Linear Optimization Problem*

From theorem 3.3.1 and the lemma 3.3.1, it is observed that the problem of optimization of the performance criterion in a class of stationary control strategies $\mathbf{U}^a$ may be considered as a problem for determination of matrix function $W_2(\lambda)$ which minimizes the quadratic functional of (3.15) in a class of stable matrix F.R.F. of dimension $m \times n$. The peculiarity of this external problem lies in the fact that the minimum is not searched in a class of arbitrary stable F.R.F.'s $W_2(\cdot)$ and that for every $W_2(\cdot)$ there exists a stable F.R.F. $W_1(\cdot)$ related to the former through the relationship in (3.9). This condition has a distinct geometrical meaning. In fact, for a fixed vector $x_0 \neq 0$ the formula

$$< W'(\cdot), \quad W''(\cdot) > x_0 = \frac{1}{2\pi i} \oint x_0^* [W'(\lambda)]^{\nabla} N W''(\lambda) \frac{d\lambda}{\lambda} x_0 \tag{3.20}$$

determines the scalar product, which converts the set $\mathbf{W}$ of all analytical matrix functions, in the unit circle, having a dimension of $(n+m) \times n$ in the Hilbert space. $\mathbf{W}_0$, the set of stable F.R.F. $W(\lambda)$ satisfying the relationship (3.9), is an affine variety* in space $\mathbf{W}$.

It is required to find out the element $W^{opt}(\cdot) \in \mathbf{W}_0$ on which the distance between the null element of the space $\mathbf{W}$ and set $\mathbf{W}_0$ is to be realized.

The problem of determining the distance up to affine set in Hilbert space has been studied well for the set that permits appropriate parametrization.

3.3.4 *Weiner - Kolmogorov Method for Conditional Minimization of a Quadratic Functional*

Let the affine set $\mathbf{W}_0$ in the Hilbert space $\mathbf{W}$ be given by the relationship

$$W_1(\lambda) = W_1^0(\lambda) + R_1(\lambda)\Psi(\lambda), \tag{3.21}$$

$$W_2(\lambda) = W_2^0(\lambda) + R_2(\lambda)\Psi(\lambda), \tag{3.22}$$

where $W^0(\lambda) = \begin{Vmatrix} W_1^0(\lambda) \\ W_2^0(\lambda) \end{Vmatrix} \in \mathbf{W}_0, \quad R_1(\cdot), \quad R_2(\cdot)$

are the given stable matrix F.R.F.'s of dimensions $n \times m$ and $m \times m$ respectively, $\Psi(\lambda)$ is an arbitrary stable matrix F.R.F. of dimension $m \times m$**. The formulae (3.21) and (3.22) perform 'parametrization' of set $\mathbf{W}_0$: for each element $W(\cdot) \in \mathbf{W}_0$ the form (3.21) and (3.22) is true (for its own function $\Psi(\lambda)$) and in that case dependence of $W(\cdot)$ on $\Psi(\cdot)$ is linear.

When (3.22) is put in (3.15), J is determined as a (quadratic) functional of $\Psi(\cdot)$. In the theory of Weiner - Kolmogorov optimal filtration for stationary processes, this problem of minimization of functions, in a class of stable F.R.F.'s, has been studied well. The solution to this problem is given by the following assertion.

---

*This means that the set of elements $W(\cdot) - W^0(\cdot)$ is a linear set. Here $W^0(\cdot)$ is a fixed element from $\mathbf{W}_0$ and $W(\cdot) \in \mathbf{W}_0$.

**For the real values of λ, the matrix functions $R_1(\lambda)$ $R_2(\lambda)$ and $\Psi(\lambda)$ are assumed to be real.

Theorem 3.3.2 *Let us assume that for* $|\lambda| = 1$ *the matrix functions* $G(\lambda)$, $L(\lambda)$ *do not possess any singularity and*

$$\det G(\lambda) \neq 0, \quad \det R_2(\lambda) \neq 0.$$

Then the least distance up to set $\mathbf{W}_0$ *is attained on the stable F.R.F.* $\Psi^{opt}(\cdot)$ *which can be computed following the procedure given below.*

1. *The* $m \times m$ *matrix function* $\Pi(\lambda)$ *with a stable inverse matrix F.R.X.* $\Pi^{-1}(\lambda)$ *is determined from the following relationship*

$$R_2^{\nabla}(\lambda) G(\lambda) R_2(\lambda) = \Pi^{\nabla}(\lambda)\Pi(\lambda). \tag{3.23}$$

2. *The function* $R(\lambda)$ *is separated, where*

$$R(\lambda) = [\Pi^{-1}(\lambda)]^{\nabla} R_2^{\nabla}(\lambda)\,[G(\lambda) W_2^0(\lambda) + L^{\nabla}(\lambda)]. \tag{3.24}$$

In other words

$$R(\lambda) = R_+(\lambda) + R_-(\lambda), \tag{3.25}$$

where the matrix F.R.F.'s $R_+(\lambda)$, $R_-(\lambda^{-1})$ *are stable and*

$$\lim_{|\lambda| \to \infty} R_-(\lambda) = 0. \tag{3.26}$$

The function $W^{opt}(\cdot)$ *is determined from the relations* (3.21) *and* (3.22), *where*

$$\Psi(\lambda) = -\Pi^{-1}(\lambda) R_+(\lambda). \tag{3.27}$$

Then the square of the distance up to set $\mathbf{W}_0$ *is equal to*

$$< W^{opt}(\cdot),\quad W^{opt}(\cdot)\rangle_{x_0} = \frac{1}{2\pi i} \oint x_0' \left[\begin{Vmatrix} (I_n - \lambda A)^{-1}\lambda B W_2^0(\lambda) + I_n \\ W_2^0(\lambda) \end{Vmatrix}^{\nabla} \right.$$

$$\times N \begin{Vmatrix} (I_n - \lambda A)^{-1}\lambda B W_2^0(\lambda) + I_n \\ W_2^0(\lambda) \end{Vmatrix} - R^{\nabla}(\lambda) R(\lambda)$$

$$\left. + R_-^{\nabla}(\lambda) R_-(\lambda)\right] x_0 \frac{d\lambda}{\lambda}. \tag{3.28}$$

It follows from (3.23) to (3.27) that the element $W^{opt}(\cdot)$ (on which the distance up to $\mathbf{W}_0$ is realized) does not depend on the choice of x_0.

3.3.5 *Parametrization of the Set of Transfer Functions*

The procedure for finding out the optimal transfer function $W(.)$ with known conditions in (3.21) and (3.22), has been discussed in the theorem 3.3.2. Realization of these conditions for transfer functions that satisfy (3.9) will now be discussed.

Lemma 3.3.2. *Let the* $\{A,B\}$ *be stabilizable and* K *be a matrix that makes the matrix* (2.21) *stable. Then the set* $\mathbf{W}_0$ *of all stable F.R.F.'s of dimension* $(m+n)\times n$ *satisfying the identity* (3.9) *is determined by the relations of* (3.21) *and* (3.22), *with the choice as given below*

$$R_1(\lambda) = \lambda W_1^0(\lambda)B, \quad R_2(\lambda) = I_m + \lambda K W_1^0(\lambda)B, \tag{3.29}$$

$$W_1^0(\lambda) = [I_n - \lambda(A + BK)^{-1}, \quad W_2^0(\lambda) = KW_1^0(\lambda), \tag{3.30}$$

The relationship (3.9) is satisfied by the function

$$W^0(\lambda) = \begin{Vmatrix} W_1^0(\lambda) \\ KW_1^0(\lambda) \end{Vmatrix}$$

and for the function $\tilde{R}(\lambda) = \begin{Vmatrix} R_1(\lambda) \\ R_2(\lambda) \end{Vmatrix}$

the following equality is valid

$$\| I_n - \lambda A, \quad -\lambda B \| \tilde{R}(\lambda) = 0. \tag{3.31}$$

Hence, the formulae (3.21), (3.22), (3.29) and (3.30) are the general solution $W(\cdot)$ of the equation (3.9). It is presented as a sum of particular solution $W^0(\cdot)$ and the general solution $\tilde{W}(\cdot) = G(\cdot)W(\cdot)$ of the corresponding homogeneous equation.

3.3.6 *Design of the Optimal Regulator for the Object Expressed in the Standard Form*

(*a*) *Summary of the Results.* The above results of optimization using transfer function method will be summarised through the following assertion.

Theorem 3.3.3 *Let the pair {A,B} in the linear quadratic problem of* (3.1) *and* (3.2) *be stabilizable and*

$$\det(I_n - \lambda A) \neq 0, \tag{3.32}$$

$$\det G(\lambda) \neq 0 \quad \text{for} \quad |\lambda| = 1. \tag{3.33}$$

Here matrix $G(\lambda)$ *is determined by formula* (3.17).

Then minimization of the functional (3.2) *in the class of linear regulators* (3.3), (3.4) *(for given initial conditions as in* (3.14)) *is possible and the solution can be obtained following the procedure laid down in the following steps.*

1. *A matrix K is selected so that the matrix (2.21) is stable.**
2. *The matrix function* $W_1^0(\lambda)$ *and* $R_2(\lambda)$ *are computed using formula in* (3.30) *and* (3.29).
3. *Function* $\Psi(\lambda)$ *is determined in accordance with theorem* 3.3.2.

The optimal regulator has the expression as in (3.3). *Here the matrix polynomials* $\alpha(\lambda)$, $\beta(\lambda)$ *are determined using formulae* (3.13), (3.11) *and* (3.12) *in terms of functions* (3.21) *and* (3.22). *The minimum value of the performance criterion is determined by the formula* (3.28).

---

* Existence of matrix K is supported by stabilizability of the pair (A, B) (vide section 3.3.3c).

(b) Observations on solution to Linear Optimization Problem

1. Though computation of function $\Psi(\lambda)$ following the theorem 3.3.2 is related to a quite arbitrary function $W_1^0(\lambda)$ (to be precise to the matrix K - vide (3.30)), the transfer function $W_1^0(\lambda)$ (which is found out using $\Psi(\lambda)$), does not depend on the choice of $W_1^0(\lambda)$. It is so, because $W^{opt}(\lambda)$ is determined only by the set $\mathbf{W}_0$ and not by its concrete parametrization.

2. The condition $R_2^{\nabla}(\lambda)G(\lambda)R_2(\lambda) > 0$ (for $|\lambda|= 1$) that supports the spectral method (vide (3.23)) taking into account the formulae (3.29), (3.30) and (3.17) appears to be equivalent to frequency condition (A.2). This means that the regulator obtained in theorem 3.3.3 is optimal both in the class of linear stationary control strategies and the class of arbitrary unpredicting control strategies (2.5) that satisfy the inequality (2.6). In fact as was shown in section 3.2.3 the frequency condition (A.2) ensures solvability of optimal problem in a class of such control strategies. The optimal strategy results in a linear stabilizing regulator as synthesized in the theorem 3.3.3.

In case the frequency condition is not satisfied the above condition gives $R_2^{\nabla}(\lambda_0)G(\lambda_0)R_2(\lambda_0) = 0$ for some λ_0, $|\lambda_0|= 1$. As per (3.23) det $\pi(\lambda_0) = 0$ and it means that in the class of linear stabilizing regulators the optimal problem can not be solved. In fact, replacing the matrix N by $N + \varepsilon I$, where ε is a small positive parameter, we may be able to synthesize an optimal regulator. The characteristic polynomial $g(\lambda, \varepsilon)$ is proportional to det (λ, ε) (vide the theorem 3.4.6 and formula (4.86)). As $\varepsilon \to 0$ we obtain a regulator that is unstable. Hence, choice of $\varepsilon \neq 0$ leads to synthesis of a suboptimal control with the given level of suboptimality.

3. For $N > 0$ the condition $det\, G(\lambda) > 0$, (when $|\lambda|= 1$) is deliberately implemented. However, it is also implemented when $R > 0$, $S = 0$, $Q = 0$ or $R = 0$, $S = 0$, $Q > 0$. When the eigenvalues of matrix A are not present in the unit circle this condition is equivalent to the frequency condition (A.2).

4. For $Q = 0$, $S = 0$, $R > 0$ the least value of the performance criterion is not necessarily equal to zero. The control strategy $u_t \equiv 0$ is not admissible if the control object is unstable (*i.e.* if even one eigenvalue of the matrix A is with $|\lambda|\geq 1$). If the matrix A has even one eigenvalue for $|\lambda|= 1$ then the frequency condition is violated. Then the optimization problem is not solvable at least in the class of linear stationary control strategies.

(b) Example : Two dimensional object with scalar input. Theorem 3.3.3 will be illustrated taking the example of (3.1) and (3.2) where

$$A = \begin{Vmatrix} -a, & -1 \\ 1 & 0 \end{Vmatrix}, \quad B = \begin{Vmatrix} 1 \\ 0 \end{Vmatrix}, \quad N = \begin{Vmatrix} 0 & 0 & 0 \\ 0 & 0 & 0 \\ 0 & 0 & 1 \end{Vmatrix} \tag{3.34}$$

and $-a_1 > 2$. (3.35)

Condition (3.35) indicates instability of the control object of (3.1) and (3.34). The eigenvalues of the matrix A are :

$$\lambda_1=\left(\frac{1}{2}\right)\left(-a_1+\sqrt{a_1^2-4}\right),$$

$$\lambda_2=\left(\frac{1}{2}\right)\left(-a_1-\sqrt{a_1^2-4}\right) \tag{3.36}$$

and $\lambda_1>1$ (since $-a_1>2$). Even though the least value of the performance criterion

$$J=\sum_{t=0}^{\infty} u_t^2 \tag{3.37}$$

in the class of stabilizing feedback appears to be a non-zero value, the control strategy $u_t\equiv 0$ is not admissible. (For $|a_1|<2$ this strategy is admissible and $\min J=0$, and for $|a_1|=2$ the optimal regulator does not exist.)

The efficacy of the algorithm in the theorem 3.3.3 will be demonstrated. The matrix K is chosen as a row matrix

$$K=\|a_1,\quad 1\|. \tag{3.38}$$

Then the matrix $A+BK=\left\|\begin{matrix}0 & 0\\ 1 & 0\end{matrix}\right\|$ will be stable. Following (3.30), (3.29), (3.17), (3.18) and (3.34) we determine

$$W_1^0(\lambda)=\left\|\begin{matrix}1 & 0\\ \lambda & 1\end{matrix}\right\|,\quad R_2(\lambda)=1+\lambda a_1+\lambda^2=(\lambda-\lambda_1)(\lambda-\lambda_2),$$

$$R_1(\lambda)=\left\|\begin{matrix}\lambda\\ \lambda^2\end{matrix}\right\|,\quad G(\lambda)\equiv 1,\quad L(\lambda)\equiv 0. \tag{3.39}$$

Following the theorem 3.3.2 and the symbols in (3.39) we find

$$R_2^{\nabla}(\lambda)G(\lambda)R_2(\lambda)=(\lambda-\lambda_1)(1-\lambda\lambda_2)(1-\lambda^{-1}\lambda_2)(\lambda^{-1}-\lambda_1).$$

Consequently $\Pi(\lambda)$ is to be chosen as the polynomial

$$\Pi(\lambda)=(\lambda-\lambda_1)(1-\lambda\lambda_2) \tag{3.40}$$

(here the F.R.F. $\Pi^{-1}(\lambda)$ will be stable). The following row matrix is found out using formula (3.24) along with (3.39) and (3.40)

$$R(\lambda)=\left\|\frac{(1-\lambda\lambda_2)(\lambda_1+\lambda)}{\lambda-\lambda_2},\quad \frac{1-\lambda\lambda_2}{\lambda-\lambda_2}\right\|.$$

In the present case separation (3.25) leads to division of the complete parts into fractions which have been obtained. $R_+(\lambda)=\|2-\lambda_2\lambda,\quad -\lambda_2\|$ is obtained by dividing the polynomials. It permits computation of the row matrix $\Psi(\lambda)$ corresponding to the optimal transfer function

$$\Psi(\lambda)=\frac{1}{(\lambda-\lambda_1)(1-\lambda\lambda_2)}\|\lambda\lambda_2-2,\quad \lambda_2\|.$$

According to (3.21) and (3.22) we have

$$W_1(\lambda)=\frac{1}{\lambda-\lambda_1)(1-\lambda\lambda_2)}\left\|\begin{matrix}-\lambda_1 & \lambda\lambda_2\\ -\lambda\lambda_1, & 2\lambda-\lambda_1\end{matrix}\right\|,$$

$$W_2(\lambda)=\frac{1}{1-\lambda\lambda_2}\|a_1+2\lambda_2,\quad 1-\lambda_2^2\|.$$

Performing elementary operations on (3.10) we find out optimal feedback polynomials $\alpha(\lambda)$, $\beta(\lambda)$:

$$\alpha(\lambda)=\lambda_1-\lambda,$$

$$\beta(\lambda)=\|1-\lambda_1^2,\quad 0\| -\lambda\| 2a_1+3\lambda_2+\lambda_1,\quad 0\|, \tag{3.41}$$

λ_1 and λ_2 are determined by the formula (3.36). The characteristic polynomial of the control system thus obtained is $g(\lambda)=\lambda-\lambda_1$. For $a_1\to -2$, we have $\lambda_1\to 1$ and the regulator ceases to be stabilizing.

3.3.7 *Correspondence between transfer function method and method of Lur'e solving equation*

Solvability of Lur'e equations (2.37) with respect to matrices K, Γ, H for partial conditions in (A.2) was discussed in section 3.2.3. For such a case the regulator has the expression as in (3.6). The same result will be got now from theorem 3.3.3.

Lemma 3.3.3 *If the matrix K in theorem* 3.3.3. *is chosen so that it is determined by equations (2.37), then*

$$R_2^{\nabla}(\lambda)G(\lambda)R_2(\lambda)=\Gamma, \tag{3.42}$$

$$R(\lambda)=\Gamma^{-1/2}\left([\lambda^{-1}I_n-(A+BK)]^{-1}B\right)^{\nabla}H, \tag{3.43}$$

where matrices Γ *and* H *are also determined by* (2.37)

It follows from (3.43) that

$$R(\lambda)=R_-(\lambda), \tag{3.44}$$

i.e., according to theorem 3.3.2 we get $\Psi(\lambda)\equiv 0$ and so

$$W^{opt}(\lambda)=\begin{Vmatrix} W_1^0(\lambda) \\ KW_1^0(\lambda)\end{Vmatrix}.$$

Function $W_1^0(\lambda)$ is determined by the formula (3.30). Then in accordance with (3.10), the optimal transfer function has the form $\alpha^{-1}(\lambda)\beta(\lambda)=K$. It is the same as the regulator in (3.6).

3.3.8 *Design of the optimal regulator for control object equations expressed through 'input–output' variables*

Let the control object be expressed as

$$a(\nabla)y_t=b(\nabla)u_t,\quad t=0,1,\ldots, \tag{3.45}$$

$$a(\lambda)=I_n+\lambda a_1+\ldots+\lambda^r a_r,$$

$$b(\lambda)=\lambda^k b_k+\ldots+\lambda^r b_r \tag{3.46}$$

are the polynomials with matrix coefficients ($a_j - n\times n$ matrices, $b_j - n\times m$ matrices), k - control time delay $1\le k\le n$. The control objective is to minimize the functional

$$J[U_0^\infty(\cdot)] = \sum_{t=0}^{\infty}\begin{pmatrix} y_t \\ u_t \end{pmatrix}^* N\begin{pmatrix} y_t \\ u_t \end{pmatrix}, \tag{3.47}$$

where N is the non negative matrix of the form (2.19). The class of admissible control strategies $\mathbf{U}^a$ is chosen as the linear stationary strategies that result in feedback

$$\alpha(\nabla)u_t = \beta(\nabla)y_t \tag{3.48}$$

and provide the inequality

$$\sum_{t=0}^{\infty}(|u_t|^2 + |y_t|^2) < \infty \tag{3.49}$$

independent of the choice of initial conditions in control systems of (3.45), (3.48). The polynomials $\alpha(\lambda)$, $\beta(\lambda)$ are expressed as

$$\alpha(\lambda) = I_m + \lambda a_1 + \ldots + \lambda^p a_p,$$

$$\beta(\lambda) = \beta_0 + \lambda\beta_1 + \ldots + \lambda^p\beta_p, \tag{3.50}$$

where α_j are $m\times m$ matrices and β_j are $m\times n$ matrices. The regulators of (3.48) satisfying the inequality (3.49) independent of choice of initial conditions will be called as *stabilizing* (for the control object (3.45)). The necessary and sufficient conditions for the regulator (3.48) to be stabilizing are the stability of the characteristic equation of the control system (3.45), (3.48),

$$g(\lambda) = \det\begin{Vmatrix} a(\lambda) & -b(\lambda) \\ \beta(\lambda) & -\alpha(\lambda) \end{Vmatrix}. \tag{3.51}$$

Optimization of the functional (3.47) in a class of stabilizing regulators (3.48) can be reduced to the control problem already studied. However sufficiency will be absent if the transfer function $W(\lambda)$ for the control systems (3.45), (3.48) is introduced. The matrix function $W(\lambda)$ is determined by the formula

$$W(\lambda) = \begin{Vmatrix} a(\lambda) & -b(\lambda) \\ \beta(\lambda) & -a(\lambda) \end{Vmatrix}^{-1}\begin{Vmatrix} I_n \\ 0 \end{Vmatrix}. \tag{3.52}$$

The relationship (3.9) is then expressed as

$$\| a(\lambda),\ \ -b(\lambda) \|\, W(\lambda) = I_n. \tag{3.53}$$

The subsequent steps will be similar to those presented in sections 3.3.2 to 3.3.4. For the present method parametrization of the set of stable F.R.F. statisfying equation (3.53) and construction of polynomials $\alpha(\lambda)$, $\beta(\lambda)$ of the stabilizing regulator for the given transfer function are important

$$\alpha^{-1}(\lambda)\beta(\lambda) = W_1(\lambda)W_1^{-1}(\lambda), \tag{3.54}$$

$W_1(\lambda)$ and $W_2(\lambda)$ are the constituents of the function $W(\lambda)$:

$$W(\lambda) = \begin{Vmatrix} W_1(\lambda) \\ W_2(\lambda) \end{Vmatrix}. \tag{3.55}$$

Equation (3.45) with matrix coefficients a_j, b_j can be examined in the same manner. But our attention will be restricted to refinement of the problem for a particular case that leads to a relatively simple solution.

(*a*) *Control object with scalar input and output.* The polynomials $a(\lambda)$, $b(\lambda)$ are scalar and the transfer function $W(\lambda)$ (vide 3.52) is a matrix with two rows. The equation (3.9) has the form

$$a(\lambda)W_1(\lambda) - b(\lambda)W_2(\lambda) = 1 \tag{3.56}$$

where $W_1(\lambda)$, $W_2(\lambda)$ are the rows of the matrix $W(\lambda)$. Following the procedure in section 3.3.2 one obtains the following result.

Lemma 3.3.4 *Let the polynomials $\alpha(\lambda)$, $\beta(\lambda)$ in equation (3.45) be scalar. Then :*

(1) For the existence of a stabilizing regulator (3.48) the necessary and sufficient condition is the solvability of equation (3.53) in a class of analytically in the unit disk F.R.F. $W_1(\lambda)$ and $W_2(\lambda)$. For any such solution $W_1(\lambda)$, $W_2(\lambda)$ of the equation (3.53), the polynomials $\alpha(\lambda)$ and $\beta(\lambda)$ of the stabilizing regulator (3.48) will be uniquely determined from the relationship (3.54) as unreducible polynomials.*

(2) *If the initial conditions of the regulator** are given by*

$$y_{-1} = \ldots = y_{-p} = 0, \quad u_{-1} = \ldots = u_{-p} = 0 \tag{3.57}$$

then the functional (3.47) *can be expressed as*

$$J = z^* \left[\frac{1}{2\pi i} \oint \hat{N}^\nabla(\lambda) \tilde{G}[W_2(\lambda), \quad \lambda] \hat{N}(\lambda) \frac{d\lambda}{\lambda} \right] z, \tag{3.58}$$

where the Hermition form $\tilde{G}(W_2, \lambda)$ is determined by the formulae (3.16) *to* (3.19) *and $W_2(\lambda)$ is a constituent of function $W(\lambda)$* (vide 3.55),

$$\hat{N}(\lambda) = \| \sum_{s=1}^{r} \lambda^{s-1} a_s, \quad \sum_{s=2}^{r} \lambda^{s-2} a_s, \quad \ldots, \quad a_r,$$

$$- \sum_{s=k}^{r} \lambda^{s-k} b_s, \; \ldots, \; -b_r \| \tag{3.59}$$

$$\text{and } z = \text{col}\,(y_{-1}, \; \ldots, \; y_{-k}, \quad u_{-1}, \; \ldots, \; u_{-r}) \tag{3.60}$$

are the initial conditions of the control object (3.45).

(3) *The set of all stable functions (3.55) which satisfies equation (3.53) can be determined as*

$$W(\lambda) = W^0(\lambda) + \left\| \begin{matrix} b(\lambda)\nu^{-1}(\lambda) \\ a(\lambda)\nu^{-1}(\lambda) \end{matrix} \right\| \Psi(\lambda) \tag{3.61}$$

where $W^0(\lambda)$ is any double rowed stable F.R.F. satisfying equation (3.53), $\nu(\lambda)$ is the G.C.D. of the polynomials $a(\lambda)$, $b(\lambda)$ and $\Psi(\lambda)$ is an arbitrary scalar stable F.R.F.

It is evident from the lemma 3.3.4 that the optimal control problem leads to minimization of the functional (3.58) as a class of stable F.R.F's which satisfy the

---

*Henceforth such F.R.F's will be known as *stable*.

**In equation (3.45) it is not necessary to assume null initial conditions.

condition in (3.53).

Solvability of equation (3.53) in the class of stable F.R.F's $W_1(\lambda)$ and $W_2(\lambda)$, is evidently equivalent to the absence of general roots of the polynomials $a(\lambda)$ and $b(\lambda)$ for $|\lambda| \leq 1$. (This means that their general divisor $\nu(\lambda)$ must be a stable polynomial.) We can choose the transfer function of the control system in equations (3.45) and (3.48) as $W^0(\lambda)$. The polynomials $\alpha(\lambda)$ and $\beta(\lambda)$ for above will be determined from the relationship

$$a(\lambda)\alpha(\lambda) - b(\lambda)\beta(\lambda) = \nu(\lambda) \tag{3.62}$$

where $\deg \alpha(\lambda) < \deg b(\lambda) - \deg \nu(\lambda)$. This will give the stabilizing regulator of (3.48)*.

Applicability of the theorem 3.3.2 to the case under consideration is related with factorizations (vide (3.58), (3.16), (3.17), (3.13) and (3.61))

$$a^{\nabla}(\lambda)G(\lambda)a(\lambda) = \pi_1(\lambda^{-1})\pi_1(\lambda), \tag{3.63}$$

$$[\hat{N}(\lambda)z]^{\nabla}\hat{N}(\lambda)z = \pi_2(\lambda^{-1})\pi_2(\lambda), \tag{3.64}$$

$\pi_1(\lambda)$, $\pi_2(\lambda)$, $\pi_1^{-1}(\lambda)$ and $\pi_2^{-1}(\lambda)$ must be stable F.R.F's. When such functions are found out the optimal regulator as per section 3.3.4 can be determined as follows : The function

$$R(\lambda) = \frac{\pi_2(\lambda)a(\lambda^{-1})\,[G(\lambda)W_2^0(\lambda) + L(\lambda^{-1})]}{\pi_1(\lambda^{-1})} \tag{3.65}$$

is separated out, *i.e.* we have to find out its representation in the form of (3.25) with stable F.R.F's $R_+(\lambda)$, $R_-(\lambda^{-1})$. The function $R_-(\lambda)$ also satisfied the condition (3.26)**.

The stable F.R.F. $\Psi(\lambda)$ resulting in optimal transfer function following the formula (3.61) can be determined as

$$\Psi(\lambda) = -\frac{R_+(\lambda)\nu(\lambda)}{\pi_1(\lambda)\pi_2(\lambda)}. \tag{3.66}$$

The function (3.61) determines the polynomials $\alpha(\lambda)$ and $\beta(\lambda)$ of the optimal regulator using relationship (3.54).

It may be observed that the function $W^{opt}(\lambda)$ depends on the choice of the initial states (3.60) of the control object (3.45). (But for any arbitrary choice the regulator is stabilizing.) The factorisation (3.63) results in an expansion into elementary products of polynomials in the numerator and denominator of the function $G(\lambda)$ when the matrix N is non negative. Here the latter division of products correspond to roots and poles of the function located outside the unit circle. If $G(\lambda) \neq 0$ when $|\lambda| = 1, \pi_1^{-1}(\lambda)$ will be a stable F.R.F. Similar observation can be made in respect of the relationship (3.64). The F.R.F. $\pi_2^{-1}(\lambda)$ is stable if $\hat{N}(\lambda)z \neq 0$ when $|\lambda| = 1$. The possibility and complexity of factorisation may depends on choice of initial conditions z of the control object (3.45).

---

*It can be shown that the regulator obtained by this method is optimal for the minimum phase control object (3.45) if $R = O$, $S = O$, $Q = I$ in the matrix (2.19).

**It is to be remembered that the scalar functions $G(\lambda)$ and $L(\lambda)$ in equation (3.65) are determined by the formulae (3.17) and (3.18).

(*b*) *Example : Scalar Object.* Let the scalar control object be

$$Y_t = a y_{t-1} + b u_{t-k}, \quad t = 0, 1, \quad \dots \tag{3.67}$$

Let $a > 1$, $b > 0$, and the initial conditions be $|\, y_{-1} \,| = 1$, $u_{-k} = u_{-k+1} = \dots = u_{-1} = 0$, *i.e.* the control 'starts' from the k-th instant*. The performance criterion is in the form of (3.47) with

$$N = \begin{Vmatrix} Q & 0 \\ 0 & R \end{Vmatrix}, \quad Q + R > 0. \tag{3.68}$$

The 'unknown' transfer function $W^0(\lambda)$ is chosen as the one that results in the feedback

$$F(\nabla) u_t = -\frac{a^k}{b} y_t, \quad t = 0, 1, \quad \dots \tag{3.69}$$

$$F(\lambda) = 1 + \lambda a + \quad \dots \quad + \lambda^{k-1} a^{k-1}. \tag{3.70}$$

Using formula (3.52) the corresponding transfer function** is determined

$$W^0(\lambda) = \begin{Vmatrix} F(\lambda) \\ -\dfrac{a^k}{b} \end{Vmatrix}. \tag{3.71}$$

In this case $\hat{N}(\lambda) z = a y_{-1}$ and consequently $\pi_2(\lambda) = |\, a y_{-1} \,| = a$. The polynomial $\pi_1(\lambda)$ is determined from the relationship (3.63) taking into account of the formulae (3.17) and (3.68).

$$a(\lambda^{-1}) G(\lambda) a(\lambda) = Q b^2 + R (1 - \lambda a)(1 - \lambda^{-1} a)$$

$$= \pi_1(\lambda) \pi_1(\lambda^{-1})$$

and it is then of the form

$$\pi_1(\lambda) = \gamma_1 - \lambda \gamma_2, \tag{3.72}$$

where

$$2\gamma_1 = \sqrt{R(1+a)^2 + Qb^2} + \sqrt{R(1-a)^2 + Qb^2}\},$$

$$2\gamma_2 = \sqrt{R(1+a)^2 + Qb^2} - \sqrt{R(1-a)^2 + Qb^2}. \tag{3.73}$$

The formula (3.65) is then expressed as

$$R(\lambda) = \frac{a^k \gamma_2}{b} - \frac{a^k (a\gamma_1 - \gamma_2)}{b(1 - \lambda_a)} + \frac{abQ}{\lambda^{k-1}(1 - \lambda a)(\lambda\gamma_1 - \gamma_2)}.$$

Since $a > 1$, $\gamma_1 > \gamma_2$, then

---

* For any other initial conditions control problem is solved in a similar manner.

** The feedback in (3.69), (3.70) is chosen from equation (3.62), (in the present case $b_+ = 1$, $b_- = \lambda^k b$, $\nu(\lambda) = 1$); a simple computation gives polynomials $\alpha(\lambda) = F(\lambda)$, $\beta(\lambda) = -a^k / b$.

$$R_+(\lambda) = -\frac{a^k}{b}\gamma_2, \quad R_-(\lambda) = -\frac{a^k(a\gamma_1 - \gamma_2)}{b(1-\lambda a)}$$

$$+ \quad \frac{abQ}{\lambda^{k-1}(1-\lambda a)(\lambda\gamma_1 - \gamma_2)},$$

$$\Psi(\lambda) = \frac{a^{k-1}\gamma_2}{b(\gamma_1 - \lambda\gamma_2)}.$$

It is now possible to determine optimal transfer function $W^{opt}(\lambda)$ using (3.61). Again using relationship (3.54) we can find out polynomials $\alpha(\lambda)$ and $\beta(\lambda)$ for the optimal regulator. The regulator has the expression

$$[F(\nabla)(1-\rho\nabla) + a^{k-1}\rho\nabla^k]u_t = \frac{a^{k-1}}{b}(\rho - a)y_t, \tag{3.74}$$

where
$$\rho = \gamma_2\gamma_1^{-1}. \tag{3.75}$$

It is not difficult to compute the characteristic polynomial of the control system (3.67) and (3.74), viz.

$$g(\lambda) = 1 - \lambda\rho.$$

Since $\rho < 1$, the regulator of (3.74) is stabilizing. (The result agrees with the theory.)

We observe that in (3.68) when $R = 0$, then $\rho = 0$ because of (3.73) and (3.75). The regulator of (3.74) is the same as that in (3.69).

For the minimum delay in control ($k = 1$) we have $F(\lambda) = 1$ and the feedback in (3.74) is expressed as

$$u_t = \frac{\rho - a}{b}y_t.$$

The result agrees with that in section 3.3.7.

§ 3.4 Limiting Optimal Control of Stochastic Processes

When the control object is subjected to random noises, the oscillations in it do not die out and it is not possible to obtain the type of inequality as in (2.6) by selecting any control strategy. The admissible control strategies will be those which satisfy the following inequality irrespective of the choice of initial conditions :

$$\overline{\lim_{t\to\infty}}\, t^{-1}\sum_{s=1}^{t} \mathbf{M}(|x_s|^2 + u_s|^2) < \infty. \tag{4.1}$$

The performance criterion is chosen as

$$J[U^\infty(\cdot)] = \overline{\lim_{t\to\infty}}\, t^{-1}\sum_{s=1}^{t} \mathbf{M}q_s(x_s, u_s), \tag{4.2}$$

where $q_s(x, u)$ is the nonnegative objective function*.

---

*It is assumed that the mathematical expectations in (4.1) and (4.2) exist for the admissible control strategies.

The speciality of infinite time period control as regards stability of the system will be once more highlighted. The methods of finite time period optimal control may appear to be unsuitable with unlimited time of control if the stability requirement is also present. Thus, the optimal feedback was synthesized (vide formula (3.27) in Chapter 2) in section 2.3.4 for terminal control of linear stationary objects. The characteristic polynomial of the control system (3.26) and (3.37) in Chapter 2 can be easily computed as

$$g(\lambda) = a(\lambda)\alpha(\lambda) - b(\lambda)\beta(\lambda)$$

$$= a(\lambda)F(\lambda)b(\lambda)\lambda^{-k} - b(\lambda)G(\lambda)$$

$$= \lambda^{-k}b(\lambda)\,[a(\lambda)F(\lambda) - \lambda^{k}G(\lambda)].$$

For a non minimum phase control object the polynomial $b(\lambda)$ (and so $g(\lambda)$) will be unstable, *i.e.* the control strategy determined by the feedback signal will not be admissible. Hence the inequality in (4.1) is not obtained.

It may be observed that for a solvable limiting optimal control problem that it usually degenerates: from equation (4.2) it is evident that the choice of control actions in any finite time period control does not affect the quality of control. In other words the constrained optimization displays 'useful properties'of the controlled process which is true for the 'transient response'. This unique solution of the limiting optimization often allows easy inclusion of additional requirements related to constraints on transient response in the control objective.

3.4.1 *Sufficient Conditions for Optimality of Admissible Control Strategies*

A few generalised conditions for optimality of control strategy expressed in terms of Bellman - Liapunov function will be discussed before optimization of linear control objects is taken up.

Let us assume that the control object is expressed as

$$x_{t+1} = X_{t+1}(x_t, u_t, v_{t+1}), \quad t = 0, 1, \ldots, \tag{4.3}$$

whereas the output y_t is given by

$$y_t = Y_t(x_t), \quad t = 0, 1, \ldots. \tag{4.4}$$

The system is free of observational noise and here $x_t \in \mathbf{R}^n$, $u_t \in \mathbf{R}^m$, $y_t \in \mathbf{R}^l$. It is also assumed that the state x_t is uniquely determined by the previous values of output and control variables*, *i.e.*,

$$x_t = \tilde{X}_t(y^t, u^{t-1}). \tag{4.5}$$

The possibility of using Bellman's equation to study infinite time period control problems has been discussed in section 2.1.5. However, in the presence of nondecaying noise the equation obtained in section 2.1.5 is unsolvable. From the analysis of limiting optimal control problems, the corresponding analog of Bellman's equation can be found out in stochastic problems.

---

*This assumption is usually not constrained when observational noise is absent. For a wider class of control x_t is a linear combination of variables $y^t_{t-r}, u^{t-1}_{t-r}, r$ is a fixed number, or even $x_t =$ col $(y_{t-r}, \ldots, y_t)$.

Let us assume there edxists such a real random function $V_t(x)$. that the quantity $\mathbf{M}\{V_{t+1}(x_{t+1}) \mid y^t\}$ is meaningful for every admissible control strategy. Also

$$\lim_{t\to\infty} \frac{\mathbf{M}\{V_{t+1}(x_{t+1})\}}{t} = 0. \tag{4.6}$$

Using equation (4.3) we find that

$$\tilde{V}_{t+1}(x,u) = \mathbf{M}\,\{V_{t+1}[X_{t+1}(x_t, U_t, v_{t+1})] + q_t(x_t, U_t) \mid x_t = x, \quad U_t = u\} \tag{4.7}$$

Theorem 3.4.1 *Let the class of admissible control strategies* $\mathbf{U}^a$ *be not empty and let it consist of all possible regular strategies* $U_0^\infty(\cdot)$ *which provide inequality (4.1) and let the limitation of the functional (4.2) be independent of the choice of initial state* x_0 *of the control object (4.3). The control strategy* $U_0^\infty(\cdot)$ *will be optimal if for* $\tilde{U}_0^\infty(\cdot) \in \mathbf{U}^a$, *which is independent of choice of* x_0, *the following equality (with probability 1) is satisfied :*

$$V_t(x_t) = \min_{u\in\mathbf{R}^m} \tilde{V}_{t+1}(x_t, u) - r_t, \tag{4.8}$$

Here r_0^∞ *is the sequence of random quantities independent of choice of* $U^\infty(\cdot) \in \mathbf{U}^a$ *with mean* $\mathbf{M}r_t$. *Then the least value of the functional* (4.2) *is*

$$J[\tilde{U}_0^\infty(\cdot)] = \lim_{t\to\infty} t^{-1} \sum_{s=0}^{t-1} \mathbf{M}r_s. \tag{4.9}$$

Equation (4.8) is the stochastic version of the Bellman's equation for optimal control with infinite time period. The solution given by the function $V_t(x)$ will be termed as *Bellman - Liapunov function.* Thus, the sufficient condition for solvability of the optimization problem is given by the existence of Bellman - Liapunov function which satisfies the condition (4.6) for every admissible control strategy. For a known Bellman - Liapunov function, the optimal strategy $U_0^\infty(\cdot)$ is given by the relationship

$$\tilde{U}_t(x_t) = \operatorname*{argmin}_{u\in\mathbf{R}^m} \tilde{V}_{t+1}(x_t, u), \tag{4.10}$$

where the function $\tilde{V}_{t+1}(x,u)$ must be determined using formula (4.7).

3.4.2 *Statement of the Limiting Linear Quadratic Optimal Control Problem*

Let the control object be given by

$$x_{t+1} = Ax_t + Bu_t + v_{t+1}, \quad t = 0, 1, \ldots, \tag{4.11}$$

here $x_t \in \mathbf{R}^n$, $u_t \in \mathbf{R}^m$, $v_t \in \mathbf{R}^n$. A and B are constant matrices of appropriate dimensions. The noise v^∞ is assumed to be a random stationary process obtained from a linear filter, *i.e.*,

$$d(\nabla)v_t = e(\nabla)w_t, \tag{4.12}$$

where $d(\lambda) = I_n + \lambda d_1 + \ldots + \lambda^q d_q$,

$$e(\lambda) = e_0 + \lambda e_1 + \ldots + \lambda^q e_q$$

are the polynomials with matrix coefficients (matrices e_j may be rectangular), w^∞ is a random (vector) process with independent values and having the following properties

$$\mathbf{M}w_t = 0, \quad \mathbf{M}w_t w_s^* = R_w \delta_{t,s}. \tag{4.13}$$

The random process w^∞ with the above properties are known as *discrete white noise* or a *white noise process.*

It is well-known that the equation (4.12) with a white noise process w^∞ will give rise to a sole centred stationary process v^∞ (within the equivalence) having a spectral density

$$\rho_v(\lambda) = d^{-1}(\lambda)e(\lambda)R_w[d^{-1}(\lambda)e(\lambda)]^\nabla.$$

The polynomial det $d(\lambda)$ is assumed to be stable.

The optimal control problem is now to find the admissible control strategy that minimizes the functional

$$J[U_0^\infty(\cdot)] = \overline{\lim_{t\to\infty}} \ (1/t)\sum_{s=0}^{t-1} \mathbf{M}q(x_s, u_s). \tag{4.14}$$

Here $q(x,u) = \begin{pmatrix} x \\ u \end{pmatrix}^* N \begin{pmatrix} x \\ u \end{pmatrix}$, (4.15)

$$N = \begin{vmatrix} Q & S \\ S* & R \end{vmatrix}, \tag{4.16}$$

$N =$ nonnegative matrix, ($N \geq 0$).

The class of admissible control strategies $\mathbf{U}^a$ will be chosen as a set of all control strategies $U_0^\infty(\cdot)$. Here $u_t = U_t(u^{t-1}, y^t)$ are the unpredictable functions of the corresponding set of control and output variables and they satisfy the inequality (4.1) irrespective of choice of initial state x_o of the control object (4.11).

3.4.3 *Solvability of the Optimization Problem*

The existence of the optimal control strategy will be studied for observed data under different conditions.

(*a*) *Completely deterministic noise.* If the noise v_0^∞ is known then its random characteristic does not play any important role. The known noise condition can be represented as

$$y_t = (x_t, \quad w^\infty), \tag{4.17}$$

i.e. the present and future values of noise signals are known at any instant t along with the state x_t.*

Attempts will now be made to express Bellman - Liapunov function as

---

*When the quantities w_t are known the noise signals v_t are obtained from equation (4.12). The initial conditions are given uniquely by the stationary condition of the process v^∞.

$$V_t(x) = x^*H_x + x^*\xi_t. \tag{4.18}$$

here H = non-negative matrix and ξ_0^∞ random sequence to be chosen so as to satisfy Bellman's equation.

Theorem 3.4.2 *If equation (2.17) is satisfied by the non-negative matrix H and matrix $R + B^* HB$ is nondegenerative while matrix*

$$A_0 = A + BK, \tag{4.19}$$

(K is in the form given by (2.25)) is stable, i.e. does not have eigenvalues for $|\lambda| \geq 1$) then the function (4.18) with,

$$\xi_t = 2\sum_{l=t}^{\infty}(A_0^*)^{l-t+1} H v_{l+1}. \tag{4.20}$$

will satisfy Bellman's equation

$$V_t(x) = \min_{u \in \mathbf{R}^m} \mathbf{M}\{V_{t+1}(Ax + Bu + v_{t+1}) + \begin{pmatrix} x \\ u \end{pmatrix}^* N \begin{pmatrix} x \\ u \end{pmatrix} | y_t, \quad x_t = x\} = r_t, \tag{4.21}$$

where

$$r_t = v_{t+1}^* H v_{t+1} + v_{t+1}^* \xi_{l+1} - \frac{1}{4}\pi_t^* BLB^* \pi_t \tag{4.22}$$

and $\pi_t = 2Hv_{t+1} + \xi_{t+1}$

$$= 2\sum_{l=t}^{\infty}(A_0^*)^{l-t} H v_{l+1}. \tag{4.23}$$

The optimal control strategy is given by the feedback

$$u_t = -(B^*HB + R)^{-1}\left[(B^*HA + S^*)^* x_t + \frac{B^*}{2}\pi_t\right]. \tag{4.24}$$

The performance criterion has the least value, with the admissible value $\mathbf{U}^a$, equal to

$$\min_{U^\infty(\cdot) \in \mathbf{U}^a} J[U_0^\infty(\cdot)] = \overline{\lim_{t \to \infty}}\ t^{-1}\sum_{s=0}^{t-1} \mathbf{M} r_s. \tag{4.25}$$

The formula (4.20) determines the satisfactory process ξ^∞ with $\mathbf{M}\xi_t = 0$,

$$\mathbf{M}\xi_t\xi_s^* = 4\sum_{l=t}^{\infty}\sum_{l'=s}^{\infty}(A_0^*)^{l-t+1} HR_v(l-l')HA_0^{l'-s+1}, \tag{4.26}$$

here $R_v(l-l') = \mathbf{M}v_l v_{l'}^*$ is the covariance of the process v^∞. As the matrix (4.19) is stable, the series in (4.26) converges. Again since the process $\{\xi_t\}$ is bounded in a mean square sense, the right hand side in equation (4.25) is bounded.

From the formulae (4.24), (4.23) and (4.20) it follows that for an optimal control strategy the input u_t depends on the future values of the noise, and optimal strategy is linear with respect to states and noise.

In accordance with the discussions in section 3.2.3a, conditions in the theorem 3.4.2 will be satisfied if the matrix (4.16) is positive. The existence of the optimal strategy for $N \geq 0$ depends on the frequency condition (A.2) of the function $G(x, u)$.

(*b*) *The past and present values of noise are known.* If $y_t = (x_t, w^t)$ then the feedback (4.24) is not admissible (future values of noise signals are not known). However, the function (4.18) will continue to be the solution of Bellman's equation with a different choice of sequence r^∞ and reduction to admissible control strategy. We shall express the general scheme that leads to such a result. The vector x_{t+1} is excluded from the function $\mathbf{M}\{V_{t+1}(x_{t+1}) \mid y^t\} + G(x_t, u_t)$ because of equation (4.11). Then the square terms of u_t are separated out

$$\mathbf{M}\{(Ax_t + Bu_t + v_{t+1})^* H(Ax_t + Bu_t + v_{t+1})$$
$$+ (Ax_t + Bu_t + v_{t+1})^* \xi_{t+1} + x_t^* Q x_t + 2u_t^* S^* x_t$$
$$+ u_t^* R u_t \mid y_t\} = \mid L^{-1/2} u_t + L^{1/2}[(B^* HA + S^*)x_t$$
$$+ \frac{B^*}{2} \mathbf{M}(\pi_t \mid y_t)] \mid^2 + x_t^* [Q + A^* HA - (B^* HA + S^*)^*$$
$$\times L(B^* HA + S^*)] x_t + x_t^* [A^* - (B^* HA + S^*)^* LB^*]$$
$$\times \mathbf{M}(\pi_t \mid y_t) + r_t. \tag{4.27}$$

Here the designation (4.23) has been used for brevity's sake end

$$L = (B^* HB + R)^{-1}, \tag{4.28}$$

$$r_t = \mathbf{M}\left\{v_{t+1}^* H v_{t+1} + v_{t+1}^* \xi_{t+1} = \pi_t^* \frac{BLB^*}{4} \pi_t \mid y_t\right\} +$$
$$+ \mathbf{M}\left\{[\pi_t - \mathbf{M}(\pi_t \mid y_t)]^* \frac{BLB^*}{4} [\pi_t - \mathbf{M}(\pi_t \mid y_t)] \mid y_t\right\}. \tag{4.29}$$

We assume that the matrix H satisfies the equation (2.17) and ξ_t is determined by formula (4.20). Then considering (4.18), the expression (4.27) is transformed to

$$\mathbf{M}\{V_{t+1}(x_{t+1}) \mid y_t\} + \begin{pmatrix} x_t \\ u_t \end{pmatrix}^* N \begin{pmatrix} x_t \\ u_t \end{pmatrix} = \mid L^{-1/2} u_t$$
$$+ L^{1/2}\left[(B^* HA + S^*)x_t + \frac{B^*}{2} \mathbf{M}(\pi_t \mid y_t)\right] \mid^2 V_t(x_t) + r_t. \tag{4.30}$$

It follows from equation (4.30) that the linear quadratic form (4.18) will be the Bellman function for equation (4.20), if the following conditions are fulfilled :

(1) the mean values of random quantities (4.29) must not depend on the choice of the control strategy,

(2) the control strategy that provides with minimum value of the right hand side of (4.30) w.r.t. u_t, *e.g.*

$$u_t = -(B^* HB + R)^{-1}\left[(B^* HA + S^*)x_t + \frac{B^*}{2} \mathbf{M}(\pi_t \mid y_t)\right], \tag{4.31}$$

must also give the inequality in (4.1) independent of the choice of the initial conditions of the system (4.11), (4.31).

Comparison of formulae (4.31) and (4.24) shows that the synthesis of optimal control, with future values of noise unknown, can be done by replacing the latter by the optimal values at the corresponding instant of time*.

The above gives a generalised picture as to application of Bellman's equation to synthesis of optimal control strategy. The methods of computing 'given observations' $\mathbf{M}\{\pi_t \mid y_t\}$ and matrix H will be shown. It will permit us to consider the feedback (4.31) as an algorithm for computation of control signals. An accurate result is presented.

Theorem 3.4.3 *Let the noise v^∞ be a stationary process obtained from a filter (4.12), and a white noise signal w^∞ with independent values having properties as in (4.13)** acts upon the input of the filter. Then according to the conditions in theorem 4.3.2, the optimal control strategy is given by formulae :*

$$u_t = -(B^*HB+R)^{-1}[(B^*HA+S^*)x_t$$

$$+B^*\sum_{l=t}^{\infty}(A_0^*)^{l-t}H\sum_{t'=l+1-t}^{\infty}h(t')w_{l+1-t'}], \tag{4.32}$$

$$h(t)=\frac{1}{2\pi i}\oint \lambda^{-t}d^{-1}(\lambda)e(\lambda)\frac{d\lambda}{\lambda}.$$

The performance criterion has the form given by (4.25) for optimal control. Here the random quantities r_t are given by formulae (4.29).

(c) *Unknown noise.* We now consider the case when the noise $y_t = x^t$ *i.e.*, it is not known at any instant past and future. The control strategy of (4.32) turns out to be not admissible. But it can be transformed into a form not containing random quantities w_t and with this the optimal strategy can be formulated.

Theorem 3.4.4 *Let us assume that the noise v^∞ is obtained from the filter (4.12), where*

$$d(\lambda)\equiv I_n, \tag{4.33}$$

*and polynomial $e(\lambda)$ is such that the polynomial det $e^T(\lambda)e(\lambda)$ is stable***. Then the optimal control for unobservable noise and under the conditions given by theorem 3.4.3 will be given by the relationship*

$$[\det e^*(\nabla)e(\nabla)]u_t = -(B^*HB+R)^{-1}\{(B^*HA+S^*)\times$$

$$\times[\det e^*(\nabla)e(\nabla)]x_t + B^*{}_\gamma(\nabla)\tilde{e}(\nabla)e^*(\nabla)\times$$

---

*It is well-known that the random quantities $\mathbf{M}\{\pi_s \mid y_t\}$ minimize the functional $\mathbf{M}|\pi_s - f(u_t)|^2$ in the class of arbitrary Borel functions $f(\cdot)$. Hence it is known as the *optimal value* of the random quantities π_s (in the mean square sense). If $s > t$ the optimal value is known as *optimal prediction*.

**The polynomial det $d(\lambda)$ is assumed to have no roots for $|\lambda| \le 1$.

***This partially means that the matrix $e_0^* e_0$ is not degenerate, *i.e.* dimension of vector w_t must not exceed n.

$$\times[x_t - Ax_{t-1} - Bu_{t-1}]\}. \quad (4.34)$$

Here $\tilde{e}(\lambda)$ *and* $\gamma(\lambda)$ *are the matrix polynomials determined by formulae*

$$\tilde{e}(\lambda) = [e^T(\lambda)e(\lambda)]^{-1}\det[e^T(\lambda)e(\lambda)], \quad (4.35)$$

$$\gamma(\lambda) \equiv 0 \quad for \quad q \equiv 0,$$

$$\gamma(\lambda) = \gamma_0 + \lambda\gamma_1 + \ldots + \gamma^{q-1}\gamma_{q-1} \quad (4.36)$$

$$for \quad q > 0,$$

$$\gamma_s = \sum_{l=0}^{q-s-1} (A_0^*)^l \ He_{l+s+1}. \quad (4.37)$$

Thus, the optimal strategy is linear w.r.t. the data of observation and takes the form of a regulator $\alpha(\nabla)u_t = \beta(\nabla)x_t$ with matrix polynomials $\alpha(\lambda)$, $\beta(\lambda)$ which can be easily obtained from relationship (4.34).

It can be shown that the optimal control strategy under conditions of theorem 4.3.2 is linear when the arbitrary matrix polynomial $d(\lambda)$ has its polynomial det $d(\lambda)$ a stable one. For this we re-write the equation (4.11) as

$$d(\nabla)x_{t+1} = d(\nabla)\,[Ax_t + Bu_t] + e(\nabla)w_{t+1}$$

and the equation (4.39) thus obtained is transformed to a standard form by introducing a new state vector. In the latter equation the noise signal will appear as a linear combination of random quantities w_t and again the theorem 4.3.4 can be made use of.

In order to use regulator (4.34), it is necessary to have H, the solution to Lur'e equation (2.17). A generalised method developed in section 3.3 and based on transfer function method is presented below. It permits synthesis of an optimal regulator (4.38) without using Lur'e equation.

In the particular case $e(\nabla) \equiv I_n$, *i.e.*, when the notice v^∞ is white noise with independent values, the relationship of (4.34) becomes considerably simpler. In fact, since $\gamma(\lambda) \equiv 0, q = 0$ so (4.34) is reduced to the form of (3.6) and (2.25). Thus, the optimal regulators with or without white noise become one and the same.

It may be observed that the assumption regarding independence of random quantities w_t, obtained from the equation (4.12), is essential for linearity of the optimal strategy. When the independence is not granted (but the properties of (4.13) are retained), there are possibilities of having optimal prediction $\mathbf{M}(\pi_t \mid w^t)$ as a non-linear function of observational data. Obviously, the optimal strategy will also be not a linear* one.

(*d*) *Performance criteria dependent on randomness.* The optimization problem for performance criteria which are random variables will now be studied. Such functionals are more real than the unconditional** ones. Formu-

---

*For a Gaussian noise the condition in (4.13) guaranties its independence. Hence the example must be formulated for random quantities which are not Gaussian.

**Averaging of realizations of noise and the sequence of control signals are carried out in the functional of (4.14) for the adaptive version of the control problem. The control signals are often formed by using identification procedures, which may include test algorithms. In such cases a more realistic approach is to ensure the required control objective at every realization of the test procedure and control process. Such a consideration leads to the requirement for performance criteria which may depend on realization of control signals.

lation and solution of such optimization problems will be shown below.

Once again we consider the control object expressed by the equation (4.11). The performance criteria are given by

$$J[U_0^\infty(\cdot),\quad w_0^\infty] = \overline{\lim_{t\to\infty}}\ t^{-1}\sum_{s=0}^{t-1} q(x_s, u_{s-1}), \tag{4.38}$$

where $q(x,u)=\begin{pmatrix}x\\u\end{pmatrix}^* N\begin{pmatrix}x\\u\end{pmatrix}$, and N is a nonnegative matrix of the from (4.16). The admissible control strategies $\mathbf{U}^a$ will be the set of functions $U^\infty(\cdot)$ which give the control u_0^∞ as per formula (2.5). They ensure the following relations with a probability of 1 :

$$\overline{\lim_{t\to\infty}}\, t^{-1}\sum_{s=0}^{t-1}(|u_s|^2+|x_s|^2)<\infty,$$

$$\lim_{t\to\infty} t^{-1}(|u_t|^2+|x_t|^2)=0,$$

$$\sup_t \mathbf{M}(|x_t|^2+|u_t|^2)<\infty. \tag{4.39}$$

The speciality of the functional (4.38) compared to (4.14) lies in the fact that, for a fixed strategy $U_0^\infty(\cdot)$, it depends on realization of noise v_0^∞, *i.e.*, random quantities. However, it does not disturb the formulation of the optimal control problem. This is exactly so, that if we put $\tilde{U}_0^\infty(\cdot) < U_0^\infty(\cdot)$ when the inequality $J[\tilde{U}^\infty(\cdot),\quad v^\infty] \le J[U^\infty(\cdot),\quad v^\infty]$ with probability 1 is fulfilled, then the set of control strategies $\mathbf{U}^a$ may be semiregulated. Then the optimal control problem is to locate $U_{opt}^\infty(\cdot)$ such that the condition $U_{opt}^\infty < U^\infty(\cdot)$ is satisfied for any arbitrary strategy $U^\infty(\cdot)$. Any such control strategy that may exist can be really called as *optimal* in the set $\mathbf{U}^a$. Again the control $\hat{u}^\infty$ determined from (2.5) using the optimal strategy will be called as *limited optimal control.*

Theorem 3.4.5 *Let us assume that the following conditions are satisfied :*

(1) the noise v^∞ is formed by the filter $v_t = e(\nabla)w_t$ where $e(\lambda) = I + \lambda e_1 + \ldots + \lambda^n e_n$ = a matrix polynomial with the stable polynomial det $e^T(\lambda)e(\lambda)$. A random process w^∞ with independent values and following properties is fed to the input of the filter :

$$\mathbf{M}w_t = 0,\quad \mathbf{M}w_t w_t^* = R_w,\quad \sup_t \mathbf{M}\,|\,w_t\,|^4 < \infty;$$

(2) the pair (A,B) is stabilizable;

(3) the matrix of (4.16) is nonnegative and for quadratic form $q(x,u)$ the frequency condition of (A.2) is satisfied.

Then the optimal control strategy $U_{opt}^\infty(\cdot)$ exists in the set $\mathbf{U}^a$ and it can be realized using the linear feedback as given in the theorem 3.4.4.

Thus, the solutions to optimization problem for functions (4.14) and (4.38) become the same.

The following is the expression for the random performance criterion with the control object given in (4.11) :

$$J[U^\infty(\cdot), v^\infty] = \overline{\lim_{t\to\infty}}(1/t)\sum_{s=1}^{t} \mathbf{M}\{q(x_s, u_{s-1}) \mid x^{s-1}, u^{s-1}\}.$$

It contains the ensemble operation of a conditional mathematical expectation. Here $q(x, u)$ represents a quadratic nonnegative form for which fulfilment of the frequency condition is assumed. The optimization problem has again a solution determined by a linear stationary control strategy. The proof is the same as given in theorem 3.4.5.

3.4.4 *Design of Optimal Linear Regulators through Transfer Function Method*

If the class $\mathbf{U}^a$ is confined to linear stationary control strategies, then the constrained optimal control may be realized using the unimportant generalisation of the method laid in Section 3.3. The scheme of the method suitable for random stationary noises with fractionally rational spectral density will be discussed.

(*a*) *Statement of the problem of linear optimal control of stochastic processes.* Let the control object be expressed by equations

$$a(\nabla)x_t = b(\nabla)u_t + v'_t, \tag{4.40}$$

$$c(\nabla)y_t = d(\nabla)x_t + v''_t. \tag{4.41}$$

Here x is the state variable, $x \in \mathbf{R}^n$; u is the control variable, $u \in \mathbf{R}^m$; y is the observable variable, $y \in \mathbf{R}^l$; and $v =$ col (v', v'') is the noise variables (in the object as well as in the observation path), $v \in \mathbf{R}^{n+l}$. Also $a(\lambda)$, $b(\lambda)$, $c(\lambda), d(\lambda)$ are the polynomials of the form

$$a(\lambda) = I_n + \lambda a_1 + \ \dots\ + \lambda^p a_p,\ \ b(\lambda) = \lambda b_1 + \ \dots\ \lambda^p b_p. \tag{4.42}$$

$$c(\lambda) = I_l + \lambda c_1 + \ \dots\ + \lambda^p c_p,\ \ d(\lambda) = d_0 + \lambda d_1 + \ \dots\ \lambda^p d_p. \tag{4.43}$$

Here the coefficients are the matrices of proper dimension, I_n, I_l are the unit matrices of dimensions $n \times n$ and $l \times l$. The noise signal

$$v_t = \text{col}\,(v_t', v_t'') \tag{4.44}$$

is considered to be random quantities, the aggregate v^∞ of which becomes a stationary process (in the general sense) with the properties

$$\mathbf{M}v_t = 0,\ \mathbf{M}v_t v_s^* = R_v(t-s). \tag{4.45}$$

It also possesses a fractionally rational spectral density $S_v(\lambda)$. The covariance $R_v(t)$of the stationary process and its spectral density $S_v(\lambda)$ are related by formulae

$$S_v(\lambda) = \sum_{t=-\infty}^{\infty} R_v(t)\lambda^t, \tag{4.46}$$

$$R_v(t) = \frac{1}{2\pi i}\oint \lambda^{-t} S_v(\lambda)\frac{d\lambda}{\lambda}. \tag{4.47}$$

It is assumed that the series (4.46) converges in some neighbourhood of the unit circumference, *i.e.* the matrix function $R(t)$ decays quite rapidly for $t \to \infty$.

In the class of admissible control strategies we choose the set of all possible sequences of functions $U^\infty(\cdot)$ that can be obtained by feedbacks, *i.e.*,

$$\alpha(\nabla)u_t = \beta(\nabla)y_t. \tag{4.48}$$

They also ensure fulfilment of the inequality

$$\overline{\lim_{t\to\infty}}(1/t)\sum_{s=0}^{t-1}\mathbf{M}(|y_s|^2+|u_s|^2+|x_s|^2)<\infty. \tag{4.49}$$

It is independent of the choice of initial conditions in the control system (4.40), (4.41), (4.48).

Using discrete Laplace transforms it can be shown that the inequality (4.49) is satisfied if the characteristic polynomial of the control system (4.40), (4.41), (4.48)* is stable. This polinominal is

$$g(\lambda)=\det\begin{Vmatrix} a(\lambda) & -b(\lambda) & 0\\ -d(\lambda) & 0 & c(\lambda)\\ 0 & \alpha(\lambda) & -\beta(\lambda)\end{Vmatrix}$$

$$=\det a(\lambda)\det c(\lambda)\det[\alpha(\lambda)-\beta(\lambda)c^{-1}(\lambda)d(\lambda)a^{-1}(\lambda)b(\lambda)]. \tag{4.50}$$

The regulator (4.48) for a stable polynomial (4.50) will be known as *stabilizing* for the control object (4.40), (4.41).

The performance criterion to be minimized in a class of control strategy $\mathbf{U}^a$ and resulting in stabilizing regulators (4.48), can be expressed as

$$J[U^\infty(\cdot)]=\overline{\lim_{t\to\infty}}\frac{1}{t}\sum_{s=0}^{t-1}\mathbf{M}\begin{pmatrix}y_s\\u_s\end{pmatrix}^*N\begin{pmatrix}y_s\\u_s\end{pmatrix}. \tag{4.51}$$

Here N = nonnegative matrix of the form (4.15).

(*b*) *Transition to control problem without observational noise*. It is possible to simplify the algebraic transformation of the control system. Let us consider F.R.F. $W(\lambda)=d(\lambda)a^{-1}(\lambda)b(\lambda)$. We shall say that the F.R.F. has a pole at $\lambda=\lambda_0$ if there is any matrix element possessing a *pole* at that point. The multiplicity of F.R.F. pole λ_0 will be set as maximum multiplicity pole λ_0 among all possible matrix ($W(\lambda)$) minors. Let all the F.R.F. ($\alpha(\lambda)a^{-1}(\lambda)b(\lambda)$) poles $\lambda_1, ..., \lambda_q$ be of multiplicity $r_1, ..., r_q$ and let

$$p(\lambda)=\prod_{k=1}^{q}(\lambda-\lambda_k)^{r_k}$$

be a scalar polynomial whose index does not exceed the multiplicity of all poles of the F.R.F. $a^{-1}(\lambda)$ (here $p(0)\neq 0$, since $a^{-1}(0)=I_n$). Then the function

$$\tilde{b}(\lambda)=p(\lambda)d(\lambda)a^{-1}(\lambda)b(\lambda) \tag{4.52}$$

will be a polynomial function. Thus, a polynomial function is introduced as

$$\tilde{a}(\lambda)=p(\lambda)c(\lambda). \tag{4.53}$$

It is not difficult to see that by virtue of (4.40) and (4.41) the following relationship is satisfied**

---

*The well-known properties of the determinants were used :

$$\det\begin{vmatrix}A B\\ C D\end{vmatrix}=\det\begin{vmatrix}A & 0\\ C & D-CA^{-1}B\end{vmatrix}=\det A\det[D-CA^{-1}B].$$

**This relationship can be obtained by using Laplace transform in equations (4.40) and (4.41) with exclusion of transformed state variable at a later step.

$$\bar{a}(\nabla)y_t = \bar{b}(\nabla)u_t + \bar{v}_t, \tag{4.54}$$

$$\bar{v}_t = v_t'' + p(\nabla)d(\nabla)a^{-1}(\nabla)v_t'. \tag{4.55}$$

The equation (4.54) looks same as the equation (4.40). But here the 'state' y_t (of dimension l) is 'being observed' without any noise. The formula (4.55) gives rise to a noise $\bar{v}_t^{\infty}$ is also a centered stationary process with fractionally rational spectral density

$$S_{\bar{v}}(\lambda) = \| p(\lambda)d(\lambda)a^{-1}(\lambda),\ I_n \| \, S_v(\lambda) \, \| p(\lambda)d(\lambda)a^{-1}(\lambda)I_n \|^{\nabla}. \tag{4.56}$$

Here $A^{\nabla}(\lambda) = A^T(\lambda^{-1})$. The characteristic polynomial of system (4.54), (4.48) is given by

$$\bar{g}(\lambda) = \det \begin{Vmatrix} \bar{a}(\lambda) & -\bar{b}(\lambda) \\ -\beta(\lambda) & \alpha(\lambda) \end{Vmatrix} = \det \bar{a}(\lambda) \det [\alpha(\lambda)$$

$$- \beta(\lambda)\bar{a}^{-1}(\lambda)\bar{b}(\lambda)] = \det a(\lambda) \det [p(\lambda)a^{-1}(\lambda)] \det c(\lambda)$$

$$\times \det [\alpha(\lambda) - \beta(\lambda)c^{-1}(\lambda)d(\lambda)a^{-1}(\lambda)b(\lambda)]. \tag{4.57}$$

Polynomials (4.50) and (4.57) are simultaneously stable or unstable. In fact from the stability of the polynomial $\bar{g}(\lambda)$(due to formulae (4.57) and (4.50)), the stability of the polynomial $g(\lambda)$ follows. The stability of the polynomial $g(\lambda)$ is equivalent to fulfilment of the inequality (4.49) for any arbitrary initial conditions of the control system (4.40), (4.41). When the inequality (4.49) is satisfied, the polynomial (4.57), as is evident, can not be unstable. In this manner, the optimal control problem of (4.40) with observations determined by equation (4.41) appeared to be equivalent* to corresponding problem with 'control object' given by (4.54) and with 'states' y_t being observed without any noise. Hence the following linear optimization problem will be investigated to minimize the performance criterion (4.51) in the presence of 'linear relationship' between y and u, *i.e.*,

$$a(\nabla)y_t = b(\nabla)u_t + v_t. \tag{4.58}$$

The polynomials $a(\lambda)$, $b(\lambda)$ are of the form (4.42). As before, minimization is realized in the class of control strategy $\mathbf{U}^a$ that gives rise to stabilizing regulators (4.48).

(*c*) *Control system transfer functions and their properties.* We introduce matrix F.R.F. $W(\lambda)$ of complex argument λ using the formula

$$W(\lambda) = G^{-1}(\lambda) \begin{Vmatrix} I_n \\ 0 \end{Vmatrix}, \tag{4.59}$$

and

$$G(\lambda) = \begin{Vmatrix} a(\lambda) & -b(\lambda) \\ -\beta(\lambda) & \alpha(\lambda) \end{Vmatrix}. \tag{4.60}$$

The matrix (4.59) is the transfer function of the control system (4.58), (4.48).

Lemma 3.4.1 *If $W(\lambda)$ is the control system transfer function with a stabilizing regulator*

---

*We note that the functional (4.51) is not dependent on state variables x_t.

(4.48), then the functional (4.51) may be written in the form

$$J[U^{\infty}(\cdot)] = \frac{1}{2\pi i}\oint \mathrm{Sp}[N^{1/2}W(\lambda)S_v(\lambda)W^{\nabla}(\lambda)N^{1/2}]\frac{d\lambda}{\lambda}. \tag{4.61}$$

Here $S_v(\lambda)$ is the spectral density of noise v^{∞}.

It follows from the lemma 3.4.1 that the linear optimization problem leads to optimization of a quadratic functional (4.61) in a class of stable control system transfer functions. The latter problem has been already discussed for special form of polynomials $a(\lambda)$, $b(\lambda)$ in section 3.3.

The structure of the transfer functions $W(\lambda)$ will be examined in more details. Let $W_1(\lambda)$ and $W_2(\lambda)$ be the constituent matrices of dimensions $n \times n$ and $m \times n$ related to the transfer function $W(\lambda)$:

$$W(\lambda) = \begin{Vmatrix} W_1(\lambda) \\ W_2(\lambda) \end{Vmatrix}. \tag{4.62}$$

From the formula (4.59), we have the relationship between these constituents as

$$a(\lambda)W_1(\lambda) - b(\lambda)W_2(\lambda) = I_n, \tag{4.63}$$

$$\alpha(\lambda)W_2(\lambda) - \beta(\lambda)W_1(\lambda) = 0. \tag{4.64}$$

The latter can uniquely determine the regulator (4.48) transfer function $\alpha^{-1}(\lambda)\beta(\lambda)$.

It appears that any stable F.R.F. (4.62) fulfilling (4.63) generates transfer function of a stabilizing regulator (4.48) because of (4.64). Similar assertion is made in section 3.3 for special choice of polynomials $a(\lambda)$, $b(\lambda)$. The general case of polynomial matrices $a(\lambda)$, $b(\lambda)$ is discussed below.

Definition 3.4.1 The pair of matrix polynomials $a(\lambda)$, $b(\lambda)$ is known as *stabilizable*, if

$$\det\,[a(\lambda^{-1})a^{*}(\lambda^{-1}) + b(\lambda^{-1})b^{*}(\lambda^{-1})]\,|\lambda|^{2p} \neq 0 \tag{4.65}$$

for all $|\lambda| \geq 1$.* If the inequality (4.65) is satisfied for all complex λ, then the pair $\{a(\lambda), b(\lambda)\}$ is known as controllable.

For $a(\lambda) = I_n - \lambda A$, $b(\lambda) = \lambda B$ the definition 3.4.1 leads to the concepts of stabilizable and controlled pairs (A, B) mentioned in section 1.3.3c.

Definition 3.4.2 The sum of multiplicities of all poles of the matrix $W(\lambda)$ located in the set $|\lambda| \leq 1$ will be called as the *Number $N_{-}(W)$ unstable poles*** *of matrix* $W(\lambda)$. The sum of multiplicities of all poles of this matrix will be known as the *Number $N(W)$ of poles of matrix* $W(\lambda)$.

Lemma 3.4.2 *The following statements are equivalent :*

---

*In the definition above it is assumed that the numbers of rows in matrices $a(\lambda)$ and $b(\lambda)$ are the same.

**This means, of the poles located in the set $|\lambda| \leq 1$.

(1) the pair $\{a(\lambda), b(\lambda)\}$ *is stabilizable (controllable);*

(2) the number $N_{-}[a^{-1}(\lambda)b(\lambda)]$ $(N[a^{-1}(\lambda)b(\lambda)]$ *is equal to the number of roots (taking into account of their multiplicity) of the polynomial* det $a(\lambda)$ *in the set* $|\lambda| \leq 1$ *(correspondingly in the complex plane);*

(3) the $n \times n$ *polynomial matrix* $A(\lambda)$ *having its inverse for all* λ *such that the matrices* $A^{-1}(\lambda)a(\lambda)$, $A^{-1}(\lambda)b(\lambda)$ *are polynomials and det* $A(\lambda)$ *is an unstable polynomial correspondingly det* $A(\lambda)$ *is a zero degree polynomial) does not exist.*

If follows from the lemma 3.4.2. that the stabilizability of the pair $\{a(\lambda), b(\lambda)\}$ is a necessary condition for solvability of the equation (4.63) in a class of stable F.R.F's. If the pair $\{a(\lambda), b(\lambda)\}$ is not stabilizable, then the following can be presented

$$a(\lambda) = A(\lambda)a_0(\lambda), \quad b(\lambda) = A(\lambda)b_0(\lambda). \tag{4.66}$$

Here $A(\lambda)$, $a_0(\lambda)$, $b_0(\lambda)$ are polynomial matrices and det $A(\lambda)$ is unstable. Then equation (4.63) can be rewritten as

$$a_0(\lambda)W_1(\lambda) - b_0(\lambda)W_2(\lambda) = A^{-1}(\lambda).$$

Since the matrix $A^{-1}(\lambda)$ has a pole for $|\lambda| \leq 1$, then at least one of the F.R.F's $W_1(\lambda)$ or $W_2(\lambda)$ is unstable.

It is quite justified to present (4.66) for the stabilizable pair $\{a(\lambda), b(\lambda)\}$ (which follows from lemma 3.4.2). Here $A(\lambda)$ is the polynomial matrix with a stable polynomial det $A(\lambda)$ and the pair of polynomial matrices $\{a_0(\lambda), b_0(\lambda)\}$ is controllable. If stable F.R.F's $W_1(\lambda)$ and $W_2(\lambda)$ satisfy the equation (4.63), then it is evident that the following F.R.F's will be stable

$$\tilde{W}_1(\lambda) = W_1(\lambda)A^{-1}(\lambda), \ \tilde{W}_2(\lambda) = W_2(\lambda)A^{-1}(\lambda). \tag{4.67}$$

Again by virtue of (4.66) they will satisfy equation

$$a_0(\lambda)\tilde{W}_1(\lambda) - b_0(\lambda)\tilde{W}_2(\lambda) = I_n; \tag{4.68}$$

conversely, any pair of stable F.R.F's $W_1(\lambda)$ and $W_2(\lambda)$ related by equation (4.68) will generate, following (4.67), a pair of stable F.R.F's $W_1(\lambda)$ and $W_2(\lambda)$ that satisfy equation (4.63). Hence the investigation of the set of stable solutions of equation (4.63) is considerably restricted to controllability of the pair $\{a(\lambda), b(\lambda)\}$.

(d) Parametrization of the solutions of matrix equation for transfer functions. We assume that the pair of polynomial matrices $\{a(\lambda), b(\lambda)\}$ is stabilizable and the equation (4.63) is solvable in the class of stable F.R.F's (4.62). If $W^0(\lambda)$ is the partially stable solution of the equation (4.63), $R(\lambda)$ is a stable F.R.F. of $(n+m)\times m$ dimension, then parametrization of the set of all such solutions, as mentioned, will be in the form

$$W(\lambda) = W^0(\lambda) + R(\lambda)\Psi(\lambda). \tag{4.69}$$

Here $R(\lambda)$ satisfies the homogeneous solution

$$||a(\lambda), \ -b(\lambda)||R(\lambda) = 0, \tag{4.70}$$

and $\Psi(\lambda)$ is an arbitrary stable F.R.F. of $m \times n$ dimension, which plays the role of a parameter. Let $\mathbf{W}_0(W^0, R)$ be a family of all matrices (4.62) obtained from (4.69) with all possible stable F.R.F's $\Psi(\lambda)$. The family $\mathbf{W}_0(W^0, R)$ will be known as *complete*, if it coincides with the set $\mathbf{W}_0$ of all stable F.R.F's (4.62) of dimension $(m+n)\times m$ that satisfy equation (4.63).

Theorem 3.4.6 *If the pair $\{a(\lambda),\ b(\lambda)\}$ is stabilizable, then the following statements are equivalent :.*

(1) Family $\mathbf{W}_0(W^0,\ R)$ is complete, det $R_2(\lambda)\not\equiv 0$;

(2) The stable F.R.F. $W^0(\lambda)$ satisfies (4.63). The function $R(\lambda)$ satisfies (4.70) and can be presented as

$$R(\lambda)=R^0(\lambda)Q(\lambda), \tag{4.71}$$

here $R^0(\lambda)=\begin{Vmatrix} R_1^0(\lambda) \\ R_2^0(\lambda) \end{Vmatrix}$ *is the polynominal matrix, $Q(\lambda)$ and $Q^{-1}(\lambda)$ are stable F.R.F's of dimension $n\times n$ and*

$$\deg r_-^0(\lambda)=\ \deg a_-(\lambda). \tag{4.72}$$

Here $a_-(\lambda)$ and $r_-^0(\lambda)$ are unstable pair of the polynominals det $a(\lambda)$, det $R_1^0(\lambda)$.*

With the conditions in (2) of the theorem the pair $\left([R_2^0(\lambda)]^*,\ [R_1^0(\lambda)]^*\right)$ is stabilizable and

$$r_-^0(\lambda)=Ca_-(\lambda),\ C\neq 0. \tag{4.73}$$

If the pair $\{\ (a(\lambda),\ b(\lambda)\}$ is controllable, then the equivalence of conditions (1) and (2) is justified provided $a_-(\lambda)$, $r_-^0(\lambda)$ signify det $a(\lambda)$ and det R_2^0, respectively. Then $\left([R_2^0(\lambda)]^*,\ [R_1^0(\lambda)]^*\right)$ is a controllable pair.

The equation (4.70) is solvable in a class of polynomial functions $R(\lambda)$, det $R_2(\lambda)\neq 0$. Let $P_1(\lambda)$, $P_2(\lambda)$ be the polynomial matrices, $\det P_1(\lambda)=\det P_2(\lambda)=1$; $L(\lambda)$ be the diagonal matrix with elements on the diagonal $L_{jj}(\lambda)=\ b_j(\lambda)/a_j(\lambda)$, $b_j(\lambda)$, $a_j(\lambda)$, $j=1,\ \ldots,\ \min(n,\ m)$, are scalar polynomials. Then we diagonalise the F.R.F. $a^{-1}(\lambda)b(\lambda)$ with the help of above elementary matrix functions $P_1(\lambda)$, $P_2(\lambda)$ and receive

$$a^{-1}(\lambda)b(\lambda)=P_1(\lambda)L(\lambda)P_2(\lambda). \tag{4.74}$$

Let $\tilde{a}(\lambda)$, $\tilde{b}(\lambda)$ be the polynomial matrices of degree $m\times m$ and $m\times n$ and $\tilde{a}_{ij}(\lambda)=a_j(\lambda)\delta_{ij}$, $\tilde{b}_{ij}(\lambda)=b_j(\lambda)\delta_{ij}$ be their elements. (Here, $\delta_{ij}=0(i\neq j)$, $\delta_{jj}=1$.) Obviously, the equality (4.70) holds good for the polynomial matrix

$$R(\lambda)=\begin{Vmatrix} P_1^{-1}(\lambda)\tilde{b}(\lambda) \\ P_2^{-1}(\lambda)\tilde{a}(\lambda) \end{Vmatrix}.$$

The polynomials $R(\lambda)$ generate the complete family $\mathbf{W}_0(W^0,\ R)$. They can be obtained

---

*This means det $a(\lambda)=a_+(\lambda)a_-(\lambda)$ where the polynominal $a_+(\lambda)$ does not possess any root for $|\lambda|\leq 1$ and polynominal $a_-(\lambda)$ does not have any root for $|\lambda|>1$. It is analogous for $R_2^0(\lambda)$.

from the solution of the system of linear equations (4.70) related to the coefficients of the matrix polynomial $R(\lambda)$ and taking into account of the statement (2) of the theorem 3.4.6, for example, with $Q(\lambda) \equiv I_n$.

(e) *Examples of perfect classes of transfer functions.* Let us examine the possible particular cases.

1. If the control object (4.58) is stable w.r.t. output, *i.e.*, the polynomial $\det a(\lambda)$ is stable, then the complete family $\mathbf{W}_0(W^0, R)$ gives the function

$$R(\lambda) = \begin{Vmatrix} a^{-1}(\lambda)\, b(\lambda) \\ I_m \end{Vmatrix}.$$

Let $W(\lambda)$ be the arbitrary stable F.R.F. satisfying the equation (4.63) and let (4.69) be its parametrization, then the F.R.F. $\Psi(\lambda) = W_2(\lambda) - W_2^0(\lambda)$ is stable and so $W(\lambda) \in \mathbf{W}_0(W^0(\lambda), R(\lambda))$.

2. Let K be a matrix, for which $(A + BK)$ is a stable matrix. Then for the equation in a standard form : $a(\lambda) = I_n - \lambda A$ and $b(\lambda) = \lambda B$, we have, the complete family $\mathbf{W}_0(W^0, R)$, by virtue of the lemma 3.3.2, is generated by the functions

$$R(\lambda) = \begin{Vmatrix} \lambda W^0(\lambda) B \\ I_n + \lambda K W^0(\lambda) B \end{Vmatrix} \quad \text{and} \quad W^0(\lambda) = \begin{Vmatrix} [I_n - \lambda(A + BK)]^{-1} \\ K[I_n - \lambda(A + BK)]^{-1} \end{Vmatrix}.$$

3. If $m = n$ and $b(\lambda) = \lambda^k b_+(\lambda)$, $\det b_+(\lambda)$ is a stable polynomial (*i.e.*, the control object is stable in respect of control), then the family $\mathbf{W}_0(W^0, R)$ will be complete when

$$R(\lambda) = \begin{Vmatrix} \lambda^k I_n \\ b_+^{-1}(\lambda) a(\lambda) \end{Vmatrix}, \quad W^0(\lambda) = \begin{Vmatrix} W_1^0(\lambda) \\ W_2^0(\lambda) \end{Vmatrix},$$

here $W_1^0(\lambda)$ is a polynomial matrix such that $W_1^0(\lambda) = \sum_{j=0}^{k-1} W^{(j)} \lambda^j$, and can be determined from

$$W_1^0(0) = I_n, \frac{d^s[a(\lambda) W_1^0(\lambda)]}{d\lambda^s}\Big|_{\lambda=0} = 0, \quad s = 1, \ldots, k-1. \tag{4.75}$$

Also $W_2^0(\lambda) = b_+^{-1}(\lambda) \lambda^{-k} [a(\lambda) W_1^0(\lambda) - I_n]$.

It may be observed that the stability of F.R.F. $W_2^0(\lambda)$follows polynomiality of the matrix $\lambda^{-k}[a(\lambda) W_1^0(\lambda) - I_n]$ as given in (4.75). The completeness of the family $\mathbf{W}_0(W^0, R)$ is evident, if we note that for the function $W(\lambda)$ satisfying (4.63), the expression (4.70) holds good when $\Psi(\lambda) = \lambda^{-k}[W_1(\lambda) - W_1^0(\lambda)]$. This F.R.F. is stable, if the F.R.F. $W_1(\lambda)$ is stable.

4. From the second statement of the theorem a complete family $\mathbf{W}_0(W^0, R)$ (for an arbitrary choice of a stable F.R.F. $W^0(\lambda)$ satisfying equation (4.63)) generates the following function for $m = 1$ (scalar control)

$$R(\lambda) = \left\| \begin{matrix} a^{-1}(\lambda) b(\lambda) \det a(\lambda) \\ \det a(\lambda) \end{matrix} \right\|.$$

$R(\lambda)$ can also be chosen as follows

$$R(\lambda) \left\| \begin{matrix} a^{-1}(\lambda) b(\lambda) a_{-}(\lambda) \\ a_{-}(\lambda) \end{matrix} \right\|.$$

Its computation is related to location of the unstable portion $a_{-}(\lambda)$ of the polynomial $\det a(\lambda)$.

5. Let $\{a(\lambda),\ b(\lambda)\}$ be a stabilizable pair of matrix polynomials (vide definition 3.4.1). According to lemma 6.3.3, there exists matrix polynomials $\alpha(\lambda)$ and $\beta(\lambda)$ of proper dimensions such that polynomial (3.51) is stable. Then the following relationship determines the complete family $\mathbf{W}_0(W^0, R)$

$$\left\| \begin{matrix} W_1(\lambda) \\ W_2(\lambda) \end{matrix} \right\| = \left\| \begin{matrix} a(\lambda) & -b(\lambda) \\ \beta(\lambda) & -\alpha(\lambda) \end{matrix} \right\|^{-1} \left\| \begin{matrix} I \\ \Psi(\lambda) \end{matrix} \right\|.$$

The matrices $W^0(\lambda)$ and $R(\lambda)$ can be easily written when the above relationship is presented in the form (4.69).

Various parametrizations of the set $\mathbf{W}_0$ may differ by the computational complexities of the functions $W^0(\lambda)$ and $R(\lambda)$.

(*f*) *Solution of the linear optimization problem.* Assuming that the parametrization (4.69) of the set $\mathbf{W}_0$ is chosen the functional (4.61) can be looked as a quadratic one with respect to the function $\Psi(\lambda)$. Minimization of such functionals in a class of stable F.R.F's was investigated in section 3.3 (vide section 3.3.3 and 3.3.4). The general result will be now laid down.

Theorem 3.4.7 *Let the following conditions be fulfilled :*

(1) the pair of matrix polynomials $\{a(\lambda),\ b(\lambda)\}$ *be stabilizable,*

(2) the stable F.R.F. $R(\lambda)$ *satisfies the condition (2) of theorem 3.4.6,*

(3) the function $R^{\nabla}(\lambda) N R(\lambda)$ *is positive for* $|\lambda| = 1$,

(4) the noise v^{∞} *of (4.58) is a centered process possessing (matrix) spectral density*

$$S_0(\lambda) = \Gamma(\lambda) \Gamma^{\nabla}(\lambda), \tag{4.76}$$

where $\Gamma(\lambda)$ *is the stable F.R.F. of dimension* $n \times 1 (1 \le n)$ *and is such that there exists a stable F.R.F.* $\Gamma^{+}(\lambda)$ *that satisfies the identity* $\Gamma^{+}(\lambda) \Gamma(\lambda) = I_l$.*

Then the following statements are true.

1. The linear optimization problem has a solution. In this case

$$\min_{U^{\infty}(\cdot) \in \mathbf{U}^{a}} J[U^{\infty}(\cdot)] = \frac{1}{2\pi i} \oint \mathrm{Sp}\, \{[W^0(\lambda) \Gamma(\lambda)]^{\nabla} N W^0(\lambda) \Gamma(\lambda)$$

---

*For a quadratic nondegenerative matrix $\Gamma(\lambda)$ ($\det \Gamma \neq 0$) the function $\Gamma^{+}(\lambda)$ coincides with the inverse $\Gamma^{-}(\lambda)$. If $\Gamma(\lambda)$ has a quasi-inverse matrix $\tilde{\Gamma}^{+}(\lambda)$ possessing F.R.F. then $\tilde{\Gamma}^{+}(\lambda) = \Gamma^{+}(\lambda)$.

$$-L^{\nabla}(\lambda)L(\lambda)+L_{-}^{\nabla}(\lambda)L_{-}(\lambda)\}\,\frac{dy}{\lambda}. \tag{4.77}$$

Here $W^{0}(\lambda)$ *is an arbitrary stable F.R.F. that satisfies (4.63), i.e.*

$$L(\lambda)=[R(\lambda)\Pi^{-1}(\lambda)]^{\nabla}NW^{0}(\lambda)\Gamma(\lambda). \tag{4.78}$$

Here the F.R.F. $\Pi(\lambda)$ *is stable along with its inverse* $\Pi^{-1}(\lambda)$ *and is determined from the factorisation* condition below*

$$R^{\nabla}(\lambda)NR(\lambda)=\Pi^{\nabla}(\lambda)\Pi(\lambda), \tag{4.79}$$

$L_{-}(\lambda)$ *is obtained by separating the function* $L(\lambda)$, *i.e., presenting it in the form*

$$L(\lambda)=L_{+}(\lambda)+L_{-}(\lambda), \tag{4.80}$$

where $L_{+}(\lambda)$ *and* $L_{-}(\lambda)^{-1}$ *are the stable F.R.F's and*

$$\lim_{|\lambda|\to\infty} L_{-}(\lambda)=0. \tag{4.81}$$

2. *The optimal transfer function* $W^{opt}(\lambda)$ *of the control system has the expression*

$$W^{opt}(\lambda)=\{I_n-R(\lambda)\,[\Pi^{\nabla}(\lambda)\Pi(\lambda)]^{-1}R^{\nabla}(\lambda)N\}W^{0}(\lambda)$$

$$+R(\lambda)\Pi^{-1}(\lambda)\Phi(\lambda). \tag{4.82}$$

Here $\Phi(\lambda)=L_{-}(\lambda)\Gamma^{+}(\lambda)+\Psi_{0}(\lambda)\,[I_n-\Gamma(\lambda)\Gamma^{+}(\lambda)]$ (4.83)

and $\Psi_0(\lambda)$ *is an arbitrary F.R.F. that ensures stability of the following F.R.F.*

$$(\Psi_0(\lambda)-[\Pi^{-1}(\lambda)]^{\nabla}R^{\nabla}NW^{0}(\lambda)\,[I_n-\Gamma(\lambda)\Gamma^{+}(\lambda)]. \tag{4.84}$$

3. *The function* $\Psi^{opt}(\lambda)$ *determines the following polynomials* $\alpha(\lambda)$ and $\beta(\lambda)$ *of the stabilizing regulator (4.48) :*

$$\alpha(\lambda)=p(\lambda)\left\{\Phi(\lambda)b(\lambda)+[R(\lambda)\Pi^{-1}(\lambda)]^{\nabla}N\begin{Vmatrix}0\\ I_m\end{Vmatrix}\right\}, \tag{4.85}$$

$$\beta(\lambda)=p(\lambda)\left\{\Phi(\lambda)a(\lambda)-[R(\lambda)\Pi^{-1}(\lambda)]^{\nabla}N\begin{Vmatrix}I_n\\ 0\end{Vmatrix}\right\}.$$

Here $p(\lambda)$ *is the scalar polynomial equal to the product of the denominators of the matrix elements of F.R.F's as shown in braces. The fractionally rational functions within braces are stable and hence the polynomial* $p(\lambda)$ *is stable. The characteristic polynomial of the control system of (4.58), (4.85) is given by*

$$g(\lambda)=\det\begin{Vmatrix}a(\lambda) & -b(\lambda)\\ -\beta(\lambda) & \alpha(\lambda)\end{Vmatrix}=C\,\frac{p^{m}(\lambda)a_{+}(\lambda)\det\Pi(\lambda)}{r_{+}^{0}(\lambda)\det Q(\lambda)}. \tag{4.86}$$

Here $a_{+}(\lambda)$, $r_{+}^{0}(\lambda)$ *are stable parts of the polynomials* $\det a(\lambda)$, $\det R_2^0(\lambda)$. *The functions* $R_2^0(\lambda)$, $Q(\lambda)$ *and the constant C are determined from the condition (2) of the theorem 3.4.6.*

Since the F.R.F. $Q^{-1}(\lambda)$ is stable (as follows from the statement of the theorem 3.4.6) and F.R.F. $\det\Pi(\lambda)$ is stable (because of the choice of matrix $\Pi(\lambda)$ from the factorisation

---

*Existence of such a $\Pi(\lambda)$ is ensured by the condition (3) of the theorem.

condition (4.79)), it follows from formula (4.86) that the regulator of (4.48), (4.85) is stabilizing for the control object (4.58).

The function $\Psi_0(\lambda)$is determined uniquely. The only requirement in its presentation is the stability of the F.R.F. (4.84). Thus, $\Psi_0(\lambda)$ is determined within the stable F.R.F. The uniqueness of the choice of the regulator for the noise v^∞ (*i.e.*, if det $S_v(\lambda) \equiv 0$), is also linked up with $\Psi_0(\lambda)$. When det $S_v(\lambda) \not\equiv 0$, we have $\Gamma^+(\lambda) = \Gamma^{-1}(\lambda)$ and from (4.83) it follows that $\Phi(\lambda) = L_-(\lambda)\Gamma^{-1}(\lambda)$. Then in such a case, it is not necessary to introduce the function $\Psi_0(\lambda)$.

The theorem 3.4.7 is introduced verbatim to continuous control objects. The only change is done in replacement of ∇^{-1} by d/dt and consequently instead of analysis in the unit circle we have analysis in the left-half of the plane, etc.

The positiveness of the function $R^\nabla(\lambda)NR(\lambda)$ (for $|\lambda|= 1$) is equivalent to, as is laid down in section 3.3.6b, fulfilment of the frequency condition (A.2) for equation (4.58) when expressed in the form of (4.11). Hence it can be now stated that for a noise v^∞ (theorem 3.4.7) generated by a stable filter (4.12) of the process w^∞ with independent values, then the regulator of (4.48) and (4.85) gives an optimal strategy in a class $\mathbf{U}^a$ of arbitrary strategies. The strategies also satisfy inequality in (4.1) independent of the choice of the initial states x_0 of the control object (4.58).

The F.R.F's $L_-(\lambda)$ and $\Psi_o(\lambda)$ of (4.83) can be determined from the following conditions : (1) the poles of the function $L_-(\lambda)$ can be only unstable poles of the matrix $[R(\lambda)\Pi^{-1}(\lambda)]^\nabla$, it is evident from the formulae (4.78) and (4.80); (2) $L_-(\lambda)$ is the correct F.R.F. (vide (4.81)); (3) the elements of the matrices in the braces of (4.85) are stable F.R.F's.

The functions $L_-(\lambda)$ and $\Psi(\lambda)$ are determined by the method of undetermined coefficients. The functions $L_-(\lambda)$ and $\Psi(\lambda)$ are expressed as $\sum_{i=1}^{N} A_i(\lambda-\lambda_i)^{-1}$ where $\lambda_i, \ldots, \lambda_N$ are unstable poles of the matrix $[R(\lambda)\Pi^{-1}(\lambda)]^\nabla$ (if these poles are simple). The coefficients A_i are determined from the system of linear equations by equating the coefficients for $(\lambda-\lambda_i)^{-1}$ to zero $\left(i = 1, \ldots, N \text{ in } \frac{\alpha(\lambda)}{p(\lambda)} \text{ and } \frac{\beta(\lambda)}{p(\lambda)}\right)$. According to theorem 3.4.6 the system thus obtained will be solvable.

(*g*) *Example.* Theorem 3.4.7 will be applied to solve the following optimization problem. The control object is written by the scalar equation

$$y_t + a_1 y_{t-1} + y_{t-2} = b_1 u_{t-1} + b_2 u_{t-2} + v_t + c_1 v_{t-1}. \tag{4.87}$$

Here the object parameters satisfy the conditions

$$a_1 > -2, \quad b_1 b_2^{-1} > 1, \quad |c_1| < 1,$$

$$b_1^2 + 2a_1 b_1 b_2 + b_2^2 \neq 0. \tag{4.88}$$

The noise v^∞ is a stationary process with properties as below

$$\mathbf{M}v_t = 0, \quad \mathbf{M}v_t v_s = \delta_{ts}. \tag{4.89}$$

The conditions in (4.88) signify that the control object is unstable with respect to the output but it is stable for control and noise signals (vide section 1.3.3d).

The control objective is to minimize the performance criterion in a class of linear stabilizing regulators :

$$J = \overline{\lim_{t\to\infty}}\ \mathbf{M}(Qy_t^2 + Ru_t^2). \tag{4.90}$$

Here $Q \geq 0$, $R > 0$ are the weighting coefficients.

To solve the problem we choose a column matrix (following section 3.4.4.d)

$$R(\lambda) = \begin{vmatrix} b(\lambda) \\ a(\lambda) \end{vmatrix}. \tag{4.91}$$

It will help us in generating a complete family $\mathbf{W}_0(W^0, R)$ for any arbitrary stable solution $W^0(\lambda)$ of equation (4.63). Thus, control stability of the control object will permit us to select

$$W^0(\lambda) = \begin{vmatrix} 1 \\ \dfrac{a_1 + \lambda}{b_1 + \lambda b_2} \end{vmatrix}. \tag{4.92}$$

In theorem 3.4.7 the central point of the optimal regulator synthesis is determination of a stable F.R.F. $\Pi(\lambda)$ (along with its inverse) using the relationship in (4.79). In the present case $N = \begin{vmatrix} Q & 0 \\ 0 & R \end{vmatrix}$, and considering (4.91), we have

$$R^{\nabla}(\lambda) N R(\lambda) = b(\lambda^{-1}) b(\lambda) Q + a(\lambda^{-1}) a(\lambda) R. \tag{4.93}$$

The polynomials $a(\lambda)$ and $b(\lambda)$ do not have any general root (as per last condition of (4.88)) in the unit circle. Hence the right hand side of (4.93) is stable for $|\lambda| = 1$. Consequently, there exists a stable polynomial $\Pi(\lambda)$ that satisfies the relationship of (4.79). We now find out this polynomial.

Let $q = QR^{-1}$ and $Z = \lambda + \lambda^{-1}$. We can express (4.93) as

$$\frac{R^{\nabla}(\lambda) N R(\lambda)}{R} = z^2 + 2\left(a_1 + q\frac{b_1 b_2}{2}\right) z + a_1^2 + q(b_1^2 + b_2^2). \tag{4.94}$$

Let us denote* $2\lambda_1 = z_1 - \sqrt{z_1^2 - 4}$, $2\lambda_2 = z_2 - \sqrt{z_2^2 - 4}$

and $z_{1,2} = -\left(a_1 + q\dfrac{b_1 b_2}{2}\right) \pm \sqrt{\left(a_1 + q\dfrac{b_1 b_2}{2}\right)^2 - a_1^2 - q(b_1^2 + b_2^2)}$.

The roots z_1, z_2 are determined from the right hand side of (4.94) and z_1, z_2 are the roots of the equations $\lambda + \lambda^{-1} = z_{1,2}$. Then we have

$$\Pi(\lambda) = \sqrt{R \lambda_1^{-1} \lambda_2^{-1}} (\lambda - \lambda_1)(\lambda - \lambda_2). \tag{4.95}$$

Remembering that $\Gamma(\lambda) = (1 + \lambda c_1)^{-1}$, we get, according to formulae (4.78), (4.91) and (4.95),

---

*Here it is assumed that $|z_1 - \sqrt{z_1^2 - 4}| > 2$, otherwise $2\lambda_1 = z_1 + \sqrt{z_1^2 - 4}$.

$$L(\lambda)=\sqrt{\frac{\lambda_1\lambda_2}{R}}\ \frac{Q(b_1\lambda+b_2)(b_1+\lambda b_2)+R\,(a_1+\lambda)(1+a_1\lambda+\lambda^2)}{(1-\lambda\lambda_1)(1-\lambda\lambda_2)(1+\lambda c_1)(b_1+\lambda b_2)}. \tag{4.96}$$

Using (4.80) and (4.81), the function $L_-(\lambda)$ can be expressed as

$$L_-(\lambda)=\frac{l_0+\lambda l_1}{(1-\lambda\lambda_1)(1-\lambda\lambda_2)}.$$

The coefficients l_0, l_1 are uniquely determined by the condition so that the F.R.F. $L(\lambda)-L_-(\lambda)$ does not have any singularity for $\lambda=\lambda_1^{-1}$ and $\lambda=\lambda_2^{-1}$. This means that the following function must reduce to zero for $\lambda=\lambda_1^{-1}$ and $\lambda=\lambda_2^{-1}$:

$$\mu(\lambda)=\sqrt{\frac{\lambda_1\lambda_2}{R}}\ Q(b_1\lambda+b_2)(b_1+\lambda b_2)+R(a_1+\lambda)(1+a_1\lambda+\lambda^2)$$

$$-(l_0+\lambda l_1)(1+\lambda c_1)(b_1+\lambda b_2).$$

If $\lambda_1\neq\lambda_2$, then the linear system of equation $\mu(\lambda_1^{-1})=0$ and $\mu(\lambda_2^{-1})=0$ uniquely determine the coefficients l_0 and l_1. For $\lambda_1=\lambda_2$, the coefficients l_0, l_1 are uniquely determined by the system of equations

$$\mu(\lambda_1^{-1})=0,\quad \left.\frac{d\mu(\lambda)}{d\lambda}\right|_{\lambda=\lambda_1^{-1}}=0.$$

Thus, the polynomials $\alpha(\lambda)$ and $\beta(\lambda)$ of the optimal regulator are determined by using the coefficients l_0 and l_1 for computation of the function as per formulae (4.85).

3.4.5 *Formulation of the Limited Optimal Control Problems using Riccati Equation*

Formulation of the optimal control of linear quadratic problem has been investigated with sufficient generalisation. The complexity of the computational algorithms in its realization is related to factorisation of matrix functions. Again it has been observed that the constrained optimal control problems do not have unique solutions. Hence, computationally simpler algorithms for synthesis of constrained optimal control problems are suggested. They are realized with the help of non-stationary control strategies.

(*a*) *Initial Premises.* Let the control object be expressed in the standard form as

$$x_{t+1}=Ax_t+Bu_t+v_{t+1},\quad t=0,1,\quad \ldots . \tag{4.97}$$

The performance criterion is given by

$$J[U_0^\infty(\cdot)]=\overline{\lim_{t\to\infty}}\ t^{-1}\sum_{s=1}^{t}\mathbf{M}q(x_s,u_{s-1}). \tag{4.98}$$

Here the matrix N in the quadratic for a $q(x,u)=\begin{pmatrix}x\\u\end{pmatrix}^* N\begin{pmatrix}x\\u\end{pmatrix}$ is nonnegative and is given by (4.16) with $R>0$. The class of admissible strategies $\mathbf{U}^a$ for minimization of the functional in (4.98) is formed out of the sequence of functions $U_0^\infty(\cdot)$. The latter gives the control law in the form $u_t = U_t(u^{t-1}, x^t)$ and ensures the inequality (4.1) independent of the choice of the initial state x_0 of the control object (4.97).

A similar problem has been already studied in section 3.4.2c and it was shown that the optimal strategy may be realised in the form of a linear stabilizing feedback signal that can be determined from solution $H \geq 0$ of the Lur'e equation (theorem 3.4.4). Such a matrix computation is also related to the requirement for factorisation of polynomial matrices. However, to determine the necessary solutions of the Lur'e equation, we can use iterative processes which do not require any factorisation. If the iterative process converges, then the independence of the functional (4.98) of the transient processes permits us to ascertain that the control in the present case will be a constrained optimal one. From the point of view of computation, the iterative process may be considerably simpler than the method of computation of coefficients of the optimal regulator.

(*b*) *Recursive procedure for finding out the nonnegative solution of Lur'e equation.* The questions of convergence of the solutions of the stationary Riccati equation to the solution of Lur'e equation were discussed in theorem 3.2.2. The following statement will be introduced as a simple corollary of this theorem.

Lemma 3.4.3 *Let the following conditions be satisfied :*

(1) the pair of matrices (A,B) in equation (4.97) is stabilizable,

(2) the matrix N in the performance criterion (4.98) is nonnegative and of the form as in (4.16) for $R > 0$,

(3) the pair of matrices $\{A - BR^{-1}S^*, \quad (Q - SR^{-1}S^*)^{1/2}\}$ *is detectable.*

Then there exists $\lim_{t\to\infty} H_t = H_\infty$, *a unique nonnegative solution of the Lur'e equation in the expression for determination of the sequences of matrices* H_t *:*

$$H_{t+1} = A^*H_tA - (A^*H_tB + S)(B^*H_tB + R)^{-1}(A^*H_tB + S)^* + Q,$$

$$H_0 \geq 0. \tag{4.99}$$

Then for $H = H_\infty$, *the matrix* $\Phi(H)$ *is stable where*

$$\Phi(H) = A - B(B^*HB + R)^{-1}(A^*HB + S)^*. \tag{4.100}$$

It may be noted that in the lemma 3.4.3 positiveness of the matrix N is not assumed but the conditions in the lemma lead to the fulfilment of the frequency condition (A.2).

(*c*) *Formulation of the constrained optimal control.* Combining the results of theorem 3.4.4 and lemma 3.4.3 the following algorithm for formulation of the control signals is obtained.

$$[\det\ e^*(\nabla)\,e\,(\nabla)]u_t = -(B^*H_tB + R)^{-1}\{(B^*H_tA + S^*)[\det\ e^*(\nabla)\,e\,(\nabla)]x_t +$$

$$+B^*\gamma(t,\nabla)\,\tilde{e}\,(\nabla)\,e^*(\nabla)\,[x_t - Ax_{t-1} - Bu_{t-1}]\}. \tag{4.101}$$

Here $\tilde{e}(\lambda)$ and $e(\lambda)$ are matrix polynomials as also in (4.34) :

$\gamma(t,\lambda) \equiv 0$ for $q = 0$,

$\gamma(t,\lambda) = \gamma_0(t) + \lambda\gamma_1(t) + \cdots + \lambda^{q-1}\gamma_{q-1}(t)$,

$$\gamma_s(t) = 2\sum_{l=0}^{q-s-1} [\Phi^*(H_t)]^{1} H_t e_{l+s+1} \text{ for } q \geq 1;$$

the matrices H_t are determined by the conditions in (4.99) for any $H_0 \geq 0$.

Algorithm in (4.99) and (4.101) is in the form of a nonstationary feedback signal. On fulfilment of the conditions laid down in the theorem 3.4.4. and lemma 3.4.3, this determinable feedback control is also optimal in the same sense as that of synthesis of stationary strategy in the theorem 3.4.4.

§ 3.5 Minimax Control

The optimal control design becomes considerably complex when the noise is not a stationary process. The problem of stabilization in the presence of noise was studied in section 3.1, but only the constraint on the noise was considered. Ensemble operation is usable during computation of performance criterion. It is the maximization of the objective function for all possible values of noise (vide, for example, (1.6)). It is a nonlinear operation, whereas the averaging operation applied in stochastic control problems is a linear one. As a rule, the non-linear ensemble operation does not permit construction of the Bellman - Liapunov function in the quadratic form.* Determination of the solutions of corresponding Bellman equation in a class of other functions is a complex and little studied problem.

The results of optimization of linear system with constrained noise which does not have any useful stochastic property will be shown below. It is assumed that the admissible class of control strategies $\mathbf{U}^a$ will generate linear stabilizing regulators. The performance criterion depends on the behaviour of the controllable process for $t \to \infty$. This is also true for constrained optimal control. However, unlike the stochastic version of control problems, the control objective requires the best control for the least noise. It is for this reason that the control problem as laid down is known as *minimax problem*. The method of minimax control design is a modification of the transfer function method of sections 3.3. and 3.4.4. It uses the techniques of convex analysis.

3.5.1 *Statement of the Minimax Control Problem*

The linear dynamic control object is given through 'input-output' relationship as

$$a(\nabla)y_t = b(\nabla)u_t + v_t. \tag{5.1}$$

Here y is the output variable, u is the control variable and v is the noise variable. All these variables are scalar. Let k be the control delay, $k \geq 1$. Then the polynomials $a(\lambda)$ and $b(\lambda)$ can be expressed as

$$a(\lambda) = 1 + \lambda a_1 + \cdots + \lambda^n a_n,$$

$$b(\lambda) = \lambda^k b_k + \cdots + \lambda^n b_n. \tag{5.2}$$

If ρ is a given constant, then the noise v^∞ belongs to a class of noise $\mathbf{V}_\rho$, such that

---

*It can be done only in exceptional cases, one such case has been discussed in 3.1.2a.

$|v_t| \le \rho$.

The admissible control strategies result in linear feedback such as

$$\alpha(\nabla)u_t = \beta(\nabla)y_t \tag{5.3}$$

and ensures the inequality irrespective of the choice of initial conditions of the control system in (5.1) and (5.3) :

$$\overline{\lim_{t\to\infty}} \; (|y_t| + |u_t|) < \infty. \tag{5.4}$$

The regulators of (5.3) having above property will be known as *stabilizable*. Obviously in order that the regulator of (5.3) is *stabilizable*, it is necessary and sufficient to have the following characteristic polynomial of the control system ((5.1) and (5.3)) stable :

$$g(\lambda) = a(\lambda)\alpha(\lambda) - b(\lambda)\beta(\lambda). \tag{5.5}$$

The degree p of the polynomials $\alpha(\lambda)$ and $\beta(\lambda)$ in the regulator of (5.3) may be an arbitrary one, $\alpha(0) \neq 0$,

$$\alpha(\lambda) = \alpha_0 + \lambda\alpha_1 + \cdots + \lambda^p\alpha_p,$$

$$\beta(\lambda) = \beta_0 + \lambda\beta_1 + \cdots + \lambda^p\beta_p. \tag{5.6}$$

An additional constraint may be given for the class U^a, *e.g.*, $\beta_0 = \cdots = \beta_{s-1} = 0$. The number s is known as *measurement delay*.

Let the performance criterion be of the form*

$$J(\alpha, \beta) = \sup_{v^\infty \in V_\rho} \overline{\lim_{t\to\infty}} \; |y_t|. \tag{5.7}$$

The objective of the minimax control is to minimize the functional (5.7) in a class of stabilizing regulators (5.3).**

3.5.2 *Control System Transfer Function and its Properties*

Let $g(\lambda)$ be the polynomial as in (5.5). Then following 3.3.8 the transfer function of the control system (5.1) and (5.3) can be written as (vide (3.52))

$$W(\lambda) = \begin{vmatrix} \alpha(\lambda)g^{-1}(\lambda) \\ \beta(\lambda)g^{-1}(\lambda) \end{vmatrix}. \tag{5.8}$$

For a stabilizing regulator of (5.3), the F.R.F. in (5.8) is stable, *i.e.*, it does not have any pole for $|\lambda| \le 1$. From (5.8) it follows that the constituents of the transfer function

$$W_1(\lambda) = \alpha(\lambda)g^{-1}(\lambda), \quad W_2(\lambda) = \beta(\lambda)g^{-1}(\lambda) \tag{5.9}$$

are related as follows

$$a(\lambda)W_1(\lambda) - b(\lambda)W_2(\lambda) = 1, \tag{5.10}$$

$$\alpha(\lambda)W_2(\lambda) - \beta(\lambda)W_1(\lambda) = 0. \tag{5.11}$$

The latter relationship determines the transfer function of the regulator (5.3), *i.e.*, $\alpha^{-1}(\lambda) \times \beta(\lambda)$.

---

*Including in the functional (5.7) of the control signals sharply complicates the control problem and interesting examples of its solutions are still not known.

**For a functional (5.7) the problem with measurement delay s is easily converted to the problem without measurement delay, but with a control delay of $k+s$.

If the pair of polynomials $\{a(\lambda), b(\lambda)\}$ is stabilizable, *i.e.*, polynomials $b(\lambda)$ and $a(\lambda)$ do not have roots for $|\lambda| \le 1$, then the set $\mathbf{W}_o$ of all stable F.R.F 's, *e.g.*, $W(\lambda) = \begin{vmatrix} W_1(\lambda) \\ W_2(\lambda) \end{vmatrix}$ (of dimension 2×1) satisfying (5.10) is not empty set. In accordance with the theorem 3.4.6, the set $\mathbf{W}_o$ permits parametrization, such as

$$W(\lambda) = W^0(\lambda) + \begin{vmatrix} b(\lambda) \\ a(\lambda) \end{vmatrix} \Psi(\lambda). \tag{5.12}$$

Here $W^0(\lambda)$ is any stable F.R.F. satisfying the relationship of (5.10) and $\Psi(\lambda)$ is an arbitrary stable F.R.F.

The component $W_1(\lambda)$ of the transfer function $W(\lambda)$ plays an important role. The stable function $W_1(\lambda)$ is expanded in a series that converges for $|\lambda| \le 1$. Thus,

$$W_1(\lambda) = \sum_{j=0}^{\infty} W^{(j)}\lambda^j, \quad \sum_{j=0}^{\infty} |W^{(j)}| < \infty. \tag{5.13}$$

In addition to that, the function $W_1(\lambda)$ determines the dependence of y_t on initial conditions and noise signals. Since in the case of a stabilizing regulator (5.3) of a stable transfer function (5.8), the choice of initial conditions does not effect the value of the performance criterion (5.7) and the inequality of (5.4), then we can assume them to be equal to zero and write (vide, for example, section 3.P.19)

$$y_t = \frac{1}{2\pi i} \oint \lambda^{-t} W_1(\lambda)\hat{v}(\lambda)\frac{d\lambda}{\lambda}. \tag{5.14}$$

Here $\hat{v}(\lambda) = \sum_{t=0}^{\infty} v_t \lambda^t$. Putting (5.14) in (5.13) and using inverse Laplace transform

$$v_t = \frac{1}{2\pi i} \oint \lambda^{-t} \hat{v}(\lambda)\frac{d\lambda}{\lambda}, \text{ we have}$$

$$y_t = \sum_{k=0}^{t} W^{(k)} v_{t-k}. \tag{5.15}$$

The relationship of (5.15) permits us to express the functional (5.7) through the transfer function $W_1(\lambda)$:

$$J(\alpha, \beta) = \rho \sum_{j=0}^{\infty} |W^{(j)}| = J_*(W_1)\rho. \tag{5.16}$$

Thus we arrive at the same problem of minimization of the functional (5.16) in the class of stable F.R.F.'s. However, unlike the linear optimization problem the functional (5.16) here is not quadratic with respect to $W_1(\cdot)$ and the method evolved in section 3.4.4 cannot be applied for minimization of this functional. Nevertheless dependence of J_* on $W_1(\cdot)$ is evident and it permits necessary modification of the transfer functions.

3.5.3 *Geometrical Interpretation of the Minimax Control Problem*

Let $\mathbf{W}$ represent the set of all analytical scalar functions in the unit circle and $\mathbf{W}_0$ be its subset obtained from the relationship

$$W(\lambda) = W^0(\lambda) + b(\lambda)\Psi(\lambda). \tag{5.17}$$

Here $W^0(\lambda)$ is a fixed and $\Psi(\lambda)$ is an arbitrary analytical function in the unit circle. The set $\mathbf{W}_o$ is an affine variety in $\mathbf{W}$. Then the minimax control problem may be interpreted as the problem of finding out the minimizing $J_*(W)$ (vide (5.16)) in the affine variety $\mathbf{W}_o$ of element $W^{opt}(\cdot)$.

If scalar product is introduced to set $\mathbf{W}$, then it can be seen as a Hilbert Space (Hardy space) :

$$\langle W', W''\rangle = \frac{1}{2\pi i}\oint W'(\lambda^{-1})W''(\lambda)\frac{d\lambda}{\lambda}. \tag{5.18}$$

The set $\mathbf{W}_o$ becomes a closed subset. The element $W^{opt}(\cdot)$ that minimizes the functional J_* is not responsible for realization of the distance between the null element and the set $\mathbf{W}_o$.

However, the set $\mathbf{K}_\mu$ of elements $\mathbf{W}$ is convex and symmetrical for any μ, where

$$J_*(W) \le \mu. \tag{5.19}$$

If $\mathbf{K}_\mu \cap \mathbf{W}_0 = \varnothing$, then the minimum of the functional J_* is greater than μ. Hence,

$$\min J_* = \min\{\mu \ge 0 : \mathbf{K}_\mu \cap \mathbf{W}_0 \ne \varnothing\} = \mu_0. \tag{5.20}$$

This means that the least value of the functional J_* is equal to μ_0 and is determined by the point of tangency of the closed convex set $\mathbf{K}_{\mu_0}$ with the linear closed variety $\mathbf{W}_0$. If this leads to a condition that the point at the tangent $W^{opt}(\cdot) \in \mathbf{W}_0$ is a fractionally rational function, then it will be the first component of the optimal transfer function of the control system*. Second component of the transfer function can be found out from (5.10).**

The case of the tangency of sets $\mathbf{K}_{\mu_0}$ and $\mathbf{W}_0$ is equivalent to that of the tangency of their orthogonal projections $P_{\mathbf{W}_\perp}[\mathbf{K}_{\mu_0}]$ and $P_{\mathbf{W}_\perp}[\mathbf{W}_0]$ on the orthogonal supplement $\mathbf{W}_\perp$ of the set $\mathbf{W}_o$. Since $P_{\mathbf{W}_\perp}[\mathbf{W}_0] = \mathbf{W}_0 \cap \mathbf{W}_\perp$ consists of a unique element $\overline{W}^{opt}(\cdot)$, on which the distance between the null element and set W_o is realized, then this element is the first component of the optimal transfer function in the linear quadratic problem with white noise***. The point of tangency of sets $\mathbf{K}_{\mu_0}$ and $\mathbf{W}_0$ and not necessarily coincides with $\overline{W}^{opt}(\cdot)$.

The notable speciality of the optimization problem under consideration lies in finiteness of the set $\mathbf{W}_\perp$. This leads to a simple expression. The projection of the set $\mathbf{K}_\mu$ on $\mathbf{W}_\perp$ is a convex polyhedron. Its vertices can be very easily determined. The projection of the point of tangency on $\mathbf{W}_\perp$, *i.e.*, the point $P_{\mathbf{W}_\perp}\mathbf{W}^{opt}(\cdot)$ is a convex combi-

---

*If $W^{opt}(\cdot)$ is not an F.R.F., then the finite difference order optimal regulator (regulator with limited memory) does not exist. If $W^{opt}(\cdot)$ can be approximated as fractionally rational functions, then suboptimal control with high degree of suboptimality can be synthesised.

**Because of the completeness of parametrization (5.12) of the set of stable transfer functions, this component will be a stable F.R.F.

***This element is a F.R.F. (vide results of section 3.3. and section 3.4).

nation of end points of the set $P_{\mathbf{w}_l}[\mathbf{K}_{\mu_0}]$. Evidently, the point of tangency itself is a combination of end points of the set $\mathbf{K}_{\mu_0}$. Their projections coincide with end points of set $P_{\mathbf{w}_l}[\mathbf{K}_{\mu_0}]$. Although $\mathbf{K}_{\mu_0}$ is an infinite dimensional set, the number of end points included in the convex combination is finite. It is because the number of end points of the set $P_{W_1}[\mathbf{K}_{\mu_0}]$ is finite. Since the end points of the set $\mathbf{K}_{\mu_0}$ can be expressed in a simple form and they happen to be F.R.F.'s the above condition permits location of tangency of the sets $\mathbf{K}_{\mu_0}$, $\mathbf{W}_0$, and at the same time computation of the optimal transfer function in the minimax control problem. An accurate formulation of the problem will be presented.

3.5.4 *Properties of the Sets in Geometrical Interpretation of the Optimization Problem*

It will be shown that the set $\mathbf{W}_l$ which is orthogonal to set $\mathbf{W}_0$ in $\mathbf{W}$ is finite dimensional.

Actually the elements $W(\cdot) \in \mathbf{W}_l$ satisfy the orthogonality condition for an arbitrary function $\Psi(\lambda)$ as given below :

$$\langle W, b\Psi \rangle = \frac{1}{2\pi i} \oint W(\lambda^{-1}) b(\lambda) \Psi(\lambda) \Psi(\lambda) \frac{d\lambda}{\lambda} = 0. \tag{5.21}$$

Here $\Psi(\lambda)$ is analytical in the unit disk $D_1 = \{\lambda : |\lambda| \leq 1\}$. From (5.21) it follows that $W(\lambda^{-1})b(\lambda)$ is analytical in D_1. Let us assume that

$$b(\lambda) \neq 0 \text{ when } |\lambda| = 1. \tag{5.22}$$

Now $b(\lambda)$ is expressed as

$$b(\lambda) = b_+(\lambda) \cdot b_-(\lambda), \tag{5.23}$$

where the polynomial $b_+(\lambda)$ does not have any root in D_1 and the polynomial $b_-(\lambda)$ does not have any root for $|\lambda| > 1^*$. Then analyticity of the function $W(\lambda^{-1})b(\lambda)$ in D_1 means that the function $W(\lambda^{-1})$ is analytical and bounded outside the set D_1 (since $W(\lambda)$ is an analytical function in D_1) and may possess poles only at the roots of the polynomial $b_-(\lambda)$. The multiplicity of the poles does not excel that of corresponding roots. The number of linear independent analytical functions with the above property is finite and equal to $l = \deg b_-(\lambda) \leq k$.

The following F.R.F.'s are an example of such functions

$$f_k(\lambda) = \frac{\lambda^k b_-^{(*)}(0)}{b_-^{(*)}(\lambda)}, \quad k = 0, 1, \ldots, l-1. \tag{5.24}$$

Here $b_-^{(*)}(\lambda) = \lambda^l b_-(\lambda^{-1})$ (5.25)

and it is the stable modification of the polynomial $b_-(\lambda)^{**}$.

---

* *i.e.*, $b_+(\lambda)$ and $b_-(\lambda)$ are the stable and unstable constituents of the polynomial $b(\lambda)$.

** If $b-(\lambda) = \bar{b}_0 + \ldots + \lambda^l \bar{b}_l$, then $b_-^{(*)}(\lambda) = \lambda^l \bar{b}_l$. So the constituent of polynominal $b_-^{(*)}(\lambda)$ is stable.

Let us now examine the set $\mathbf{K}_1$ that can be obtained from the inequality of of (5.19) putting $\mu = 1$ in it. It may be recalled that the set consists of the functions $W(\lambda)$ which are analytical in D_1. The coefficients $W^{(j)}$ of the expansion in (5.13) related to the said functions satisfy the inequality

$$\sum_{j=0}^{\infty} |W^{(j)}| \leq 1. \tag{5.26}$$

It follows from (5.26) that the functions $\pm z_j(\lambda)$ belong to the set $\mathbf{K}_1$ and they are the end points of this set. This means that $\mathbf{K}_1$ is the convex hull (in $\mathbf{W}$) of functions $\{\pm z_j\}$. Here,

$$z_j(\lambda) = \lambda^j, \; j = 0, 1, \; \dots . \tag{5.27}$$

We compute the function $P_{\mathbf{W}_l} z_j$ which is the projection of the function $z_j(\lambda)$ on the set $\mathbf{W}_l$. For this $\{g_k(\lambda)\}$ is chosen as the base in $\mathbf{W}_0 (k = 0, \dots, l-1)$, which is a conjugate in $\mathbf{W}$ to the base $\{f_k(\lambda)\}$:

$$\langle g_k(\cdot), f_j(\cdot) >= \delta_{kj}. \tag{5.28}$$

Then

$$P_{\mathbf{W}_l} z_j(\cdot) = \sum_{k=0}^{l-1} \langle z_j(\cdot), f_k(\cdot) \rangle g_k(\lambda). \tag{5.29}$$

On expansion of the function $P_{\mathbf{W}_l} z_j$ with $\{g_k(\cdot)\}$ as the base, we get coefficients $\langle z_j, f_k \rangle$ which can be computed easily as follows :

$$\begin{aligned} < z_j, f_k > &= \frac{1}{2\pi i} \oint \lambda^{-j} \; \frac{\lambda^k b_-^{(*)}(0)}{b_-^{(*)}(\lambda)} \frac{d\lambda}{\lambda} \\ &= \frac{1}{2\pi i} \oint \lambda^{-(j-k)} \; \frac{b_-^{(*)}(0)}{b_-^{(*)}(\lambda)} \frac{d\lambda}{\lambda} \\ &= \gamma_{j-k} \end{aligned} \tag{5.30}$$

It follows from (5.30) that

$$\gamma_s = \begin{cases} < z_s(\cdot), f_0(\cdot) > & \text{for } s \geq 0, \\ 0 & \text{for } s < 0 \end{cases} \tag{5.31}$$

Thus

$$P_{\mathbf{W}_l}(\pm z_j) = \pm \sum_{k=0}^{\min(j, l-1)} < z_{j-k}(\cdot), f_0(\cdot) > g_k(\lambda). \tag{5.32}$$

The convex hull of the points $P_{\mathbf{W}_l}(\pm z_j)$ coincides with the set of projections $P_{\mathbf{W}_l}[\mathbf{K}_1]$ on $\mathbf{W}_l$ of the set $\mathbf{K}_1$. The set $P_{\mathbf{W}_l}[\mathbf{K}_1]$ is convex and all its end points are included in the points (5.32). It can be shown that because of the condition in (5.22)

$$< P_{\mathbf{W}_l} z_j, P_{\mathbf{W}_l} z_j > \to 0 \text{ for } j \to \infty \tag{5.33}$$

and since the points of (5.32) are symmetrical with respect to zero, then the number of end points of the set $P_{\mathbf{W}_l}[\mathbf{K}_1]$ is finite, *i.e.* $P_{\mathbf{W}_l}[\mathbf{K}_1]$ is a polyhedron.

Without limiting our attention to the problem of choosing the end points from the points of (5.32), we assume that the choice is already made and then designate them accordingly :

$$P_{W_\perp} z_{j_1}(\lambda) \ldots, P_{W_\perp} z_{j_q}(\lambda). \tag{5.34}$$

Here q is the number of end points* and z_{j_k} are the corresponding end points in the set $\mathbf{K}_1$.

Let us now find out $P_{W_\perp}[\mathbf{W}_0]$. For this it is sufficient to compute $P_{W_\perp}[W_1(\cdot)]$, where $W_1(\cdot)$ is the first component of the transfer function of any stabilizing regulator (5.3). Let

$$W_1(\lambda) = \alpha(\lambda) g^{-1}(\lambda) = \sum_{j=0}^{\infty} W_1^{(j)} \lambda^j. \tag{5.35}$$

Then by virtue of (5.32) we have

$$\overline{W}_1^{opt}(\lambda) = P_{W_\perp}[W_1(\cdot)]$$

$$= \sum_{j=0}^{\infty} W_1^{(j)} P_{W_\perp} z_j(\cdot)$$

$$= \sum_{k=0}^{l-1} \sum_{j=k}^{\infty} W_1^{(j)} < z_{j-k}, f_0 > g_k(\lambda)$$

$$= \sum_{k=0}^{l-1} < W_{1,k}, f_0 > g_k(\lambda). \tag{5.36}$$

Here

$$W_{1,k}(\lambda) = \sum_{j=k}^{\infty} W_1^{(j)} \lambda^{j-k}. \tag{5.37}$$

($W_1^{(j)}$ have been considered to be real numbers).

The least value μ_0 of the functional J_* is determined by the condition

$$\mu_0 = \min\{\mu > 0 : \overline{W}_1^{opt}(\cdot) \in \mu P_{W_\perp}[\mathbf{K}_1]\}. \tag{5.38}$$

Evidently, the function $\mu_0^{-1}\overline{W}^{opt}(\cdot)$ must be a convex combination of end points of (5.32). Let

$$\mu_0^{-1}\overline{W}^{opt}(\cdot) = \sum_{s=1}^{q} \delta_s P_{W_\perp} z_{j_s}(\cdot), \quad \sum_{s=1}^{q} |\delta_s| = 1, \tag{5.39}$$

be the expansion along end points. Since it is assumed that the F.R.F. $\overline{W}^{opt}(\cdot), P_{W_\perp} z_{j_1}(\cdot), \ldots, P_{W_\perp} z_{j_q}(\cdot)$ and the number μ_0 are known, then the relationship of (5.39) determines a set of coefficients $\{\delta_s\}$ of the expansion. The element $F(\lambda) = W_1^{opt}(\lambda)$ of the tangency between the sets $\mathbf{K}_{\mu_0}$ and $\mathbf{W}_0$ then takes the form of a polynomial

*It can be shown that the number of end points may be very large if the polynomial $b_-(\lambda)$ possesses roots in the sufficiently small neighbourhood of the unit circle - (vide section 3, P.27).

$$F(\lambda) = \sum_{s=1}^{q} \delta_s \lambda^{j_s}, \quad \delta_s = \mu_0 \tilde{\delta}_s. \tag{5.40}$$

Here $\tilde{\delta}_s$, j_s are the coefficients and indices from (5.39).

Thus, the first component of the optimal transfer function is determined by the formula (5.40) and it has the expression of a polynomial. The second component $W_2^{opt}(\cdot)$ is determined from the equation

$$a(\lambda)F(\lambda) - b(\lambda)W_2^{opt}(\lambda) = 1 \tag{5.41}$$

and it is solvable. From (5.41) it follows that $W_2^{opt}(\lambda)$ has the expression

$$W_2^{opt}(\lambda) = G(\lambda)/b_+(\lambda). \tag{5.42}$$

Here $G(\lambda)$ is the polynomial that can be determined from the equation

$$a(\lambda)F(\lambda) - b_-(\lambda)G(\lambda) = 1 \tag{5.43}$$

when the polynomial (5.40) is given. The equation (5.43) is solvable*. The optimal regulator polynomials $\alpha(\lambda)$ and $\beta(\lambda)$ are then determined from the relationship (5.11). The optimal control strategy is determined by the regulator

$$F(\nabla)b_+(\nabla)u_t = G(\nabla)y_t. \tag{5.44}$$

Thus, the minimax control has the expression of a linear feedback.

3.5.5 *Statement of the Basic Result*

The results of analysis of minimax control problems obtained in earlier sections will be formulated as follows.

Theorem 3.5.1 *Let the condition (5.22) be satisfied for the polynomial $b(\lambda)$ and the pair $(a(\lambda), b(\lambda))$ be stabilizable. Then the solution of the minimax control problem exists and it is determined by the following conditions.*

1. We determine the sequence γ_0^∞ as the solution of the equation

$$b_-^{(*)}(\nabla)\gamma_t = 0, \quad t = 1, \quad 2, \quad \dots, \tag{5.45}$$

with initial conditions

$$\gamma_s = 0 \text{ for } s < 0, \quad \gamma_0 = 1. \tag{5.46}$$

Here $b_-^{(*)}(\lambda)$ is a stable polynomial (5.25). It can be determined by using the unstable part $b_-(\lambda)$ of the polynomial $b(\lambda)$ (vide (5.23)).

2. We express Γ_t the phase vector of the equation (5.45) :

$$\Gamma_t = \text{col}\,(\gamma_t, \quad \dots, \quad \gamma_{t-l+1}) \tag{5.47}$$

and define the set $\mathbf{K} \subseteq \mathbf{R}^l$ as the convex symmetrical hull of the points $\{\pm\Gamma_j\}$, $j = 0, 1, \dots$.

The set **K** *is a polyhedron and so it is a convex hull of the finite number of points*

---

*Since $b_-(0) = 0$ then from the solvability of the equation (5.43) we get $F(0) = 1$, *i.e.* in the expansion (5.40) the constant component (related to $j_s = 0$) is different from zero.

$\{\pm\Gamma_j\}$, $0 \le j \le j_0$, we introduce the matrix B in terms of the coefficients $\bar{b}_s$ of the polynomial $b_-(\lambda)$,

$$b_-(\lambda) = \bar{b}_0 + \lambda\bar{b}_1 + \cdots + \lambda^l \bar{b}_l, \tag{5.48}$$

$$B = \left\| \begin{matrix} -\bar{b}_{l-1}\bar{b}_l^{-1} & \dots & -\bar{b}_0\bar{b}_l^{-1} \\ 1 & \dots & 0 \\ 0 & 1.. & 0 \\ \cdot & \cdot & \cdot \\ \cdot & \cdot & \cdot \\ \cdot & \cdot & \cdot \\ 0 & 0 & 1\,0 \end{matrix} \right\|$$

We also designate

$$a(B) = I_l + a_1 B + \cdots + a_l B^l. \tag{5.50}$$

Let the ray

$$\{\mu^{-1}a^{-1}(B)\Gamma_0,\ \ \mu > 0\} \tag{5.51}$$

intersect the boundary of the set **K** *at the point $\Gamma \in \mathbf{R}^l$ for $\mu = \mu_0$,*

$$\Gamma = \sum_{s=1}^{q} \delta_s \Gamma_{j_s},\quad \sum_{s=1}^{q} |\delta_s| = 1, \tag{5.52}$$

is the expansion of the point Γ in terms of end points of the set **K**.

Then the least value of the function (5.16) will be $\mu_0\rho$, and the optimal control strategy will be given by the regulator (5.44). Here the polynomial $F(\lambda)$ is determined by the formula (5.40) and the polynomial $G(\lambda)$ is uniquely determined by the relations in (5.43).

3.5.6 *The Properties of the Optimal Regulator*

We shall list a few properties of the optimal regulator that follow its design and the proof of the theorem 3.5.1.

1. The characteristic polynomial (5.5) of the control system (5.1) and (5.44) is easily computed as

$$\begin{aligned} g(\lambda) &= a(\lambda)F(\lambda)b_+(\lambda) - b(\lambda)G(\lambda) \\ &= b_+(\lambda)\,[a(\lambda)F(\lambda) - b_-(\lambda)G(\lambda)] \\ &= b_+(\lambda). \end{aligned} \tag{5.53}$$

The condition of unpredictability in (5.44) is satisfied since
$F(0)b_+(0) \ne 0.$

2. If in (5.1) noise is absent and the performance criterion is selected in the form of

$$J = \sum_{t=0}^{\infty} y_t^2,$$

then the optimal regulator may be realized by the methods of analytical construction (vide section 3.2). If the control object (5.1) is of minimal phase, *i.e.*, $b_-(\lambda) = b_+(\lambda)^k$, then the above regulator is the same as that of (5.44). For nonminimal phase control object the optimal regulators are as a rule different. From the proof of the theorem 3.5.1, the geometrical interpretation of this fact in terms of spheres in Banach spaces l_2 (for a quadratic functional) and l_1 (for quasi-linear functional of the form as in (5.7)) follows.

3. The degree of the optimal polynomials $\alpha(\lambda)$ and $\beta(\lambda)$ are estimated from the number of the end points of the polyhedron **K**. A rough estimate of this number is $j_0 \le C\rho_0^{-1}$, where ρ_0 is the distance of the $b_-(\lambda)$ roots from the unit circumference. This follows from the linear independence of the first l vectors Γ_t, $t = 0, \ \ldots, \ l-1$, and from the properties of the linear equation given in (5.45). The degree of the polynomials $\alpha(\lambda)$) and $\beta(\lambda)$ determine the 'memory' of the regulator (5.44). It may increase indefinitely if the roots of the polynomial $b_-(\lambda)$ approach the unit circumference.

4. If the control object (5.1) is of minimal phase then $b(\lambda) = \lambda^k b_+(\lambda)$, where k is the control delay. Then it is evident that $b_-^{(*)}(\lambda) = b_k$ and $\Gamma_t = 0$ for $t \ge k$. This means that all the vector Γ_t (for $t \ge k$) are included in the convex symmetrical hull of the previous ones. Hence deg $F(\lambda) \le k-1$ and the pair $\{F(\lambda), G(\lambda)\}$ is uniquely found out from the following equation applying lemma 2.3.1

$$a(\lambda)F(\lambda) - \lambda^k G(\lambda) = 1. \tag{5.54}$$

This method of determining the optimal regulator will be extended to the condition when the polynomial $b_-(\lambda)$ is strongly unstable : $|\bar{b}_l| > |\bar{b}_0| + \cdots + |\bar{b}_{l-1}|$. In fact, due to the obvious equality $b_-^{(*)}(\nabla)\Gamma_t = 0$ all the vectors Γ_t for $t \ge k$ are included in the convex symmetrical hull of the previous ones. Hence, the degree of the polynomial $F(\lambda)$ is not more than $l-1$ and the regulator is formed from the equation (5.54) in the same manner as that for a minimal phase control object.

5. The degenerated case of condition (5.22) not being fulfilled will now be examined. In a linear-quadratic problem, the greatest lower bound of the performance criterion is not achieved when such conditions are present, and so the optimal regulator does not exist. In the minimax control problem related to existence of the optimal regulator, both the answers are possible depending on polynomials $a(\lambda)$ and $b(\lambda)$. This has been illustrated by examples.

3.5.7 *A Few Generalisations*

Analysis carried out in sections 3.5.3 and 3.5.4 show how the results of the theorem 3.5.2 can be extended closely to the problem formulation. A few of them will be now examined.

(*a*) *Equation of the control object in the standard form.* Let the control object be expressed by the equations

$$x_{t+1} = Ax_t + Bu_t + Cv_{t+1}, \tag{5.55}$$

$$y_t = D^* x_t. \tag{5.56}$$

Here A is an $n \times n$ matrix and B, C, D are n-vectors, $v^\infty \in \mathbf{V}_\rho$. Let the performance criterion be expressed in the form of (5.7). We introduce the polynomials

$$a(\lambda) = \det(I_n - \lambda A),$$

$$b(\lambda) = D^*(I_n - \lambda A)^{-1} B \det a(\lambda),$$

$$c(\lambda) = D^*(I_n - \lambda A)^{-1} C \det a(\lambda).$$

Then the system of (5.55) and (5.56) can be transformed to

$$a(\nabla)y_t = b(\nabla)u_t + c(\nabla)v_t. \tag{5.57}$$

The equation (5.57) differs from the equation (5.1) in the fact that in the former noise has been 'filtered'.

Minimization of the performance criterion (5.7) for the control object (5.57) in a class of stabilizing regulators (5.3) can be implemented by the methods same as those of minimax control. We explain the necessary changes for a few more generalised problems.

(*b*) *Object with scalar input and output, and smooth noise.* Let the control object be expressed by the equation (5.1) where the noise v^∞ is determined as the output of the filter

$$c(\nabla)v_t = d(\nabla)e_t. \tag{5.58}$$

The input to the filter is an irregular noise $e^\infty \in \mathbf{V}_\rho$.*

The performance criterion, subjected to minimization in the class of stabilizing regulators (5.3), can be expressed as (5.7). Assuming the polynomials $c(\lambda)$, $d(\lambda)$ to be mutually unreductible and the polynomial $c(\lambda)$ to be stable, we can formulate the optimal control problem solution in the same terms as those of a minimax control problem.

Let the polynomial $d(\lambda)$ be separated into stable and unstable components as

$$d(\lambda) = d_+(\lambda)d_-(\lambda). \tag{5.59}$$

Let $d_-^{(*)}(\lambda) = \lambda^r d_-(\lambda^{-1})$

where $r = \deg d_-(\lambda)$. (5.60)

Also let γ_t be the solution of the equation

$$d_-^{(*)}(\nabla)b_-^{(*)}(\nabla)\gamma_t = 0, \quad t = 1, 2, \quad \ldots, \tag{5.61}$$

with initial conditions given in (5.46), we define the phase vector $\Gamma_t = \mathrm{col}\,(\gamma_t, \ldots, \gamma_{t-r-l+1})$ in $\mathbf{R}^{r+l}$. By virtue of (5.61) Γ_t satisfies the equation

$$\Gamma_{t+1} = B\Gamma_t, \quad t = 0, 1, \ldots, \tag{5.62}$$

and also the initial condition

$$\Gamma_t = \mathrm{col}\,(1, 0, \ldots, 0).$$

The stable matrix B is determined by the standard technique, *e.g.* through coefficients of the equation (5.61).

---

*We recall that $\mathbf{V}_\rho$ includes all the possible sequences e^∞ that satisfy the unique condition $|e_t| \le \rho$.

Let **K** represent the convex hull of the vectors $\{\pm\Gamma_t\}$.

Theorem 3.5.2 *Let there be no root of the polynomials* $d(\lambda)$ and $b(\lambda)$ *in the unit circle. Then according to the conditions in theorem 3.5.1 the optimal regulator exists. Let the ray*

$$\{\mu^{-1}C(B)a^{-1}(B)\Gamma_0,\quad \mu>0\}$$

interset the border of the set K^* *at the point* Γ *for* $\mu=\mu_0$. *Again the expansion* Γ *along end points of the set* K *is given by*

$$\Gamma=\sum_{j=1}^{r+l}\delta_j\Gamma_{k_j},\quad \sum_{j=1}^{r+l}|\delta_j|=1.$$

Then –

(1) the polynomial $F(\lambda)=\mu_0\sum_{j=0}^{r+l}\delta_j\lambda^{k_j}$, *is divided as* $f(\lambda)=m(\lambda)d_-(\lambda)$ *where in* (λ) *is a polynomial ;*

(2) the optimal value of the performance criterion is $J_{\min}=\mu_0\rho$;

(3) the polynomial $\alpha(\lambda)$ *of the optimal regulator (5.3) is determined by the formula* $\alpha(\lambda)=m(\lambda)b_+(\lambda)c(\lambda)$. *The polynomial* $\beta(\lambda)$ *is given by the relationship* $m(\lambda)a(\lambda)c(\lambda)-b_-(\lambda)\beta(\lambda)=d_+(\lambda)$ *and it is solvable ;*

(4) the characteristic polynomial of the optimal control system is equal to $g(\lambda)=b_+(\lambda)d_+(\lambda)$.

(c) *Finite time period control.* Let us examine the control object (5.1) in the time interval $[0, T]$ with zero initial conditions. The performance criterion is expressed as

$$J_T(\alpha,\beta)=\sup_{v_0^T\in V_{\rho,T}}\ \sup_{0\le t\le T}|y_t|, \tag{5.63}$$

where

$$V_{\rho,T}=\{v_0^T: |v_t|\le\rho\}.$$

The minimization of the performance criterion (5.63) in the class of stabilizing regulators (5.3)** will be the subject of our study.

Then the formulation of theorem 3.5.1 remains valid if the performance criterion (5.7) in it is replaced by (5.63) and the set **K** by $\mathbf{K}_T$ (convex hull of points $\pm\Gamma_t$, $t=0,1,\dots,T-1$).

(d) *Minimaxity of control.* A similar solution of the control problem is achieved when the performance criterion *has the form given below and it is to be minimized in the class of stabilizing regulators (5.5) :*

---

*When the condition of the theorem of polynomials $d(\lambda)$ and $b(\lambda)$ is satisfied the set **K** becomes a polyhedron.

**For finite time period control, the condition for stabilizability of regulators is introduced so that the control signal does not become very large when T is large.

$$J(\alpha,\beta) = \sup_{v_0^{-\infty} \in V_\rho} \overline{\lim_{t\to\infty}} \; |u_t|. \qquad (5.64)$$

If the control object (5.1) is output stable (polynomial $\alpha(\lambda)$ is stable) then the optimal strategy is evident : $u_t \equiv 0$. For an unstable control object the problem of minimization of the performance criterion (5.7) is interesting.

The solution can be obtained by using the same reasoning as used for solution of minimax problems with functional (5.7). The only difference is that the functional (5.7) is expressed through the second component of the transfer function. Thus, if

$$W_2(\lambda) = \sum_{j=0}^{\infty} W_2^{(j)}\lambda^j$$

then

$$J(\alpha, \quad \beta) = \rho \sum_{j=0}^{\infty} |W_2^{(j)}|.$$

The relations of (5.10) and (5.11) are retained and along with the parametrization formula (5.17) the following formula is to be used (vide (5.12)) :

$$W(\lambda) = W^0(\lambda) + a(\lambda)\Psi(\lambda).$$

Here $W^0(\lambda)$ and $\Psi(\lambda)$ are fixed and arbitrary analytical functions, respectively. In this manner the formulation of the theorem 3.5.1 remains valid, if, in the light of the conditions of the theorem, the polynomials $b_-(\lambda)$ and $a_-(\lambda)$ (unstable component of the polynomial $a(\lambda)$ are examined together).

§ 3.A Appendix

3.A.1 *Frequency Theorem*

While analyzing the solutions of Riccati equation and establishing the existence of Liapunov quadratic functions, we have always referred to the frequency theorem of Kalman-Iakubovich. We shall introduce the version of the theorem [177] which will be convenient for future statements.

Let $q(x,u)$ be the symmetric real form of n-vectors x and m-vectors u :

$$q(x,u) = \begin{pmatrix} x \\ u \end{pmatrix}^* N \begin{pmatrix} x \\ u \end{pmatrix}, \qquad (A.1)$$

where matrix N has the expression as in (2.19). Let A, B be the real matrices of dimensions $n \times n$ and $n \times m$ and the F.R.F. of the complex argument be given by

$$W(\lambda) = \lambda(I_n - \lambda A)^{-1}B.$$

Theorem 3.A.1. *Let the pair* (A, B) *be stabilizable and let for some* $\varepsilon > 0$ *and for all* λ, $|\lambda| = 1$, *(for which det* $(I_n - \lambda A) \neq 0$) *the following inequality be satisfied :*

$$\begin{vmatrix} W(\lambda) \\ I_m \end{vmatrix}^{\nabla} N \begin{vmatrix} W(\lambda) \\ I_m \end{vmatrix} \geq \varepsilon[W^{\nabla}(\lambda)W(\lambda) + I_m]. \qquad (A.2)$$

Then there exist such real matrices $H = H^*$, $\Gamma > 0$, K that will make the following identity valid for all $x \in \mathbf{R}^n$, $u \in \mathbf{R}^m$:

$$x^*Hx - (Ax + Bu)^*H(Ax + Bu)$$

$$= \begin{pmatrix} x \\ u \end{pmatrix}^* N \begin{pmatrix} x \\ u \end{pmatrix} - (Kx - u)^* \Gamma(Kx - u). \qquad \text{(A.3)}$$

Here the matrix $A + BK$ is stable (does not have any eigenvalue for $|\lambda| \geq 1$)

Condition of (*A*.2) is known as *frequency condition.*

Results close to the above are given by the following assertion (generalisation of Kalman-Sage lemma (137)).

Theorem 3.A.2 *If the pair $\{A, B\}$ is controllable and det $(I_n - \lambda A) \neq 0$ for $|\lambda| = 1$, then the non-negativeness of the matrix (A.2) is the necessary and sufficient condition for solvability of the system of Lure' equations in respect of matrices $H = H$, $\Gamma > 0, K$, i.e.,*

$$Q = K^*\Gamma K + H - A^*HA,$$

$$S = -K^*\Gamma - A^*HB,$$

$$R = \Gamma - B^*HB$$

§ 3.P. Proofs of Lemmas and Theorems

3.P.1. *Proof of the Theorem 3.1.1*

By virtue of (1.7) and (1.8), the following inequality is satisfied for the functions $U_t(x)$ which are the solution of the equation (1.9), for all x_t that satisfies equation (1.1)

$$V_{t+1}[X_{t+1}(x_t, U_t(x_t), v_t)] - V_t(x_t) + q_t(x_t, \quad U_t(x_t)) \leq r^{-1}. \qquad \text{(P.1)}$$

Adding up the inequality of (P.1) upto the time T we have

$$V_T[X_T(x_{T-1}, U_{T-1}(x_{T-1}), v_{T-1})] - V_0(x_0)$$

$$+ \sum_{t=0}^{T-1} q_t(x_t, U_t(x_t)) \leq r^{-1}T.$$

Added to the above inequality is the non-negativeness of the function $V_t(x)$ and thus we get,

$$T^{-1} \sum_{t=0}^{T-1} q_t(x_t, U_t(x_t)) \leq r^{-1} + T^{-1}V_0(x_0).$$

In the above inequality, it can be shown that the performance criterion of (1.6) is satisfied, when the upper bound is $T \to \infty$.

3.P.2 *Proof of the Theorem 3.1.2*

In accordance with the relations in (1.7), (1.10), (1.16) and (1.17) we get

$$\Psi_t(\tilde{x}, \quad \tilde{u}) = \sup_{w \in W} \quad |A^T\tilde{x} + \tilde{B}\tilde{u} + w|^2 = \rho_W(A^T\tilde{x} + \tilde{B}\tilde{u}).$$

Then because of (1.12), (1.11), (1.14) and (1.16), the expression in (1.8) takes the form

$$\inf_{\tilde{u}} \ [\rho_W(A^T\tilde{x}+\tilde{B}\tilde{u})] - |\tilde{x}|^2 + |\tilde{x}|^2 \le r^{-1}$$

or $\rho_W(w_W) \le r^{-1}$

(by virtue of (1.19)). Consequently $\tilde{u}$ must be determinable from the relationship $A^T\tilde{x}+\tilde{B}\tilde{u} = w_W$. This result is also given by the conditions 1 and 2 of the theorem. The solution of the latter equation leads to (1.20) which concludes the proof of the theorem 3.1.2.

3.P.3 *Proof of the Theorem 3.1.3*

We re-write the inequality of (1.30)

$$x_t^*\left\{(1+\rho)\left[A^*HA - \frac{A^*HBB^*HA}{B^*HB}\right] - H + Q\right\}x_t$$

$$+\Lambda_H C_v^2(1+\rho^{-1}) \le r^{-1}. \tag{P.2}$$

The expression in the square brackets can be transformed by elementary operations into

$$A^*HA - \frac{A^*HBB^*HA}{B^*HB} = (A+BK)^*H(A+BK)$$

$$-(B^*HB)^{-1}[(A+BK)^* \quad HBB^*H(A+BK)] \tag{P.3}$$

Here K is a row vector from the first condition of the theorem. Since the matrix in the square bracket of the right hand side of the identity (P.3) is non-negative, the inequality of (P.2) (taking into consideration of (1.32)) can be re-written as

$$x_t^*(\tilde{A}^*H\tilde{A} + Q - H)x_t + \Lambda_H C_v^2(1+\rho^{-1}) \le r^{-1}.$$

By virtue of (1.33), and (1.34) this inequality (and consequently (1.30)) is satisfied for all x_t. According to the theorem 3.1.1 this ensures stabilizability of the control system of (1.10) and (1.31) and is affirmed. Theorem 3.1.3 is proved.

3.P.4 *Proof of the Theorem 3.1.4*

Referring to (1.44), (1.48) and (1.53) we have

$$V_2(\varepsilon_{t+1}) = \left[(A_L\varepsilon_t + v_t' + Lv_t'')^* \ H_2(A_L\varepsilon_t + v_t' + Lv_t'')\right]^{1/2}$$

$$\le [(A_L\varepsilon_t)^* \ H_2(A_L\varepsilon_t)]^{1/2} + [(v'_t)^* \ H_2 v_t']^{1/2}$$

$$+\left[(v_t'')^* L^* H_2 L v_t''\right]^{1/2} \le \beta^{1/2} V_2(\varepsilon_t) +$$

$$+ C_v(|H_2|^{1/2} |L^* H_2 L|^{1/2}).$$

From above we get the inequality of (1.50).

By virtue of (1.45), (1.47) and (1.49) we similarly get for the function $V_1(x)$

$$V_1(x_{t+1}) = [(A_k x_t - BK\varepsilon_t + v'_t)^* \ H_1(A_k x_t - BK\varepsilon_t + v'_t)]^{1/2}$$

$$\le \alpha^{1/2} V_1(x_t) + \gamma V_2(\varepsilon_t) + C_v |H_1|^{1/2}.$$

Hence the second inequality of (1.50) follows and the theorem is proved.

3.P.5 *Proof of the Theorem 3.2.1*

Since $f(0) = Q - SQ^{-1}S^*$ the implication of (2.39) follows from (2.40). We shall prove (2.40) for $H_1 > 0$. Then the case of $H_1 > 0$ will be justified by the limiting transition. For $H > 0$ the following formula is valid

$$f(H) = (A - BR^{-1}S^*)^*(H^{-1} + BR^{-1}B^*)^{-1}(A - BR^{-1}S^*)$$

$$+Q - SR^{-1}S^*. \tag{P.4}$$

The above formula is easily verified using the identity

$$(H^{-1} + BR^{-1}B^*)^{-1} = H - HB(R + B^*HB)^{-1}B^*H.$$

From (P.4) it follows that

$$f(H_2) - f(H_1) = (A - BR^{-1}S^*) \quad [(H_2^{-1} + BR^{-1}B^*)^{-1}$$

$$-(H_1^{-1} - BR^{-1}B^*)^{-1}] (A + BR^{-1}S^*). \tag{P.5}$$

But

$$H_2 \geq H_1 \Rightarrow H_1^{-1} \geq H_2^{-1} \Rightarrow H_1^{-1} + BR^{-1}B^*$$

$$\geq H_2^{-1} + BR^{-1} + B^* \Rightarrow (H_1^{-1} + BR^{-1}B^*)^{-1}$$

$$\leq (H_2^{-1} + BR^{-1}B^*)^{-1}.$$

Hence (2.40) follows from (P.5).

The inequality of (2.41) follows directly from the relationship

$$f(H_1) - f(H_2) = \Phi^*(H_1)(H_1 - H_2)\Phi(H_1)$$

$$+(K_{H_1} - K_{H_2})^*(B^*H_2B + R)(K_{H_1} - K_{H_2}). \tag{P.6}$$

We shall establish (P.6) using (2.42) and the equalities

$$f(H) = A^*HA + Q + (A^*HB + S)K_H,$$

$$K_H^*(B^*HB + R) = -(A^*HB + S),$$

$$K_H^*(B^*HA + S^*) = -K_H^*(B^*HB + R)K_H,$$

The above equalities are obtained from (2.43) and (2.38).

$$f(H_1) - f(H_2) - \Phi^*(H_1)(H_1 - H_2)\Phi(H_1) = f(H_1) - f(H_2)$$

$$-A^*(H_1 - H_2)A - A_1^*(H_1 - H_2)BK_{H_1} - K_{H_1}^*B^*(H_1 - H_2)A$$

$$-K_{H_1}^*B^*(H_1 - H_2)BK_{H_1} = (A^*H_1B + S)K_{H_1} - (A^*H_2B + S)K_{H_2}$$

$$-[A^*H_1B + S - (A^*H_2B + S)]K_{H_1}$$

$$-K_{H_1}^*[B^*H_1A + S^* - (B^*H_2A + S^*)] - K_{H_1}^*(B^*H_1B + R)K_{H_1}$$

$$+K_{H_1}^*(B^*H_2B + R)K_{H_1} = K_{H_2}^*(B^*H_2B + R)K_{H_2}$$

$$-K_{H_2}^*(B^*H_2B + R)K_{H_1} - K_{H_1}^*(B^*H_2B + R)K_{H_2}$$

$$+K_{H_1}^*(B^*H_2B + R)K_{H_1} = (K_{H_1} - K_{H_2})^*(B^*H_2B + R)(K_{H_1} - K_{H_2}).$$

The equality of (P.6) is established and the theorem is proved.

3.P.6 *Proof of the Theorem 3.2.2*

Let the sequence matrices H_t be determined by the equation (2.45) and the initial condition $H_0=0$. Then, due to the non-negativeness of the matrix N, we have $H_1=f(0)\geq Q-SR^{-1}S^*\geq 0$. Consequently, $H_1\geq H_0$. Since $H_2=f(H_1)\geq f(H_0)=H_1$, consideration of the theorem 3.2.1 gives us $H_2\geq H_1$. Hence $H_{t+1}>H_t$ for all t. Thus, the sequence of non-negative matrices $\{H_t\}$ does not decay monotonously. We shall show that the sequence of $\{H_t\}$ is bounded. We shall preliminarily estabilish the auxiliary assertions which will be later useful.

Lemma 3.P.1 *The following presentation follows the matrix function of (2.38):*

$$f(H)=\tilde{A}^*H\tilde{A}-\tilde{A}^*HBL(H)B^*H\tilde{A}+\tilde{Q}, \tag{P.7}$$

$$f(H)=(\tilde{A}+\tilde{B}K_H)^*H(\tilde{A}+\tilde{B}K_H)+\tilde{K}_H^*R\tilde{K}_H+\tilde{Q}, \tag{P.8}$$

where

$$L(H)=(B^*HB+R)^{-1},\ \tilde{K}_H=-L(H)B^*H\tilde{A},$$

$$\tilde{A}=A-BR^{-1}S^*,\ \tilde{Q}=Q-SR^{-1}S^*. \tag{P.9}$$

Proof We have,

$$\tilde{A}^*H\tilde{A}-\tilde{A}^*HBLB^*H\tilde{A}+\tilde{Q}=A^*HA-A^*HBR^{-1}S^*$$
$$-SR^{-1}B^*HA+SR^{-1}B^*HBR^{-1}S^*-A^*HBLB^*HA$$
$$-SR^{-1}B^*HBLB^*HBR^{-1}S^*+SR^{-1}B^*HBLB^*HA$$
$$+A^*HBLB^*HBR^{-1}S^*+\tilde{Q}. \tag{P.10}$$

Using the formula $LB^*HB=1+LR$ in (P.10) and then making the similar terms shorter we have

$$\tilde{A}^*H\tilde{A}-\tilde{A}^*HBLB^*H\tilde{A}+\tilde{Q}=A^*HA-A^*HBLB^*HA$$
$$-SR^{-1}S^*-SLS^*-SLB^*HA-A^*HBLS^*+\tilde{Q}$$
$$=A^*HA-(A^*HA+S)L(BHA^*+S^*)+Q=f(H).$$

It is the same as (P.7). Using the formulae

$$\tilde{K}_H^*(B^*HB+R)=-\tilde{A}^*HB,$$

$$\tilde{K}_H^*B^*H\tilde{A}=-\tilde{K}_H(\tilde{B}^*HB+R)^{-1}\tilde{K}_H$$

and the definition of $\tilde{K}_H$ (vide (P.9)), we get

$$f(H)=(\tilde{A}+B\tilde{K}_H)^*H(A+B\tilde{K}_H)-\tilde{A}^*HBLB^*HA+\tilde{Q}$$
$$-\tilde{K}_H^*B^*H\tilde{A}-\tilde{A}^*HB\tilde{K}_H-\tilde{K}_H^*B^*HB\tilde{K}_H$$
$$=(\tilde{A}+B\tilde{K}_H)^*H(\tilde{A}+B\tilde{K}_H)-\tilde{K}_H^*(B^*HB+R)\tilde{K}_H+\tilde{Q}$$
$$-2\tilde{K}_H^*(B^*HB+R)\tilde{K}_H-\tilde{K}_HB^*HB\tilde{K}_H.$$

It will lead to (P.8) proving the lemma 3.P.1.

Lemma 3.P.2 *For the matrix function*

$$\phi(H,K)=(\tilde{A}+BK)^*H(\tilde{A}+BK)+K^*RK+\tilde{Q}$$

with an arbitrary choice of the matrix K of appropriate dimension the following inequality is valid

$$f(H) = \phi(H, \tilde{K}_H) \le \phi(H, K). \qquad (P.11)$$

Proof We choose an arbitrary n-vector $a \ne 0$ and examine the expression

$$a^*(\tilde{A} + BK)^* H(A + BK)a + a^*K^*RKa + a^*\tilde{Q}a$$

$$= (z - Bu)^* \quad H(z - Bu) + u^*RU + a^*\tilde{Q}a$$

where $z = Aa$ and $u = -Ka$.

The least value with respect to u is obtained for $u = -(B^*HB + R)^{-1}BHz$. Since the vector a is arbitrary, it leads to matrix $\tilde{K}_H$. This establishes the inequality (P.11) and proves the lemma 3.P.2.

We re-examine the proof of the theorem 3.2.2. The Riccati equation corresponding to the lemma 3.P.1 may be transformed as

$$H_{t+1} = \left(\tilde{A} + B\tilde{K}_{H_t}\right)^* \quad H_t\left(\tilde{A} + B\tilde{K}_{H_t}\right) + \tilde{K}^*_{H_t}R\tilde{K}_{H_t} + \tilde{Q}.$$

The following inequality holds in accordance with the lemma 3.P.2 :

$$H_{t+1} \le (\tilde{A} + BK)^* \quad H_t(\tilde{A} + BK) + K^*RK + \tilde{Q}. \qquad (P.12)$$

Here the matrix K is arbitrarily chosen with an appropriate dimension. Since the pair (A, B) is stabilizable this property is also attributed to the pair $(A - BR^{-1}S^*, B)$. Hence there exists a matrix K such that the matrix $\tilde{A}_K = \tilde{A} + BK$ is stable. We first fix up such a matrix K and then obtain the estimate from (P.12)

$$H_{t+1} \le \sum_{s=1}^{\infty} (A^*_K)^s (K^*RK + \tilde{Q})A^s_K.$$

It establishes the boundedness of the sequence of matrices $\{H_t\}$. Consequently there exists a finite $\lim_{t \to \infty} H_t = H_\infty$. It is a non-negative matrix and it can be shown that it is a stationary point of the function $f(H) : f(H_\infty) = H_\infty$. The first assertion of the theorem is established.

Lemma 3.P.3 *For the matrix $H > 0$ under the conditions of the theorem 3.2.2 there exists $\alpha > 1$ such that $f(\alpha H) \le \alpha H$.*

Proof From the stabilizability of the pair (A, B) follows the stabilizatbility of the pair $(\tilde{A}, B)$, where $\tilde{A}$ is given by (P.9). Let the matrix K be such that the matrix $\Phi = \tilde{A} + BK$ is stable. By virtue of (P.11) we have

$$f(H) \le \phi(H, K) = \Phi^* H\Phi + K^*RK + \tilde{Q}$$

for any $H > 0$. Since the matrix $\Phi^*H\Phi - H$ is negative, we may choose $\alpha > 1$ such that $\alpha(\Phi^*H\Phi - H) \le -(K^*RK + \tilde{Q})$. For this value of α we have

$$f(\alpha H) \le \alpha\Phi^*H\Phi + K^*RK + \tilde{Q} \le \alpha H.$$

Hence the proof.

We continue the proof of the theorem. It is well-known that the number of different solutions of Lur'e equation is finite. Hence, there exists a matrix $\tilde{H} > 0$, such that for any non-negative solution H of the Lur'e equation, the condition $H \le \tilde{H}$ is satisfied. Without limiting the generalisation we can consider (by virtue of the lemma 3.P.3) that $\tilde{H}$ satisfies the inequality $f(\tilde{H}) \le \tilde{H}$. Let $\tilde{H}_t$ be the sequence of matrices determined by the equation (2.45) for $\tilde{H}_0 = \tilde{H}$. Then $\tilde{H}_1 = f(\tilde{H}_0) \le \tilde{H}_0$ and $f(\tilde{H}_1) \le f(\tilde{H}_0)$ (because of theorem 3.2.1), *i.e.*, $\tilde{H}_2 \le \tilde{H}_1$. Continuing this argument, we can establish the monotonicity of the sequence $\{\tilde{H}_t\}$, $\tilde{H}_{t+1} \le \tilde{H}_t$. Since $\tilde{H}_t \ge 0$, there exists $\lim_{t\to\infty} \tilde{H}_t = \tilde{H}_\infty$ and it satisfies the Lur'e equation. Again for any solution $H \ge 0$ of the Lur'e equation, we have $H \le \tilde{H}_0$ and so $H \le \tilde{H}_t$. Consequently, $H \le \tilde{H}_\infty$, *i.e.*, $\tilde{H}_\infty = H_{max}$, the maximum of the solutions of the Lur'e equation. Arguments similar to above, will prove that the solution H_∞ obtained from the proof of the assertion 1 coincides with the minimum. In fact for any solution $H \ge 0$ we have $H_0 = 0 \le H$ and so $H_t \le H$. Consequently, $H_\infty = H_{min}$. This establishes the second assertion of the theorem.

We assume that for some solution $H \ge 0$ of the Lur'e equation, a stable matrix (2.42) and (2.43) is obtained. Then, in accordance with the theorem 3.2.1, we have

$$H_{max} - H = f(H_{max}) - f(H) \le \Phi^*(H)(H_{max} - H)\Phi(H)$$

$$\le [\Phi^*(H)]^t (H_{max} - H)[\Phi(H)]^t$$

for any $t > 0$. Since the matrix $\Phi(H)$ is stable, it follows from here that $H_{max} - H \le 0$, *i.e.*, $H = H_{max}$. Thus, the assertion 3 is proved.

By virtue of (P.8), we have

$$H = \tilde{\Phi}_H^* H \tilde{\Phi}_H + \tilde{K}_H^* R \tilde{K}_H + \tilde{Q}. \qquad \text{Here } \tilde{\Phi}_H = \tilde{A} + B\tilde{K}_H \tag{P.13}$$

has been introduced for brevity's sake.

For any $t > 1$ we can find out by iterations :

$$H = (\tilde{\Phi}_H^{t+1})^* H \tilde{\Phi}_H^{t+1} + \sum_{s=0}^{t} (\tilde{\Phi}_H^s)^* (\tilde{Q} + \tilde{K}_H^* R \tilde{K}_H) \tilde{\Phi}_H^s. \tag{P.14}$$

We assume that the matrix Φ_H possesses an eigen value for λ_0, $|\lambda_0| \ge 1$. Let x_o be the corresponding eigen vector. Then from (P.14) and remembering the non-negativeness of the matrix $\tilde{Q}$, we have

$$x_0^* H x_0 = |\lambda_0|^{2(t+1)} x_0^* H x_0 + \sum_{s=0}^{t} |\lambda_0|^{2s} |(\tilde{Q} + \tilde{K}_H^* R \tilde{K}_H)^{1/2} x_0|^2. \tag{P.15}$$

Consequently, $(\tilde{Q} + \tilde{K}_H^* R \tilde{K}_H)^{1/2} x_0 = 0$, since $\tilde{K}_H x_0 = 0$ (because $R > 0$). But then

$$\lambda_0 x_0 = \tilde{\Phi}_H x_0 = (\tilde{A} + B\tilde{K}_H) x_0 = (A - BR^{-1}S^*) x_0.$$

For the positive matrix H, we arrive at a contradiction since we have, from (P.15), $|\lambda_0|^{2t} \le 1$, *i.e.* $|\lambda_0| = 1$. Thus, under the conditions of the assertion 4, the maxtrix $\tilde{\Phi}_H = \tilde{A} + B\tilde{K}_H$ is stable. A simple calculation (vide P.9)) shows that

$$\tilde{A} + B\tilde{K}_H = A - BR^{-1}S^* - BL(H)B^*HA + BL(H)B^*HBR^{-1}S^*$$

$$= A - BR^{-1}S^* - BL(H)B^*HA + BR^{-1}S^*$$

$$- BL(H)S^* = A + BK_H.$$

The assertion 4 is established.

The pair $\left(\tilde{\Phi}_H, (\tilde{Q} + \tilde{K}_H^* R \tilde{K}_H)^{1/2}\right)$ can not be detectable if the matrix $\tilde{\Phi}_H$ has eigenvalues λ_0 outside the unit circle $|\lambda_0| \geq 1$. In fact, on the corresponding eigenvector x_0 of the matrix $\tilde{\Phi}_H$ we have $\tilde{\Phi}_H x_0 = \lambda_0 x_0$ and $(\tilde{Q} + \tilde{K}_H^* R \tilde{K}_H)^{1/2} x_0 = 0$ (vide (P.15)). We also note that for any matrix D, the matrix $\tilde{\Phi}_H + D(\tilde{Q} + \tilde{K}_H^* R \tilde{K}_H)^{1/2}$ can not be stable, because

$$\left[\tilde{\Phi}_H + D(\tilde{Q} + \tilde{K}_H^* R \tilde{K}_H)^{1/2}\right] x_0 = \tilde{\Phi}_H x_0 = \lambda_0 x_0, \quad |\lambda_0| \geq 1.$$

Hence for the proof of the assertion 5, it is sufficient to show that the detectability of the matrices $\left(\tilde{\Phi}_H, (\tilde{Q} + \tilde{K}_H^* R \tilde{K}_H)^{1/2}\right)$ follows from the detectability of the matrices $(\tilde{A}, \tilde{Q}^{1/2})$.

Let D be an arbitrary n x n matrix. Since the region enclosing the values of the symmetric matrix $(\tilde{Q} + \tilde{K}_H^* R \tilde{K}_H)^{1/2}$ also contains the region enclosing the values of the matrix $D\tilde{Q}^{1/2} - B\tilde{K}_H$, the following equation is solvable with respect to the $n \times n$ matrix $\tilde{D}$

$$\tilde{D}(\tilde{Q} + \tilde{K}_H^* R \tilde{K}_H)^{1/2} = DQ^{1/2} - B\tilde{K}_H. \tag{P.16}$$

In fact, if $(\tilde{Q} + \tilde{K}_H^* R \tilde{K}_H)^{1/2} a = 0$ for an n-vector a, then because of the non-negativeness of the matrix $\tilde{Q}$ and positiveness of the matrix R, we get

$$a^* \tilde{Q} a = 0, \quad \tilde{K}_H a = 0, \text{ i.e. } (D\tilde{Q}^{1/2} - B\tilde{K}_H) a = 0.$$

Hence it is possible to affirm that the equality of (P.16) permits determination of the matrix $\tilde{D}$ for any known matrix D. The equality of (P.16) is then rewritten as

$$\tilde{\Phi}_H + \tilde{D}(\tilde{Q} + \tilde{K}_H^* R \tilde{K}_H)^{1/2} = \tilde{A} + D\tilde{Q}^{1/2}.$$

We choose a D matrix such that the matrix $\tilde{A} + D\tilde{Q}^{1/2}$ is stable. (It is possible to choose so if the pair $(\tilde{A}, \tilde{Q}^{1/2})$ is detectable.) But this means that the matrix $\tilde{\Phi}_H + \tilde{D}(\tilde{Q} + \tilde{K}_H^* R \tilde{K}_H)^{1/2}$ is stable, *i.e.*, the pair $\left[\tilde{\Phi}_H, (\tilde{Q} + \tilde{K}_H^* R \tilde{K}_H)^{1/2}\right]$ is detectable. Thus, the stability of the matrix $\tilde{\Phi}_H = \tilde{A} + B\tilde{K}_H = A + BK_H$ and the assertion 5 are established.

The assertion 5 is established when the assertion 1 is proved. Theorem 3.2.2 is proved.

3.P.7 *Proof of the Theorem 3.3.1*

1. Let (3.3) be the stabilizing regulator. We use it to determine the transfer function $W(\lambda)$ in accordance with the formulae (3.7) and (3.8). For convenience of writing we introduce the symbols

$$a(\lambda) = I_n - \lambda A, \quad b(\lambda) = \lambda B. \tag{P.17}$$

Using the matrix identity

$$(a - b\alpha^{-1}\beta)^{-1} = a^{-1} - a^{-1}b(\alpha^{-1} - \beta a^{-1}b)\beta\alpha^{-1}$$

we have from (3.17)

$$a(\lambda)W_1(\lambda) - b(\lambda)W_2(\lambda) = I_n - b(\lambda)\left[-\alpha(\lambda) + \beta(\lambda)a^{-1}(\lambda)b(\lambda)\right]^{-1}$$

$$\times\beta(\lambda)\alpha^{-1}(\lambda) - b(\lambda)\left[\alpha(\lambda) - \beta(\lambda)a^{-1}(\lambda)b(\lambda)\right]^{-1}\beta(\lambda)\alpha^{-1}(\lambda) = I_n.$$

It is the same as the expression (3.9). Thus, the transfer function $W(\lambda)$ satisfies the relationship in (3.9). Further we use again (3.7) and get

$$W_2(\lambda)W_1^{-1}(\lambda) = [\alpha(\lambda) - \beta(\lambda)a^{-1}(\lambda)b(\lambda)]^{-1}\beta(\lambda)a^{-1}(\lambda)$$

$$\times[a(\lambda) - b(\lambda)\alpha^{-1}(\lambda)\beta(\lambda)] = [\alpha(\lambda) - \beta(\lambda)a^{-1}(\lambda)b(\lambda)]^{-1}\beta(\lambda)$$

$$\times[I_n - a^{-1}(\lambda)b(\lambda)\alpha^{-1}(\lambda)\beta(\lambda)] = [\alpha(\lambda) - \beta(\lambda)a^{-1}(\lambda)b(\lambda)]^{-1}$$

$$\times[I_m - \beta(\lambda)a^{-1}(\lambda)b(\lambda)\alpha^{-1}(\lambda)]\beta(\lambda) = \alpha^{-1}(\lambda)\beta(\lambda). \quad \text{(P.18)}$$

This means that the transfer function $W(\lambda)$ representing the control system of (3.1) and (3.3) uniquely determines the regulator transfer function (3.3). The following formula is established

$$\left\| \begin{matrix} a(\lambda) & -b(\lambda) \\ \beta(\lambda) & -\alpha(\lambda) \end{matrix} \right\|^{-1} = \left\| \begin{matrix} W_1(\lambda) & W_1(\lambda)b(\lambda)\alpha^{-1}(\lambda) \\ W_2(\lambda) & [W_2(\lambda)b(\lambda) + I_m]\alpha^{-1}(\lambda) \end{matrix} \right\|. \quad \text{(P.19)}$$

The expressions (3.9) and (P.18) are sufficient to establish the equality

$$a(\lambda)W_1(\lambda)b(\lambda)\alpha^{-1}(\lambda) - b(\lambda)\,[W_2(\lambda)b(\lambda) + I_m]\alpha^{-1}(\lambda) = 0,$$

$$\beta(\lambda)W_1(\lambda)b(\lambda)\alpha^{-1}(\lambda) - \alpha(\lambda)\,[W_2(\lambda)b(\lambda) + I_m]\alpha^{-1}(\lambda) = I_m.$$

The first equation is obtained from (3.9). Thus,

$$b(\lambda)\,[W_2(\lambda)b(\lambda) + I_m] = [b(\lambda)W_2(\lambda) + I_n]b(\lambda).$$

The second equation follows the relationship valid for (P.18), *i.e.*,

$$\beta(\lambda)W_1(\lambda)b(\lambda)\alpha^{-1}(\lambda) = \alpha(\lambda)W_2(\lambda)b(\lambda)\alpha^{-1}(\lambda).$$

Thus, the identity of (P.18) involving matrix functions (3.7) is established. But

$$\left| \begin{matrix} a(\lambda) & -b(\lambda) \\ \beta(\lambda) & -\alpha(\lambda) \end{matrix} \right|^{-1} = g^{-1}(\lambda)S(\lambda), \quad \text{(P.20)}$$

where $S(\lambda)$ is the polynomial matrix and $g(\lambda)$ is the polynomial of (3.5). It follows from (P.19) that $g(\lambda)\,W_1(\lambda)$ and $g(\lambda)W_2(\lambda)$ are polynomial matrices. This means that for a stabilizing regulator as is (3.3) the matrix functions $W_1(\lambda)$ and $W_2(\lambda)$ can not have features for $|\lambda| \leq 1$, *i.e.*, they are analytical within the unit circle. Thus, the solvability of the equation (3.9) in a class of analytical functions $W(\lambda)$ belonging to the unit disk is established.

2. Let $W(\lambda)$ be the stable F.R.F. that is satisfied by the relationship (3.9). We rewrite it in the form

$$I_n = [I_n - \lambda(A + BK) + \lambda BK]\,W_1(\lambda) - \lambda B\,W_2(\lambda)$$

$$= \left([W_1^0(\lambda)]^{-1} + \lambda BK\right)W_1(\lambda) - \lambda B\,(\Psi(\lambda) + KW_1(\lambda)]$$

$$= [W_1^0(\lambda)]^{-1}\;\; W_1(\lambda) - \lambda B\Psi(\lambda). \quad \text{(P.21)}$$

Here the notations of (3.11) have been used and

$$W_1^0(\lambda) = [I_n - \lambda(A + BK)]^{-1}. \quad \text{(P.22)}$$

The matrices $W_1^0(\lambda)$ and $\Psi(\lambda)$ are stable F.R.F's. Solving (P.21) with respect to W_1 we get

$$W_1(\lambda) = W_1^0(\lambda)\,[I_n + \lambda B\Psi(\lambda)]. \tag{P.23}$$

Using (3.11), (P.23) and (3.12) we have

$$\begin{aligned}
W_2(\lambda)W_1^{-1}(\lambda) &= K + \Psi(\lambda)W_1^{-1}(\lambda)\\
&= K + \Psi(\lambda)\,[I_n + \lambda B\Psi(\lambda)]^{-1}\,[W_1^0(\lambda)]^{-1}\\
&= K + \Psi(\lambda)\,[I_n - \lambda B\,(I_m + \Psi\lambda B)^{-1}\Psi(\lambda)]\,[W_1^0(\lambda)]^{-1}\\
&= K + [\Psi(\lambda) - \Psi(\lambda)\lambda B\,(I_m + \Psi\lambda B)^{-1}\Psi(\lambda)]\,[W_1^0(\lambda)]^{-1}\\
&= K + [I_m - \Psi\lambda B\,(I_m + \Psi\lambda B)^{-1}]\,\Psi(\lambda)\,[W_1^0(\lambda)]^{-1}\\
&= K + (I_m + \Psi(\lambda)\lambda B)^{-1}\Psi(\lambda)\,[W_1^0(\lambda)]^{-1}\\
&= K + (I_m + \psi_1^{-1}(\lambda)\psi_2(\lambda)\lambda B)^{-1}\psi_1^{-1}(\lambda)\psi_2(\lambda)\,[W_1^0(\lambda)]^{-1}\\
&= K + [\psi_1(\lambda) + \psi_2(\lambda)B]^{-1}\psi_2(\lambda)\,[W_1^0(\lambda)]^{-1}\\
&= [\psi_1(\lambda) + \psi_2(\lambda)\lambda B]^{-1}\,\{[\psi_1(\lambda) + \psi_2(\lambda)\lambda B]K\\
&\quad + \psi_2(\lambda)\,[I_n - \lambda(A + BK)]\}\\
&= [\psi_1(\lambda) + \psi_2(\lambda)\lambda B]^{-1}\,[\psi_1(\lambda)K + \psi_2(\lambda) - \psi_2(\lambda)\lambda A].
\end{aligned}$$

It leads to (3.11) and (3.13).

It can be shown that for the regulator of (3.3) and (3.13) the function $W(\lambda)$ will be the transfer function, *i.e.*, the formulae of (3.7) are valid for the regulator. We shall show that the retulator of (3.3) and (3.13) is stabilizing. In view of the formulae (P.19) it is sufficient to ensure analyticity of the matrix functions

$$W_1(\lambda)b(\lambda)\alpha^{-1}(\lambda), \quad [W_2(\lambda)b(\lambda) + I_m]\alpha^{-1}(\lambda). \tag{P.24}$$

Taking (3.13) and (P.13) into account, we have

$$\begin{aligned}
&W_1(\lambda)b(\lambda)\,[\psi_1(\lambda) + \psi_2(\lambda)\lambda B]^{-1}\\
&= W_1(\lambda)\lambda B\,[I_m + \Psi(\lambda)\lambda B]^{-1}\psi_1^{-1}(\lambda)\\
&= W_1^0(\lambda)\,[I_n + \lambda B\Psi(\lambda)]\,\lambda B\,[I_m + \Psi(\lambda)\lambda B]^{-1}\psi_1^{-1}(\lambda)\\
&= W_1^0(\lambda)\,[\lambda B + \lambda B\Psi(\lambda)\lambda B]\,[I_m + \Psi(\lambda)\lambda B]^{-1}\psi_1^{-1}(\lambda)\\
&= W_1^0(\lambda)\lambda B\,[I_m + \Psi(\lambda)\lambda B]\,[I_m + \Psi(\lambda)\lambda B]^{-1}\psi_1^{-1}(\lambda)\\
&= W_1^0(\lambda)\psi_1^{-1}(\lambda).
\end{aligned}$$

Since the functions $W_1^0(\lambda)$ and $\psi_1^{-1}(\lambda)$ are analytical in the unit circle, the first function of (P.24) is analytical in the unit circle. Further, (3.11) and (3.13) give

$$[W_2(\lambda)b(\lambda)+I_m]\,[\psi_1(\lambda)+\psi_2(\lambda)\lambda B]^{-1}$$

$$=[KW_1(\lambda)b(\lambda)+\Psi(\lambda)\lambda B+I_m]\,[I_m+\Psi\lambda B]^{-1}\psi_1^{-1}(\lambda)$$

$$=KW_1(\lambda)b(\lambda)\,[I_m+\Psi(\lambda)\lambda B]^{-1}\psi_1^{-1}(\lambda)+\psi_1^{-1}(\lambda).$$

Again from above, the second function of (P.24) is analytical in the unit circle. Thus, the matrix in the right hand side of (P.19) is analytical in unit circle, and because of (P.20), it shows stability of the polynomial (3.5). Theorem 3.3.1 is proved.

3.P.8 *Proof of the Lemma 3.3.1*

Let us introduce vector functions $\hat{x}(\lambda)$, $\hat{u}(\lambda)$ of complex argument λ:

$$\begin{pmatrix}\hat{x}(\lambda)\\ \hat{u}(\lambda)\end{pmatrix}=\sum_{t=0}^{\infty}\begin{pmatrix}x_t\\ u_t\end{pmatrix}\lambda^t. \tag{P.25}$$

For $|\lambda|=1$, the formula (P.25) determines the Fourier transform of the vector sequence col (x_t, u_t), $t=\ldots,\ -1,\ 0,\ +1,\ldots$. Here $x_t\equiv 0$, $u_t\equiv 0$ when $t<0$. From (P.25) 'Parseval formula' can be obtained

$$\frac{1}{2\pi i}\oint\begin{pmatrix}\hat{x}(\lambda)\\ \hat{u}(\lambda)\end{pmatrix}\begin{pmatrix}\hat{x}(\lambda)\\ \hat{u}(\lambda)\end{pmatrix}^{\nabla}\frac{d\lambda}{\lambda}=\sum_{t=0}^{\infty}\begin{pmatrix}x_t\\ u_t\end{pmatrix}\begin{pmatrix}x_t\\ u_t\end{pmatrix}^{*}.$$

In particular,

$$\frac{1}{2\pi i}\oint N^{1/2}\begin{pmatrix}\hat{x}(\lambda)\\ \hat{u}(\lambda)\end{pmatrix}\begin{pmatrix}\hat{x}(\lambda)\\ \hat{u}(\lambda)\end{pmatrix}^{\nabla}N^{1/2}\frac{d\lambda}{\lambda}=\sum_{t=0}^{\infty}N^{1/2}\begin{pmatrix}x_t\\ u_t\end{pmatrix}\begin{pmatrix}x_t\\ u_t\end{pmatrix}^{*}N^{1/2},$$

where $N^{1/2}$ = non-negative root of matrix N. Considering that

$$\mathrm{Sp}\,N^{1/2}\begin{pmatrix}\hat{x}\\ \hat{u}\end{pmatrix}\begin{pmatrix}\hat{x}\\ \hat{u}\end{pmatrix}^{\nabla}N^{1/2}=\begin{pmatrix}\hat{x}\\ \hat{u}\end{pmatrix}^{\nabla}N\begin{pmatrix}\hat{x}\\ \hat{u}\end{pmatrix},$$

we get

$$J=\sum_{t=0}^{\infty}\mathrm{Sp}\,N^{1/2}\begin{pmatrix}x_t\\ u_t\end{pmatrix}\begin{pmatrix}x_t\\ u_t\end{pmatrix}^{*}N^{1/2}$$

$$=\frac{1}{2\pi i}=\oint\begin{pmatrix}\hat{x}(\lambda)\\ \hat{u}(\lambda)\end{pmatrix}^{\nabla}N\begin{pmatrix}\hat{x}(\lambda)\\ \hat{u}(\lambda)\end{pmatrix}\frac{d\lambda}{\lambda}. \tag{P.26}$$

We compute col $(\hat{x}(\lambda), \hat{u}(\lambda))$. Multiplying the equation $a(\nabla)x_t=b(\nabla)u_t$ by λ^t and summing up along t, we get*

$$\sum_{t=1}^{\infty}\lambda^t a(\nabla)x_t=\sum_{t=1}^{\infty}\lambda^t b(\nabla)u_t.$$

---

*Here the matrix polynomials $a(\lambda)$ and $b(\lambda)$ are determined by the formulae of (P.17).

But because of (P.17) and (P.25)

$$\sum_{t=1}^{\infty} \lambda^t a(\nabla)x_t = \sum_{t=1}^{\infty} \lambda^t (I_n - A\nabla)x_t = \sum_{t=1}^{\infty} \lambda^t (x_t - Ax_{t-1})$$

$$= \sum_{t=1}^{\infty} \lambda^t x_t - A\sum_{t=1}^{\infty} \lambda^t x_{t-1} = \hat{x}(\lambda) - x_0 - \lambda A\hat{x}(\lambda) = a(\lambda)\hat{x}(\lambda) - x_0.$$

Thus

$$a(\lambda)\hat{x}(\lambda) = b(\lambda)\hat{u}(\lambda) + x_0. \tag{P.27}$$

We have similarly

$$\alpha(\lambda)\hat{u}(\lambda) = \beta(\lambda)\hat{x}(\lambda) + r(\lambda, x_{-p}^{-1}, u_{-p}^{-1}), \tag{P.28}$$

where

$$r(\lambda, x_{-p}^{-1}, u_{-p}^{-1}) = \sum_{s=1}^{p} \begin{pmatrix} x_{-s} \\ u_{-s} \end{pmatrix}^{*} \begin{Vmatrix} -\sum_{k=s}^{p} \alpha_k \lambda^{k-s} \\ \sum_{k=s}^{p} \beta_k \lambda^{k-s} \end{Vmatrix}.$$

From (P.27) and (P.28) we have

$$\begin{pmatrix} \hat{x}(\lambda) \\ \hat{u}(\lambda) \end{pmatrix} = \begin{Vmatrix} a(\lambda) & -b(\lambda) \\ \beta(\lambda) & -\alpha(\lambda) \end{Vmatrix}^{-1} \begin{pmatrix} x_0 \\ r(\lambda, \; x_{-p}^{-1}, \; u_{-p}^{-1}) \end{pmatrix}.$$

Since $r(\lambda, x_{-p}^{-1}, \; u_{-p}^{-1}) = 0$ we have

from (3.14) and (P.19)

$$\hat{x}(\lambda) = W_1(\lambda)x_0 \text{ and } \hat{u}(\lambda) = W_2(\lambda)x_0.$$

From (P.27) we have

$$\hat{x}(\lambda) = a^{-1}(\lambda)b(\lambda)\hat{u}(\lambda)a^{-1}(\lambda)x_0.$$

Finally,

$$\begin{pmatrix} \hat{x}(\lambda) \\ \hat{u}(\lambda) \end{pmatrix} = \begin{Vmatrix} a^{-1}(\lambda)b(\lambda)W_2(\lambda) + a^{-1}(\lambda) \\ W_2(\lambda) \end{Vmatrix} x_0. \tag{P.29}$$

Putting (P.29) in (P.26) we get (3.15) and the lemma under consideration is proved. Inverse Laplace transform of (P.29) gives x_t and u_t :

$$\begin{pmatrix} x_t \\ u_t \end{pmatrix} = \frac{1}{2\pi i} \oint \begin{Vmatrix} a^{-1}(\lambda)b(\lambda)W_2(\lambda) + a^{-1}(\lambda) \\ W_2(\lambda) \end{Vmatrix} \lambda^{-t} \frac{d\lambda}{\lambda} x_0.$$

3.P.9 ***Proof of the Theorem 3.3.2***

Since $N \geq 0$, the matrix function $G(\lambda)$ is non-negative for $|\lambda| = 1$. The expression in (3.23) for a stable F.R.F. $\pi(\lambda)$ is true as per theorem 3.P.6 in [177]. Again for $|\lambda| < 1$ the inverse of the matrix $\pi(\lambda) = \pi^{-1}(\lambda)$ exists and it does not have any specificity in an open unit circle. Since $\det G(\lambda) \neq 0$ and $\det R_2(\lambda) \neq 0$ for $|\lambda| = 1$ then $\pi^{-1}(\lambda)$ is a stable F.R.F. Then the functional of (3.15) can be written as (using the notations of (3.21) and (3.22)).

$$\langle W(\cdot), W(\cdot)\rangle x_0 = x_0^{\bullet}\frac{1}{2\pi i}\oint [\Pi(\lambda)\Psi(\lambda)+R(\lambda)]^{\nabla}$$

$$\times[\Pi(\lambda)\Psi(\lambda)+R(\lambda)]\frac{d\lambda}{\lambda}x_0$$

$$+x_0^{\bullet}\frac{1}{2\pi i}\oint [\left\|\begin{matrix}(I_n-\lambda A)^{-1}\lambda B W_2^0(\lambda)+I_n\\ W_2^0(\lambda)\end{matrix}\right\|^{\nabla} N$$

$$\times\left\|\begin{matrix}(I_n-\lambda A)^{-1}\lambda B W_2^0(\lambda)+I_n\\ W_2^0(\lambda)\end{matrix}\right\| - R^{\nabla}(\lambda)R(\lambda)]\frac{d\lambda}{\lambda}x_0. \qquad \text{(P.30)}$$

Next we observe that

$$\frac{1}{2\pi i}\oint R_-^{\nabla}(\lambda)\,[\Pi(\lambda)\Psi(\lambda)+R_+(\lambda)]\frac{d\lambda}{\lambda}=0. \qquad \text{(P.31)}$$

In the fact the matrix function $R_-^{\nabla}(\lambda)\,[\Pi(\lambda)\Psi(\lambda)+R_+(\lambda)]$ is analytical in unit circle and so according to Cauchy's theorem the integral on the left hand side of (P.31) is equal to the value of this maxtrix function for $\lambda=0$. Since the function $\Pi(\lambda)\Psi(\lambda)+R_+(\lambda)$ is finite for $\lambda=0$, then because of (3.26) this deduction is equal to zero. Thus, (P.31) is proved. Based on this argument

$$\frac{1}{2\pi i}\oint [\Pi(\lambda)\Psi(\lambda)+R_+(\lambda)]^{\nabla}R_-(\lambda)\frac{d\lambda}{\lambda}=0.$$

Hence consideration of (3.25) gives

$$\frac{1}{2\pi i}\oint [\Pi(\lambda)\Psi(\lambda)+R(\lambda)]^{\nabla}\,[\Pi(\lambda)\Psi(\lambda)+R(\lambda)]\frac{d\lambda}{\lambda}$$

$$=\frac{1}{2\pi i}\oint [\Pi(\lambda)\Psi(\lambda)+R_+(\lambda)]^{\nabla}\,[\Pi(\lambda)\Psi(\lambda)+R_+(\lambda)]\frac{d\lambda}{\lambda}+\frac{1}{2\pi i}\oint R_-^{\nabla}(\lambda)R_-(\lambda)\frac{d\lambda}{\lambda}.$$

The last item in the right hand side of the above equality does not depend on choice of $\Psi(\lambda)$ and the first one is a non-negative matrix. Hence the optimal choice of $\Psi(\lambda)$must be realized from the relationship

$$\Pi(\lambda)\Psi(\lambda)+R_+(\lambda)=0. \qquad \text{(P.32)}$$

It leads to (3.27). Considering (P.32), the formula (P.30) takes the expression as in (3.28). The theorem is proved.

3.P.10 *Proof of the Lemma 3.3.2*

It follows from (3.29) and (3.30) that

$$\left\|\begin{matrix}W_1^0(\lambda)\\ KW_1^0(\lambda)\end{matrix}\right\| \in \mathbf{W} \text{ and matrix } \left\|\begin{matrix}R_1(\lambda)\\ R_2(\lambda)\end{matrix}\right\| \text{ is analytical}$$

in the unit circle. Hence

$$W(\lambda)=\left\|\begin{matrix}W_1(\lambda)\\ W_2(\lambda)\end{matrix}\right\| \in \mathbf{W}.$$

We observe that the function $W(\lambda)$ satisfies the relationship in (3.9) for any $\Psi(\lambda)$. From (3.30) we have

$$\| I_n - \lambda A, \quad -\lambda B \| \, W^0(\lambda) = (I_n - \lambda A) W_1^0(\lambda) - \lambda B K W_1^0(\lambda) =$$

$$= [I_n - \lambda(A + BK)] \, W_1^0(\lambda) = I_n,$$

i.e., for $W(\lambda)$ the equality of (3.9) is satisfied.
Further

$$\| I_n - \lambda A, \quad -\lambda B \| \begin{Vmatrix} R_1 l(\lambda) \\ R_2(\lambda) \end{Vmatrix} = (I_n - \lambda A) \lambda W_1^0(\lambda) B$$

$$- \lambda B \, [I_m + \lambda K W_1^0(\lambda) B] = (I_n - \lambda A) W_1^0(\lambda) \lambda B$$

$$- \lambda B K W_1^0(\lambda) \lambda B - \lambda B = [I_n - \lambda(A + BK)] \, W_1^0(\lambda) \lambda B - \lambda B = 0.$$

Hence $\| I_n - \lambda A, -\lambda B \| \, W(\lambda) = \| I_n - \lambda A, -\lambda B \| \, W_1^0(\lambda)$

$$+ \| I_n = \lambda A, -\lambda B \| \, G(\lambda) \Psi(\lambda) = I_n,$$

irrespective of the choice of the matrix function $\Psi(\lambda)$ (of dimension mxn). Thus, it is established that $W(\cdot) \in \mathbf{W}_0$. It will be shown that the expression in (3.17) is true for any function $W(\cdot) \in \mathbf{W}_0$. Here $\Psi(\lambda)$ is a function which is analytical in the unit disk. Let $W_1(\lambda)$ and $W_2(\lambda)$ be the constituent functions of $W(\lambda)$ (vide (3.8)). Then the mxn matrix function

$$\Psi(\lambda) = W_2(\lambda) - K W_1(\lambda), \tag{P.33}$$

where K is a constant matrix that determines the function $W_1^0(\cdot)$ (vide (3.30), will be analytical in the unit circle. We express the functions (determinable by relations in (3.21) and (3.22) through $W_1^{\Psi}(\lambda)$) and $W_2^{\Psi}(\lambda)$ and use them in (P.33). Then

$$\Psi(\lambda) = W_2^{\Psi}(\lambda) - K W_1^{\Psi}(\lambda). \tag{P.34}$$

Since the functions $W^{\Psi}(\cdot) = \begin{Vmatrix} W_1^{\Psi} \\ W_2^{\Psi} \end{Vmatrix}$ and $W^0(\cdot)$ satisfy the relationship in (3.9), then

$\| I_n - \lambda A, -\lambda B \| \, \Delta W(\lambda) = 0$ is true for

$$\Delta W(\lambda) = W^{\Psi}(\lambda) - W(\lambda).$$

Alternatively,

$$(I_n - \lambda A) \Delta W_1(\lambda) = \lambda B \, \Delta W_2(\lambda). \tag{P.35}$$

Also from (P.33) and (P.34), it follows that $\Delta W_2 = K \Delta W_1$ and hence from (P.35) we have

$$[I_n - \lambda(A + BK)] \Delta W_1 = 0.$$

This means that $\Delta W_1 = 0$ and $W_1^{\Psi}(\lambda) = W_1(\lambda)$. Consequently, $W_2 \Psi(\lambda) = W_2(\lambda)$ and thus $W^{\Psi}(\lambda) = W(\lambda)$.

The formula of (P.33) shows how to determine the function $\Psi(\lambda)$ resulting from the formulae of (3.21) and (3.22) with the given transfer function $W(\lambda)$. The lemma is proved.

3.P.11 *Proof of the Theorem 3.3.3*

It may be observed that the conditions of the theorem 3.3.2 have been fulfilled. In view of the inequality in (3.33) the matrices $G(\lambda)$ and $L(\lambda)$ do not have features for $|\lambda|=1$. Further we have from (3.29)

$$R_2^{-1}(\lambda) = [I_n + \lambda K W_1^0(\lambda) B]^{-1}$$

$$= I_n - K\left([W_1^0(\lambda)]^{-1} + \lambda BK\right)^{-1}\lambda B$$

$$= \lambda B - K(I_n - \lambda A)^{-1}\lambda B.$$

For $|\lambda|=1$ this function does not have any feature as conditioned by the inequality of (3.33). Hence, when $|\lambda|=1$, $\det R_2(\lambda) \neq 0$. Thus the conditions of the theorem 3.3.2 are verified and verification of the theorem 3.3.3 follows the theorems of 3.3.1 and 3.3.2.

3.P.12 *Proof of the Lemma 3.3.3*

Using the formulae of (3.17), (3.29) and (3.30) we have

$$R_2^{\nabla}(\lambda)G(\lambda)R_2(\lambda) = \left\| \begin{matrix} W_1^0(\lambda)\lambda B \\ I_m + KW_1^0(\lambda)\lambda B \end{matrix} \right\|^{\nabla} N \left\| \begin{matrix} W_1^0(\lambda)\lambda B \\ I_m + KW_1^0(\lambda)\lambda B \end{matrix} \right\|. \tag{P.36}$$

For $N > 0$ the identity of (2.20) is true. Applying this in (P.36), we have

$$R_2^{\nabla}(\lambda)G(\lambda)R_2(\lambda) = [I_m + KW_1^0(\lambda)\lambda B - KW_1^0(\lambda)\lambda B]^{\nabla}\Gamma$$

$$\times[I_m + KW_1^0(\lambda)\lambda B - KW_1^0(\lambda)\lambda B] + [W_1^0(\lambda)\lambda B]^{\nabla}H$$

$$\times[W_1^0(\lambda)\lambda B] - [\lambda AW_1^0(\lambda)\lambda B + \lambda BKW_1^0(\lambda)\lambda B$$

$$+\lambda B]^{\nabla}H[\lambda AW_1^0(\lambda)\lambda B + \lambda BKW_1^0(\lambda)\lambda B + \lambda B]$$

$$= \Gamma + [W_1^0(\lambda)\lambda B]^{\nabla}HW_1^0(\lambda)\lambda B - [W_1^0(\lambda)\lambda B]^{\nabla}HW_1^0(\lambda)\lambda B = \Gamma.$$

The formula of (3.42) is established. Thus $\pi(\lambda) = \Gamma^{1/2}$. In a similar manner we get from (3.24) (using (3.17), (3.18),(3.29) and (3.30)) as shown below

$$\Gamma^{1/2}R(\lambda) = \left\| \begin{matrix} W_1^0(\lambda)\lambda B \\ I_m + KW_1^0(\lambda)\lambda B \end{matrix} \right\|^{\nabla} N \left\| \begin{matrix} W_1^0(\lambda)\lambda BK(I_n - \lambda A)^{-1} \\ KW_1^0(\lambda) \end{matrix} \right\|$$

$$+\left\| \begin{matrix} W_1^0(\lambda)\lambda B \\ I_m + KW_1^0(\lambda)\lambda B \end{matrix} \right\|^{\nabla} N \left\| \begin{matrix} (I_n - \lambda A)^{-1} \\ 0 \end{matrix} \right\|$$

$$= \left\| \begin{matrix} W_1^0(\lambda)\lambda B \\ I_m + KW_1^0(\lambda)\lambda B \end{matrix} \right\|^{\nabla} N \left\| \begin{matrix} W_1^0(\lambda) \\ KW_1^0(\lambda) \end{matrix} \right\|.$$

Considering the relations in (2.37), the following computation is carried out :

$$\Gamma^{1/2}R(\lambda) = [\lambda W_1^0(\lambda)B]^{\nabla} QW_1^0(\lambda) + [\lambda W_1^0(\lambda)B]^{\nabla} SKW_1^0(\lambda)$$

$$+[I_m + \lambda K W_1^0(\lambda)B]^{\nabla} S^* W_1^0(\lambda) + [I_m + \lambda K W_1^0(\lambda)B]^{\nabla} RKW_1^0(\lambda)$$

$$= [\lambda W_1^0(\lambda)B]^{\nabla} (K^*\Gamma K + H - A^*HA)W_1^0(\lambda)$$

$$-[\lambda W_1^0(\lambda)B]^{\nabla} (K^*\Gamma + A^*HB)KW_1^0(\lambda) - (\Gamma K + B^*HA)W_1^0(\lambda)$$

$$-[\lambda W_1^0(\lambda)B]^{\nabla} K^*(\Gamma K + B^*HA)W_1^0(\lambda) + (\Gamma - B^*HB)KW_1^0(\lambda)$$

$$+[\lambda W_1^0(\lambda)B]^{\nabla} K^*(\Gamma - B^*HB)KW_1^0(\lambda) = [\lambda W_1^0(\lambda)B]^{*} HW_1^0(\lambda)$$

$$-\left\{[\lambda W_1^0(\lambda)B]^{\nabla} \lambda^{-1}(A+BK)^*H - \lambda^{-1}B^*H\right\}\lambda(A+BK)W_1^0(\lambda)$$

$$= [\lambda W_1^0(\lambda)B]^{\nabla} HW_1^0(\lambda)] - [\lambda W_1^0(\lambda)B]^{\nabla} H\lambda(A+BK)W_1^0(\lambda)$$

$$= [\lambda W_1^0(\lambda)B]^{\nabla} H[I_n - \lambda(A+BK)]W_1^0(\lambda) = [\lambda W_1^0(\lambda)B]^{\nabla} H.$$

Hence the formula of (3.43) is established and the lemma proved.

3.P.13 ***Proof of the Lemma 3.3.4***

(1) It follows from (3.52) that

$$W(\lambda) = \begin{Vmatrix} \alpha(\lambda)/g(\lambda) \\ \beta(\lambda)/g(\lambda) \end{Vmatrix},$$

where $g(\lambda)$ is the characteristic polynomial of the control system. The F.R.F's $\alpha(\lambda)g^{-1}(\lambda)$ and $\beta(\lambda)g^{-1}(\lambda)$ will be stable if the equation (3.56) is solvable in the class of stable F.R.F 's $W_1(\lambda)$ and $W_2(\lambda)$. Because of the nonreductibility of the polynomials $\alpha(\lambda)$ and $\beta(\lambda)$, the stability of polynomial $g(\lambda)$ is assumed, *i.e.* the regulator of (3.48) is stabilizing. The converse is obvious.

(2) Reasons similar to those given for introducing the formula (3.15) (vide section 3.P.8) will be used to establish the formula of (3.59).

(3) Let $W(\lambda) = \begin{Vmatrix} W_1(\lambda) \\ W_2(\lambda) \end{Vmatrix}$ be an arbitrary stable F.R.F. that satisfies (3.53). Since

$W^0(\lambda) = \begin{Vmatrix} W_1^0(\lambda) \\ W_2^0(\lambda) \end{Vmatrix}$, and it also satisfies (3.53), then the functions $\Delta W_1 = W_1(\lambda) - W_1^0(\lambda)$ and

$\Delta W_2 = W_2(\lambda) - W_2^0(\lambda)$ satisfy the relation $a(\lambda)W_1 - b(\lambda)\Delta W_2 = 0$. Hence,

$$\frac{\Delta W_1(\lambda)}{\Delta W_2(\lambda)} = \frac{b(\lambda)}{a(\lambda)},$$

i.e., $\nabla W_1 = \Psi(\lambda)b(\lambda)\nu^{-1}(\lambda)$

and $\nabla W_2 = \Psi(\lambda)a(\lambda)\nu^{-1}(\lambda)$

where $\Psi(\lambda)$ is F.R.F. The expression in (3.61) follows from this. As the polynomials $a(\lambda)\nu^{-1}(\lambda)$ and $b(\lambda)\nu^{-1}(\lambda)$ are nonreductible and the F.R.F.'s $W_1(\lambda), W_2(\lambda), W_1^0(\lambda)$ and $W_2^0(\lambda)$ are stable, the stability of the F.R.F. will then follow from (3.61). The lemma 3.3.4 is proved.

3.P.14 ***Proof of the Theorem 3.4.1***

The following equality is true for the control strategy $\tilde{U}^\infty(\cdot)$ determined with the help of (4.8) :

$$V_t(x_t) = \mathbf{M}\{V_{t+1}[X_{t+1}(x_t, \tilde{U}_t, v_t) + q_t(x_t, \tilde{U}_t)] \mid y^t\} - r_t.$$

Taking average of this equality and summing up from $t = 0$ to $t = T$, we get

$$V_0(x_0) = \mathbf{M}V_{T+1}(x_{T+1} + \sum_{t=0}^{T} \mathbf{M}q_t(x_t, \tilde{U}_t) - \sum_{t=0}^{T} \mathbf{M}r_t.$$

Taking into account that $V_t(\cdot) \geq 0$ and $\lim_{t\to\infty} \; t^{-1}MV_t(x_t) = 0$ we get from this

$$\overline{\lim_{T\to\infty}} \; T^{-1} \sum_{t=0}^{T} \mathbf{M}q_t(x_t, \; \tilde{U}_t) = \overline{\lim_{t\to\infty}} T^{-1} \sum_{t=0}^{T} \mathbf{M}r_t. \tag{P.37}$$

Moreover, for any admissible strategy $U^\infty(\cdot)$ the following inequality is satisfied :

$$\mathbf{M}V_t(x_t) \leq \mathbf{M}\{V_{t+1}[X_{t+1}(x_t, U_t, v_t)] + q_t(x_t, \; U_t)\} - \mathbf{M}r_t.$$

Here we have assumed that the values of Mr_t do not depend on the choice of the admissible control strategy. The inequality is time averaged and taking into account that $\lim_{t\to\infty} t^{-1}V_t(x_t) = 0$ (as per the assumption it is valid for every admissible control strategy), we get

$$\overline{\lim_{\tau\to\infty}} \; T^{-1} \sum_{t=0}^{T} \mathbf{M}q_t(x_t, U_t) \geq \overline{\lim_{\tau\to\infty}} \; T^{-1} \sum_{t=0}^{T} \mathbf{M}r_t.$$

Comparing the inequality thus obtained with the equality in (P.37), optimality of the control strategy $\tilde{\mathbf{U}}^\infty(\cdot)$ is affirmed. The theorem is proved.

3.P.15 ***Proof of the Theorem 3.4.2***

It can be easily shown that the expression

$$V_{t+1}(Ax_t + Bu_t + v_{t+1}) + \begin{pmatrix} x_t \\ u_t \end{pmatrix}^* N \begin{pmatrix} x_t \\ u_t \end{pmatrix} + (Ax_t + Bu_t + v_{t+1})^* \xi_{t+1} \tag{P.38}$$

attains the least (w.r.t. U_t) value when the value of u_t is equal to that determined by the formula (4.24). Putting (4.24) in (P.38) we rewrite this expression in the form $x_t^* H x_t + x_t^* \xi_t + r_t$. This is possible if H satisfies the equation (2.17), r_t has the form as in (4.22) and ξ_t and v_{t+1} are related as

$$\xi_t = A_0^*[\xi_{t+1} + 2Hv_{t+1}]. \tag{P.39}$$

Because of the stability of the matrix (4.19), it follows from (P.39) that the formula (4.20) determines the stationary process bounded by the mean square. For a control (4.24) the equation of (4.11) has the expression

$$x_{t+1} = A_0 x_t - B(B^* HB + R)^{-1} B^*(Hv_{t+1} - \xi_{t+1}/2).$$

The inequality follows

$$\overline{\lim_{T\to\infty}}\ T^{-1} \sum_{t=0}^{T-1} \mathbf{M} \mid x_t \mid^2 < \infty.$$

The latter inequality and (4.24) together lead to the inequality of (4.1), *i.e.*, the formulated control strategy is admissible. The theorem 3.4.2 is proved.

3.P.16 ***Proof of the Theorem 3.4.3***

It will be shown that the average values of the noise signals (4.29) do not depend on the choice of the admissible control strategy. It may be observed that, because of the stability of the polynomial det $d(\lambda)$, the stationary process v^∞ by the relationship (4.12) may be expressed in the form

$$v_t = \sum_{t'=0}^{\infty} h(t') w_{t-t'}, \tag{P.40}$$

where $h(t)$ is the weighting function of the filter (4.12) :

$$h(t) = \frac{1}{2\pi i} \oint \lambda^{-1} d^{-1}(\lambda) e(\lambda) \frac{d\lambda}{\lambda}. \tag{P.41}$$

The formula (P.40) is useful in determining the optimal value v_s, $s \geq t$ in respect of known values of w^t. Actually, if we assume

$$v_s = \sum_{t'=0}^{s-t-1} h(t') w_{s-t'} + \sum_{t'=s-t}^{\infty} h(t') w_{s-t'} \tag{P.42}$$

and consider centering and independence of random quantities w_t, then we can locate from (P.42)

$$\mathbf{M}\{v_s \mid w^t\} = \sum_{t'=s-t}^{\infty} h(t') w_{s-t'}, \tag{P.43}$$

since

$$\mathbf{M}\{w_{s-t'} \mid w^t\} = \mathbf{M} w_{s-t'} = 0, \quad t' = 0. \quad \dots, \quad s-t-1.$$

For the covariance of the estimation error we have, from (P.43) and (P.42), the following expression

$$\mathbf{M}\{[v_s - \mathbf{M}(v_s \mid w^t)]\, [v_s{'} - \mathbf{M}(v_s{'} \mid w^t)]^* \mid w^t\}$$

$$= \sum_{t'=0}^{\min(s,s')-t-1} h(t') R_w h^*(t'). \tag{P.44}$$

We consider that by virtue of (4.23) and (4.20),

$$\pi_t = 2 \sum_{l=t}^{\infty} (A_0^*)^{l-t} H v_{l+1},$$

and with the help of (P.43) and (P.44) we can find out

$$\mathbf{M}(\pi_t \mid w^t) = 2 \sum_{l=t}^{\infty} (A_0^*)^{l-t} H \mathbf{M}(v_{l+1} \mid w^t)$$

$$= 2 \sum_{l=t}^{\infty} (A_0^*)^{l-t} H \sum_{t'=l+1-t}^{\infty} h(t') w_{l+1-t'}. \tag{P.45}$$

$$\mathbf{M}\{[\pi_t-\mathbf{M}(\pi_t\mid w^t)]\,[\pi_t-\mathbf{M}(\pi_t\mid w^t)]^*\mid w^t\}$$

$$=4\sum_{l=t}^{\infty}\sum_{l'=t}^{\infty}(A_0^*)^{l-t}H\sum_{t'=0}^{\min(l,l')-t}h(t')R_w=h^*(t')Ha_0^{l'-t}. \qquad \text{(P.46)}$$

This means that the mean value and the covariance of the random variables $\{\pi_t\}$ do not depend on the choice of the admissible control strategy. Again, as because σ-algebras of the random quantities w^t and y_t coincide (we may recall that $y_t=x_t, w^t$), then

$$\mathbf{M}\{[\pi_t-\mathbf{M}(\pi_t\mid y_t)]^*BLB^*[\pi_t-\mathbf{M}(\pi_t\mid y_t)]\mid y_t\}$$

$$=\mathrm{Sp}\,(BLB^*)^{1/2}\mathbf{M}\{[\pi_t-\mathbf{M}(\pi_t-\mathbf{M}(\pi_t\mid y_t)]$$

$$\times[\pi_t-\mathbf{M}(\pi_t\mid y_t)]^*\mid y_t\}\,(BLB^*)^{1/2}. \qquad \text{(P.47)}$$

The second item in (4.29) does not depend on the choice of the control strategy. Further

$$\mathbf{M}\{\pi_t^*BLB^*\pi_t\mid y_t\}=[\mathbf{M}(\pi_t\mid y_t)]^*BLB^*\mathbf{M}(\pi_t\mid y_t)$$

$$+\mathbf{M}\{[\pi_t-\mathbf{M}(\pi_t\mid y_t)]^*BLB^*[\pi_t-\mathbf{M}(\pi_t\mid y_t)]\mid y_t\}.$$

Consequently because of (P.45), (P.47) and (P.46) the quantity $\mathbf{M}\{\pi_t^*BLB^*\pi_t\mid y_t\}$ will also be independent of the choice of the control strategy. Lastly the following quantities will not depend on the choice of the control strategy by virtue of the formulae (P.43) and (P.44)

$$\mathbf{M}\{v_{t+1}^*Hv_{t+1}\mid y_t\}=\mathrm{Sp}\,H^{1/2}h(0)R_wh^*(0)H^{1/2},$$

$$\mathbf{M}\{v_{t+1}^*\xi_{t+1}\mid y_t\}=2\sum_{l=t}^{\infty}(A_0^*)^{l-t}\mathbf{M}v_{t+1}^*Hv_{l+1}$$

$$=2\sum_{l=t}^{\infty}(A_0^*)^{l-t}[\mathbf{M}(v_{t+1}\mid y_t]^*H\mathbf{M}(v_{l+1}\mid y_t)$$

$$+2\sum_{l=t}^{\infty}(A_0^*)^{l-t}\mathbf{M}\{[v_{l+1}-\mathbf{M}(v_{l+1}\mid y_t)]^*$$

$$\times H[v_{l+1}-\mathbf{M}(v_{l+1}\mid y_t)]y_t\}.$$

This shows that the mean values of the random quantity (4.29) do not depend on the choice of the control strategy. It may be observed that the regulator (4.31) ensures fulfilment of the inequality (4.16) independent of the initial conditions of the control system. For the control system of (4.31) the equation of (4.11) takes the following expression

$$x_{t+1}=A_0x_t-B(B^*HB+R)^{-1}B^*\mathbf{M}(\pi_t\mid y_t)+v_{t+1}. \qquad \text{(P.48)}$$

In view of the stability of the matrix A_0, the formula (P.45) determines the process bounded by mean square. Again because of (P.48), the process x^∞ will also be bounded by the mean square independent of initial conditions. The limit in the mean square for the processs u^∞ then results from (4.31). This means that the inequality of (4.16) is satisfied independent of the choice of the initial values of the control system.

The identity of the formulae (4.31) and (4.32) is obtained when (P.45) is taken into account.

The functional in the optimal control easily results from the relationship of (4.30). The theorem is proved.

3.P.17 *Proof of the Theorem 3.4.4*

When the condition in (4.33) is satisfied, it results from (P.41) that $h(t) = e_t, t = 0, \dots, q$, and $h(t) \equiv 0$ for other values of t. Hence the formula in (P.43) is transformed into

$$\mathbf{M}(v_s \mid w^t) = \sum_{t'=s-q}^{\min(t,s)} e_{s-t'-1} w_{t'}.$$

It enables us to transform the formula in (P.45) into

$$\mathbf{M}(\pi_t \mid w^t) = 2 \sum_{s=0}^{q-1} \left[\sum_{l=0}^{q-s-1} (A_0^*)^l H e_{l+s+1} \right] w_{t-s} = \gamma(\nabla) w_t. \tag{P.49}$$

Here the polynomial $\gamma(\lambda)$ is determined by the formulae (4.36) and (4.37). Thus, the regulator (4.31) has the expression

$$u_t = -(B^*HB + R)^{-1} [(B^*HA + S^*)x_t + B^*\gamma(\nabla)w_t]. \tag{P.50}$$

The equation (4.11), when (4.33) is taken into consideration, can be rewritten as

$$e(\nabla)w_t = x_t - Ax_{t-1} - Bu_{t-1}. \tag{P.51}$$

Becuase of (4.35), we can express the following equality from (P.51)

$$\det\,[e^*(\nabla)e(\nabla)]\,w_t = \tilde{e}(\nabla)e^*(\nabla)\,[x_t - Ax_{t-1} - Bu_{t-1}]. \tag{P.52}$$

The operation $\det\,[e^*(\nabla)e(\nabla)]$ is applied to the relationship (P.50). Again considering (P.52) we get the formula (4.34) that determines the admissible control strategy. It results from the theorem 3.4.3 that this strategy is optimal even in a wider class of control strategy $\mathbf{U}^a$ which uses the previous values of the noise signals. The theorem 3.4.4. is proved.

3.P.18 *Proof of the Theorem 3.4.5*

The proof does not differ much from the proof of the theorem 3.4.4. Let us assume that the noise signals v_0^t at time t are known and available for synthesis of control strategy.

The Bellman-Liapunov function $V_t(x, \xi_t) = x^*Hx + \xi_t^*x$ is now examined. Here ξ_t is defined by the formula (4.20), H is the non-negative solution of the Lur'e equation (2.17), for which the matrix (4.19) is stable (such a solution of H exists by virtue of the theorem 3.P.1). Using the identity transformations from (4.15) and (4.11), we can show the validity of the relationship

$$q(x_t, u_t) = V_t(x_t) - V_{t+1}(x_{t+1}) + v_{t+1}^*(\xi_{t+1} - Hv_{t+1})$$

$$- \mid 2Hv_{t+1} + \xi_{t+1} \mid^2_{B\Gamma^{-1}B^*} + \mid Kx_t - u_t - \Gamma^{-1}B^*(2Hv_{t+1} + \xi_{t+1}) \mid^2_\Gamma .$$

Here the following notation has been used

$$\mid f \mid^2_\Gamma = f^*\Gamma f,$$

$$\mid f \mid^2_{B\Gamma^{-1}B^*} = f^*B\Gamma^{-1}B^*f.$$

It has been also considered that by virtue of (4.20)

$$\xi_t = A_0^*[\xi_{t+1} + 2HV_{t+1}].$$

From (4.20) results the formula

$$2Hv_{t+1} + \xi_{t+1} = 2\sum_{s=t}^{\infty}(A_0^*)^{s-t}HV_{s+1},$$

which can be used to rewrite the above relationship in the form shown below :

$$q(x_t,\quad u_t) = V_t(x_t) - V_{t+1}(x_{t+1}) + \zeta_t + \eta_t$$

$$+ | Kx_t - u_t - 2\Gamma^{-1}B^* \sum_{s=t}^{\infty}(A_0^*)^{s-t}H\mathbf{M}(v_{s+1} | F^t) |_\Gamma^2,$$

where

$$\zeta_t = v_{t+1}^*(Hv_{t+1} + \xi_{t+1}) - 4\left|\sum_{s=0}^{\infty}(A_0^*)^s H v_{s+t+1}\right|_{B\Gamma^{-1}B^*}^2$$

$$+4\sum_{s=0}^{\infty}(A_0^*)^s H[v_{s+t+1} - \mathbf{M}(v_{s+t+1} | F^t)] |_{B\Gamma^{-1}B^*}^2,$$

$$\eta_t = 2\left[u_t - Kx_t + \sum_{s=0}^{\infty}B^*(A_0^*)^s H\mathbf{M}(v_{s+t+1} | F^t)\right]^* \sum_{s=0}^{\infty}B^*(A_0^*)^s$$

$$\times H[v_{s+t+1} - \mathbf{M}(v_{s+t+1} | F^t)]$$

and F^t is the σ-algebra resulting from random quantities x^t, u^t and w^t. Thus,

$$t^{-1}\sum_{s=0}^{t-1}q(x_s,\quad u_s) = t^{-1}[V_{t+1}(x_{t+1}) - V_t(x_t)] + t^{-1}\sum_{s=0}^{t-1}\zeta_s$$

$$+t^{-1}\sum_{s=0}^{t-1}\eta_s + t^{-1}\sum_{s=0}^{t-1} | Rx_s - u_s$$

$$-2\Gamma^{-1}B^* \sum_{s'=s}^{\infty}(A_0^*)^{s'-s}H\mathbf{M}(v_{s'+1} | F^s) |_\Gamma^2.$$

We shall show that the random quantities $\{\zeta_t\}$ do not depend on the choice of the admissible control strategy and for the random quantities η_t (with probability 1) the limiting equality $\lim_{t\to\infty} \quad t^{-1}\sum_{s=0}^{t-1}\eta_s = 0$. is satisfied. Actually by virtue of the centering and independence of the random quantities w^t the following is satisfied

$$\mathbf{M}(w_k | F^t) = \begin{cases} w_k, & k \le t, \\ 0, & k > t. \end{cases}$$

Hence, because of the condition 1 of the theorem, we have

$$\mathbf{M}(v_{s+t+1} | F^t) = \sum_{j=0}^{n} e_j\mathbf{M}(w_{s+t+1-j} | F^t)$$

$$= \sum_{j=0}^{n} e_j w_{s+t+1-j}\theta(j - s - 1).$$

Here n = degree of the polynomial $e(\lambda)$,

$$\theta(s)=\begin{cases}1, & s\ge 0,\\ 0, & s<0.\end{cases}$$

Thus, the random quantity ξ_t along with the random quantity ζ_t does not depend on the choice of the admissible control strategy. Further, the random quantity $v_{s+t+1}-M(v_{s+t+1}\,|\,F^t)$ appears as the linear combination of the random quantities $w_{t+1}, w_{t+2},\ldots$. Consequently, the random variables

$$\eta'_t = u_t - Kx_t + \sum_{s=0}^{\infty} B^*(A_0^*)^s H\mathrm{M}(v_{s+t+1}\,|\,F^t)$$

$$= u_t - Kx_t + \gamma(\nabla)w_t$$

and

$$\eta''_t = \sum_{s=0}^{\infty} B^*(A_0^*)^s H[v_{s+t+1}-\mathrm{M}(v_{s+t+1}\,|\,F^t)]$$

$$= \sum_{s=0}^{\infty}\left[\sum_{j=0}^{n} B^* e_j^*(A_0^*)^{s+j} H w_{t+s+1}\right]$$

$$= \sum_{s=0}^{\infty} \tilde{\gamma}_s w_{t+s+1}$$

are stochastically independent. Here $\gamma(\lambda)$ is the polynomial of finite degree with matrix coefficients γ_j. Considering the fact that the matrix coefficients $\tilde{\gamma}_s$ decay exponentially to the null matrix (when $s\to\infty$), it can be shown that the random quantities $\eta_t^{\cdot}$ have second moments evenly bounded along t. Existence of such moments for the random quantities $\eta_t^{\cdot}$ results from the determination of the control strategies. For random quantities $\eta_t = \eta'_t\eta''_t$. Hence it can be easily shown that the conditions of the theorem 2.P.3 are satisfied and so $\lim_{t\to\infty} t^{-1}\sum_{s=0}^{t-1}\eta_s = 0$ with probability 1. In an analogous manner we can show the existence of the limit $\lim_{t\to\infty} t^{-1}\sum_{s=0}^{t-1}\xi_s = \xi$ (with probability 1). It does not depend on the choice of the control strategy as the random quantity ξ_s does not depend on this strategy. Lastly, we observe that, because of the admissible control strategy $\lim_{t\to\infty} t^{-1}V_t = 0$ with probability 1, we have

$$\overline{\lim_{t\to\infty}}\, t^{-1}\sum_{s=0}^{t-1} q(x_s,\ \ u_s) = \xi + \overline{\lim_{t\to\infty}}\ \ t^{-1}\sum_{s=0}^{t-1} |Kx_s - u_s$$

$$-2\Gamma^{-1}B^*\sum_{s'=0}^{\infty}(A_0^*)^{s'} H\mathrm{M}(v_{s+s'+1}|F^s)|_{\Gamma}^2.$$

The optimality of the strategy is obvious from above, *i.e.*,

$$u_t = Kx_t - 2\Gamma^{\cdot}B^*\sum_{s=0}^{\infty}(A_0^*)^s H\mathrm{M}(v_{t+s+1}\,|\,F^t).$$

Then min $J[U^{\infty}(\cdot), v_0^{\infty}] = \xi$. Using the assertions, as in the proof of the theorem 3.4.4, the linear feedback, thus obtained, can be transformed to a form independent of noise signals. This leads to the relationship in (4.34). (It may be observed that K and Γ can be expressed through H - vide (2.37).) The feedback, thus obtained, gives a stable characteristic polynomial of the closed loop control system. Hence the admissibility of the control strategy resulting from this feedback is obvious. The theorem 3.4.5 is proved.

3.P.19 ***Proof of the Lemma 3.4.1***

The equation of (4.58) and (3.48) is rewritten applying discrete Laplace transform (vide section 3, P.8) :

$$a(\lambda)\hat{y}(\lambda) - b(\lambda)\hat{u}(\lambda) = \hat{v}(\lambda) + w_1(\lambda),$$

$$\alpha(\lambda)\hat{u}(\lambda) - \beta(\lambda)\hat{y}(\lambda) = w_1(\lambda). \tag{P.53}$$

Here $\hat{v}(\lambda)$ is the Laplace transform of the noise v^{∞} and the polynomials $w_1(\lambda)$ and $w_2(\lambda)$ are conditioned by the initial values of the control system. If we assume, for simplicity's sake, zero initial conditions, then $w_1 = 0$, $w_2 = 0$. Solving (P.53) with respect to $\hat{y}$ and $\hat{u}$ and using (4.59), we find out col $(\hat{y}(\lambda), \hat{u}(\lambda)) = W(\lambda)\, \hat{v}(\lambda)$. Applying inverse Laplace transform we get

$$\begin{pmatrix} x_t \\ u_t \end{pmatrix} = \frac{1}{2\pi i} \oint \lambda^{-t} W(\lambda)\hat{v}(\lambda) d\lambda/\lambda.$$

Consequently,

$$\mathbf{M}\begin{pmatrix} x_t \\ u_t \end{pmatrix}\begin{pmatrix} x_t \\ u_t \end{pmatrix}^{*} = \frac{1}{2\pi i} \oint W(\lambda) S_v(\lambda) W^{\nabla}(\lambda) \frac{d\lambda}{\lambda}.$$

It is the same as (4.61).

3.P.20 ***Proof of the Lemma 3.4.2***

Let the pair $\{a(\lambda), b(\lambda)\}$ be stabilizable. The matrix F.R.F. $a^{-1}(\lambda)$ can be expressed in the diagonal form with the help of elementary matrix functions (vide (46))

$$a^{-}1(\lambda) = P_1(\lambda)\tilde{a}^{-1}(\lambda)P_2(\lambda), \tag{P.54}$$

where $\tilde{a}(\lambda)$ is the diagonal matrix with polynomials as its elements in the diagonal, *i.e.*, $\tilde{a}_{ii}(\lambda) = a_i(\lambda)$, $P_1(\lambda)$ and $P_2(\lambda)$ are the polynomial matrices having det $P_1(\lambda) =$ Const. $\neq 0$ and det $P_2(\lambda) =$ Const $\neq 0$. (It means that $P_1^{-1}(\lambda)$ and $P_2^{-1}(\lambda)$ are also polynomial matrices.)

Let λ_0 be the unstable root of the polynomial det $a(\lambda)$ multiplicity of r. It is assumed that the multiplicity of the $a^{-1}(\lambda)b(\lambda)$ pole is less than r. Without the loss of generalisation, it can be considered that $a_i(\lambda_0) = 0$ when $1 \le i \le k$, and $a_i(\lambda_0) \neq 0$ when $i > k$. Since det $a^{-1}(\lambda) = \prod_{i=1}^{n} a_i^{-1}(\lambda)$ the multiplicity of the root λ_0 in $\prod_{i=1}^{k} a_i(\lambda)$ is equal to r.

Let us examine the polynomial matrix

$$Q(\lambda) = P_2(\lambda)b(\lambda) = P_2(\lambda)a(\lambda)\,[a^{-1}(\lambda)b(\lambda)]$$

$$= \tilde{a}^{-1}(\lambda)P_1^{-1}(\lambda)\,[a^{-1}(\lambda)b(\lambda)]. \qquad \text{(P.55)}$$

The first k rows of the matrix $Q(\lambda)_0)$ are linearly dependent. Actually, if it is not so, then the k-th order nonzero minor can be determined in these rows. But then the corresponding minor in the matrix $P_1^{-1}(\lambda)\,[a^{-1}(\lambda)b(\lambda)]$ possessess pole of multiplicity r. Since $P_1^{-1}(\lambda)$ is a polynomial matrix, it means that the multiplicity of the pole λ_0 of the matrix $a^{-1}(\lambda)b(\lambda)$ is not less than r. It contradicts the assumption made. Hence, the first k rows of the matrix $Q(\lambda_0)$ are linearly dependent and hence there exists a nonzero vector d such that

$$d^*Q(\lambda_0) = 0, \quad d^*\tilde{a}_i(\lambda_0) = 0. \qquad \text{(P.56)}$$

Then by virtue of (P.55) for the vector $c = P_2^*(\lambda)\overline{d}$ we have $c^*b(\lambda_0) = 0$. Moreover, by virtue of (P.54), $c^*a(\lambda_0) = d^*P_2(\lambda_0)a(\lambda_0) = d^*\tilde{a}(\lambda_0)P_1^{-1}(\lambda_0) = 0$, *i.e.*, the matrix $a(\lambda_0)a^*(\lambda_0) + b(\lambda_0)b^*(\lambda_0)$ appears to be degenerated, which contradicts the stabilizability condition of the pair $\{a(\lambda), b(\lambda)\}$. This transition from (1) to (2) is shown. Let there exist an invertible matrix $A(\lambda)$ such that $a_0(\lambda) = A^{-1}(\lambda)a(\lambda)$ and $b_0(\lambda) = A^{-1}(\lambda)b(\lambda)$, which are polynomial matrices and let the polynomial det $A(\lambda)$ be unstable. We designate unstable parts of the polynomial det $a(\lambda)$ and det $a_0(\lambda)$ through $a^-(\lambda)$ and $a_0^-(\lambda)$ respectively. Then it is evident that deg $a_0^-(\lambda) <$ deg $a^-(\lambda)$ and $a_0^{-1}(\lambda)b_0(\lambda) = a^{-1}(\lambda)b(\lambda)$. From the expansion of (P.54) we get deg $N_-[a_0^{-1}(\lambda), b_0(\lambda)] < a^-(\lambda)$for the matrix $a_0^{-1}(\lambda)$ and this condition contradicts the assertion (2) of the lemma. Thus the transition from (2) to (3) is established. Let us now consider the assertion (3) of the lemma. For any complex λ_0, $|\lambda_0| = 1$, and non-zero n-vector c, $|c| = 1$, the equation $c^*a(\lambda_0) = 0$ and $c^*b(\lambda_0) = 0$ are satisfied. The following notations are introduced :

$$a(\lambda) = a(\lambda_0) + (\lambda - \lambda_0)a_1(\lambda),$$

$$b(\lambda) = b(\lambda_0) + (\lambda - \lambda_0)b_1(\lambda),$$

$$A(\lambda) = I_n + (\lambda - \lambda_0 - 1)cc^*,$$

$$a_0(\lambda) = a(\lambda) - (\lambda - \lambda_0 - 1)cc^*a_1(\lambda),$$

$$b_0(\lambda) = b(\lambda) - (\lambda - \lambda_0 - 1)cc^*b_1(\lambda).$$

Evidently, $c^*A(\lambda_0) = 0$ and det $A(\lambda) = \lambda - \lambda_0$.

It then gives $A(\lambda)a_0(\lambda) = a(\lambda)$ and $A(\lambda)b_0(\lambda) = b(\lambda)$.

This contradicts the assertion (3) of the lemma, namely,

$$A(\lambda)a_0(\lambda) = a(\lambda) - (\lambda - \lambda_0 - 1)cc^*a_1(\lambda)$$
$$+(\lambda - \lambda_0 - 1)cc^*a(\lambda) - (\lambda - \lambda_0 - 1)^2cc^*a_1(\lambda).$$

But, as because $c^*a(\lambda_0) = 0$, we have $c^*a(\lambda) = (\lambda - \lambda_0)c^*a_1(\lambda)$. Hence, $A(\lambda)a_0(\lambda) = a(\lambda)$ and analogously $A(\lambda)b_0(\lambda) = b(\lambda)$.

The transition from (3) to (1) is established and at the same time the lemma has been proved for a stabilizing pair $\{a(\lambda), b(\lambda)\}$. The version with the controllable pair is proved in a similar manner.

3.P.21 ***Proof of the Theorem 3.4.6***

Let $R(\lambda)$ be a polynomial matrix that satisfies (4.70). In view of the stabilizability of the pair of polynomial matrices $\{a(\lambda), b(\lambda)\}$, condition in (4.70) and the lemma 3.4.2, we have

$$\deg a_-(\lambda) = N_-[a^{-1}(\lambda)b(\lambda)] = N_-[R_1(\lambda)R_2^{-1}(\lambda)]$$

$$= N\left([R_2^*(\lambda)]^{-1},\quad R_1^*(\lambda)\right) \leq \deg r_-(\lambda). \tag{P.57}$$

Here $r_-(\lambda)$ is the negative part of the polynomial det $R_2(\lambda)$. If the pair $\left\{[R_2^*(\lambda)]^{-1}, R_1^*(\lambda)\right\}$ is stabilizable, then the equality of (P.57) is achieved. Since every root of the polynomial $a_-(\lambda)$ in (P.57) appears to be a pole of the function $R_1(\lambda)R_2^{-1}(\lambda)$, then in case of equality (P.57) we must have (4.73) satisfied.

Let $R(\lambda)$ be an arbitrary stable F.R.F. satisfying (4.70) and for which det $R_2(\lambda) \not\equiv 0$. Let $r(\lambda)$ be the product of all denominators of matrix elements of $R(\lambda)$ so that the matrix

$$R'(\lambda) = R(\lambda)r(\lambda) \tag{P.58}$$

and it is a polynomial matrix. Let us designate the negative part of the polynomial det $R'(\lambda)$ by $r'(\lambda)$. Now if the following equality is satisifed

$$\deg r'_-(\lambda) = \deg a_-(\lambda), \tag{P.59}$$

then because of the condition (2) in the lemma 3.4.2 and the relationship of (P.57), we have the pair $\{[R'_2(\lambda)]^*, [R'_1(\lambda)]^*\}$ stabilizable. So the expression in (4.71) exists when $R^0(\lambda) = R'(\lambda)$ and $Q(\lambda) = \text{Im}\, r^{-1}(\lambda)$. If the equality of (P.59) does not exist, then by virtue of the condition (3) of the lemma 3.4.2, we may lower the degree $r'(\lambda)$ and once more repeat the above procedure till the polynomial matrix $R^0(\lambda)$ is found out, for which the equality $deg\, r_-^0(\lambda) = deg\, a_-(\lambda)$ is verified. In that case $[R'(\lambda)]^* = A(\lambda)\,[R^0(\lambda)]^*$, where $A(\lambda)$ is a polynomial matrix, deg det $A(\lambda) > 0$, and the pair $\left([R_2^0(\lambda)]^*, [R_1^0(\lambda)]^*\right)$ is stabilizable. It remains now to prove that the completeness of the family $\mathbf{W}_0(W^0, R)$ is equivalent to the stability of F.R.F's $Q(\lambda)$ and $Q^{-1}(\lambda)$.

Let the family $\mathbf{W}_0(W^0, R)$ be complete. It will be shown that the F.R.F's $Q(\lambda)$ and $Q^{-1}(\lambda)$, as given in (4.71), are stable. Since $W^0(\lambda)$ satisfies (4.69) and its parameter $\Psi(\lambda) \equiv 0$ is a stable F.R.F. then from the completeness of $\mathbf{W}_0(W^0, R)$ results the stability of the F.R.F. $W^0(\lambda)$. The stability of the F.R.F. $R^0(\lambda)\ Q(\lambda)\ \Psi(\lambda)$ results from the stability of any arbitrary F.R.F. $W(\lambda) \in W_0\,(W^0, R)$. If we can show that the stability of the F.R.F. $Q'(\lambda)$ results from the stability of the F.R.F. $R_0(\lambda)\ Q'(\lambda)$ then by virtue of the arbitrariness of $\Psi(\lambda)$ the F.R.F. $Q'(\lambda)$ must be stable. Actually, if it is not so and λ_0 is the unstable pole of the function $Q'(\lambda)$, *i.e.*, $Q'(\lambda) = Q'(\lambda) + (\lambda - \lambda_0)^{-k}\, Q'_2(\lambda)$, $\quad k > 0, \quad Q'_2(\lambda_0) \neq 0$ and

$Q'_1(\lambda)$ is analytical in some neighbourhood of the point λ_0 then $R^0(\lambda_0)$ and $Q'_2(\lambda_0)=0$. It contradicts the stabilizability of the pair $([R_2^0(\lambda)]^*, [R_1^0(\lambda)]^*)$. Thus, $Q'(\lambda)$ (along with it $Q(\lambda)$) is a stable F.R.F. if the F.R.F. $R^0(\lambda)Q'(\lambda)$ is stable. Let C be any arbitrary matrix of dimension $m \times n$. Then the function $W(\lambda) = W^0(\lambda)+R^0(\lambda)C$ satisfies the relationship in (4.63), and consequently, the relationship in (4.69) which possesses a stable $\Psi(\lambda)$. But in the present case $\Psi(\lambda)=Q^{-1}(\lambda)C$. It proves the stability of $Q^{-1}(\lambda)$.

Let $W(\lambda)$ be an arbitrary stable solution of the equation (4.63). It may be observed that (4.69) is satisfied with

$$\Psi(\lambda)=R_2^{-1}(\lambda)\,[W_2(\lambda)-W_2^0(\lambda)]. \qquad \text{(P.60)}$$

Evidently, it is sufficient to assert the validity of the expression

$$W_1(\lambda)=W_1^0(\lambda)+R_1(\lambda)R_2^{-1}(\lambda)\,[W_2(\lambda)-W_2^0(\lambda)].$$

But the function $W(\lambda)-W^0(\lambda)$ satisfies the equation

$$a(\lambda)\,[W_1(\lambda)-W_1^0(\lambda)]-b(\lambda)\,[W_2(\lambda)-W_2^0(\lambda)]=0.$$

Hence we shall affirm the validity of this expression taking into account (4.70). Thus, the expression (4.64) for an arbitrary solution $W(\lambda)$ of the equation (4.63) (with F.R.F. $\Psi(\lambda)$ in the form as in (P.60)) is established. It now remains to show the stability of this function.

Because of (4.71), we have $R_2^0(\lambda)Q(\lambda)\Psi(\lambda)=[W_2(\lambda)-W_2^0(\lambda)]$ from (P.60). This means that the F.R.F. $R_2^0(\lambda)Q(\lambda)\Psi(\lambda)$ is stable. It has been shown above that the stability of the F.R.F. $R_2^0(\lambda)\tilde{Q}(\lambda)$ brings forth the stability of the F.R.F. $\tilde{Q}(\lambda)$. Hence the F.R.F. $Q(\lambda)\Psi(\lambda)$ is stable. But because $[Q(\lambda)]^{-1}$ is a stable F.R.F., stability of the F.R.F. $\Psi(\lambda)$ is established and the theorem is proved.

3.P.22 *Proof of the Theorem 3.4.7*

1. The assertion (1) of the theorem results from the assertions (2) and (3).

(2) For arbitrary matrices a, b the products of which (ab and ba) exist the integrand in (4.61) is transformed, as shown below, taking into account of relations in (4.77) and (4.69) and the equality $\mathrm{Sp}\,\{ab\}=\mathrm{Sp}\,\{ba\}$

$$\mathrm{Sp}\,\{[N^{1/2}W(\lambda)S_v(\lambda)W^\nabla(\lambda)N^{1/2}]\} = \mathrm{Sp}\,\{[\Gamma^\nabla(\lambda)W^\nabla(\lambda)NW(\lambda)\Gamma(\lambda)]\}$$

$$=\mathrm{Sp}\,\{[\Gamma^\nabla(\lambda)\,\{\Psi^\nabla(\lambda)\Pi^\nabla(\lambda)\Pi(\lambda)\Psi(\lambda)+\Psi^\nabla(\lambda)R^\nabla(\lambda)NW^0(\lambda)$$

$$+[W^0(\lambda)]^\nabla NR(\lambda)\Psi(\lambda)+[W^0(\lambda)]^\nabla NW^0(\lambda)\}\,\Gamma(\lambda)]\}$$

$$=\mathrm{Sp}\,\{[\Pi(\lambda)\Psi(\lambda)\Gamma(\lambda)+L(\lambda)]^\nabla\,[\Pi(\lambda)\Psi(\lambda)\Gamma(\lambda)+L(\lambda)]\}$$

$$+\mathrm{Sp}\,\{\{[W^0(\lambda)\Gamma(\lambda)]\}^\nabla NW^0(\lambda)\Gamma(\lambda)-L^\nabla(\lambda)L(\lambda)\}.$$

Here the F.R.F. $L(\lambda)$ is defined by the formulae (4.78), (4.79) and (4.76). The functional (4.61) is transformed, as shown below, after taking into account (4.80) and (4.81) (as done in the proof of the theorem 3.3.2) :

$$J[U^{\infty}(\cdot)] = J_0 + \frac{1}{2\pi i}\oint \mathrm{Sp}\,\{[\Pi(\lambda)\Psi(\lambda)\Gamma(\lambda)$$

$$+L_+(\lambda)\}^{\nabla}[\Pi(\lambda)\Psi(\lambda)\Gamma(\lambda)+L_+(\lambda)]\}\,\frac{d\lambda}{\lambda}. \qquad \text{(P.61)}$$

Here the quantity J_0 is defined by the formula

$$J_0 = \frac{1}{2\pi i}\oint \mathrm{Sp}\,\{L_-^{\nabla}(\lambda)L_-(\lambda)$$

$$+[W^0(\lambda)\Gamma(\lambda)]^{\nabla}NW^0(\lambda)\Gamma(\lambda) - L^{\nabla}(\lambda)L(\lambda)\}\,\frac{d\lambda}{\lambda}$$

and it does not depend on the choice of $\Psi(\lambda)$. Since the first item in (P.61) is non-negative for any choice of $\Psi(\lambda)$, then the least value of the functional $J[U^{\infty}(\cdot)]$ is attained for a choice of $\Psi(\lambda)$ from the condition

$$\Pi(\lambda)\Psi(\lambda)\Gamma(\lambda)+L_+(\lambda)=0. \qquad \text{(P.62)}$$

It is equal to J_0 and thus the formula of (4.77) is proved. Solving the equation (P.62) with respect of $\Psi(\lambda)$, we determine

$$\Psi(\lambda) = -\Pi^{-1}(\lambda)L_+(\lambda)\Gamma^+(\lambda) + \Psi_0(\lambda)\,[I_n - \Gamma(\lambda)\Gamma+(\lambda)]. \qquad \text{(P.63)}$$

Here $\Psi_0(\lambda)$ is an arbitrary F.R.F. of dimension $m\times n$, which has stability of the F.R.F. $\Psi(\lambda)$ $[I_n\ -\Gamma(\lambda)\Gamma+(\lambda)]$ as the only condition imposed on it. The choice of $\Psi_0(\lambda)$ does not affect the magnitude of the functional (P.61). This is because

$\Psi_0(\lambda)\ [I_n - \Gamma(\lambda)\Gamma^+(\lambda)]\ .\Gamma(\lambda) = 0.$

Because of (4.80) and (4.78) we can transform (P.63) as

$$\Psi(\lambda) = \Pi^{-1}(\lambda)\Phi(\lambda) - [\Pi^{\nabla}(\lambda)\Pi(\lambda)]^{-1}\,R^{\nabla}NW^0(\lambda). \qquad \text{(P.64)}$$

Here the function $\Phi(\lambda)$ is defined by the formula (4.83). The optimal transfer function $W^{opt}(\lambda)$ is determined by the formulae of (4.69), (P.64) and has the expression as in (4.82).

3. The formulae of (4.85) can be written as follows :

$$\|\,\beta(\lambda),\ -\alpha(\lambda)\| = p(\lambda)\,[\Phi(\lambda)\,\|\,a(\lambda),\ -b(\lambda)\| - [\Pi^{-1}(\lambda)]^{\Delta}R^{\nabla}(\lambda)N].$$

We recall that $W^0(\lambda)$ and $R(\lambda)$ satisfy the relations in (4.63) and (4.70) and then find out the following by virtue of (4.69) and (P.64) :

$$\|\,\beta(\lambda),-\alpha(\lambda)\|\ W^{opt}(\lambda) = \|\,\beta(\lambda),\ -\alpha(\lambda)\|\ [W^0(\lambda)+R(\lambda)\Psi(\lambda)]$$
$$= p(\lambda)\,\{\Phi(\lambda) - [\Pi^{-1}(\lambda)]^{\nabla}R^{\nabla}(\lambda)NW^0(\lambda)$$
$$-[\Pi^{-1}(\lambda)]^{\nabla}R^{\nabla}(\lambda)NR(\lambda)\Pi^{-1}(\lambda)\,[\Phi(\lambda)$$
$$-[\Pi^{-1}(\lambda)]^{\nabla}R^{\nabla}(\lambda)NW^0(\lambda)]\}. \qquad \text{(P.65)}$$

But because of (4.79) we have

$$[\Pi^{-1}(\lambda)]^{\nabla}R^{\nabla}(\lambda)NR(\lambda)\Pi^{-1}(\lambda) = I_m$$

and the right hand side of (P.65) is equal to zero.

This indicates that $W^{opt}(\lambda)$ satisfies the relationship of (4.64), *i.e.* it is the transfer function of the control system given by (4.58), (4.48) and (4.85).

We introduce the F.R.F's $\tilde{\alpha}(\lambda)$ and $\tilde{\beta}(\lambda)$ using the formula

$$\| \tilde{\beta}(\lambda), -\tilde{\alpha}(\lambda) \| = \Phi(\lambda) \| a(\lambda), -b(\lambda) \| - [R(\lambda)\Pi^{-1}(\lambda)]^{\Delta} N$$

and establish its stability. It results from (P.65) that

$$\| \tilde{\beta}(\lambda), -\tilde{\alpha}(\lambda) \| W^{opt}(\lambda) = 0. \tag{P.66}$$

Using (P.64), we get $\Pi(\lambda)\,\Psi(\lambda) = \| \tilde{\beta}(\lambda) - \tilde{\alpha}(\lambda) \| .W^{o}(\lambda)$.

Again this along with (4.69) and (P.66) gives $0 = \| \tilde{\beta}(\lambda), -\tilde{\alpha}(\lambda) \| W^{o}(\lambda) + \| \tilde{\beta}(\lambda), -\tilde{\alpha}(\lambda) \| R(\lambda)\,\Psi(\lambda) = [\Pi(\lambda) + \| \tilde{\beta}, -\tilde{\alpha} \| R]\Psi$, *i.e.*,

$$\| \tilde{\beta}(\lambda), -\tilde{\alpha}(\lambda) \| R(\lambda) = -\Pi(\lambda). \tag{P.67}$$

We determine $\| \tilde{\beta}, -\tilde{\alpha} \| = \| 0, -\Pi \| \, \| W^{opt}, R \|^{-1}$ using (P.66) and (P.67). Since $\Pi(\lambda), W^{opt}(\lambda)$ and $R(\lambda)$ are stable F.R.F.'s then it is sufficient to establish the stability of the F.R.F. $(\det \| W^{opt},\ R \|)^{-1}$. We have

$$\det \| W^{opt},\ R \| \det \begin{Vmatrix} W_1 & R_1 \\ W_2 & R_2 \end{Vmatrix} = \det R_2 \det [W_1 - R_1 R_2^{-1} W_2].$$

But we consider (4.70) and (4.63) to find out $\det [W_1 - R_1 R_2^{-1} W_2] = \det [W_1 - a^{-1} b W_2]$ $= \det a^{-1} \det [aW_1 - bW_2] = \det a^{-1}$ and $(\det || W^{opt}(\lambda),\ R(\lambda) ||)^{-1} = \det a(\lambda)\, [\det R_2(\lambda)]^{-1}$.

Because of (4.72) and (4.71), the R.R.F. $\det a(\lambda)\, [\det R_2(\lambda)]^{-1}$ is stable and it proves the stability of the F.R.F's $\tilde{\alpha}(\lambda)$ and $\tilde{\beta}(\lambda)$. They also establish the formulae (4.85). Using them in (P.67), we have

$$g(\lambda) = \det \begin{Vmatrix} a(\lambda) & -b(\lambda) \\ \beta(\lambda) & -\alpha(\lambda) \end{Vmatrix} = \det a(\lambda) \det \alpha(\lambda) - \beta(\lambda) a^{-1}(\lambda) b(\lambda)]$$

$$= \det a(\lambda) \det [\alpha(\lambda) - \beta(\lambda) R_1(\lambda) R_2^{-1}(\lambda)]$$

$$= \det a(\lambda) \det R_2^{-1}(\lambda) \det [\alpha(\lambda) R_2(\lambda) - \beta(\lambda) R_1(\lambda)]$$

$$= p(\lambda) \det a(\lambda) \det \Pi(\lambda)\, [\det R_2(\lambda)]^{-1}.$$

Because of the condition (2) of the theorem 3.4.6, we have the following conditions $R_2(\lambda) = R_2^{0}(\lambda)\, Q(\lambda)$ and $r_-^{0}(\lambda) = Ca_-(\lambda)$ and thus (4.86) is obtained. The theorem 3.4.7 is proved.

3.P.23 *Proof of the Lemma 3.4.3*

According to the theorem 3.2.2. the Lur'e equation of (2.17) possesses a unique non-negative matrix solution and the stable matrix of (4.100) relates to this solution.

Let $\{\tilde{H}_t\}$ denote the sequence of functions which can be determined by the equation (4.99) when $\tilde{H}_0 = 0$. According to the theorem 3.2.2, the sequence $\{\tilde{H}_t\}$ converges to the solution of Lure' equation when $t \to \infty$. Since $H_0 \geq \tilde{H}_0 = 0$, then according to the theorem 3.2.1 :

$$H_1 = f(H_0) \geq f(\tilde{H}_0) = f(0) = Q - SR^{-1}S^{*} = \tilde{H}_1$$

and by induction $H_t \geq \tilde{H}_t$ (for all t).

Because of (2.41),

$$H_{t+1}-\bar{H}_{t+1}=f(H_t)-f(\bar{H}_t)\le\Phi^*(\bar{H}_t)(H_t-\bar{H}_t)\Pi(\bar{H}_t)$$
$$\le\prod_{s=0}^{t}\Phi^*(\bar{H}_s)H_0\Phi(\bar{H}_s),\ \Phi(\bar{H})+A+BK(\bar{H}).$$

Since $\bar{H}_s\to H$ and the matrix $\Phi(\bar{H})$ is stable the $H_t-\bar{H}_t\le 0$ for sufficiently large t, *i.e* $H_t\to H_\infty=H$, as $t\to\infty$. Hence the proof.

3.P.24 *Proof of the Theorem 3.5.1*

When the sequence γ_t (vide 5.31) is operated by $b^*(V)$ (for $t=1,2,\ldots$), then

$$b_-^{(*)}(\nabla)\gamma_t=<b_-^{(*)}(\nabla)z_t(\cdot),\ f_0(\cdot)>=\frac{1}{2\pi i}\oint b_-^{(*)}(\nabla)\lambda^{-t}\frac{b_-^{(*)}(0)}{b_-^{(*)}(\lambda)}$$
$$\times\frac{d\lambda}{\lambda}=\frac{1}{2\pi i}\oint b_-^{(*)}(\lambda)\lambda^{-t}\frac{b_-^{(*)}(0)}{b_-^{(*)}(\lambda)}\frac{d\lambda}{\lambda}=\frac{b_-^{(*)}(0)}{2\pi i}\oint\frac{d\lambda}{\lambda^{t+1}}=0. \qquad \text{(P.68)}$$

Consequently, the coefficients (5.31) coincide with the sequence defined by the relations (5.45) and (5.46).

Let us transform (5.34). We observe that the vectors of (5.47) satisfy the following equation (because of (P.68)) :

$$\Gamma_{t+1}=B\Gamma,\quad t=0,\quad\ldots,\quad 1,\quad \Gamma_0=\text{col}\,(1,\,0,\quad\ldots,\quad 0). \qquad \text{(P.69)}$$

Here the stable matrix has the form as in (5.49).

Introducing the notation $G(\cdot)=$ col $(g_0(\cdot).\ \ \ldots,\ \ g_{p-t}(\cdot))$, the expression of (5.32) can be rewritten as

$$\Gamma_t=B^t\Gamma_0. \qquad \text{(P.70)}$$

Because of (5.34) and (P.71) we have

$$P_{W_\perp}(\pm z_j)=\pm\Gamma_j^*G(\lambda)=\pm(B^j\Gamma_0)^*G(\lambda). \qquad \text{(P.71)}$$

Because of (5.34) and (P.71) we have

$$W_1^{opt}(\lambda)=\sum_{j=0}^{\infty}(W_1^{(j)}B^j\Gamma_0)^*G(\lambda).$$

Since $\sum_{j=0}^{\infty}W_1^{(j)}\lambda^j=\alpha(\lambda)\,[(\lambda)\alpha(\lambda)-b(\lambda)\beta(\lambda)]^{-1}$

then $\sum_{j=0}^{\infty}W_1^{(j)}B^j=\alpha(B)\,[a(B)\alpha(B)-b(B)\beta(B)]^{-1}.$ (P.72)

It can be easily shown that $b_-(\lambda)=\det(B-\lambda I)$ and so $b_-(B)=0$ (because of the Hamilton - Caley theorem). Consequently, $b(B)=0$ and $\sum_{j=0}^{\infty}W_1^{(j)}B^j=a^{-1}(B)$. Thus, $W_1^{opt}(\lambda)=(a^{-1}(B)\Gamma_0)G(\lambda)$. By the same procedure it can be proved that the coefficients $P_{W_\perp}(K_1)$ and $P_{W_\perp}[W_1(\cdot)]$ (on some base $W_\perp$) possess the same form as in the said theorem.

As per (5.41), $F(\lambda)$ happens to be the first component of the optimal transfer function. It remains now to show that **K** is a polyhedron. This results from the stability of the matrix in (P.70) and from the fact that the zero is the interior point of the convex envelope of points $\pm\Gamma_0$, $\pm\Gamma_1$, ..., $\pm\Gamma_{p-1}$ and especially of the set **K**. The remaining assertions are established in section 3.5.4.

3.P.25 ***Proof of the Theorem 3.5.2***

It is almost word by word repetition of the proof of the theorem 3.5.1. The main difference here is that the role of the polynomial $b_-(\lambda)$ is replaced by that of $b_-(\lambda)\, d_-(\lambda)$. The transfer function of the control system possesses the components

$$W_1(\lambda) = \alpha(\lambda)d(\lambda)\,[c(\lambda)g(\lambda)]^{-1} \text{ and } W_2(\lambda) = \beta(\lambda)d(\lambda)\,[c(\lambda)g(\lambda)]^{-1},$$

where $g(\lambda)$ is defined by the formula (5.5). The relationship of (5.10) is replaced by $a(\lambda)\,c(\lambda)\,W_1(\lambda) - b(\lambda)\,c(\lambda)\,W_2(\lambda) = d(\lambda)$,and (5.17) by $W(\lambda) = W^0(\lambda) + b_-(\lambda)\,d_-(\lambda)\Psi(\lambda)$, where $W^0(\lambda)$ and $\Psi(\lambda)$ are analytical functions in D_1, while $W^0(\lambda)$ is fixed, $\Psi(\lambda)$ is arbitrary. The set $\mathbf{W}_\perp$ orthogonal to W_0 has the dimension $p+k = \deg b_-(\lambda)d_-(\lambda)$. The basis here are the functions $p(\lambda)$, $\lambda p(\lambda)$, ..., $\lambda^{p+k-1}p(\lambda)$, $p(\lambda) = [b_-^{(*)}(\lambda)c_-^{(*)}(\lambda)]^{-1}$. When the relationship analogous to (P.72) is introduced, we have to consider that,

$$\alpha(B)d(B)\,[a(B)\alpha(B) - b(B)\beta(B)]^{-1}d(B)a^{-1}(B)$$
$$= a^{-1}(B)\,[a(B)\alpha(B) - b(B)\beta(B)]^{-1}d(B)b(B)\beta(B) = 0.$$

For the optimal transfer function $W_1(\lambda) = F(\lambda)$, the equation for $\alpha(\lambda)$ and $\beta(\lambda)$ (as per conclusion of the theorem) has the form $W_1(\lambda)\,c(\lambda)g(\lambda) - \alpha(\lambda)d(\lambda) = 0$.

4

Adaptive Linear Control Systems with Bounded Noise

The genesis of adaptive control lies in the attempts to solve the practical problems which could not be solved by traditional methods.

The present control theory penetrates into the ideas of optimization. This often reveals that there is a nonconformity between the objective of optimization prescribed for the control system and the achievements. The primary source of such a nonconformity is usually the absence of information about the states, structure and parameters of the control object and the influence of the external environment on the control object. All these are required to be known for optimal control. The noise acting on the control system is usually *a priori* not known and the parameters of the object may vary with time (drift) in an unknown manner, etc. During the initial stages of development of the control theory, similar problems were overcome by lowering the requirements of the control objective, for example, optimal control required to be only on the average and the objective functions contained in them averaging operations. The theory of stochastic (statistical) control was developed in this direction.

The stochastic optimization requires a proper expression for the probabilistic characteristic of the noise acting on the object. Under this condition the proposed theories of optimal control strategy are often critical to the deviations of the computed values of the system parameters and the noise from their real values. (In such cases we call it to be the absence of coarseness (robustness) with respect of small variations of parameters.) Attempts to change over to minimax control, computed on the worst variant, may appear to be inconsistent because of the excessive deterioration of the control objective.

Under these conditions of insufficient *a priori* information about the structure, control object parameters and noise, the idea of its accomplishment in the control process is presented in an interesting manner. Such a control assumes existence of feedback but the feedback principle is used here with a greater load. The accessible observed data for control may serve as the basis for reorganization of the control strategy. The task of reorganizing the control strategy is solved by the method of adaptation (or adjustment) of the control system to *a priori* unknown conditions of functioning. The control systems which work on this principle (and their control strategies) are called adaptive systems.

Adaptive strategies are often (but not always) based on possibility of identification of the control object and (or) the noise present in the system. In this case identification may be carried out to the extent necessary for achieving the control objective.

The subjective understanding of adaptation is as follows : As soon as the control strategy ensures achievement of the control objective, the interpretation of its adaptive action becomes conditional in nature.* It means that one and the same control strategy ensures achievement of the control objective for some class of control object and the noise acting on it. The advantages of adaptive control can be explained only when the methods not possessing above property are well studied.**

Nevertheless, the adaptive control concept appeared to be useful and received a wide recognition amongst specialists. Methods were worked out and adaptive strategies for various classes of adaptation were suggested (related to the *a priori* uncertainty and universality of adaptive control strategy).

Another speciality of the adaptive control, which one can rarely think of, is that through the adaptive control methods all the optimization problems of the classical control theory can be solved. Thus, the solution of a usual time optimal problem or any terminal control problem requires a detailed knowledge of the control object structure and its parameters. The very idea of 'adaptation' or 'identification' contradicts the possibility of ensuring the control objective. The time lost in collection of unachieved information is not made up. So adaptation in such problems can be achieved only at the cost of the optimization in the control objective.

In adaptive control, optimal control objective (in the general sense) can be incorporated only when the functional of the control quality does not depend on the transient response of the control system.

In a sense it means that the control process does not have any time bound. Similar optimization problems were named limiting optimal in Chapter 3.

The control of the transient processes is of considerable interest for the supplements. This problem is little studied within the limits of this summed up theory. It has not been discussed any further.

In this chapter we shall examine the basic concepts related to the adaptive control and discuss a few results of adaptive control when the noise acting on the control object is bounded. The stochastic version of the control problem in case of parametric uncertainty is discussed in Chapter 6.

---

*The adaptive control system is sometimes defined as the one which automatically searches out the necessary control law by the method of analysis of the system response to the current control. If studied formally, this type of concept is meaningless, because the process of searching out the necessary control law is also a control law itself. This can be explained as follows : When a set of "control laws" is separated out, a subsequent division of this set can not be called as the control law from the partitioned set. From this point of view adaptive control is a control law of higher level.

**In non-degenerated control problems, when the optimal strategy is determined by a unique method, the latter, as a rule, ensures optimization of the control objective only for a definite control object.

§ 4.1 Fundamentals of Adaptive Control

4.1.1 *Adaptive Control Strategy*

We shall present the generalised adaptive control problems, distinct from those discussed in Chapters 2 and 3. As before, we consider discretized time, t = 0, 1, ... , and the control process is assumed to be for an infinite period.

The classes of control object and noise will be considered later. We introduce an abstract parameter ξ (uncertainty parameter) from a set Σ. If ξ is fixed, the control object becomes definite and we can express it as

$$x_t = X_t(x_0^{t-1}, u_0^{t-1}, v_1^t, \xi), \quad t = 1, 2, \dots, \tag{1.1}$$

with the initial condition

$$x_0 = X_0(\xi). \tag{1.2}$$

As before, x represents the state variable, $x \in R^n$; u is the control variable, $u \in R^m$; and v is the noise variable, $v \varepsilon R^q$. We assume the sequence $X_0^\infty(\cdot)$ consisting of the function $X_t(\cdot)$ to be fixed, and the functions $X_t(\cdot)$ to be known, whereas the parameter ξ is unknown. Thus, unlike the equations presented in Chapter 1, we have here the control object equation which is dependent on the uncertainty parameter ξ. This is the 'parametrizing' class of the control object. The noise or disturbance v_1^∞ in equations (1.1) and (1.2) are assumed to be unknown (unobservable). It is convenient to assume that v_1^∞ is completely described by the parameter ξ. We call

$$v_1^\infty = v_1^\infty(\xi) \tag{1.3}$$

as a realization of noise.* The control object is characterized by the output. The corresponding equation is

$$y_t = Y_t(x_0^t, u_0^{t-1}, v_1^t, \xi), \quad t = 1, \dots, y_0 = Y_0(x_0). \tag{1.4}$$

Here y is the output variable, $y \in R^l$. y_t is identified with a sensor, *i.e.*, with the collection of data related to the control object and noise, known at the instant t. In (1.4) $Y_0^\infty(\cdot)$ is the fixed sequence of functions of corresponding arguments. The values may be time variant.

Let us assume that the class of control strategy is defined. There will be a differentiation made between the control strategies which depend on ξ and the strategies which are independent of ξ. The latter will be called as realizable control strategies different from the admissible one which can be dependent on ξ. Thus, a class **U**^{r} will be separated out to represent the realizable control strategies. This class consists of the functions

$$U_t = U_t(u_0^{t-1}, y_0^t), \quad t = 1, 2, \dots, \quad U_0 = U_0(y_0)$$

and determines the control u_0^∞ according to the rule

$$u_t = U_t(u_0^{t-1}, y_0^t), \quad t = 1, 2, \dots, u_0 = U_0(y_0) \tag{1.5}$$

---

*The uncertainty parameter ξ may be considered as the aggregate of two parameters ξ^0 and ξ^n. The first parameter 'gives' the control object and the second realizes the noise. In the expressions of (1.1) to (1.3), such a possibility is not excluded. The functions $x_t(\cdot)$ and $v_t(\cdot)$ may depend only on some values expressed through the uncertainty parameter ξ.

when the strategy $U_0^\infty(\cdot) \in U^r$ is fixed, it enables realization of the states $x_0^\infty(\xi)$, control $u_0^\infty(\xi)$ and output $y_0^\infty(\xi)$ related to the control object. Let the functional $J[U_0^\infty(\cdot), \xi]$ be defined on the sets of $x_0^\infty(\cdot)$, $u_0^\infty(\cdot)$ and $y_0^\infty(\cdot)$ as

$$J[U_0^\infty(\cdot), \quad \xi] = \overline{\lim_{t \to \infty}} \quad \tilde{M} q_t(x_0^t(\cdot), u_0^t(\cdot), y_0^t(\cdot), \xi). \tag{1.6}$$

Here $\tilde{M}$ is the ensemble operation (vide section 1.2.3) and $q_0^\infty(\cdot)$ is the sequence of non-negative objective functions.* The relationship in (1.6) determines the functional $J(\cdot)$ as the function of fixed strategy and may be of uncertainty parameter ξ.

The control objective is to ensure the inequality

$$J[U_0^\infty(\cdot), \xi] \le r^{-1}(\xi), \tag{1.7}$$

Here $r(\xi)$ is the level of control quality. It may also be dependent on ξ.

Definition 4.1.1 The control strategy $u_0^\infty(\cdot) \in U^r$ will be called *adaptive with respect to the control objective defined by the functional J with quality level r, in a class* Σ, if for every $\xi \in \Sigma$, the inequality *(1.7)* is satisfied.

We shall again point out the speciality of adaptive control that though the strategy does not make use of the values of uncertainty parameter ξ, it still ensures the control objective for each value of this parameter.

Thus the adaptive control problem means formulation of an adaptive control strategy.**

The content of the formulated problem depends on how extensive is the class of adaptation (*i.e.*, how extensively generalised is the set of uncertainty which the adaptive strategy uses for 'correction'), what form is the quality functional and how is the level of control quality that has been chosen. It is now possible to classify the control objectives as done before in section 1.4.

The controlling system realizing the adaptive control is known as adaptive regulator. The aggregate of the control object and the adaptive control is sometimes known as the adaptive control system.***

4.1.2 *Identification Method in Adaptive Control*

Another interpretation of the adaptive strategy used for a larger group of control problems is to determine the 'structure' of the feedback.

In many cases the uncertainty set Σ is actually broken into direct multiplication of the subsets $\mathcal{T}$ and Σ', $\Sigma = \mathcal{T} \otimes \Sigma'$. Again the set $\mathcal{T}$ is a finite dimensional vector space or its

---

*The ensemble operation $\tilde{M}$ may be applied only to a portion of the set of realizing sequence $x_0^\infty(.)$, $u_0^\infty(\cdot)$ and $y_0^\infty(\cdot)$. (In a stochastic case, such an operation may be a conditioned average.)

**If in (1.1) the functional J and the quality index r are not dependent on ξ then the problem becomes an usual control problem (vide section 1.2.3).

***The concept of adaptive control system lies in the cases where the regulator adaptability appears in its capability to ensure achievement of the control objective with the possible variations of external conditions of the system.

part. The vector parameter $\tau \in \mathcal{T}$ describes the parameters of the control object and then as the parameter $\xi' \in \Sigma'$, it can realize the noise v_1^∞, etc. The class $\mathbf{U}^a$ of the admissible control strategies consisting of the sequence $U_0^\infty(\cdot)$ of functions $U_t = U_t(y_o^t, u_0^{t-1}, \tau)$ is now introduced. The functions may depend on τ, but not on ξ'. Minimization of the function (1.6) in the admissible class of strategies $\mathbf{U}^a$, leads us to the optimal control design problem under 'sufficiently complete' certainty conditions as regards control object parameters. Similar problems have been studied in Chapter 3. τ may be called as an essential parameter (its value is necessary for determination of the optimal strategy in the class $\mathbf{U}^a$), and the parameter ξ' may be called a non-essential parameter.*

Let us assume that the optimization problem is solved in the class $\mathbf{U}^a$ for every $\tau \varepsilon \mathcal{T}$ and the corresponding control u_0^∞ is determined with the help of the optimal strategy given by

$$u_t = U_t(y_0^t, u_0^{t-1}, \tau), \quad U_0^\infty(\cdot) \in \mathbf{U}^a. \tag{1.8}$$

Synthesis of optimal strategy is the basis of the extremal control theory. It is interesting and complex. It has been studied well in Chapter 3. Now the functions $U_t(\cdot)$ in equation (1.8) will be considered to be known. If the parameter τ is not known then the control (1.8) can not be realized. The realizable control will be formulated according to the law

$$u_t = U_t(y_0^t, u_0^{t-1}, \tau_t), \quad t = 1, 2, \ldots, u_0 = U_0(y_0, \tau_0), \tag{1.9}$$

Here τ_t are the tunable parameters to be chosen from the values of the set $\mathcal{T}$.

The equation (1.9) gives the regulator with tunable parameters and the aggregate of the equations (1.1) and (1.9) gives the generalised tunable object. The problem consequently pointed to formation of suitable tuners τ_0^∞. It is now possible to mention admissible and realizable strategies of parameter tuning. So the admissible tuning strategy (*i.e.*, predicting and ensuring achievement of the control objective) may be served by the strategy $\tau_t = \tau$. Realizable tuning strategies will be defined by the sets (sequences) of functions $T_0^\infty(\cdot)$, realizing parameter tuning as per law given by

$$\tau = T_t(y_0^t, u_0^t, \tau_0^{t-1}), \; t = 1, 2, \ldots, \tau_0 = T_0(y_0). \tag{1.10}$$

We assume separation of a set $\mathbf{T}^r$ of realizable parameter tuning strategies.

Then we introduce all the necessary ideas so as to determine the adaptive system.

It may be then observed that since the control strategy $U_0^\infty(\cdot)$ in (1.9) is determined uniquely and within the task of parameter tuning strategy ($T_0^\infty(\cdot)$ in (1.10), the functional of (1.6) can be considered to be a function of the strategy $T_0^\infty(\cdot)$:

$$J[U_0^\infty(\cdot), \xi] = \tilde{J}[T_0^\infty(\cdot), \xi]. \tag{1.11}$$

Definition 4.1.2 The control system *(1.1), (1.9)* and *(1.10)* will be called *adaptive in a class* $\Sigma = \mathcal{T} \otimes \Sigma'$ *with respect to the control objective*

---

*The role of ξ' is played by the elementary event ω in stochastic control problems. Its knowledge is not required for minimization of 'average' control objective functional.

$$\tilde{J}[T_0^\infty(\cdot),\xi] \le r^{-1}(\xi) \tag{1.12}$$

if in (1.10) the realizable strategy ($T_0^\infty(\cdot) \in \mathbf{T}^r$) is used and then the inequality of (1.12) is satisfied for every $\xi \in \Sigma$.

The special optimization problem of adaptive control is presented separately. Here the level of control quality is given by the relationship

$$r^{-1}(\xi) = \inf_{\tilde{U}_0^\infty \in \mathbf{U}^a} J[\tilde{U}_0^\infty(\cdot),\xi], \tag{1.13}$$

where the minimum is attained in the class of admissible control strategies $\mathbf{U}^a$.

The adaptive control strategy given by the definition of 4.1.2 is uniquely determined by the problem of functional J, and classes $\mathbf{U}^a$ and $\mathbf{T}^r$. Hence, such a strategy will be known, for brevity's sake, as (J, $\mathbf{U}^a$, $\mathbf{T}^r$) adaptive in a class Σ.

Here introduction of the classes $\mathbf{U}^a$ not only determines the level of control quality, but it leads the synthesis of realizable control strategy to the synthesis of realizable parameter tuning strategy. Of course, it does not exclude the possibility of the condition $\mathbf{U}^a = \mathbf{U}^r$.

The idea of adaptive control strategy is also applied to other (not necessarily to optimization) control objectives. In this case $r(\xi)$ does not coincide with the maximum possible level of quality control.

In the stochastic variant of the adaptive control, probability 1 may be assigned to the inequality (1.12).

The equation (1.10) in the adaptive control system describes the block of parameter tuner known as adaptor. The aggregate of the equations (1.9) and (1.10) describes the controlling system. When the inequality (1.12) is satisfied for every $\xi \in \Sigma$, this system is naturally known as adaptive in the class Σ (or the adaptive regulator). The control u_0^∞ given by the adaptive optimal regulator is known to be limiting optimal.

Synthesis of the realizable parameter tuning strategies which determine the adaptor functioning is usually based on identification methods. With the help of these methods the tuner τ_0^∞ coinciding with the unknown parameter can be obtained. As a rule the identification methods assume application of recursive estimation procedures.

The basis of most of the identification methods in feedback systems puts forth considerable difficulties, because one of the basic conditions of convergence of tuner is the requirement of a uniform limit (in the time domain) on the states of the control system. Also for the control, with *a priori* uncertainty of control object parameters, this characteristic can not be guaranteed beforehand. Its fulfilment assumes formulation of the control objective. The above speciality of closed loop control systems appeared to be so much important that there was a considerable development of the recursive estimation methods. The examples below will illustrate this.

§ 4.2 Existence of adaptive control strategy in a minimax control problem

Identification technique is usually not applied to problems where the noise in the control object is uncontrolled or does not possess any useful statistical properties (centricity, stationarity, etc.). The answer to the question about the existence of a universal adaptive control strategy is important under such conditions. The example of a linear control object with time invariant (but unknown) coefficients is given below. Here the disturbance is adaptable, arbitrary and bounded, and a universal adaptive control strategy exists.

4.2.1 *Statement of the Problem*

We examine the object with scalar input and output described by the equation

$$a(\nabla)y_t = b(\nabla)u_t + v_t, \quad t = 0, 1, \dots . \tag{2.1}$$

Here $a(\lambda)$, $b(\lambda)$ are the polynomials with scalar coefficients

$$a(\lambda) = 1 + \lambda a_1 + \quad \dots \quad + \lambda^n a_n,$$

$$b(\lambda) = \lambda b_1 + \quad \dots \quad + \lambda^n b_n \tag{2.2}$$

and v_0^∞ is the disturbances of the class $\mathbf{V}_\rho$ determined by the condition

$$| v_t | \leq \rho. \tag{2.3}$$

Here ρ is the given level of possible disturbances.

The control object of (2.1) is realized as a feedback system. It is assumed that the set $\mathbf{U}^r$ of realizable control strategies $U_0^\infty(\cdot)$ is chosen. The control strategies determine the feedback given by

$$u_t = U_t(u_0^{t-1}, y_0^t) \tag{2.4}$$

and ensure fulfilment of the following inequality independent of the choice of initial condition in (2.1) and (2.4)

$$\lim_{t \to \infty}(| y_t | + | u_t |) < \infty. \tag{2.5}$$

It is assumed that for every realizable strategy $U_0^\infty \in \mathbf{U}^r$ the functional $J[y_0^\infty, u_0^\infty]$ is determined for the control system trajectory y_0^∞, u_0^∞. The functional does not depend on the choice of initial conditions of the control system.* The control objective consists of the choice of realizable strategy that ensures fulfilment of the inequality

$$J[y_0^\infty, u_0^\infty] \leq r^{-1}. \tag{2.6}$$

Here the level of control quality is given.

A similar control problem was studied in section 3.5 with the following functional which was minimized in the class of linear stabilizable feedbacks $\mathbf{U}^s$.

$$J = \overline{\lim_{t \to \infty}} | y_t | . \tag{2.7}$$

If the coefficients of the control object (2.1) are known, then theorem 3.5.1 permits synthesis of the optimal strategy.

We are interested in control of the system (2.1) under the conditions that all or a part of the polynomial coefficients (2.2) is not known and the admissible strategies are to be used. As the admissible strategies can be determined by these coefficients, this problem does not appear to be a possible one. Due to this reason, the control strategies are chosen in the class of realizable strategies $\mathbf{U}^r$. Here the feedbacks (2.4) do not make use of the

* The left hand side of the inequality (2.5) is an example of such a functional.

control object (2.1) parameters. In other words, the discussion is now on the formulation of adaptive control strategy in a sense given in the definition 4.1.1.

Let the level of control quality r be chosen from the condition

$$r^{-1} = \inf_{U^{\infty}(\cdot) \in U^{a}} \sup_{v^{\infty} \in V_{\rho}} J[y_0^{\infty}, u_0^{\infty}], \tag{2.8}$$

where U^a is the class of linear strategies for the stabilizable regulators (vide 3.5) given by

$$\alpha(\nabla)u_t = \beta(\nabla)y_t, \tag{2.9}$$

$$\alpha(\lambda) = 1 + \lambda\alpha_1 + \quad \dots \quad +\lambda^p\alpha_p,$$

$$\beta(\lambda) = \beta_0 + \quad \dots \quad +\lambda^p\beta_p. \tag{2.10}$$

We shall now solve the formulated adaptive control problem in the optimal frame. Let τ and δ be the sets of coefficients corresponding to polynomials (2.2) and (2.10),* *i.e.*,

$$\tau = \text{col}\,(a_1, a_2, \quad \dots, \quad b_1, b_2, \quad \dots),$$

$$\delta = \text{col}\,(\alpha_1, \alpha_2, \quad \dots, \quad \beta_0, \beta_1, \quad \dots). \tag{2.11}$$

The quality functional is a function of τ, δ and v_0^{∞}, when the admissible control strategy is fixed

$$J(y_0^{\infty}, u_0^{\infty}) = \tilde{J}(\tau, \delta, v_0^{\infty}). \tag{2.12}$$

Let us designate the quality functional of the control system (2.1) and (2.9) having the 'worst' disturbance through

$$Q(\tau, \delta, \rho) = \sup_{v^{\infty} \in V_{\rho}} \tilde{J}(\tau, \delta, v_0^{\infty}). \tag{2.13}$$

Optimization of this functional in a class of stabilizing regulators (2.9) leads to the function

$$q(\tau, \rho) \min_{\delta \in \Delta} Q(\tau, \delta, \rho). \tag{2.14}$$

Here the minimum is taken from the set Δ of stabilizing regulators (2.9). Thus, the adaptive control problem in a minimax case consists of formulation of the control strategy from the class U^r that ensures fulfilment of the inequality

$$J(y_0^{\infty}, u_0^{\infty}) \le q(\tau, \rho). \tag{2.15}$$

Here, $\tau \in \mathcal{T}$, where $\mathcal{T}$ is the given set of probable control objects (*i.e.*, $\mathcal{T}$ is the class of adaptation).

The following is the solution of the suboptimal control problem, when the inequality is satisfied

$$J(y_0^{\infty}, u_0^{\infty}) \le \rho' q(\tau, \rho) \tag{2.16}$$

for a given $\rho' > 1$.

---

*Depending on the choice of the control object and the regulator, the dimension of these sets may be different. Also the sets of τ and δ will be identified with the control object (2.1) and regulator (2.10).

4.2.2 *Synthesis of an Adaptive Control Strategy*

The realizable control strategy corresponding to the section 4.1.2 will be synthesized in the form

$$\alpha(\nabla, \tau_t)u_t = \beta(\nabla, \tau_t)y_t. \tag{2.17}$$

Here τ_t are the tunable parameters, determinable by the adaptation algorithm; and $\alpha(\lambda, \tau)$ and $\beta(\lambda, \tau)$ are the polynomials

$$\alpha(\lambda, \tau) = 1 + \lambda\alpha_1(\tau) + \ldots + \lambda^p\alpha_p(\tau),$$

$$\beta(\lambda, \tau) = \beta_0(\tau) + \ldots + \lambda^p\beta_p(\tau). \tag{2.18}$$

The aggregate of the coefficients of these polynomials is computed using the formula $\delta(\tau) = \text{argmin}\, Q\,(\tau, \delta, \rho)$, *i.e.*, it is searched in the solution of the corresponding minimax problem*

$$\delta(\tau) = \text{ col } (a_1(\tau), \ldots, a_p(\tau), \beta_0(\tau), \ldots, \beta_p(\tau). \tag{2.19}$$

a) Information sets in the control object parametric space : Let the control object, realizable control strategy and control system initial conditions (2.1) and (2.4) be fixed in such a manner that the sequences y_0^∞ and u_0^∞ are determined as unpredictable functions (vide section 1.2) of disturbance $v\infty$, object parameter τ and the adaptable strategy $U^\infty(\cdot)$. Thus,

$$y_t = y_t[v_0^t, \tau, U_0^t(\cdot)], \; u_t = u_t[v_0^t, \tau, U_0^{t-1}(\cdot)]. \tag{2.20}$$

Let us designate

$$\Phi_{t-1} = (y_{t-1}, \ldots, y_{t-n}, -u_{t-1}, \ldots, -u_{t-n}). \tag{2.21}$$

Then the equation (2.1) may be written in the form

$$y_t + \Phi_{t-1}\tau = v_t. \tag{2.22}$$

Assuming that the realizable strategy is determined by the equation (2.17) and the parameter tuning τ_0^∞, the following sets are introduced in the parametric space with the conditions as shown

$$\mathcal{T}'_t(\varepsilon) = \{\tilde{\tau} \in \mathcal{T} : | \, y_t + \Phi_{t-1}\tilde{\tau} \, | \leq \rho + \varepsilon \, | \, \Phi_{t-1} \, | \}, \tag{2.23}$$

$$\mathcal{T} = \begin{cases} \mathcal{T}_{t-1}, & \text{if } \tau_t \in \mathcal{T}'_t(\varepsilon), \\ \mathcal{T}_{t-1}, \cap \mathcal{T}'_t(0), & \text{if } \tau_t \notin \mathcal{T}'_t(\varepsilon), \end{cases} \tag{2.24}$$

$t = 1, 2, \quad \ldots, \mathcal{T}_0 = \mathcal{T}$ and ε is a non-negaitve parameter.

For $\varepsilon = 0$ the set $\mathcal{T}$ may be treated as the set of systems with respective initial conditions which are conformable with the observations and satisfy the limit (2.3) on the disturbance v_0^∞ in the interval $[0, t]$.

---

*The use of the regulator (2.17) with the tunable parameters makes the assumption that the dependence of $\delta = \delta(\tau)$ coefficients of the optimal regulator (for the control object with the parameter τ) is known for every $\tau \in \mathcal{T}$ and $\tau_t \in \mathcal{T}$ is included for all t. It may be observed that the probable dependence on the parameter ρ is not mentioned for simplicity's sake. For synthesis of an adaptive regulator, the parameter ρ is assumed to be fixed. Moreover, for functionals $J(\cdot)$ of one family y^∞ and u^∞, the function $q(\tau, \rho)$ degenerates: $q(\tau, \rho) = q'(\tau)q''(\rho)$ and the optimal regulator does not depend on the level of disturbance, ρ.

The sets $\mathcal{T}$ can be truly called as *information sets* in the control object parametric space. They are degenerated by the control u_0^∞, observation y_0^∞ and tuning τ_0^∞, all considered at the same instant. Evidently, $\mathcal{T}_0^\infty$ is a non-increasing sequence of sets (polyhedrons, if the initial set $\mathcal{T}_0 = \mathcal{T}$ is a polyhedron).

b) Adaptation algorithm. It has been assumed that the tuner τ_0^∞ is fixed. It is now considered that the tuning parameters τ_t are formed according to the rule

$$\tau_{t+1} = \underset{\tau \in \mathcal{T}_t}{\operatorname{argmin}}\ q(\tau, \rho), \tag{2.25}$$

i.e., from this minimization condition we determine the functions (2.14)*. The relations of (2.25) and (2.24) (adaptation algorithm) recursively determine the sequence τ_0^∞ and $\mathcal{T}_0^\infty$ for a given initial parameter τ_0.

Lemma 4.2.1 *Let $\mathcal{T}$ be a bounded space such that for all $\tau \in \mathcal{T}$, there exists an optimal regulator determined by a set of coefficients (2.19).*

Then for all disturbances and every control object (2.1), determinable by parameter $\tau \in \mathcal{T}$ and the set of initial conditions ϕ_{-1}, there exists (for all $\varepsilon > 0$) a finite moment of time $t_* = t_*(v_0^\infty, \ \tau, \ \Phi_{-1})$. In that case, for the adaptation algorithm of (2.25) and (2.24), the following conditions are satisfied for $t \geq t_*$

$$\mathcal{T}_t = \mathcal{T}_{t_*}, \quad \tau_t = \tau_{t_*}, \quad \delta(\tau_t) = \delta(\tau_{t_*}). \tag{2.26}$$

Thus, the adaptation algorithm (2.24) and (2.25) is finitely convergent.

c) Suboptimality of the adaptive control strategy. The following assertions about the properties of the control strategies In (2.17), (2.24) and (2.25), are presented.

Theorem 4.2.1 *Let the following conditions be satisfied :-*

1) $\mathcal{T}$ is a compact set of an Euclidean space such that, for all $\tau \in \mathcal{T}$, there exists an optimal regulator with coefficients in (2.19);

2) the functional $J(y_0^\infty, \ u_0^\infty)$ for every admissible control strategy that is degenerated by the stabilizing regulator of (2.9), does not depend on the choice of the initial conditions in the control system of (2.1) and (2.9);

3) the function $q(\tau, \ \tilde{\rho})$ is continuously on the right along $\tilde{\rho}$, at the point $\tilde{\rho} = \rho$ uniformly along $\tau \in \mathcal{T}$.

Then for any number $\rho' > 1$, there exists $\varepsilon > 0$, such that the control strategy determinable by relations of (2.17), (2.24) and (2.25) is adaptive in relation to the control objective (2.16) (in a class $\mathcal{T} \otimes V_\rho$).

It may be observed that realization of a finite 'memory' (storage of the finite number of output and control signals) is necessary for the adaptation algorithm of (2.24) and (2.25). But this value may appear to be too large to be a practical one.

---

*If the function $q(\tau, \rho)$ attains the minima at a few points, then τ_t is chosen from any one of them. We shall not again mention the dependence of τ_t on ρ (vide the foot note on page no. 146).

4.2.3 *Examples*

a) Minimum phase object with control and measurement delay. The control object equation is expressed by (2.1) when the polynomial

$$b(\lambda) = \lambda^k b_+(\lambda). \tag{2.27}$$

Here $k \geq 1$ is the control delay and $b_+(\lambda)$ is the stable polynomial. The minimax solution of the functional $J = \overline{\lim_{t\to\infty}} |y_t|$, as obtained in section 3.5, for control strategies degenerated from stabilizing regulations of (2.9), and for an additional condition $\beta_0 = \ldots. = \beta_{s-1} = 0$ (s is the measurement delay) it leads to the optimal regulator

$$F(\nabla)b_+(\nabla)u_t = \nabla^s G(\nabla)y_t. \tag{2.28}$$

Here the polynomials $F(\lambda)$ and $G(\lambda)$ are uniquely determined by the conditions (vide lemma 2.3.1)

$$F(\lambda)a(\lambda) - \lambda^{k+s}G(\lambda) = 1, \ \deg F(\lambda) < k + s. \tag{2.29}$$

The identity in λ in (2.29) degenerates into a linear system in relation to polynomial coefficients $F(\lambda)$ and $G(\lambda)$.

For the control in (2.28), the control object of (2.1) and (2.27) can be expressed in the form $y_t = F(\nabla)v_t$ (by virtue of (2.29)). Again because of arbitrariness of v_t, it is not difficult to obtain the formula for the function $q(\tau, \rho)$:

$$q(\tau, \rho) = \rho \sum_{l=0}^{k+s-1} |F_l(\tau)|. \tag{2.30}$$

Here $F_l = F_l(\tau)$ are the coefficients of the polynomial $F(\lambda) = F(\lambda, \tau)$, and $F_0 = 1$, τ is the set of control object coefficients as in (2.1) and (2.27).

If the control object of (2.1) and (2.27) (with the parameter τ) is controlled by a stabilizing regulator (2.9) (with parameter δ), then the function (2.13) can be easily computed. As a matter of fact, let

$$\alpha(\lambda)/[a(\lambda)\alpha(\lambda) - b(\lambda)\beta(\lambda)] = \sum_{k=0}^{\infty} \lambda^k W^{(k)}, \tag{2.31}$$

where $W^{(k)} = W^{(k)}(\tau, \delta)$ are the exponentially decaying (up to zero, for $k \to \infty$) coefficients of the transfer function in the expanded form. Then from (2.31) we find out

$$Q(\tau, \delta, \rho) = \rho \sum_{k=0}^{\infty} |W^{(k)}(\tau, \delta)|. \tag{2.32}$$

It is evident that the functions $Q(\tau, \delta, \rho)$ and $q(\tau, \rho)$ are continuous with respect to the parameter ρ and for stabilizability of the polynomials $\alpha(\lambda, \tau)$ and $\beta(\lambda, \tau)$ (for $\tau \in \mathcal{T}$), the conditions in the theorem 4.2.1 appear to be satisfied.

Realization of the algorithm (2.24) and (2.25) requires computation of the minimum points of the function $q(\tau, \rho)$ with respect to the sets $\mathcal{T}_t$. This operation simplifies considerably if $\mathcal{T}$ is a polyhedron. Then all the sets $\mathcal{T}_t$ will be polyhedrons and the minimization of the function (2.30) leads to its computation at the vertices of the regular polyhedron. Let, for example, $a(\lambda) = 1 + \lambda a_1$. Then,

$$F(\lambda) = \sum_{l=0}^{k+s-1} \lambda^l (a_1)^l,$$

$$G(\lambda) = (a_1)^{k+1} \text{ and}$$

$$q(\tau, \quad \rho) = \rho \sum_{l=0}^{k+s-1} |a_1|^l .$$

If a_l is the first component of the set of τ (apart from a_l, coefficients of the polynomial $b_+(\tau)$ may be included in τ), then the realization of the algorithm (2.24) and (2.25) is very easy. When $\mathcal{T}$ is a polyhedron, the vector τ_t coincides with one of its vertices.

b) Scalar object. In the above example, the quality functional did not appear in the control signals. Sufficiently generalised forms of minimization control problems for such functionals is not yet solved. Hence we shall restrict our study to a scalar control object, for which synthesis of the minimax problem is realized easily. If the control object is expressed as

$$y_{t+1} + a y_t = b u_t + v_{t+1}, \quad \tau = \text{col}\,(a, \quad b),$$

and the admissible control strategies U^a are determined by the stabilizing feedback signals $u_t = K y_t$, $|a - bK| < 1$, then for the quality functional

$$J(y_0^\infty, \quad u_0^\infty) = \overline{\lim_{t \to \infty}} \quad (|y_t| + R\,|u_t|), \quad R \geq 0,$$

we have, $Q(a,b,K) = \rho(1 - |bK - a|)^{-1}(1 + R\,|K)$. For optimal feedback

$$K(a, \quad b) = \begin{cases} 0, & \text{if } |b < (1 - |a|)R, \\ ab^{-1}, & \text{if } |b| \geq (1 - |a|)R. \end{cases}$$

Then the value of $q(\tau) = q(a, b)$ is given by

$$q(a, \quad b) = \begin{cases} \frac{\rho}{1-|a|}, & \text{if } |b| < (1 - |a|)R, \\ \rho(1 + R\,|a/b|), & \text{if } |b| \geq (1 - |a|)R. \end{cases}$$

If $\mathcal{T}$ is a compact set not intersecting the line $b = 0$ (when $|a| \geq 1$), then the conditions of the theorem 4.2.1 are satisfied. This means that the objects from $\mathcal{T}$ must be controllable ($b \neq 0$) or else stable ($|a| < 1$). Realization of the formula (2.25) leads to the choice (from $\mathcal{T}_t$) of elements either with respect to the modulus of the first component (if $|b_t| < (1 - |a_t|R)$), or with respect to the relationship $|a_t b_t^{-1}|$ (if $|b_t| \geq (1 - |a_t|R)$).

§ 4.3 Self-Tuning Systems

Design of adaptive strategies for continuous linear control objects with unknown parameters has been well-studied using Liapunov functions for problems of dissipation and stabilization (vide section 1.5). Such control systems are known *as self-tuning systems.*

Adaptation and self-tuning are both conceptually of the same category. However, we restrain ourselves in identifying them with each other. The self- tuning systems are specific method of designing the adaptive control strategy consisting of the following two key steps :

1) the choice of suitable feedbacks within the composition of the tunable parameters (transition to the generalised tunable objects in the framework of identification technique, vide section 4.1.2) ;

2) the choice of the parameter tuning strategy from the condition that a function (usually quadratic) of state variables would decay along the control trajectories passing through the exterior of a compact set.

Such a technique usually ensures achievement of stabilizing control objectives and rarely, optimization. Then the adaptation class depends on the method taking into account of each of the above mentioned steps.

The self-tuning method can be used for discrete systems as shown below.

4.3.1 *Self-Tuning with no Disturbance*

The self-tuning method will be stated for a simpler case of a linear control object with no disturbance. The control object is given by

$$a(\nabla)y_t = bu_{t-1}, \quad t=0, \quad 1, \quad \dots, \tag{3.1}$$

where $y_t \in R^l$, $u_t \in R^m$, $l \leq m$; $a(\lambda)$ is a polynomial

$$a(\lambda) = I_l + \lambda a_1 + \quad \dots \quad + \lambda^n a_n. \tag{3.2}$$

The matrix coefficients of the polynomial are completely or partially unknown; I_l is the unit $l \times l$ matrix; b is an $l \times m$ matrix which is assumed to be known and for which

$$\det bb^* \neq 0. \tag{3.3}$$

The control objective is taken as the limited equality

$$J(y_0^\infty, u_0^\infty) = \lim_{t \to \infty} (|\, y_t \,| + |\, u_t \,|) = 0. \tag{3.4}$$

It has to ensure the realizable control strategies in the class $\mathbf{U}^r$. Let us assume that $\mathbf{U}^r$ is composed of all the probable sequences $U_0^\infty(\cdot)$ of functions $U_t(\cdot)$, independent of the unknown control object parameters, future input quantities and output variables. With the choice of the strategy $U_0^\infty(\cdot)$, the control $U_0^\infty$is given by the relations

$$u_t = U_t(u_0^{t-1}, y_0^t), \; t = 1, 2, \quad \dots, \quad u_0 = U_0(y_0). \tag{3.5}$$

The adaptive control problem in this case consists of the choice of a control strategy $U_0^\infty(\cdot) \in \mathbf{U}^r$, which, independent of unknown parameters of the polynomial (3.1) and initial conditions of the control object, will ensure the control objective (3.4).

This problem is quite simple because of the assumptions made about absence of noise, and knowledge of the matrix b. A more complicated situation will be studied later.

a) Transition to a generalised tunable object. The equation (3.1) can be conveniently transformed to a standard form. To do this the following vector function is introduced

$$x_t = \operatorname{col}(y_t, \quad \dots, \quad y_{t-n+1}). \tag{3.6}$$

The following equation is meant for x_t

$$x_{t+1} = Ax_t + B[\tau x_t + bu_t]. \tag{3.7}$$

Here

$$A = \begin{Vmatrix} \mathbf{0}_l \dots \mathbf{0}_l\mathbf{0}_l \\ \mathbf{I}_l \dots \mathbf{0}_l\mathbf{0}_l \\ \cdot \qquad \cdot \\ \cdot \qquad \cdot \\ \cdot \qquad \cdot \\ \mathbf{0}_l \dots \mathbf{I}_l\mathbf{0}_l \end{Vmatrix}, \quad B = \begin{Vmatrix} \mathbf{I}_l \\ \mathbf{0}_l \\ \cdot \\ \cdot \\ \cdot \\ \mathbf{0}_l \end{Vmatrix} \text{ and}$$

$$\tau = \| -a_1, \quad \ldots, -a_n \| . \tag{3.8}$$

Here $\mathbf{0}_l$ is the null matrix of dimension $l \times l$.

The control signal u_t can be formed as per following rule :

$$u_t = -b^*(bb^*)^{-1}\tau_t x_t, \tag{3.9}$$

where $\{ \tau_t \}$ are the tunable parameters of feedback. Thus, the self-tuning system design leads to formation of an adaptive parameter tuning strategy for a generalised tunable object (GTO)

$$x_{t+1} = [A + B(\tau - \tau_t)]x_t. \tag{3.10}$$

The parameter tuning objective is to ensure asymptotic stability of GTO. When the tuner τ_0^∞ is bounded

$$\lim_{t \to \infty} x_t = 0, \quad \sup_t \mid \tau_t \mid < \infty, \tag{3.11}$$

it is equivalent to control objective of (3.4).

b) Self tuning algorithm. The realizable parameter tuning strategy can be expressed as

$$\tau_{t+1} = \tau_t + \gamma_t \frac{B^*(x_{t+1} - Ax_t)x_t^*}{\mu + |x_t|^2}. \tag{3.12}$$

Here $\mu > 0$ and γ_t are arbitrary $l \times l$ matrices, $\gamma_t^* = \gamma_t$, and they satisfy the conditions (for $\varepsilon > 0$) as follows

$$\varepsilon I_l \le \gamma_t \le (2 - \varepsilon) I_l. \tag{3.13}$$

The matrix τ (vide (3.8)) determining the structure of the algorithm (3.12) may contain known coefficients of the polynomial (3.2). When such coefficients are available, there is no need to recalculate their values according to formulae (3.12). The algorithm of (3.12) can be rewritten into a more economical form as shown below.

Let the unknown elements of matrix τ (as given in (3.8)) be numbered in any order. Their aggregate, put to order, transforms r-vector τ'. We define $l \times r$ matrices Φ_t and l-vector ϕ_t by the relationship

$$\tau x_t = \Phi_t \tau' + \phi_t. \tag{3.14}$$

This means that the functions Φ_t and ϕ_t are determined by x_t and known elements of the matrix τ at all instants. The vector tuning algorithm, τ_t is determined by the relations

$$\tau'_{t+1} = \tau'_t + \Phi_t^* \gamma_t \frac{B^*(x_{t+1} - Ax_t)}{\mu + |\Phi_t \Phi_t^*|}. \tag{3.15}$$

Here $\mu > 0$ and $l \times l$ matrices γ_t satisfy the conditions in (3.13). Then the control strategy is given by the feedback

$$u_t = -b^*(bb^*)^{-1}(\Phi_t \tau'_t + \phi_t). \tag{3.16}$$

c) Adaptability of the self-tuning strategy. The control strategies determined by the relations (3.9), (3.12), (3.16) and (3.15) are realizable since it has the expression as in (3.5) by virtue of (3.6).

Theorem 4.3.1 *The control strategies (3.9), (3.12) are adaptive with respect to the control objective (3.11) and hence to the control objective (3.4) in an arbitrary class $\mathcal{T}$ of control object parameters τ (3.1). Then*

$$\mathrm{Sp}\,(\tau_t - \tau)(\tau_t - \tau)^* \le \mathrm{Sp}\,(\tau_0 - \tau)(\tau_o - \tau)^* \tag{3.17}$$

and for $\mu = 0$ the convergence in (3.11) is exponential. It is given by the estimate

$$|x_t| \le C\left(\frac{n-1}{n}\right)^t |x_0|^2 \tag{3.18}$$

where C is a constant and n is the degree of the polynomial (3.2).

A similar assertion is possible about adaptability of the control strategy (3.16) and (3.15) in an arbitrary class $\mathcal{T}'$ of parameters τ'. For adaptation algorithm (3.15) we have the estimate of the following nature (3.17) :

$$|\tau'_t - \tau'| \le |\tau'_0 - \tau'|,$$

Here r-vector τ' is determined by the relationship in (3.14) and is assured of exponential convergence rate in (3.11). The proof of this assertion is analogous to that of the theorem 4.3.1.

Let $l = 1$ in the algorithm (3.15). Designating by

$$c_t = \Phi_t^*, \quad \aleph_t = \Phi_t \tau'_t - B^*(x_{t+1} - Ax_t), \quad \gamma_t \equiv 1, \quad \mu = 0,$$

we have the following expression for the algorithm (3.15) :

$$\tau'_{t+1} = \tau'_t + (c_t^* c_t)^{-1}(c_t^* \tau'_t - \aleph_t)c_t. \tag{3.19}$$

It is the same as the known Kaczmarz's algorithm for the solution of the system of linear equations

$$c_t^* \tau' - x_t = 0, \quad t = 0, \quad t, \quad \ldots . \tag{3.20}$$

4.3.2 *Self-Tuning in the Presence of Disturbance*

If the control object is subjected to disturbance, then as a rule, the asymptotic stability becomes an unachievable control objective. The stabilizing control objectives are the achievable ones and then it is desirable that the output deviation from the null becomes minimum. If no *a priori* information, excepting the level mentioned above, is known about the noise, then the control problem can be considered as a minimax problem. Again for the adaptive control problem, minimum of the maximum output deviation (of the control object) is attached only for $t \to \infty$.

a) Statement of the problem. Let the GTO be expressed as

$$x_{t+1} = Ax_t + B[(\tau - \tau_t)x_t + v_{t+1}], \quad t = 0, \quad 1, \quad \ldots . \tag{3.21}$$

Here A is an $n \times n$ matrix ; B is an $n \times l$ matrix ; τ is an $l \times n$ matrix on unknown parameters of the object ; τ_t is a matrix of tunable parameters and $\{ v_t \}$ is a disturbance, about which only the following information is available, viz.,

$$v_t \in \mathbf{V}, \quad t = 0, \quad 1, \quad \ldots . \tag{3.22}$$

Here $\mathbf{V}$ is a convex symmetrical (in relation to zero) set in $\mathbf{R}^l$. All the possible sequences v_0^∞ satisfying (3.22) transform the class of disturbance $\mathbf{V}^\infty$. The matrices A and B in (3.21) are assumed to be known. Here A is a stable matrix and

$$\det B^* B \ne 0. \tag{3.23}$$

The tuning quality will be characterized by the functional

$$J(x_0^\infty) = \overline{\lim_{t\to\infty}} \ |x_t|. \tag{3.24}$$

As regards the matrix τ of unknown parameters, it is assumed that it belongs to a convex bounded set $\mathcal{T}$. Moreover, it is assumed that the projector $P_{\mathcal{T}}$ on this set is known. The projector possesses the following properties: $P_{\mathcal{T}} \ \tau' = \tau'$ and

$$\mathrm{Sp}\,(P_{\mathcal{T}}\tau'' - \tau')(P_{\mathcal{T}}\tau'' - \tau')^* \le \mathrm{Sp}\,(\tau'' - \tau')(\tau'' - \tau')^*. \tag{3.25}$$

Here $\tau' \in \mathcal{T}$ and τ'' is an arbitrary matrix of the dimension same as that of τ'.

The class of realizable parameter tuning strategies, $\mathbf{T}^r$ will be all the possible methods of forming the estimates τ_t. These do not make use of future states of GTO (3.21), unknown matrix τ and the terminal value of the functional (3.24). Also

$$\tau_t \in \mathcal{T}. \tag{3.26}$$

It will be shown that the class $\mathbf{T}^r$ is not empty and a few realizable tuning strategies will be built up.

b) Synthesis of a realizable parameter tuning strategy. Let the parameter tuning algorithm be expressed in the analogous manner as in (3.12)

$$\tau_{t+1} = P_{\mathcal{T}}[\tau_t + \gamma_t \eta_{t+1} |x_t|^{-2} x_t^*], \quad t = 0, \ 1, \ \dots . \tag{3.27}$$

Here

$$\eta_{t+1} = (B^*B)^{-1} B^*(x_{t+1} - Ax_t). \tag{3.28}$$

Here $\mu = 0$ (unlike (3.12)) and the operator $P_{\mathcal{T}}$ (so as to satisfy (3.26)) is also used. Because of (3.21), the following expression is valid

$$\eta_{t+1} = (\tau - \tau_t)x_t + v_{t+1}. \tag{3.29}$$

The presence of the disturbance in (3.29) does not permit us establish the convergence of the tuners τ_0^∞ when the matrix coefficients γ_t are chosen from the condition (3.13). Hence we assume that

$$\gamma_t = \delta_t \bar{H}. \tag{3.30}$$

Here $\bar{H}$ is a positive matrix whose choice will be discussed later; δ_t are non-negative numbers which must satisfy the following inequality for all t and any $v_{t+1} \in \mathbf{V}$:

$$\delta_t(\eta_{t+1}^* \bar{H} \eta_{t+1} - \delta_t \eta_{t+1}^* \bar{H}^2 \eta_{t+1} - v_{t+1}^* \bar{H} \eta_{t+1}) \ge 0. \tag{3.31}$$

Let $v_{t+1}^{(0)}$ be the point of tangency between the $\mathbf{V}$ and the plane with the directional vecotr $\bar{H}\eta_{t+1}$, *i.e.*

$$[v_{t+1}^{(0)}]^* \bar{H} \eta_{t+1} = \max_{v \in \mathbf{V}} v^* \bar{H} \eta_{t+1}. \tag{3.32}$$

Then (3.31) is satisfied when $\varepsilon > 0$ and

$$\delta_t = \begin{cases} \mu_{t+1} & \text{for } \mu_{t+1} \ge \varepsilon, \\ 0 & \text{for } \mu_{t+1} < \varepsilon \end{cases}. \tag{3.33}$$

Here

$$\mu_{t+1} = |\bar{H}\eta_{t+1}|^{-2} (\eta_{t+1} - v_{t+1}^{(0)})^* \bar{H} \eta_{t+1}. \tag{3.34}$$

c) Adaptability of the parameter tuning strategy. The relations (3.27), (3.28), (3.30) and (3.32) to (3.34) determine the realizable parameter tuning strategy of GTO (3.21).

It requires refinement of the choice of positive matrix $\tilde{H}$ and number ε, and the given initial matrix τ_0. The following assertion may be expressed in respect of this strategy.

Theorem 4.3.2 *Let us assume that the following conditions are fulfilled :*

1) given a positive matrix H that satisfies the inequality for any $\alpha \varepsilon (0.1)^*$

$$A^* H A \leq \alpha^2 H, \quad H \geq I; \tag{3.35}$$

2) null is the interior point of the set V, i.e.,

$$c_v^{(0)} = \inf |v| > 0, \quad v \overline{\in} V; \tag{3.36}$$

3) for the number $\varepsilon > 0$ *and matrix* $\tilde{H}$ *the following relations are true*

$$\tilde{H} = \tilde{H}^* = (B^* H B)^{1/2}, \quad \varepsilon\sqrt{|H| \quad |B^* B|} < 1. \tag{3.37}$$

Then the parameter tuning strategy GTO determined by the relations (3.27), (3.28), (3.30), (3.32) to (3.34) for any arbitrary initial condition $\tau_0 \in \mathcal{T}$ *ensures the following inequality (for t = 1,2,...)*

$$\mathrm{Sp}\,(\tau_{t+1} - \tau)(\tau_{t+1} - \tau)^* \leq \mathrm{Sp}\,(\tau_t - \tau)^*, \tag{3.38}$$

$$|x_t| \leq |H|^{1/2} |x_0| (\alpha + v/\sqrt{t})^t + C \sum_{k=0}^{t-1} (\alpha + v\sqrt{k})^k, \tag{3.39}$$

where the number α *is from the inequality in (3.35) and*

$$v = \varepsilon^{-1}\sqrt{\mathrm{Sp}\,(\tau - \tau_0)(\tau - \tau_0)^*}, \quad C = (|\tilde{H}|^{-1} - \varepsilon)^{-1} \sup_{v \in V} |v|. \tag{3.40}$$

The inequality (3.39) permits estimation of the transient process in the adaptive system. In particular, for the instant t_*, which satisfies the conditions $t_* \geq \tilde{\varepsilon}^{-2} \upsilon^2$ and $\tilde{\varepsilon} < 1 - \alpha$, we get the following estimate from (3.39) for $t \geq t_*$:

$$|x_t| \leq |H|^{1/2} |x_0| (\alpha + \tilde{\varepsilon})^t + C \sum_{k=0}^{t_*-1} (\alpha + v/\sqrt{k})^k$$

$$+ C\left[(\alpha + \tilde{\varepsilon})^{t_*} - (\alpha + \tilde{\varepsilon})^t\right] [1 - (\alpha + \tilde{\varepsilon})]^{-1}. \tag{3.41}$$

From here the inequality results as

$$\sup_t \quad |x_t| \leq |H|^{1/2} |x_0| + 4\tilde{\varepsilon}^{-2} v^2 \max_{t \geq 1} (\alpha + v/\sqrt{t})^t$$

$$+ C(\alpha + \tilde{\varepsilon})^{v^2 \tilde{\varepsilon}^{-2}} / [1 - (\alpha + \tilde{\varepsilon})].$$

This inequality signifies that this adaptive strategy is in the class $\mathcal{T}$ in relation to the dissipative control objective. When the disturbance is absent ($v_t \equiv 0$) then from (3.40) we have C = 0 and the estimate (3.41) shows that as $t \to \infty$, $x_t \to 0$ with an exponential rate (vide also 3.18).

---

* Such matrices exist. Actually, with some C and υ, $C \geq 1$ and $0 < \upsilon < 1$, and for a stable matrix A the inequality $|A^t| \leq C\,\upsilon^t$, $t = 0, 1, \ldots$, holds. If $1 > \alpha = C^{-2}(C^2 - 1 + v^2)$, then (3.35) is satisfied for a matrix $H = \sum_{k=0}^{\infty} (A^k)^* A^k$.

4.3.3 *Adaptive Control with Bounded Disturbance in the Control Object*

a) Description of the mathematical model of the object. The above self-tuning result for a GTO may be applied to the solution of the control problem stated as follows : the control is described by the equation

$$a(\nabla)y_t = b(\nabla)u_t + v_t \tag{3.42}$$

where $y_t \in \mathbf{R}^l$, $u_t \in \mathbf{R}^m$, $v_t \in \mathbf{R}^l$, $a(\lambda)$ and $b(\lambda)$ are the polynomials with matrix coefficients

$$a(\lambda) = I_l + \lambda a_1 + \quad \dots \quad + \lambda^n a_n,$$

$$b(\lambda) = \lambda b_1 + \quad \dots \quad + \lambda^n b_n.$$

The following conditions are assumed to be satisfied for the equation (3.42) :

1) the polynomial coefficients $a(\lambda)$ and $b(\lambda)$ can depend on vector parameter τ, $a(\lambda) = a(\lambda, \tau)$ and $b(\lambda) = b(\lambda, \tau)$. Here the polynomial coefficients are not known and their dependence on τ is assumed to be linear ;

2) the set $\mathcal{T}$ of possible values of τ is a convex compact subset of the space $\mathbf{R}^q$. The projector $P_{\mathcal{T}}$ on the set $\mathcal{T}$ is then assumed to be known;

3) for each point $\tau \in \mathcal{T}$ there exist matrix polynomials $\alpha(\lambda, \tau)$ and $\beta(\lambda, \tau)$ of appropriate dimensions

$$\alpha(\lambda, \tau) = I_m + \lambda\alpha_1 + \quad \dots \quad + \lambda^n \alpha_n,$$

$$\beta(\lambda, \tau) = \lambda\beta_1 + \quad \dots \quad + \lambda^n \beta_n, \tag{3.43}$$

so that the feedback

$$\alpha(\nabla, \tau)u_t = \beta(\nabla, \tau)y_t \tag{3.44}$$

is stabilizing for the control object (3.42).* It may be recalled that this signifies stability of the scalar polynomial

$$g(\lambda, \tau) = \det \begin{Vmatrix} a(\lambda, \tau) & -b(\lambda, \tau) \\ \beta(\lambda, \tau) & -\alpha(\lambda, \tau) \end{Vmatrix}; \tag{3.45}$$

4) the class $\mathbf{V}_\rho$ of the disturbances v_0^∞ in (3.42) is determined by the level $\rho > 0$: $\mathbf{V}_\rho = \{v_0^\infty : |\, v_t \,| \le \rho\}$.

Let the control objective be given by the inequality

$$J(y_0^\infty, u_0^\infty) = \overline{\lim_{t \to \infty}}(|\, y_t \,|^2 + |\, u_t \,|^2) \le r^{-1}. \tag{3.46}$$

Here the quality level r may depend on τ.

The control objective (3.46) is required to ensure stabilizing control strategies (in a class $\mathbf{U}^\tau$) determinable by feedback of the nature $u_t = U_t(u^{t-1}, y^t)$, $t = 1, 2, \dots\dots$, $u_0 = U_0(y_0)$. Here the functions $U_t(\cdot)$ do not depend on τ.

b) Transition to a generalised self-tuning object. Vector function

$$x_t = \text{col}\,(y_{t-1}, \quad \dots, \quad y_{t-n}, u_{t-1}, \quad \dots, \quad u_{t-n}) \tag{3.47}$$

---

* We note the time delay in (3.44) : $\beta(0, \tau) = 0$ (vide (3.43). In the scalar case ($l=m=1$) such a stabilizing link exists if $|\, a(\lambda, \tau) \,| + |\, b(\lambda, \tau) \,| \neq 0$, for $|\, \lambda \,| \le 1$.

satisfies the equation in the standard form

$$x_{t+1} = Ax_t + (BDx_t + v_t) + Cu_t, \tag{3.48}$$

where

$$A = \left\| \begin{matrix} E_l & 0 \\ 0 & E_m \end{matrix} \right\|, \quad E_l = \left\| \begin{matrix} 0 & 0 & \dots & 0 & 0 \\ I_l & 0 & \dots & 0 & 0 \\ \cdot & \cdot & \cdot & \cdot & \cdot \\ \cdot & \cdot & \cdot & \cdot & \cdot \\ \cdot & \cdot & \cdot & \cdot & \cdot \\ 0 & 0 & \dots & I_l & 0 \end{matrix} \right\|,$$

$$E_m = \left\| \begin{matrix} 0 & 0 & \dots & 0 & 0 \\ I_m & 0 & \dots & 0 & 0 \\ \cdot & \cdot & \cdot & \cdot & \cdot \\ \cdot & \cdot & \cdot & \cdot & \cdot \\ \cdot & \cdot & \cdot & \cdot & \cdot \\ 0 & 0 & \dots & I_m & 0 \end{matrix} \right\|, \quad B = \left\| \begin{matrix} I_l \\ 0 \\ \cdot \\ \cdot \\ \cdot \\ 0 \end{matrix} \right\|, \quad C = \left\| \begin{matrix} 0 \\ \cdot \\ \cdot \\ \cdot \\ 0 \\ I_m \\ 0 \\ \cdot \\ \cdot \\ \cdot \\ 0 \end{matrix} \right\|, \tag{3.49}$$

$$D = | -a_1, \dots, -a_n, b_1, \dots, b_n | .$$

We observe the following to be essential for further properties of the equation ((3.48) : 1) all the unknown parameters are included in $l \times n(l+m)$ matrix $D = D(\tau)$ which is linear with τ; (2) the matrix pair $(A + BD(\tau), C)$ is stabilizable for all $\tau \in \mathcal{T}$.*

For every point $\tau \in \mathcal{T}$ we assume the matrix $K(\tau)$ to be known, such that the following matrix is stable

For every point $\tau \in \mathcal{T}$ we asume the matrix $K(\tau)$ to be known, such that the following matrix is stable

$$A(\tau) = A + BD(\tau) + CK(\tau). \tag{3.50}$$

The realizable control strategy will be formed as

$$u_t = K(\tau_t)x_t, \tag{3.51}$$

where τ_t are the tunable parameters of the regulator. Putting (3.51) in (3.48), the GTO is obtained as

$$x_{t+1} = A(\tau_t)x_t + B[D(\tau - \tau_t) + v_{t+1}]. \tag{3.52}$$

---

*Stabilizability of the above pair (vide section 1.3.3) can be established by the following manner : the regulator (3.44) is given by $u_t = K(\tau)x_t$ where $K(\tau)$ is the regulator-coefficient-matrix (3.44) and the symbols used in the expression correspond to those in (3.47). Since the polynomial (3.45) is stable, then the states x_t of the equation (3.48) will be evenly bounded (in relation to t) independent of the choice of the initial conditions, and it signifies stability of the matrix $A + BD(\tau) + CK(\tau)$, i.e., stabilizability of the pair $(A + BD(\tau), C)$.

Here the matrix $A(\tau)$ is determined by the formula (3.50) for $\tau \in \mathcal{T}$.

Unlike that in the equation (3.21), the matrix $A_t = A(\tau_t)$ is non-stationary in equation (3.52) but stable for all $\tau_t \in \mathcal{T}$. It appears that for a parameter tuning strategy, in the line of strategies for GTO (3.21), it is necessary to prove the finite time convergence of the sequence τ_0^∞. Hence the non-stationarity of the matrix A_t for an established dissipative control objective does not play any role. The estimate of the control quality of a corresponding strategy permits provision of the control objective (3.46) if the given level of control quality is rather high.

c) Synthesis of adaptive strategy. We determine the $l \times q$ matrix Φ_t by the relationship

$$[D(\tau) - D(0)]x_t = \Phi_t \tau. \tag{3.53}$$

It is possible by virtue of the linear dependence of $D(\tau)$ on τ. It is evident that there exists a constant $C_\Phi > 0$, such that

$$|\Phi_t \Phi_t^*| \leq C_\Phi^2 |x_t|^2, \quad t = 1, 2, \quad \ldots, \tag{3.54}$$

$|\Phi\Phi^*|$ is the norm of the matrix $\Phi\Phi^*$. Using the symbols

$$\eta_{t+1} = (B^*B)^{-1} B^*[x_{t+1} - A(\tau_t)x_t] = \Phi_t(\tau - \tau_t) + v_t, \tag{3.55}$$

we can write the algorithm (3.27) as

$$\tau_{t+1} = P_{\mathcal{T}}\left[\tau_t + \delta_t \frac{\Phi_t^* \bar{H}_t \eta_{t+1}}{|\Phi_t \Phi_t^*|}\right], \quad t = 0, 1, \quad \ldots. \tag{3.56}$$

Here $\bar{H}_0^\infty$ is some sequence of positive matrices and δ_t are the scalar values determinable by the formula (3.33) when

$$\mu_t = \frac{\eta_{t+1}^* \bar{H}_t \eta_{t+1}}{|\bar{H}_t \eta_{t+1}|^2} - \frac{\rho}{|\bar{H}_t \eta_{t+1}|}. \tag{3.57}$$

The procedure in (3.56) is supplemented by the initial vector $\tau_0 \in \mathcal{T}$.

Theorem 4.3.3 *We assume that the following conditions are fulfilled in addition to (1) to (4).*

5) for every $\tau \in \mathcal{T}$ a positive matrix $H(\tau)$ is defined so that the following inequality is satisfied

$$A^*(\tau)H(\tau)A(\tau) \leq \alpha^2 H(\tau), \quad H(\tau) \geq I_N, \tag{3.58}$$

*Where $\alpha \in (0,1)$; $N = n(1+m)$ and $A(\tau)$ is determined from the formula (3.50);**

6) the number ε and the matrices $\bar{H}_t$ in the adaptation algorithm (3.56), (3.33) and (3.57) satisfy the conditions

$$\varepsilon \sup_{\tau \in \mathcal{T}} |H(\tau)|^{1/2} < 1, \quad \bar{H}_t = [B^* H(\tau_{t-1}) B]^{1/2}. \tag{3.59}$$

Then the formulae (3.51), (3.56), (3.33) and (3.57) determine the realizable control strategy which is adaptive in a class $\mathcal{T}$ in respect of the control objective (3.46); if

---

*Existence of the number $\alpha^2 = \alpha^2(\tau)$ and matrix $H(\tau)$ (satisfying (3.58)) is substantiated in the same manner as the inequality of (3.35). Because of the compactness of the set $\mathcal{T}$ we can consider $\sup_{\tau \in \mathcal{T}} \alpha^2(\tau) = \alpha^2 < 1$.

$$r < \rho^{-1}\left(1+\sum_{k=1}^{\infty}(\alpha+\nu/\sqrt{k})^k\right)^{-1} C_H^{-1}(1-\varepsilon C_H), \tag{3.60}$$

where $C_H = \sup|H(\tau)|^{1/2}$, $\tau \in \mathcal{T}$. *In this case the tuner* τ_0^∞ *is finitely-convergent and the magnitude* $|\tau_t - \tau|$ *is not monotonously increasing.*

4.3.4 ***Method of Recursive Objective Inequalities in an Adaptive Tracking Problem***

The adaptation algorithms used in sections 4.3.2, 4.3.3 possessed a special property : at every stage the estimate τ_t monotonously approached the unknown parameter τ. It corresponded to a definite control object or GTO. It is a nonessential property for ensuring the control objective. It permits avoiding 'deterioration' of the tuners when the control object parameters are considerably undeterminable and at the same time their identification is not possible due to arbitrary character of the disturbances. Such algorithms appear in the theory of learning systems, where the closeness of τ_t to τ often determines the degree of 'learning' of the system. The estimation algorithm realizes the correction in an estimate only in such cases, where the estimate is guaranteed to be near the estimated parameter. Otherwise, the estimate remains unchanged. Such a 'guarantee' condition is given by some inequality related to estimates already obtained and the given observations. Presence of similar inequalities permits interpretation of the estimation process as a technique for finding out their solutions and thus the purpose of estimation is defined. Due to this, the above inequalities are known as objective inequalities. The following inequalities are the objective ones in the algorithm of (3.27) :

$$\mu_{t+1}(\hat{\tau}) < \varepsilon, \quad t = 0, \quad 1, \quad \ldots . \tag{3.61}$$

When the next inequality is satisfied, for $\tilde{\tau} = \tau_t$, the estimate τ_t(vide (3.33)) did not change. The characteristic of the objective inequalities (3.61) is that the functions $\mu_{t+1}(\tilde{\tau})$ are not *a priori* given. They depend on the values of the variables x_{t+1} ; x_{t+1}, in its turn, depends on the estimates τ_s which were used for $s < t$. Hence the objective inequalities are themselves obtained by a sequence (recursive) depending on preformulation of the estimation process. Here the solution of the inequalities is also to be found out. Hence these inequalities are known as recursive and the method, for the adaptation problem, to find out the solution of the objective inequalities, is known as the method of recursive objective inequalities.

We shall now briefly discuss the method of recursive objective inequalities for the adaptive control of a scalar object.

a) Statement of the adaptive optimal tracking problem. Let us consider the following tracking problem : the control object is expressed by the equation with scalar input and output

$$a(\nabla)y_t = u_{t-1} + v_t. \tag{3.62}$$

The coefficients of the polynomial $a(\lambda) = 1 + \lambda a_1 + \ldots + \lambda^n a_n$ may be partially known.

The disturbance v_0^∞ is assumed to satisfy the condition

$$|v_t| \leq \rho \tag{3.63}$$

with a given $\rho > 0$, and otherwise it is arbitrary.

Let the sequence y_0^∞ be given. We call it as a *programmed motion.* The control objective is to ensure the inequalities

$$\overline{\lim}\ |y_t - \hat{y}_t| \leq \rho, \quad \overline{\lim}\ |u_t| < \infty, \quad t \to \infty. \tag{3.64}$$

It is not difficult to observe that because of the arbitrariness of the disturbance υ_t, the inequality $\overline{\lim}\ |y_t - \hat{y}_t| < \rho, \quad t \to \infty$, will not be achieved for any choice of control u_0^∞. Hence the control objective (3.64) can be characterized as optimal with respect to output variable y and dissipative with respect to the control variable u.

The control objective (3.64) is ensured with the help of realizable control strategies from the class U^r, resulting from all unpredictable feedbacks, which do not make use of unknown coefficients of the polynomial $a(\lambda)$.

Let τ be the uncertainty parameter (a set of unknown coefficients) of the equation (3.62)

$$\tau = \text{col}\left(a_{i_1}, \ \ldots, \ a_{i_k}\right). \tag{3.65}$$

It is considered as a vector of dimension k. We shall assume that the set $\mathcal{T} \subseteq \mathbf{R}^k$ of all possible values of the parameter τ is known.

When the paramter τ is known, the optimal feedback is determined without any difficulty and has the expression

$$u_t = \tilde{a}(\nabla)y_t + \hat{y}_{t+1},$$

$$\tilde{a}(\lambda) = \lambda^{-1}[a(\lambda) - 1]. \tag{3.66}$$

Here it is assumed that the programmed motion is known at least one step ahead and bounded in the time domain. For the control in (3.66) the equation (3.62) has the expression $y_{t+1} = \hat{y}_{t+1} + \upsilon_{t+1}$, *i.e.*, it ensures fulfilment of the first inequality in (3.64). The characteristic polynomial of the control system (3.63) and (3.66), is equal to unity, and so the second inequality in (3.64) is also fulfilled. However, the regulator of (3.66) is not realizable, if the parameter (3.65) is not known.

b) Reduction of the adaptive tracking problem to recursive objective inequalities. Let us introduce

$$\Phi_{t-1} = \text{col}\left(-y_{t-i_1}, \ \ldots, \ -y_{t-i_k}\right),$$

$$\phi_{t-1} = -\sum_{k=1}^{n}{}' \ a_k y_{t-k}. \tag{3.67}$$

Here the prime signifies that the summation is carried only up to the indices of known coefficients. In (3.67), the control object is (3.62) described by

$$y_t = \Phi^*_{t-1}\tau + u_{t-1} + \phi_{t-1} + v_t. \tag{3.68}$$

The realizable control strategy can be synthesized with the help of the feedback

$$u_t + \Phi^*_t\tau_t + \phi_t = \hat{y}_{t+1}, \tag{3.69}$$

where τ_0^∞ are the tunable parameters of the regulator.

Putting (3.69) in (3.68), we get the relationship

$$y_t - \hat{y}_t = \Phi^*_{t-1}(\tau - \tau_{t-1}) + v_t. \tag{3.70}$$

It is the initial value for the recursive inequalities. Let the function of parameter $\hat{\tau}$ be

$$\Psi_t(\hat{\tau}) = \varepsilon - |\Phi^*_{t-1}(\tau - \hat{\tau}) + v_t|. \tag{3.71}$$

The parameter τ and the number $\varepsilon > 0$ are assumed to be fixed.

It results from (3.62) that if $\varepsilon \geq \rho$, then for all t the inequality

$$\Psi_t(\hat{\tau}) \geq 0 \tag{3.72}$$

is solved.

If it is possible to find out a realizable bounded tuner τ_0^∞ that satisfies the limiting inequality

$$\underline{\lim}\ \ \Psi_t(\tau_{t-1}) \geq 0, \quad t \to \infty, \tag{3.73}$$

then the realizable strategy (3.69) ensures fulfilment of the first inequality in (3.64), when $\varepsilon = \rho$. The second inequality in (3.64) is obtained from the control object (3.62).

Thus, the synthesis of the adaptive control strategy is reduced to the problem of finding out bounded tuners τ_0^∞ fulfilling the objective inequality in (3.73).

At every instant t the inequality (3.72) determines a region in the space $\mathbf{R}^k$ of parameters $\hat{\tau}$ which opens up the possibility of forming relaxation algorithms which solve these inequalities.* One of the possible algorithms of this type will be considered below.

c) 'Extended stripe' algorithm

The gradient of the function (3.71) (with respect to $\hat{\tau}$) is easily computed as

$$\text{grade}_t \Psi_t(\hat{\tau}) = \Phi_{t-1}\ \text{sign}\ [\eta_t(\hat{\tau})], \tag{3.74}$$

where

$$\eta_t(\hat{\tau}) = \Phi^*_{t-1}(\tau - \hat{\tau}) + v_t \tag{3.75}$$

and it is assumed that $\eta_t(\hat{\tau}) \neq 0$. If the estimate τ_{t-1} is obtained for the instant t and the observation y_t becomes accessible then the estiamate τ_t can be found out in the form of

$$\tau_t = \tau_{t-1} + \gamma_{t-1}\Phi_{t-1}\ \text{sign}\ \eta_t(\tau_{t-1}). \tag{3.76}$$

Here γ_{t-1} is a non-negative number determining the shift magnitude for the estimate τ_{t-1} in the direction of the gradient of the function $\Psi_t(\hat{\tau})$ at the point $\hat{\tau} = \tau_{t-1}$. This magnitude can be determined from the condition so as to fulfill the monotonity condition, viz.,

$$|\tau_t - \tau| \leq |\tau_{t-1} - \tau|. \tag{3.77}$$

It can be shown by simple calculation that the condition in (3.77) is satisfied when γ_{t-1} satisfies the inequality

$$\gamma_{t-1}(2|\eta_t| - 2v_t\ \text{sign}\ \eta_t - \gamma_{t-1}|\Phi_{t-1}|^2) \geq 0.$$

Considering (3.63) and the worst case possible the following will be the requirement for the fulfilment of the inequality

$$\gamma_{t-1}(2|\eta_t| - 2\rho - \gamma_{t-1}|\Phi_{t-1}|^2) \geq 0. \tag{3.78}$$

---

*The solution algorithms of convex system of equations are known as relaxation algorithms for which the usual unfulfilled inequality is satisfied at every step at the cost of shifting the estimation τ_{t-1} in the direction of the gradient of the function $\Psi_t(\cdot)$. For this, the rigidity $\Psi_t\tau_{t-1} < \Psi_t(\tau_t)$ is reduced (weakened).

The condition in (3.78) is fulfilled if, for example,

$$\gamma_{t-1} = \begin{cases} (|\eta_t| - \rho) |\Phi_{t-1}|^{-2} & \text{for } |\eta_t| > \varepsilon, \\ 0 & \text{for } |\eta_t| \le \varepsilon. \end{cases} \tag{3.79}$$

The algorithm of (3.76) ensuring (3.77) has, consequently, the following expression

$$\tau_{t+1} = \tau_t + \theta_t |\Phi_t|^{-2} (\eta_{t+1} - \rho \text{ sign } \eta_{t+1})\Phi_t, \tag{3.80}$$

where $\theta_t = 1$ for $|\eta_{t+1}| \ge \varepsilon$ and $\theta_t = 0$ for $|\eta_{t+1}| < \varepsilon$. We note that the algorithm (3.80) uses only the values at the instant t, which are accessible : by virtue of (3.75), (3.86) and (3.69)

$$\eta_{t+1} = \eta_{t+1}(\tau_t) = y_{t+1} - \hat{y}_{t+1}. \tag{3.81}$$

It results from (3.80) that the estimation process is directed towards ensuring the inequality $|\eta_{t+1}(\tau_t)| \le \varepsilon$. If starting from any instant t these inequalities are satisfied, then $\tau_t = \tau_{t_*}$, for $t \ge t_*$.

The in equality of (3.72) is equivalent to the inequality below

$$|\Phi_t^* \hat{\tau} - (y_{t+1} - \hat{y}_{t+1})| \le \varepsilon. \tag{3.82}$$

It determines the stripe region of width $2\varepsilon |\Phi_t|^{-1}$in the space of parameters $\hat{\tau}$ with the middle plane $\Phi_t^* \tau = y_{t+1} - \hat{y}_{t+1}$. If the point τ_t is located in the above region (*i.e.* for $\hat{\tau} = \tau_t$ (3.82) is satisfied) then $\tau_{t+1} = \tau_t$, by virtue of the algorithm (3.80). If τ_t is located outside the region (3.82) then τ_{t+1} is obtained by projecting τ_t on the plane nearest to it, viz. $|\Phi_t^* \hat{\tau} - (y_{t+1} - \hat{y}_{t+1}| \le \rho$ Thus the point so located, namely, τ_{t+1} appears to be in the region (3.82) as because $\varepsilon \ge \rho$. The geometric interpretation of the algorithm in (3.80) serve as the basis for calling it as the 'extended stripe'.

The algorithm 'extended stripe' is a variant of the 'regional' algorithm type. Each of the successive estimates τ_t located outside this region will be shifted towards it. (Then the shift may make the hit points to be inside the region.) We recall that basic algorithm of this family is the algorithm 'stripe'. For this the projection of τ_t is obtained on the middle plane. The algorithm 'stripe' has the expression of (3.80) for $\rho = 0$. If this projection is to be done independent of the location of the point τ_t, then we arrive at the Kaczmarz's algorithm (vide 3.19).

For an adaptive control problem more prefered variant is the 'extended stripe', since this algorithm ensures finite convergence of the adaptation algorithm when $\varepsilon > \rho$ and the boundaries are sufficiently weak.

c) *Adaptability of the synthesized control strategy.* The algorithm (3.80) along with the equation (3.69) determines the realizable control strategy. The studies of adaptive properties of this strategy are carried out by the methods similar to those adopted for control strategy in section 4.3.3. The result of these studies is presented in the following assertion.

Theorem 4.3.4 *When $\varepsilon \ge \rho$, the control strategy determinable by the relations (3.69), (3.67), (3.80) and (3.81), ensures fulfilment of the following inequalities with any initial condition $\tau_0 \in \mathcal{T}$.*

$$|\tau_{t+1} - \tau| \le |\tau_t - \tau|, \tag{3.83}$$

$$\left(\sum_{k=0}^{n-1} |y_{t-k}|^2\right)^{1/2} \le \sqrt{n}\left(\sum_{k=0}^{n-1} |y_{-k}|^2\right)^{1/2} (\alpha+\nu/\sqrt{t})^t$$

$$+C\left[1+\sum_{k=1}^{t-1}(\alpha+\nu/\sqrt{k})^k\right], \tag{3.84}$$

where $\alpha=\sqrt{(n-1)/n}, \quad \nu=\sqrt{n}\,|\tau-\tau_0|,$

$$C=\sqrt{n}\left(\rho+\varepsilon+\sup_t |\hat{y}_t|\right). \tag{3.85}$$

Moreover, the following bounded conditions are satisfied and the algorithm (3.80) *converges in finite time, if* $\varepsilon>\rho$

$$\overline{\lim}\,|y_t-\hat{y}_t|\le\varepsilon, \quad \overline{\lim}\,|u_t|<\infty, \quad t\to\infty, \tag{3.86}$$

The theorem 4.3.4 indicates adaptability of the synthesized control strategy in an arbitray class $\mathcal{T}$ in respect of control objective (3.86). This control objective evidently coincides with (3.64) when $\varepsilon=\rho$ and determines the optimal property of the adaptive control strategy.

The inequality of (3.84) permits estimation of the transient process of the adaptive system.

§ 4.P Proof of the Lemmas and Theorems

4.P.1 *Proof of the Lemma 4.2.1*

Let t be the instant of variation of the set $\mathcal{T}_t$, such that

$\mathcal{T}_{t+1}\subset\mathcal{T}_t, \quad \mathcal{T}_{t+1}\ne\mathcal{T}_t$. Let $\tau\in\mathcal{T}_{t+1}$.

Then $\tau\in\mathcal{T}'_{t+1}(0)$, *i.e.*,

$$|a(\nabla,\tau)y_{t+1}-b(\nabla,\tau)u_{t+1}|\le\rho. \tag{P.1}$$

Moreover, by virtue of (2.24)

$$|a(\nabla,\tau_t)y_{t+1}-b(\nabla,\tau_t)u_{t+1}|>\rho+\varepsilon\,|\Phi_t|. \tag{P.2}$$

Since $a(\nabla,\tau)y_{t+1}-b(\nabla,\tau)u_{t+1}=y_{t+1}-\Phi_t^*\tau$

and $a(\nabla,\tau_t)y_{t+1}-b(\nabla,\tau_t)u_{t+1}=y_{t+1}-\Phi_t^*\tau_t$,

then by the virtue of (P.1) and (P.2), $|\Phi_t^*(\tau-\tau_t)|>\varepsilon\,|\Phi_t|$.

It then follows that the inequality $|\tau-\tau_t|>\varepsilon>0$ is true for any $\tau\in\mathcal{T}_{t+1}$. Consequently, for any variation of the set $\mathcal{T}_t$ the inequality

$$\inf_{\tau'\in\mathcal{T}_{t+1}} |\tau'-\tau_t|>\varepsilon>0 \tag{P.3}$$

is satisfied. Because of the boundedness of the set $\mathcal{T}_0=\mathcal{T}$ the number of variations in vector τ_t is certainly not more than what has been ascertained. The rest of the conditions

(2.26) are the consequence of the finiteness of the number of variations of the estimates τ_t. The lemma is proved.

4.P.2 Proof of the Theorem 4.2.1

Because of the lemma 4.2.1 the $\lim_{t\to\infty}\tau_t = \tau_\infty$ is achieved in the finite time period. We shall consider later only those t for which $\tau_t = \tau_\infty$. Because of (2.24) we have then

$$|y_t - \Phi^*_{t-1}\tau_\infty| = |a(\nabla, \tau_\infty)y_t - b(\nabla, \tau_\infty)u_t| \le \rho + \varepsilon|\Phi_{t-1}|. \tag{P.4}$$

Hence we get

$$y_t = \Phi^*_{t-1}\tau_\infty + \bar{v}_t$$

where $\{\bar{v}_t\}$ is the 'new noise' that satisfies the following estimate (because of (P.4))

$$|\bar{v}_t| \le \rho + \varepsilon|\Phi_{t-1}| \tag{P.5}$$

for all sufficiently large t. It permits interpretation of the observable process (y^∞, u^∞) as a control system solution, where the object is determined by the parameter τ_∞ and the regulator by a set of coefficients $\delta(\tau_\infty)$. The 'noise', with the property as in (P.5), acts on such an object. The aforesaid control system can be expressed in the form

$$x_{t+1} = A(\tau_\infty)x_t + B\bar{v}_{t+1}. \tag{P.6}$$

Here the matrix A and vector B are written in the standard form. The matrix $A(\tau)$ is stable for any $\tau \in \mathcal{T}$. Hence for every $\tau \in \mathcal{T}$ there exists a positive matrix $H = H(\tau)$ which satisfies the inequality $A^*(\tau)\,H(\tau)\,A(\tau) \le \bar{\rho}H(\tau)$. By virtue of the compactness of the set $\mathcal{T}$, the quantity $\bar{\rho}$, $(0<\bar{\rho}<1)$, can be chosen independent of vector $\tau \in \mathcal{T}$. $V(x) = [x^*H(\tau_\infty)x]^{1/2}$, the Liapunov function for the system trajectory (P.6) can be found out as

$$V(x_{t+1}) = [(A(\tau_\infty)x_t + B\bar{v}_{t+1})^* \times H(\tau_\infty)(A(\tau_\infty)x_t + B\bar{v}_{t+1})]^{1/2}$$

$$\le [x_t^*A^*(\tau_\infty)H(\tau_\infty)A(\tau_\infty)x_t]^{1/2} + |\bar{v}_{t+1}|\,[B^*H(\tau_\infty)B]^{1/2}$$

$$\le \bar{\rho}V(x_t) + C(\rho + \varepsilon|\Phi_t|).$$

It is not difficult to show that there exists a constant C_1 such that $|\Phi_t| \le C_1|x_t|$ and so

$$V(x_{t+1}) \le (\bar{\rho} + \varepsilon C_2)V(x_t) + \rho C. \tag{P.7}$$

Here C_2 is a constant. If we make a choice of ε so small that $\bar{\rho} + \varepsilon C_2 < 1$, then from (P.7) we have

$$C_\Phi = \sup_t |\Phi_t| < \infty. \tag{P.8}$$

Thus, because of (P.8), the 'noise' $\bar{v}_\infty$ appears to be bounded for sufficiently small ε (vide (P.5)). To be precise, for any $\varepsilon' > 0$ we can choose ε such that $|\bar{v}_t| \le \rho + \varepsilon'$ for sufficiently large t. In view of the conditon (3) of the theorem, we have

$$\tilde{J}[\tau_\infty, \delta(\tau_\infty), \overline{v_0^\infty}] \le Q[\tau_\infty, \delta(\tau_\infty), \quad \rho+\varepsilon']$$

$$= q(\tau_\infty, \rho+\varepsilon') \le q(\tau_\infty, \rho)+\varepsilon'',$$

Here $\varepsilon''>0$ can be chosen as small as possible for a sufficiently small value of ε'. Thus,

$$J(y_0^\infty, \quad u_0^\infty) = \tilde{J}[\tau_\infty, \delta(\tau_\infty), \overline{v_0^\infty}]$$

$$\le q(\tau_\infty, \rho+\varepsilon'') \le \rho' q(\tau_\infty, \rho). \tag{P.9}$$

Here ρ' may be chosen as close as possible to unity for sufficiently small number of ε''.

Corresponding to the algorithm (2.25) of the choice of estimates τ_t, the parameter τ_∞ appears to be the 'best' in the set $\mathcal{T}_\infty$. Since for the 'true' parameter τ inclusion of $\tau \in \mathcal{T}_t$ is justified for all t, then the inequality $q(\tau_\infty, \rho) \le q(\tau, \rho)$ must be satisfied and hence the inequality (2.10) follows from (P.9). Thus, the theorem 4.2.1 is proved.

4.P.3 Proof of the Theorem 4.3.1

Let $\eta_t = B^*(x_{t+1} - Ax_t)$. By virtue of (3.10) we have

$$\eta_t = (\tau - \tau_t)x_t. \tag{P.10}$$

Considering (P.10), (3.13) and (3.12)

$$\mathrm{Sp}\,(\tau_{t+1}-\tau)(\tau_{t+1}-\tau)^* = \mathrm{Sp}\,(\tau_t-\tau)(\tau_t-\tau)^*$$

$$-2\frac{\eta_t^*\gamma_t\eta_t}{\mu+|x_t|^2}+\frac{\eta_t^*\gamma_t^2\eta_t|x_t|^2}{(\mu+|x_t|^2)^2} \le \mathrm{Sp}\,(\tau_t-\tau)(\tau_t-\tau)^*$$

$$-2\frac{\eta_t^*\gamma_t\eta_t}{\mu+|x_t|^2}+\frac{\eta_t^*\gamma_t^2\eta_t}{\mu+|x_t|^2} \le \mathrm{Sp}\,(\tau_t-\tau)(\tau_t-\tau)^*$$

$$-\varepsilon^2\frac{\eta_t^*\eta_t}{\mu+|x_t|^2}. \tag{P.11}$$

From here we get the estimates

$$\mathrm{Sp}\,(\tau_{t+1}-\tau)(\tau_{t+1}-\tau)^* \le \mathrm{Sp}\,(\tau_0-\tau)(\tau_0-\tau)^*$$

$$-\varepsilon^2\sum_{s=0}^{t}\frac{|\eta_s|^2}{\mu+|x_s|^2} \tag{P.12}$$

$$\text{and } \sum_{s=0}^{\infty}|\eta_s|^2\left(\mu+|\overset{1}{x}_s|^2\right)^{-1} < \infty. \tag{P.13}$$

The inequality (P.12) leads to (3.17). The matrix H is defined by the equation

$$A^*HA - H = -I. \tag{P.14}$$

Because of (3.8), $A^{n-1}=0$ is satisfied and so from (P.14) we find out

$$H = \sum_{s=0}^{n-1}(A^*)^s A^s. \tag{P.15}$$

It can be easily shown that $|A|=1$ and hence from (P.15) the estimates $1 \leq |H| \leq n$ follow. We examine the Liapunov function $V(x) = x^* Hx$ for the system trajectory (3.10) with consideration of (P.10), (P.14) and (P.15). We have

$$V(x_{t+1}) - V(x_t) \leq -|x_t|^2 + 2\eta_t^* HAx_t + \eta_t^* H\eta_t$$

$$\leq -|x_t|^2(1-\rho) + (\rho^{-1}|HA|^2 + |H|)|\eta_t|^2. \tag{P.16}$$

It results from (P.13) that $|\eta_t|^2 \leq \rho_t(\mu + |x_t|^2)$, $\sum_{t=0}^{\infty} \rho_t < \infty$.

Considering these estimates, we have

$$V(x_{t+1}) \leq V(x_t) - |x_t|^2(1-\rho - C\rho_t) + C\mu\rho_t$$

$$\leq \left(1 - \frac{1-\rho-C\rho_t}{n}\right)V(x_t) + C\mu\rho_t. \tag{P.17}$$

Here, for brevity's sake, we have $C = \rho^{-1}|HA|^2 + |H|$. Since $\lim_{t\to\infty} \rho_t = 0$, it follows from (P.17) that $\lim_{t\to\infty} V(x_t) = 0$. It means fulfillment of the control objective (3.11). For $\mu = 0$, the inequality may be written more precisely as

$$V(x_{t+1}) \leq V(x_t) - |x_t|^2 + \rho'_t |x_t|^2 \leq (1 - (1-\rho'_t)/n)V(x_t).$$

Here $\rho'_t = (2|HA| + \rho_t |H|)\rho_t$. From here

$$V(x_{t+1}) \leq \prod_{s=0}^{t}\left(\frac{n-1}{n} + \frac{\rho' s}{n}\right)V(x_0)$$

$$= \left(\frac{n-1}{n}\right)^{t+1} \prod_{s=0}^{t}(1+\rho''_s)V(x_0)$$

$$\leq \left(\frac{n-1}{n}\right)^{t+1} V(x_0) \exp \sum_{s=0}^{t} \rho''_s.$$

Here $\{\rho''_s\}$ are non-negative numbers of a finite series. The expression (3.18) results from the inequality thus obtained. The theorem is proved.

4.P.4 Proof of the Theorem 4.3.2

By virtue of (3.36), (3.30), 3.25) and (3.29), we have

$$\text{Sp}\,(\tau_{t+1}-\tau)(\tau_{t+1}-\tau)^* \leq \text{Sp}\left(\tau_t - \tau + \gamma_t \frac{\eta_{t+1}^* x_t^*}{|x_t|^2}\right)$$

$$\times\left(\tau_t - \tau + \gamma_t \frac{\eta_{t+1} x_t^*}{|x_t|^2}\right) = \text{Sp}\,(\tau_t - \tau)(\tau_t - \tau)^*$$

$$-\delta_t (y_{t+1} - v_{t+1})^* \tilde{H}\eta_{t+1} |x_t|^{-2}$$

$$-\delta_t |x_t|^{-2}[\eta_{t+1}^* \tilde{H}\eta_{t+1} - \delta_t \eta_{t+1}^* \tilde{H}^2 \eta_{t+1} + v_{t+1}^* \tilde{H}\eta_{t+1}].$$

Considering the inequality (3.31) and equality (3.32), we get

$$\mathrm{Sp}\,(\tau_{t+1}-\tau)(\tau_{t+1}-\tau)^* \le \mathrm{Sp}\,(\tau_t-\tau)(\tau_t-\tau)^*$$

$$-\delta_t\,|x_t|^{-2}(\eta_{t+1}-v_{t+1})^*\tilde{H}\eta_{t+1}. \tag{P.18}$$

The inequality (3.39) follows from above and

$$\sum_{t=0}^{\infty}\delta_t\,|x_t|^{-2}(\eta_{t+1}-v_{t+1}^{(0)})^*\tilde{H}\eta_{t+1} \le \mathrm{Sp}\,(\tau_0-\tau)(\tau_0-\tau)^*. \tag{P.19}$$

Considering (3.33), (3.34) and (3.36), we find out (from (P.19))

$$\sum_{t=0}^{\infty}\theta(\mu_{t+1}-\varepsilon)\,|x_t|^{-2}|\tilde{H}\eta_{t+1}|^2 \le \varepsilon^{-2}\,\mathrm{Sp}\,(\tau-\tau_0)(\tau-\tau_0)^* \tag{P.20}$$

where $\theta(s)=1$ for $s\ge 0$, and $\theta(s)=0$ for $s<0$.

We introduce the Liapunov function $V(x)=(x^*Hx)^{1/2}$with matrix that satisfies (3.35). By virtue of (3.21) and (3.28), we have

$$V(x_{t+1})=[(Ax_t+B\eta_{t+1})^*H(Ax_t+B\eta_{t+1})]^{1/2}$$

$$\le (x_t^*A^*HAx_t)^{1/2}+(\eta_{t+1}^*\tilde{H}^2\eta_{T+1})^{1/2}$$

$$\le \alpha V(x_t)+|\tilde{H}\eta_{t+1}| \le \alpha V(x_t)$$

$$+\theta(\mu_{t+1}-\varepsilon)\,|\tilde{H}\eta_{t+1}|+\theta(\varepsilon-\mu_{t+1})\,|\tilde{H}\eta_{t+1}|. \tag{P.21}$$

Next we observe that for $\mu_{t+1}<\varepsilon$ the inequality $|\tilde{H}\eta_{t+1}|<C$ results (3.40) and (3.34). Hence we determine from (P.21)

$$V(x_{t+1}) \le (\alpha+\alpha_t)V(x_t)+C. \tag{P.22}$$

According to (P.20), the series from α_t^2 converages. The theorem of 4.3.2 then follows the simple assertion given below.

Lemma 4.P.1 *For the values* $V_t\ge 0$, *which satisfy the inequality*

$$V_{t+1} \le (\alpha+\alpha_t)V_t+C \tag{P.23}$$

with positive constants $\alpha, C\,(\alpha<1)$, and non-negative values of α_t, for which

$$\sum_{t=0}^{\infty}\alpha_t^2 \le \nu^2, \quad \nu>0, \tag{P.24}$$

we have the following estimate true for all $t=1,2,\ldots,$

$$V_t \le (\alpha+\nu/\sqrt{t})^t V_0 + +C\sum_{k=0}^{t-1}(\alpha+\nu/\sqrt{k})^k,$$

$$(\alpha+\nu/\sqrt{k})^k\,|_{k=0}=1. \tag{P.25}$$

Proof of Lemma We have from (P.23)

$$V_t \le \prod_{s=0}^{t-1}(\alpha+\alpha_s)V_0+C\left[1+\sum_{s=1}^{t-1}\prod_{k=s}^{t-1}(\alpha+\alpha_k)\right]. \tag{P.26}$$

Using the inequality

$$\left(\prod_{k=s}^{t-1} x_k\right)^{1/(t-s)} \le \left(\sum_{k=s}^{t-1} x_k\right)/(t-s)$$

we have

$$\prod_{k=s}^{t-1}(\alpha+\alpha_k) \le \left(\alpha+\sum_{k=s}^{t-1}\alpha_k/(t-s)\right)^{t-s} \le (\alpha+\nu/\sqrt{t-s})^{t-s}.$$

(P.26) leads to (P.25). The lemma is proved.

The inequality (3.39) in the same obvious manner results from (P.23) and (P.20) along with lemma 4.P.1.

4.P.5 Proof of the Theorem 4.3.3

Following the discussions in section 4.P.4, the inequality below is introduced using (3.56) and (3.55) :

$$|\tau_{t+1}-\tau|^2 \le |\tau_t-\tau|^2-\varepsilon\delta_t\,|\tilde{H}_t\eta_{t+1}|^2\,|\Phi_t\Phi_t^*|^{-1}. \tag{P.27}$$

The relations below result from above

$$\lim_{t\to\infty}(\tau_{t+1}-\tau_t)=0,\quad \sum_{t=0}^{\infty}\theta(\mu_{t+1}-\varepsilon)\,|\tilde{H}_t\eta_{t+1}|^2\,|\Phi_t\Phi_t^*|^{-1}$$

$$<\varepsilon^{-2}\,|\tau-\tau_0|^2. \tag{P.28}$$

Because of (3.58), we get for the function $V_t(x)=[x^*H(\tau_{t-1})x]^{1/2}$

$$V_{t+1}(x_{t+1}) \le \alpha[x_t^*H(\tau_t)x_t]^{1/2}+|\tilde{H}_{tt}\eta_{t+1}| \le \alpha V_t(x_t)$$

$$+\alpha\,|x_t|\cdot|H(\tau_t)-H(\tau_{t-1})|^{1/2}+|\tilde{H}_t\eta_{t+1}|.$$

We observe that for $\mu_{t+1}<\varepsilon$, the following is satisfied

$$|\tilde{H}_t\eta_{t+1}| \le \rho(C_H^{-1}-\varepsilon)^{-1}.$$

Here $C_H = \operatorname{Sup}|H(\tau)|^{1/2},\quad \tau\in\mathcal{T}.$

Hence considering (3.54), we have

$$V_{t+1}(x_{t+1}) \le (\tilde{\alpha}_t-\alpha_t)V_t(x_t)+\rho(C_H^{-1}-\varepsilon)^{-1} \tag{P.29}$$

where $\tilde{\alpha}_t=\alpha+\alpha\,|H(\tau_t)-H(\tau_{t-1})|^{1/2}$ and

$$\alpha_t=\theta(\mu_{t+1}-\varepsilon)C_\Phi\,|\tilde{H}_t\eta_{t+1}|\,|\Phi_t\Phi_t^*|^{-1/2}. \tag{P.30}$$

Because of the first relationship in (P.28) and compactness of the set $\mathcal{T}$ there exists a final instant t such that for $t\ge t_*$ the condition $\tilde{\alpha}_t\le\tilde{\alpha}<1$ is satisfied. Hence, for such t, the following inequality is valid

$$V_{t+1}(x_{t+1}) \le (\tilde{\alpha}+\alpha_t)V_t(x_t)+\rho(C_H^{-1}-\varepsilon)^{-1}. \tag{P.31}$$

Considering (P.28) and applying the lemma 4.P.1 to $V_t = V_t(x_t)$, we have

$$V_t(x_t) \le V_{t_*}(x_{t_*})(\tilde{\alpha}+\nu/\sqrt{t-t_*})^{t-t_*}$$

$$+\rho(C_H^{-1}-\varepsilon)^{-1}\sum_{k=0}^{t-t_*-1}(\tilde{\alpha}+\nu/\sqrt{k})^k, \tag{P.32}$$

where $\upsilon = \varepsilon^{-1}|\tau-\tau_0|$. The inequality $\sup|x_t|<\infty$ follows (P.32). Consideration of fulfillment of the inequalities

$$|H_t\eta_{t+1}| \ge \rho(1-\varepsilon)^{-1},$$

$$|\Phi_t\Phi_t^*|^{-1/2} \ge C_\Phi^{-2}|x_t|^{-2} \ge C_\Phi^{-2}(\sup|x_t|)^{-2} > 0.$$

(when $\mu_{t+1} \ge \varepsilon$), it can be established from (P.28) that $\sum\theta(\mu_{t+1}-\varepsilon)<\infty$, *i.e.*, the algorithm (3.56) converges in finite time. Let it be $\bar{t} = \bar{t}(x_0^\infty, \tau_0^\infty, \tau)$. When $t \ge \bar{t}$, we get from (P.30), $\alpha_t = 0$. Hence the inequality (3.60) results easily from (P.31). The montonicity of $\{|\tau_t-\tau|\}$ follows (P.27). The theorem is proved.

4.P.6 Proof of the Theorem 4.3.4

The inequality below is introduced following the section 4.P.5 and from (3.80) :

$$|\tau_{t+1}-\tau|^2 \le |\tau_t-\tau|^2 - \theta_t(|\eta_{t+1}|\Phi_t|^2$$

$$-\nu_{t+1}\operatorname{sign}\eta_{t+1})(|\eta_{t+1}|-\rho). \tag{P.33}$$

The above is a proof of the validity of (3.83). From (P.33) we find out the following taking (3.79) into account :

$$\sum_{t=0}^{\infty}\theta_t|\Phi_t|^{-2}(|\eta_{t+1}|-\rho)^2 \le |\tau-\tau_0|^2. \tag{P.34}$$

The vector $x_t =$ col $(y_t, \ldots, y_{t-n+1})$ satisfies the equation below, because of (3.62) :

$$x_{t+1} = Ax_t + B(\eta_{t+1}-\hat{y}_{t+1}). \tag{P.35}$$

Here $n \times n$ matrix A and the n-vector B are determined by the formulae (3.8) when $l = 1$. The inequality $A^*HA \le n^{-1}(n-1)H$ is satisfied by the diagonal matrix $H =$ diag $(n, \;\; n-1, \ldots, 1)$. Hence considering the formula $B^*HB = n$, we found out (vide (P.21) an (P.22))

$$V(x_{t+1}) \le (\alpha+\alpha_t)V(x_t)+C, \quad V(x) = (x^*Hx)^{1/2}. \tag{P.36}$$

Here α, C are the constant given by the formulae (3.85) and $\alpha_t = \sqrt{n}\,\theta_t|x_t|^{-1}(|\eta_{t+1}|-\rho)$. Considering the obvious inequality $|\Phi_t| \le |x_t|$ (vide (3.67)), the convergence of the series in α_t^2 is ascertained because of (P.34). Application of the lemma 4.P.1 to (P.36) leads to the inequality (3.84). From (3.84) it results that $\sup_t|\Phi_t|<\infty$ and hence from (P.34) we obtain the bounded relationship $\lim_{t\to\infty}(|\eta_{t+1}|-\rho)\theta_t = 0$. It is equivalent to the first inequality in (3.86). The second inequality of (3.86) follows (3.69).

For $\varepsilon > \rho$, the above bounded relationship signifies $\lim\theta_t = 0, \quad t\to\infty$, *i.e.*, the algorithm (3.80) converges in finite time. The theorem 4.3.4 is proved.

5

The Problem of Dynamic System Identification

The identification problem is a multifaceted one. We shall consider only some of its aspects related to the estimation of unknown parameters of dynamic systems.

In this chapter a few general estimation schemes grouped around the algorithms of Kalman-Bucy and stochastic approximation will be studied. The studies result in a sufficiently generalised situation that embraces not only the series of traditional stochastic estimation but a few minimax estimation problems as well. The basic aim of the analysis carried out below is determined by the desire to obtain the consistent (in stochastic cases strongly consistent) estimates with little assumptions about the characteristics of the estimation scheme. It may permit application of these algorithms to the adaptive control problems.

§ 5.1 Optimal Recursive Estimation

The implication of the optimal estimation is that the criterion for estimation quality is given and hints for the set of admissible estimates, in which the optimal estimate of the given quality is to be determined, are available. The set of admissible estimates is itself identified with the set of functions of attainable (observable and measurable) quantities related to the estimable quantity*. The effectiveness of the solution of the estimation problem depends, to a great extent, on the character of this relationship, and (it is understood) on the expression of the criterion and the set of admissible estimates. The theory of linear estimates is more or less well developed for cases where the estimates are linear functions of the given observation, and the criterion is to minimize the mean square error of the estimation. In a number of cases, the solution of the optimal linear estimation may be carried out up to the algorithm that can be realized simply in a computer. If the optimal estimation process is considered to be growing with time (for

*In mathematical statistics the functions of the observable quantities are known as statistical, and the optimal estimate is searched out in the given class of statistics.

example, it is necessary to consider the estimates of the process due to new observations), then it is an optimal filter problem. Partially one can speak of linear filtering, if a sequence of linear estimates is formed.

The central result of the filter theory is a recursive estimation algorithm known as the Kalman-Bucy filter. The primary optimal characteristics of this algorithm were established with the assumptions that the estimable quantity satisfies the linear stochastic equation and the linear function of this quantity is observed in the background of white noise. Various other generalisations of the estimation problem have also been obtained and these have led to the solution in the form of Kalman-Bucy filter. These results have immediate relationship to the problem of adaptive control strategy design. The recursive estimation algorithms of unknown parameters become, in many cases, constituents of the control strategies.

In this paragraph we shall study the estimation problem in a scheme used for achieving the solution in the form of Kalman-Bucy filter. The functional to determine the optimal criterion contains ensemble operation which may be, in case of a stochastic noise, a mathematical expectation. Or else it may be the maximization operation if it is not a stochastic problem. In the former case, we apply mean square error criteria of minimization, which has been well-studied. The latter is a less known case that corresponds to a minimax estimation problem.

5.1.1 *Formulation of the Estimation Problem*

The estimation problem can be formulated for the terminal states of the linear system following R.Kalman. These are estimated from the output observation.

Let the dynamic object be expressed by the equation

$$x_0 = v'_0, \quad x_{t+1} = F_t x_t + v'_{t+1}, \quad t = 0, \quad 1, \quad \ldots, \quad t_*. \tag{1.1}$$

Here $F_0, \quad \ldots, \quad F_{t_*}$ is the given sequence of matrices (dimension $n \times n$); $\{v'_t\}$ is the observable disturbance on the object. The output is written by the equation

$$y_t = \Phi_t x_t + v''_t, \quad t = 0, \quad 1, \quad \ldots, \quad t_*. \tag{1.2}$$

Here $\Phi_0, \quad \ldots, \quad \Phi_{t_*}$ is the sequence of matrices (dimension $l \times n$); $\{v''_t\}$ is the noise in the observation channel. Let the ensemble operation $\tilde{\mathbf{M}}$ be determined in the set $\left\{v_0^{t_*+1}\right\}$ of disturbance aggregates $v_0^{t_*+1}$, $v_t = \operatorname{col}(v'_t, \quad v''_{t-1})$, $v''_{-1} = 0$. The class V^{t_*+1} of uncertainty in the object and its conditions of operation will be characterized as a set of all disturbances $v_0^{t_*+1}$ that satisfy the following condition

$$\tilde{\mathbf{M}} \sum_{t=0}^{t_*+1} v_t^* N_t^{-1} v_t = r_{t_*+1}. \tag{1.3}$$

Here N_t represents the non-negative matrices, *e.g.*,

$$N_t = \begin{vmatrix} Q_t & S_t \\ S_t^* & R_t \end{vmatrix} \tag{1.4}$$

and r_{t_*+1} is a positive number. N_t may be singular matrices. Then it may be assumed that for all disturbances $v_0^{t_*+1} \in \mathbf{V}^{t_*+1}$, the vector v_t belongs to the region where matrices N_t assume values, and the expression $v_t^* N_t^{-1} v_t$ exists. For example if $Q_t = 0$, $S_t = 0$, $R_t > 0$, then the vector v_t belongs to the region of the matrices (1.4) for $v_t' = 0$. Then

$$v_t^* N_t^{-1} v_t = (v_t'')^* R_t^{-1} v''_t.$$

The estimation problem consists of achieving the estimate $\hat{x}_t$ of the state variable x_t from the observations $y_0^{t_*}$. The estimates to be found out are linear, *i.e.*

$$\hat{x}_t = \sum_{s=0}^{t_*} h_s y_s.$$

Here $h_s = h(t,s)$ is the weighting function of the linear filter. The estimate $\hat{x}_t$ must be the best in the sense of minimization of the functional

$$J_t\left[h_0^{t_*}\right] = \tilde{\mathbf{M}} \mid x_t - \hat{x}_t \mid^2 \rightarrow \min_{\{h_0^{t_*}\}}$$

where $\tilde{\mathbf{M}}$ is the above mentioned ensemble operation in the set $\mathbf{V}^{t_*+1}$.

Depending on the instant t in the estimation process, we introduce two terminologies : interpolation (smoothening), if $t \le t_*$ and extrapolation (prediction, forecasting), if $t \ge t_*$. In future only the extrapolation problems in forward direction (in time $t = t_* + 1$) will be studied. Other cases can be stuided in a similar manner. The linear estimate and the performance criterion have the following forms :

$$\hat{x}_{t_*+1} = \sum_{t=0}^{t_*} h_t y_t, \tag{1.5}$$

$$J_{t_*+1}\left[h_0^{t_*}\right] = \tilde{\mathbf{M}} \mid x_{t_*+1} - \hat{x}_{t_*+1} \mid^2 \rightarrow \min_{\{h_0^{t_*}\}} . \tag{1.6}$$

5.1.2 *Duality of the Estimation and Optimal Control Problems*

Minimization of the functional (1.6) with respect to $h_0^{t_*}$ and in the presence of the boundary conditions (1.1) to (1.3) degenerates into equivalent dual (conjugate) problem. The Z_t matrices (dimension $n \times n$) are used in the conjugate problem and they are determined by the conditions :

$$F_t^* Z_{t+1} = Z_t + \Phi_t^* h_t^*, \quad t = 0, \quad 1, \quad \ldots, \quad t_*, \tag{1.7}$$

$$Z_{t_*+1} = I_n.$$

Here F_t, Φ_t and h_t are the matrices obtained from the relations (1.1) to (1.5). The elementary schemes permit us to establish the equality

$$x_{t_*+1} - \hat{x}_{t_*+1} = L_{t_*+1} v_0^{t_*+1}, \tag{1.8}$$

where

$$L_{t_*+1} = \mid Z_0^*, 0, \quad Z_1^*, \quad -h_0, \quad Z_2^*, \quad -h_1, \quad \ldots, \quad Z_{t_*+1}^*, -h_{t_*} \mid . \tag{1.9}$$

By virtue of (1.5) it follows from (1.8) that this problem is equivalent to the problem of minimization of the functional

$$J_{t_*+1}(\cdot) = \bar{\mathbf{M}}\left\{| L_{t_*+1} v_0^{t_*+1} |^2\right\} \to \min_{\{h_0^{t_*}\}} \tag{1.10}$$

for the relations in (1.7) based on the variables Z_t and h_t.

The problem of (1.10) and (1.7) is a conjugate one of the estimation problem and it has the expression of an optimal control problem. The above is true when the matrices Z_t are interpreted as the states of the linear object (1.7), which could be determined by the matrix control h'^*_0.

5.1.3 *Solution of the Matrix Linear Quadratic Cost Optimization Problem*

$$G_{t_*+1} = G_{t_*+1}\left(Z_0^{t_*+1}, h_0^{t_*}\right) = Z_0^* Q_0 Z_0 + \sum_{t=0}^{t_*} | Z^*_{t+1}, -h_t | N_t | Z^*_{t+1}, -h_t |^* . \tag{1.11}$$

Here the matrices N_t are given by (1.4) and Q_0 is the constituent of the matrix N_0.

If the matrices Z_t and h_t vary arbitrarily satisfying the linear relations (1.7) only, then the formula (1.11) defines the set $\{G_{t^*+1}\}$ of symmetric matrices G_{t^*+1}. The matrix linear quadratic cost optimization problem consists of determination of a matrix $\tilde{G}_{t_*+1} \varepsilon \{G_{t^*+1}\}$ such that for any matrix $G_{t_*+1} \in \{G_{t^*+1}\}$, the inequality $G_{t_*+1} \geq \tilde{G}_{t^*+1}$ is satisfied. Presumably the above inequality is in the corresponding quadratic form.

It will be shown that the solution of the matrix problem exists, when degeneration of the matrices R_t, $t = 0, 1 \ldots\ldots, t_* + 1$, is not assumed. The following symbol is introduced for the non-zero n-vector

$$J_a\left[h_0^{t_*}\right] = a^* G_{t_*+1} a, \tag{1.12}$$

$$z_t = Z_{t_*-t+1} a, \quad A_t = F^*_{t_*-t},$$

$$B_t = \Phi^*_{t_*-t}, \quad u_t = -h^*_{t_*-t} a. \tag{1.13}$$

Then the linear quadratic problem is obtained as

$$z_{t+1} = A_t z_t + B_t u_t, \quad t = 1, \quad \ldots, \quad t_*, \tag{1.14}$$

$$J\left[u_0^{t_*}\right] = J_a\left[h_0^{t_*}\right] = \sum_{t=0}^{t_*} \begin{pmatrix} z_t \\ u_t \end{pmatrix}^* N_{t_*-t+1} \begin{pmatrix} z_t \\ u_t \end{pmatrix} + z^*_{t_*+1} Q_0 z_{t_*+1}. \tag{1.15}$$

A special case of the above ($S_t \equiv 0$) was studied in section 2.1.4. The optimal control strategy minimizing the quadratic functional (1.15), for linear boundaries as in (1.14), results from the results obtained in section 2.3 and can be expressed as

$$u_t = K_t^* z_t,$$

where $K_t = -\left(R_{t_*-t} + B_t^* P_{t_*-t} B_t\right)^{-1} \left(A_t^* P_{t_*-t} B_t + S^*_{t_*-t}\right)$ and P_t,
the symmetric matrices are determined by Riccati's equation

$$P_{t+1} = A^*_{t_*-t} P_t A_{t_*-t} + Q_{t+1} - \left(A^*_{t_*-t} P_t B_{t_*-t} + S_t\right)$$

$$\times\left(R_t + B^*_{t_*-t} P_t - B_{t_*-t}\right)^{-1}\left(A_{t_*-t} P_t B_{t_*-t} + S_t\right)^*. \tag{1.16}$$

The initial condition is

$$P_0 = Q_0. \tag{1.17}$$

The performance criterion of (1.15) takes the least value equal to $J(u'^*_0) = z_0^* P_{t_*+1} z_0.$

It is obvious from (1.16) and (1.17) that the matrices P_t and hence K_t do not depend on the choice of the vector A. Considering (1.14) and (1.15), and also the arbitrariness of the vector a, it may be shown that the solution of the matrix optimal control problem does exist and it has the following expression.

The sequence of the matrices h'^*_0 that minimizes the matrix $G_{t_*+1}\left(Z_0^{t_*+1}, h'^*_0\right)$, with the boundaries as in (1.7), is determined by the relations

$$h_t = Z^*_{t+1} K_t, \quad t = 0, \quad \ldots, \quad t_*, \tag{1.18}$$

$$\text{where } K_t = (F_t P_t \Phi_t^* + S_t)(R_t + \Phi_t P_t \Phi_t^*)^{-1} \tag{1.19}$$

P_t, the non-negative matrices are found out from the following conditions

$$P_{t+1} = F_t P_t F_t^* - (F_t P_t \Phi_t^* + S_t)(R_t + \Phi_t P_t \Phi_t^*)^{-1} (F_t P_t \Phi_t^* + S_t)^* + Q_{t+1}, \tag{1.20}$$

$$P_0 = Q_0. \tag{1.21}$$

Then

$$\min_{\{G_{t_*+1}\}} G_{t_*+1} = P_{t_*+1}. \tag{1.22}$$

when the matrices h_t are chosen from the condition in (1.18), the equation (1.7) can be expressed as

$$Z_t = (F_t^* - \Phi_t^* K_t^*) Z_{t+1}, Z_{(t+1)} = I_n.$$

Hence it follows that

$$Z_t = \prod_{s=t}^{t_*} (F_s - K_s \Phi_s)^*, \quad t = 0, \quad \ldots, \quad t_*, \quad Z_{t_*+1} = I_n. \tag{1.23}$$

5.1.4 *The Kalman-Bucy Filter*

From the equality of (1.8) we have

$$\left(x_{t_*+1} - \hat{x}_{t_*+1}\right)\left(x_{t_*+1} - \hat{x}_{t_*+1}\right)^* = \left(L_{t_*+1} N^{1/2}\right)\left(N^{-1/2} v_0^{t_*+1}\right)\left(N^{-1/2} v_0^{t_*+1}\right)^*\left(L_{t_*+1} N^{1/2}\right)^*,$$

where N is the arbitrary non-negative matrix of proper dimension. The expression $N^{-1/2} v_0^{t_*+1}$ exists for this matrix. Choosing N as a block-diagonal matrix, for which the blocks are given by N_t, t = 0, ... , t_*, and using the Schwartz inequality, we can find out the following (by virtue of (1.11))

$$\left(x_{t_*+1}-\hat{x}_{t_*+1}\right)\left(x_{t_*+1}-\hat{x}_{t_*+1}\right)^* \leq$$

$$\leq\left(\sum_{t=0}^{t_*+1} v_t^* N_t^{-1} v_t\right) G_{t_*+1}\left(Z_0^{t_*+1}, h_0^{t_*}\right).$$

Hence by virtue of (1.3)

$$\tilde{\mathbf{M}}\,|\,x_{t_*+1}-\hat{x}_{t_*+1}\,|^2 \leq r_{t_*+1}\,\mathrm{Sp}\,G_{t_*+1}\left(Z_0^{t_*+1}, h_0^{t_*}\right). \tag{1.24}$$

It results from (1.24) that the relations (1.18) to (1.21) and (1.23) give h'^*_0, the matrices minimizing G_{t_*+1}, and hence the quantity $\mathrm{Sp}\,G_{t_*+1}$ may serve as the approximation of the optimal weighting function of the filter (1.5) provided the quantity $\mathrm{Sp}\,P_{t_*}$ (vide (1.22)) is sufficiently small. An important fact is that for typical ensemble operations $\tilde{\mathbf{M}}$ the above matrices $h_0^{t_*}$ minimize the functional (1.10) and/or functional (1.6). Attention is now diverted, for the time being, to the characteristics of a filter (1.5), in which the weighting function is given by the relations in (1.18) to (1.21) and (1.23).

Theorem 5.1.1 *We assume that the matrices R_t in (1.4) are positive and the estimates $(\hat{x}_t)$ are formed with the help of the filter given by*

$$\hat{x}_{s+1}=\sum_{t=0}^{s} h(s,t)y_t,\ t=0,\quad 1,\quad \ldots, \tag{1.25}$$

where $h(s,t)=(F_s-K_s\Phi_s)(F_{s-1}-K_{s-1}\Phi_{s-1})\quad\ldots$

$$\times(F_{t+1}-K_{t+1}\Phi_{t+1})K_t. \tag{1.26}$$

Here the matrices K_t are given by the formulae (1.19), (1.20) and (1.21).
Then the estimates (1.25) satisfy the difference equation

$$\hat{x}_{t+1}=F_t\hat{x}_t+K_t[y_t-\Phi_t\hat{x}_t],\quad t=0,1,\quad \ldots. \tag{1.27}$$

The intial condition $\hat{x}_0$ is chosen from the condition

$$x_0=\underset{x_1}{\operatorname{argmin}}\ \tilde{\mathbf{M}}\,|\,x_0-v'_0\,|^2. \tag{1.28}$$

The recursive relations in (1.27), (1.19) and (1.20) are known as Kalman-Bucy filter and the matrix K_t is known as the Kalman gain.

5.1.5 *Optimal Properties of the Kalman-Bucy Filter*

(*a*) *Mean square estimation.* Let us assume that in the equations of (1.1) and (1.2), the disturbances v_t are random quantities (they are designated as v_t). Let them possess the following properties

$$\mathbf{M}v_t=0,\quad \mathbf{M}v_tv_s^*=N_t\delta_{ts}. \tag{1.29}$$

Here N_t is the covariance of the random signals v_t. The relationship of (1.3) determining the set of uncertainty of the control object (1.1) has the following expression

$$\mathbf{M}\sum_{t=0}^{t_*+1} v_t^* N_t^{-1} v_t = \mathrm{Sp}\sum_{t=0}^{t_*+1} N_t^{-1/2}{}_t\, \mathbf{M} v_t v_t^* N_t^{-1/2} = t_* + 1. \qquad (1.30)$$

This means that $r_{t_*+1} = t_* + 1$. The problem is to determine the linear estimate

$$\hat{x}_{t+1} = \sum_{s=0}^{t} h_s y_s, \qquad (1.31)$$

that minimizes the functional

$$J_{t+1}[h_0^t] = \mathbf{M}\,|x_{t+1} - \hat{x}_{t+1}|^2. \qquad (1.32)$$

This shows that for optimization in the mean square sense, it is necessary to predict linearly in the forward time direction, the states of the control object (1.1) from the observations of the output variable.

Lemma 5.1.1 *The estimation problem of (1.31) and (1.32) in a class of disturbances to be determined by the conditions in (1.29) and (1.30), is equivalent to the related matrix linear quadratic problem.*[*] *Consequently, the estimates* $\hat{x}_t$ *which can be obtained by using Kalman-Bucy filter of (1.27), (1.19), (1.20) and (1.21) (for initial condition* $\hat{x}_0 = 0$*) are optimal in the mean square sense in a class of linear estimates and for all* t.

If the random quantities v_t *are Gaussian, then the above estimates are optimal in a class of arbitrary unpredictable estimates.*[**] Then

$$\min_{\{h_0^t\}} \mathbf{M}\,|x_{t+1} - \hat{x}_{t+1}|^2 = \mathrm{Sp}\,P_t.$$

(*b*) *Minimax estimation.* Let the sequence of the matrices N_t (as in (1.4)) and the positive numbers r_t be given. It is assumed that the disturbances (v_t) satisfy the following inequality for any t

$$\sum_{s=0}^{t} v_s^* N_s^{-1} v_s \le r_t. \qquad (1.33)$$

$\mathbf{V}_t$, the set of all possible disturbances v_0^t, have the same property and in the present case it characterizes the uncertainty of the control objects under the conditions it operates. Let the estimate (1.31) of the states of the control object (1.1) be found out so that it is optimal in the sense of minimum of the functional

$$J_{t+1}[h_0^t] = \sup_{v_0^t \in \mathbf{V}^t} |x_{t+1} - \hat{x}_{t+1}|^2, \qquad (1.34)$$

i.e., it is necessary to determine the minimax linear prediction for the next interval of time.

Lemma 5.1.2 *The estimation problem of (1.31) and (1.34) in a class of disturbances* $\mathbf{V}^\infty$, *which is to be determined by the conditions in (1.33), is equivalent to the related linear quadratic problem. Consequently, the estimates* $\hat{x}_t$ *obtained from the Kalman-Bucy filter of (1.27), (1.19), (1.20) and (1.21) with an initial condition of (1.28) are optimal in the minimax sense in a class of linear estimates. Then*

---

[*] This means that the minima of the corresponding functionals can be achieved by choosing one and only one set of matrices h_0^t.

[**] In a class of estimates $\hat{x}_{t+1} = f_t(y_0^t)$, where $f_t(\cdot)$ is an arbitrary Borelev function.

$$\min_{\{h_j^t\}} \sup_{v_0^t \in V^t} |x_{t+1} - \hat{x}_{t+1}|^2 = r_{t+1}\Lambda_{t+1}. \tag{1.35}$$

(c) *Estimation of the parameter of a useful signal observed in the background of a noncentered noise.* The above results will be now applied to a particular case, where

$$y_t = \Phi_t \tau + v_t, \quad t = 0, \quad 1, \quad \ldots, \tag{1.36}$$

i.e., when the 'state' of the object τ does not vary with time ($\tau_{t+1} = \tau_t$). The relationship of (1.36) will be interpreted as the equation of the observation channel where the useful signal depends on the unknown vector parameter τ. The other possible interpretation of the relationship of (1.36) is that it is the expression for the dynamic object with an unknown set of coefficients τ. In either case y_t and Φ_t are assumed to be known and may be used, at any instant t, for estimation of the unknown vector τ. For simplicity's sake, we limit our attention only to a scalar y_t.

The estimation of parameter τ from (1.36) with centered stationary and independent disturbance v^∞ has been well-studied. The Kalman-Bucy filter here coincides with the well known recursive modification of the method of least squares. It has the expression as :

$$\tau_t = \tau_{t-1} + \gamma_{t-1}\Phi_t^* L_t(y_t - \Phi_t \tau_{t-1}),$$

$$\gamma_t = \gamma_{t-1} - \gamma_{t-1}\Phi_t^* L_t \Phi_t \gamma_{t-1}, \tag{1.37}$$

$$L_t = (R + \Phi_t \gamma_{t-1} \Phi_t^*)^{-1}.$$

Here $\mathbf{M}v_t^2 = R$. The convergence can be proved for a probability of 1 for the estimates (τ_t) to the vector τ independent of the choice of the initial values (vector τ_0 and the positive matrix γ_0 in the algorithm (1.37), and for natural assumptions.

It is interesting to explain how much sensitive similar results are to the assumptions made about the disturbance. To be precise, is it possible to ensure smallness of the estimation error $\Delta\tau_t = \tau - \tau_t$, if the disturbance v^∞ differs little from the white noise v^∞ ? A similar question arises, for example, in the case when the random quantities v_t are noncentered and their mean values change in some neighboured of the null.

We shall introduce the result obtained for disturbance character on the basis of 'invariance' of Kalman-Bucy filter.

Lemma 5.1.3 *Let us assume that the following conditions are fulfilled :*

(1) disturbance v^∞ in (1.36) is determined by the properties

$$v_t = \bar{v}_t + \tilde{v}_t.$$

Here $\tilde{v}_\infty$ is the centered random process with independent values and

$$(t+1)^{-1} \sum_{k=0}^{t} \bar{v}_k^2 \le C_{\bar{v}}^2, \quad \mathbf{M}\tilde{v}_t^2 \le \sigma_{\tilde{v}}^2 \tag{1.38}$$

where $C_{\bar{v}}$ and $\sigma_{\tilde{v}}$ are constants ;

2) the row vectors Φ_t for the random quantities are independent (at any instant) with random quantities $\tilde{v}_t$ and

$$\mathbf{M}|\Phi_t|^2 \le C_\Phi < \infty; \tag{1.39}$$

3) a positive random quantity λ_0 exists such that it satisfies the limiting inequality, given below, with probability 1

$$\underline{\lim}_{t\to\infty} \quad (t+1)^{-1} \quad \sum_{s=0}^{t} \Phi_s^*\Phi_s \geq \lambda_0 I. \tag{1.40}$$

Here I is the unit matrix of power dimension.

Then for estimates τ_t formed by the procedure of (1.37) with probability 1, the following inequality is true

$$\overline{\lim}_{t\to\infty} \quad |\tau_t - \tau|^2 \leq \lambda_0^{-1} C_{\bar{v}}^2. \tag{1.41}$$

From the inequality of (1.41) it results that, when $\bar{v}_t \equiv 0$ the estimates τ_t are strongly consistent.

It results from the proof of the lemma 5.1.3 that the estimate (1.41) can not be improved upon in a class of disturbances (1.38), if the inequality of (1.40) can not be improved upon.

Thus, when the Kalman-Bucy filter is examined, we observe that the estimates obtained have the characteristics of robustness with respect to the centricity of independent disturbances. In a special case of the filter, the estimates converge to the small neighborhood of the unknown parameter τ, if the deviation from the stationarity determinable by the quantity $C_{\bar{v}}^2$ is sufficiently small.

It may be observed that for identification of dynamic systems (expressed by the equation in (1.36)), the row vector Φ_t consists of a set of outputs at previous intervals of time t and the control signals. This leads to the fact that λ_0 in (1.40) may appear to be dependent on $C_{\bar{v}}$ and then the smallness of the right-hand side in the inequality (1.41) may not exist, when $C_{\bar{v}} \to 0$. This property is observed in practice, if in $\{v_t\}$ the white noise component $\{\bar{v}_t\}$ is absent.

§ 5.2 The Kalman-Bucy Filter for Tracking the Parameter Drift in Dynamic Systems

Application of the estimation scheme as developed in paragraphs 5.1.4 and 5.1.5 for dynamic system parameter estimation requires an additional foundation. When the object is written in the form of (1.1) and (1.2), we get the matrices F_t and Φ_t which may depend on previous observations. The estimation algorithms in paragraphs of 5.1.3 and 5.1.4 excluded such a possibility.

In this section it is assumed that the varying parameters of the object represent themselves as a Markovian process. The Kalman-Bucy filter is studied in the estimation problem with a parameter drift for the state observations.

Let the control object be expressed by the nonstationary equation

$$x_0 = v_0, \quad x_{t+1} = A_t(\tau_t)x_t + B_t(\tau_t)u_t + v_{t+1}, \quad t = 0, \quad 1, \quad \dots, \tag{2.1}$$

where $x_t \in R^n$, $u_t \in R^m$, $v_t \in R^n$ and $A_t(\tau)$, $B_t(\tau)$ are the time varying matrices of appropriate dimensions, which depend linearly on vector parameter $\tau \in \mathcal{T} \subseteq \mathbf{R}^p$. v^∞ the disturbance is assumed to be a white noise :

$$\mathbf{M}v_t = 0, \quad \mathbf{M}v_t v_s^* = R_v(t)\delta_{ts}. \tag{2.2}$$

The parameter $\tau = \tau_t$, on which the coefficients of the equation (2.1) depend, 'drifts' with time. Let the evolution τ_t be given by the equation

$$\tau_t = w_0, \quad \tau_{t+1} = F_t \tau_t + w_{t+1}, \quad t = 0, \quad 1, \quad \ldots . \tag{2.3}$$

Here F_t is the given quadratic matrices of dimensions $p \times p$ and w^∞ is the white noise with the following properties

$$\mathbf{M}w_t = 0, \quad \mathbf{M}w_t w_s^* = R_w(t)\delta_{ts}. \tag{2.4}$$

It is assumed that the control u^∞ in (2.1) is formed with the help of the relations

$$u_t = U_t(u^{t-1}, x^t) + e_t, \tag{2.5}$$

where $U_0^\infty(\cdot)$ is the fixed sequence of functions $U_t(\cdot)$. The number of arguments of these functions may change with time and e_t is the random quantity which is mutually as well as disturbance (v^∞, w^∞) independent.

The n-dimensional vector function Φ_t and $n \times p$ dimensional Φ_t are introduced by the relations

$$\phi_t = A_t(0)x_t + B_t(0)u_t, \tag{2.6}$$

$$\Phi_t \tau_t = A_t(\tau_t)x_t + B_t(\tau_t)u_t - \phi_t.$$

The above is permitted on the assumption of linear dependence of coefficients of the equation (2.1) on parameter τ_t. It is further assumed that for all t the values ϕ_t, Φ_t possess finite second moments

$$\mathbf{M}\phi_t\phi_t^* \; < \; \infty, \quad \mathbf{M}\Phi_t\Phi_t^* \; < \; \infty. \tag{2.7}$$

Designating

$$y_t = x_{t+1} - \phi_t, \tag{2.8}$$

we can write the control object (2.1) in the form

$$y_t = \Phi_t \tau_t + v_{t+1}. \tag{2.9}$$

Thus, we arrive at the system (1.1) and (1.2) for $v'_{t+1} = w_{t+1}$, $v''_t = v_{t+1}$, $x_t = \tau_t$. However, there exists an essential speciality of this case : the matrices Φ_t in 'the observation channel' (2.9) depend on previous 'states' τ^{t-1}(vide (2.6)) and 'observations' y^{t-1}, and consequently they are random.

5.2.1 *Optimal Tracking of the Parameter Drift in the Presence of Gaussian Disturbances*

Let at the instant t the random quantities y^t and Φ^t be known. It is required to find out the estimate τ_{t+1} with these information when τ_{t+1} is measurable with respect to the random quantities y^t and Φ^t, and also the minimizing functional given below in a class of estimates $\hat{\tau}_{t+1} = f_1(y^t, \Phi^t)$ (which are arbitrary Borelev functions of observed quantities)

$$J_{t+1}[\hat{\tau}_{t+1}] = \mathbf{M}\,|\,\tau_{t+1} - \hat{\tau}_{t+1}\,|^2. \tag{2.10}$$

By a simple rearrangement it is possible to arrive at the well-known expression for the optimal estimation

$$\hat{\tau}_{t+1} = \mathbf{M}\{\tau_{t+1} \mid y^t, \Phi^t\} = \int \tau \mu_{\tau_{t+1}}(d\tau \mid y^t, \Phi^t). \tag{2.11}$$

Here $\mu\tau_{t+1}(\cdot \mid y^t, \Phi^t)$ is the *a posteriori* distribution of random quantities (τ_{t+1}) with observations y^t and Φ^t (vide section 2.4). Thus the estimates, best in the mean square sense, are related to determination of *a posteriori* distribution of the random quantity τ_{t+1}.

The problem of determining the *a posteriori* distributions is in a general complex. It is possible to obtain, as in section 2.5, convenient recursive relations for *a posteriori* distributions, if we assume that the disturbances v^∞ and w^∞ in equations (2.1) and (2.3) are Gaussian and the disturbance e_t in (2.5) has a density function (not necessarily Gaussian) (which we designate by $p_e(\cdot \mid t)$. Similar notations are used for Gaussian density functions of random quantities v_t and w_t :

$$p_v(\cdot \mid t) \sim N[0, \quad R_v(t)], \quad p_w(\cdot \mid t) \sim N[0, \quad R_w(t)],$$

$$p_\tau(\cdot \mid 0) = p_w(\cdot \mid 0) \sim N(\bar{\tau}, \quad R_\tau). \tag{2.12}$$

The above assumption permits us to show that the density function $\mu_{\tau_{t+1}}(\cdot \mid y^t, \Phi^t)$ has a Gaussian distribution $p_{\tau_{t+1}}(\cdot \mid y^t, \Phi^t)$. The latter can be found out by the parameters $\hat{\tau}_{t+1}$ and P_{t+1}, which depend on observed data y^t, Φ^t:

$$p_{\tau_{t+1}}(\cdot \mid y^t, \quad \Phi^t) \sim N(\hat{\tau}_{t+1}, \quad P_{t+1}). \tag{2.13}$$

A recursive relation, as in section 2.5, can also be found out for $\hat{\tau}_t$ and P_t.

Theorem 5.2.1 *With uncorrelated Gaussian disturbances v_0^∞ and w_0^∞, and the random parameter, (τ depending on them with a priori Gaussian distribution (vide (2.12)), the a posteriori densities $p_{\tau_{t+1}}(\cdot \mid y^{t-1}, \Phi^{t-1})$ (of the random quantity τ_t) are Gaussian. Their statistics $\hat{\tau}_t$ and P_t are determined by the recursive relations*

$$\hat{\tau}_{t+1} = f_t \hat{\tau}_t + K_t(y_t - \Phi_t \hat{\tau}_t), \tag{2.14}$$

$$P_{t+1} = F_t P_t F_t^* - K_t \Phi_t P_t F_t^* + R_w(t+1), \tag{2.15}$$

$$K_t = F_t P_t \Phi_t^* [R_v(t+1) + \Phi_t P_t \Phi_t^*]^{-1}, \quad t = 0, \quad 1, \quad \ldots, \tag{2.16}$$

and by the initial conditions

$$\hat{\tau}_0 = \bar{\tau}, \quad P_0 = R_\tau. \tag{2.17}$$

Then the conditional covariance of the estimation error is given by

$$P_t = \mathbf{M}\{(\tau_t - \hat{\tau}_t)(\tau_t - \hat{\tau}_t)^* \mid y^{t-1}, \quad \Phi^{t-1}\}. \tag{2.18}$$

Hence (vide (2.10))

$$\min_t \quad J_{t+1}(\hat{\tau}) = \text{ SP } \mathbf{M}P_{t+1}. \tag{2.19}$$

Comparison of the formulae (2.14) to (2.16), (1.27), (1.19) and (1.20) shows that they are identical when we choose $S_t \equiv 0$, $Q_t = R_w(t+1)$ and $R_t = R_v(t+1)$. Thus, the statistics

($\hat{\tau}_t$ and P_t) of the *a posteriori* distribution (of random quantities τ_t) can be found out by Kalman-Bucy filter of (2.14) to (2.16), when the observations are given by y^{t-1} and Φ^{t-1}, and the initial conditions are as in (2.17). Then the statistics $\hat{\tau}_t$ is optimal in a mean square sense for the estimation of the random quantity τ_t in a class of arbitrary statistics (of the given observations y^{t-1} and Φ^{t-1}, or x_t and u^{t-1}).

5.2.2 *Asymptotic Properties of the Kalman-Bucy Filter*

The theorem 5.2.1 asserts optimality of the estimates as obtained by the procedure in (2.14) to (2.17). However, it is necessary to know the preciseness of tracking of the parameter τ_t. Can we ascertain sufficient closeness of the estimate $\hat{\tau}_t$ to τ_t, or at least when $t \to \infty$, in case the disturbance w^∞ in (2.13) is small? The answer to this is related to the nature of the covariance matrix P_t determined by the equation (2.16) depending on the magnitude of the 'free member' $R_w(t+1)$. This question is discussed below without going into the means of obtaining the filter equation. In addition the properties of the equation (2.16) are analysed without any presupposition about the optimality of the Kalman-Bucy filter. The same results are applied to the estimation problem without any presupposition about the Gaussian characteristics of the disturbances v^∞ and w^∞.

(a) Continuity in the small for the Riccati equation. The equation (2.15) along with (2.16) has the expression of a nonstationary Riccati equation. We shall not refer to the presupposition that $R_v(t)$ and $R_w(t)$ are the covariances of the random quantities. We assume that these are arbitrary symmetric matrices with the following properties

$$R_t = R_v(t+1) > 0, \quad Q_t = R_w(t+1) \geq 0,$$

$$\pi(Q_0^\infty) = \sup_t \mid Q_t \mid < \infty. \tag{2.20}$$

Moreover, we shall consider the case of $F_t \equiv F$ only. It is equivalent to the Riccati equation with the following expression

$$P_{t+1} = FP_tF^* - FP_t\Phi_t^*(R_t + \Phi_tP_t\Phi_t^*)^{-1}\Phi_tP_tF^* + Q_t. \tag{2.21}$$

Definition 5.2.1 The Riccati equation (2.21) is *continuous in the small* with respect to the free member Q_0^∞, if for any initial condition $P_0 \geq 0$, the amplitude below tends to zero when $\pi(Q_0^\infty) \to 0$ (vide (2.20)).

$$\overline{\lim_{t\to\infty}} \mid P_t \mid = \rho(P_0^\infty). \tag{2.22}$$

Introduction of the continuity in the small for the Riccati equation is important for the supplements. In the identification and estimation problems it ensures asymptotic (for $t \to \infty$) smallness of the estimation error when the drift speed of the parameter is not very large. In particular, if the parameter drift is absent then the continuity in the small for the Riccati equation signifies the sound basis of the estimates of corresponding Kalman-Bucy filter irrespective of the choice of the initial conditions. It opens the possibility of a well-founded use of Kalman-Bucy filter in identification (of linear systems) and adaptive control problems. It is so in spite of the fact that the parameters of the observation scheme are dependent on the preformulation of the estimation process and the Kalman-Bucy filter may not possess optimal properties. (Optimality of Kalman-

Bucy filter in identification of unknown parameters is important in exceptional cases, such as, that shown in section 5.2.1.)

If the matrix F is stable then it results directly from the inequality (as is obvious from (2.21)),

$$P_{t+1} \leq F P_t F^* + Q_t, \tag{2.23}$$

that the equation (2.21) is continuous in the small with respect to the free member. Moreover, if the matrix F possesses even one eigen value outside the unit circle, then $\rho(P_0^\infty)$ does not decrease to zero for $\pi(Q_0^\infty) \to 0$. This means that the continuity in the small with respect to the free member is apparently absent. Hence in the following section the principal attention has been given to the case when all the eigenvalues of the matrix F are spread in the closed unit circle and a few of them on the periphery of the unit circle. If the latter eigenvalues possess short elementary divisions then the matrices $\{F^t\}$ are not bounded uniformly along t. (Some elements of these matrices grow in steps with increasing time). In practice, the parameter drift model is often used in which the matrix F has the expression of a Jordanian cage corresponding to the eigen value of 1. Such a parameter variation corresponds to their stepped drift 'tracking' the disturbance w^∞.

(b) Convergence of the homogenous Riccati equation. We introduce the following 'homogenous' Riccati equation along with (2.21)

$$H_{t+1} = F H_t F^* - F H_t \Phi_t^* (R_t + \Phi_t H_t \Phi_t^*)^{-1} \Phi_t H_t F^*. \tag{2.24}$$

Theorem 5.2.2 *Let the following conditions be fulfilled*

(1) $\det F \neq 0$ and $\det (F - \lambda I) \neq 0$, when $|\lambda| > 1$;

(2) $\sup |R_t| < \infty$ and $R_t > 0$, for all $t \geq 0$;

(3) For any p-vector $a \neq 0$ the following is true

$$\lim_{t \to \infty} \left\{ |f^{-t} a|^2 + \sum_{s=0}^{t} |\Phi_s^* f^{-s} a|^2 \right\} = \infty. \tag{2.25}$$

Then the following is true for any $H_0 \geq 0$, where H_0^∞ is the sequence that can be determined by the equation (2.24)

$$\lim_{t \to \infty} H_t = 0. \tag{2.26}$$

Then the nonstationary Liapunov equation

$$G_{t+1}^0 = (F - \tilde{K}_{tt} \Phi_t) G_t^0 (F - \tilde{K}_t \Phi_t)^*, \tag{2.27}$$

$$\tilde{K}_t = F H_t \Phi_t^* (R_t + \Phi_t H_t \Phi_t^*)^{-1}, \tag{2.28}$$

gives birth to the converging sequence of matrices $\{G_t^0\}$ for any initial condition $G_0^0 > 0$, and

$$\lim_{t \to \infty} G_t^0 = 0. \tag{2.29}$$

For a stable matrix F the condition in (2.25) is intentionally made to be satisfied. For the existence of eigen values of F on the unit circle with multiple elementary divisions, the relationship in (2.29) is not trivial. This is because $\tilde{K}_t \to 0$ when $t \to \infty$ (according to (2.28)), and for $\tilde{K}_t \equiv 0$ the equation (2.27) is unstable.

(c) *Identificability of the object when parameter drift is absent.* The theorem 5.2.2 gives an immediate application to the identification of the object (2.1), where the unknown parameters are constant : $\tau_{(t+1)} = \tau_t = \tau$, *i.e.*, a special version of the drift model (2.3) is available when the disturbance w^∞ is not present. Application of Kalman-Bucy filter to estimation of unknown parameter τ leads to a homogeneous Riccati equation (2.24). The equation of (2.26) is fulfilled under the conditions of the theorem 5.2.2. Assuming that the disturbance v^∞ is a white noise and recalling the results of the section 5.1.5, we may ascertain that the estimates $\{\hat{\tau}_t\}$ obtained by the Kalman-Bucy filter are well-founded, *i.e.*, $\lim_{t\to\infty} \hat{\tau}_t = \tau$ (in the mean square sense). The basic condition to be verified is that the following properties of the information matrix, must tend to zero in 'all directions',* for $t \to \infty$

$$I_t = \sum_{s=0}^{t} \Phi_s^* \Phi_s. \tag{2.30}$$

This condition is well-known and it requires sufficiently diversified control (or disturbance) signals in the object (2.1). In some cases, to ensure such properties of the information matrix (2.30), it is convenient to randomize the control strategy.

(d) Stationary Riccati equation. The equation (2.21) will be studied with a relatively simple case where $\Phi_t \equiv \Phi$, $R_t \equiv R$ and $Q_t \equiv Q$, *i.e.*, we take the case of a stationary Riccati equation

$$P_{t+1} = FP_tF^* - FP_t\Phi^*(R + \Phi P_t \Phi^*)^{-1}\Phi P_t F^* + Q. \tag{2.31}$$

Theorem 5.2.3 *Let us assume that the following conditions are satisfied :*

1) $\det F \neq 0, \det(F - \lambda I_p) \neq 0$, when $|\lambda| > 1$;

2) $R > 0$;

3) the pair (F, Φ) is detectable.

Then the equation (2.31) is continuous in the small with respect to the free member Q. In particular, $\lim_{t\to\infty} P_t = 0$ when $Q = 0$.

e) Tracking of the drift of the nonstationary quadratic functional extremum. Let us choose the following simple problem of extremum control where the control object is inertialess. Let the control objective be minimization of the quadratic functional

$$J(u, t) = \mathbf{M}(u_t - u_t^0)^* Q(u_t - u_t^0). \tag{2.32}$$

Here Q is the positive matrix and u_t^0 is the nonstationary solution of the external problem. The quantity u_t^0 is random and unknown. Let the drift u_t^0 be given by a linear model

$$u_t^0 = L\tau_t, \quad \tau_{t+1} = F\tau_t + w_{t+1}, \tag{2.33}$$

where L and F are the unknown matrices, det F = *0*, and w_t are the disturbances. It is assumed that the observed value at any instant t is proportional to the stochastic gradient

*This means that the least eigen value λ_t of the matrix I_t must grow unbounded for $t \to \infty$.

of the functional (2.32), *i.e.*, the following is known

$$x_t = \Phi(u_t^0 - u_t) + v_t,$$

where u_t is the control signal at that instant, v_t is the observational error and Φ is the known orthogonal matrix. Let the disturbances w^∞ and v^∞ take independent values at different instants, such that they are independent of each other and

$$\mathbf{M}w_t = 0, \quad \mathbf{M}v_t = 0, \quad \mathbf{M}w_t w_t^* = R_w, \quad \mathbf{M}v_t v_t^* = R_v > 0. \tag{2.34}$$

The drifting solution (u_t^0) of the external problem is to be estimated from the observations of x_t (and presumably u_t). Designating $y_t = x_t + \Phi u_t$, the observation scheme can be written in a more common form :

$$y_t = \Phi u_t^0 + v_t. \tag{2.35}$$

This gives us the special case of the estimation problem (2.9) when the matrix Φ is independent of time.

The solutions u_{t+1}^0 obtained from the given observations (y_t and u_t) of the Kalman-Bucy filter (2.14) to (2.17) will be made use of for determination of the estimates $\hat{u}_{t+1}$. The expressions for Kalman-Bucy filter are rewritten as :

$$\hat{u}_t = L\hat{\tau}_{t+1} l, \quad \hat{\tau}_{t+1} = F\hat{\tau}_t + K_t(y_t - \Phi L\hat{\tau}_t), \tag{2.36}$$

$$P_{t+1} = FP_tF^* - K_t\Phi P_tF^* + R_w, \tag{2.37}$$

$$K_t = FP_t\Phi^*(R_v + \Phi P\Phi^*)^{-1}. \tag{2.38}$$

The initial conditions are chosen as vector τ_0 and matrix $P_0 \geq 0$.

Thus, the equation (2.37) of the Kalman-Bucy filter is a stationary Riccati equation. Suppose the conditions of the theorem 5.2.3 are fulfilled for the equation (2.37) when $R = R_w$, and the pair $(F, R_w^{1/2})$ is stabilizable. It can be ascertained that there exists $\lim_{t \to \infty} P_t = P_*$ where P_* (following the theorem 3.2.2) is the unique non-negative solution of the Lur'e equation

$$P_* = FP_*F^* - FP_*\Phi(R_v + \Phi P_*\Phi^*)^{-1}\Phi P_*F^* + R_w. \tag{2.39}$$

A simple calculation shows* that

$$\min_u J(u, t) = \mathrm{Sp}\, Q^{1/2} L P_* L^* Q^{1/2},$$

i.e., the estimates $\hat{u}_t$ can be determined by the formulae (2.36) to (2.38) and they are asymptotically optimal independent of the choice of initial conditions τ_0 and $P_0 \geq 0$. Applying the theorem 5.2.3 to the equation (2.39) we ascertain that $|P_*| \to 0$ when $|R_w| \to 0$.

f) Nonstationary Riccati equation. Let us again consider the asymptotic properties of the solutions of the equation (2.24). It results from the theorem 5.2.4 that to establish

---

*It may be recalled that the estimates $\hat{u}_t$ are searched in a class of unpredicting functions of observation variables x^{t-1} and u^{t-1} *i.e.*, these are the forecasts for the future instants.

the properties of stability in the small, it is necessary to make some assumptions regarding the properties of the sequence $\{\Phi_t\}$, which are related to the condition of detectability of the pair (F, Φ) when $\Phi_t \equiv \Phi$. Such a property is the condition of uniform observation of the matrices (F and Φ^{∞}). This means that there exist a positive number α and a natural number $k \geq 1$, such that for all t we have

$$\sum_{j=0}^{k-1} (F^*)^j \Phi_{t+j}^* \Phi_{t+j} F^j \geq \alpha I_p. \qquad (2.40)$$

The condition (2.40) for $\Phi_t \equiv \Phi$ and $k = p - 1$ signifies the observability of the pair (F, Φ). It is a stronger condition than the detectability of this pair.

Theorem 5.2.4 *The equation (2.24) is continuous in the small in relation to the free member, if the conditions (1) and (2) of the theorem 5.2.3 and the condition of uniform observability (2.40) of the matrices (F, Φ^{∞}) are satisfied.*

The proof of the theorem 5.2.4 is based on establishing majorants for solutions of the equation (2.24). These majorants satisfy the stationary Riccati equation, for which the conditions of the theorem 5.2.3 are valid.

§ 5.3 Recursive Estimation

Kalman-Bucy filter is a representative of the generalised family of recursive algorithms for estimation. It has been noted that for adaptive control synthesis based on identification approach, the consistency of estimates plays the foremost role. Recursive algorithms for stochastic approximation, presented in various forms like stochastic gradient or pseudo gradient techniques, possess wide possibility. Recursive modifications of the least square method, such as, Kalman-Bucy filter often relates to the algorithms of stochastic approximation*. The concept of recursive estimation permits the proposition of a sufficiently generalised scheme for obtaining the estimates with various models of observation.

The results based on the generalised estimation scheme are acknowledged to be simpler. For identification of dynamic systems, most development has been made in the estimation of regressive model coefficients. The results related to the estimation of coefficients in autoregression equations have become classical in the statistical mathematics. These are obtained by the least square method (LSM) with independent Gaussian noises. The correlation of noises invites displacement of the estimates of LSM. More generalised estimation schemes, developed in recent years, have achieved consistent estimates when the noises are finitely dependent. A few of these results will be discussed in the following section.

5.3.1 *Forecasting Methods of Identification*

A generalised method for estimation is based on the minimization of the error of forecasting the observed quantities. This method can be written in a general form as follows.

---

*Sometimes the terminology 'stochastic process of second degree' is used for them. It is different from the procedure laid by Robbins-Monro.

Let us assume that at any instant t the observed data y^{t-1} and u^{t-1} have been achieved. These correspond to the controllable (observable) output and input signals of an object. It is necessary to build the model related to these quantities based on cause-effect relationship, so that an answer to the question was obtained, namely, what is the expected output of the object when some excitation is given at the input? Our attention is confined only to the cases when the model is desirably a deterministic one and gives a unique correspondence between the observed input and output. It is reasonable to choose such a model from the previously determined class of models. For example, the parametrized vector parameter τ may be chosen from the set $\mathcal{T} \subseteq \mathbf{R}^p$. When the value of τ is fixed, the model presents a set of functions realizing, at the instant t, the reflection of the set of observations y^{t-1} and u^{t-1} on to the output space of the object. It can be written as

$$\hat{y}_t(\tau) = Y_t(y^{t-1}, u^{t-1}; \quad \tau), \quad t = 1, 2, \dots . \tag{3.1}$$

Here the set of functions $\{Y_t(\cdot; \tau)\}$ determines a model that fits the parameter $\tau \in \mathcal{T}$. It is convenient to interpret $\hat{Y}_t(\tau)$ as the expected (predicted by the model) value of y_t, *i.e.*, $\hat{y}_t(\tau)$ is the forecast of the system output at the next instant. Hence the model (3.1) will be called as the forecasting model.

We shall next discuss the criterion for selection of the model that gives the best forecast. For this purpose, an forecasting error will be introduced as follows:

$$\varepsilon_t(\tau) = y_t - \hat{y}_t(\tau). \tag{3.2}$$

It is determinable by the model selected. We also set a scalar evaluation function $q_t[\tau, \varepsilon_t(\tau)]$. Then we can compute the functional below that can be considered as the mean (in the interval $[1, t]$) loss of the forecasting model for parameter τ

$$V_t(\tau) = t^{-1} \sum_{s=1}^{t} q_s[\tau, \varepsilon_s(\tau)]. \tag{3.3}$$

The least value of the functional (3.3) is obtained by using the optimal model, for which the parameter $\tau = \tau_t$ is determined from the condition

$$\tau_t = \tau_t(y^t, u^{t-1}) = \underset{\tau \in \mathcal{T}}{\operatorname{argmin}} \ V_t(\tau). \tag{3.4}$$

It also shows (within the task of finding out the algorithm for searching out the minimum function $V_t(\tau)$) a method of obtaining the estimates τ_t. It is not related to any statistical property of the observed data.

The selection of definite evaluation functions $q_t(\cdot)$ leads to a variety of schemes for estimation, which can be used in mathematical statistics, such as the method of least squares (for linear (with respect to τ) models and square evaluation functions), or the method of maximum similarity (the evaluation function is proportional to the logarithm of density function of the renewed disturbance), etc.

The value in (3.3) can be, in most of the cases of mathematical statistics, seen as an empirical functional of average risk. As t increases, it tends to the functional of average risk according to the law of large numbers and determines the estimation method.

5.3.2 *Selection of Forecasting Models*

The generalised scheme of estimation, presented in section 5.3.1, may appear to be weakly founded (to be precise, unsubstantiated), if during selection of forecasting models special features of the observed data are not taken into account. Usually the choice of the forecasting model is done beforehand by a mathematical model of the object. The correspondence between the forecasting models and the mathematical models is conditioned by the statistical interpretation of the identification process and is based on presentation of the output process in the 'augmented' form. Let it be explained in details. We assume that the mathematical model of the statistical object permits expression of the output y_t in the form :

$$y_t = Y_t(y^{t-1}, u^{t-1}, \tau_*) + v_t. \tag{3.5}$$

Here τ_* is the unknown parameter. (The mathematical model is given within the knowledge of this parameter.) v_t is the centralised random quantity independent of the values y^{t-1} and u^{t-1}. The relationship in (3.5) is known as *the expression for the system output in the innovation form,* since the second part in the right-hand side of (3.5) may be seen as that part of information about y_t, which is not contained in the observed data y^{t-1} and u^{t-1}. If it is known that $\tau_* \in \mathcal{T}$ then the sets of functions $\{Y_t(\cdot;\tau)\}$, parametrized by the parameter $\tau \in \mathcal{T}$, may be chosen as the class of forecasting models. A true model that responds to the parameter τ formulates the forecast according to the formulae in (3.1). The formula below, used for mean square forecasting error, ensures the least value of the functional (3.6) when $\tau = \tau_*$:

$$J_t(\tau) = \mathbf{M} \quad |\varepsilon_t(\tau)|^2 = \mathbf{M} \quad |Y_t(y^{t-1}, u^{t-1} : \tau_*)$$

$$-Y_t(y^{t-1}, u^{t-1} : \tau)|^2 + \mathbf{M}\,|v_t|^2. \tag{3.6}$$

Moreover, for many important cases, the functional (3.6) has the unique minimum. Then its minimization permits centralization of the system model and to determine the parameter τ_*.

The functional $J_t(\tau)$ contains ensemble operation $\mathbf{M}$ and sometimes it is permissible to express it by the empirical functional that does not contain ensemble operation :

$$V_t(\tau) = t^{-1} \sum_{s=1}^{t} |\varepsilon_s|^2 .$$

The substitution is justified in cases where the limits $\lim_{t\to\infty} V_t(\tau)$ and $\lim_{t\to\infty} J_t(\tau)$ exist and they coincide (this is possible due to the law of large numbers). These statements show how the forecasting models of estimation are to be chosen and indicate the possible method of substantiating the estimates furnished by them.

Typical examples of models of objects and the corresponding forecasting models will now be taken up.

a) Regressive model. Let the control object be expressed as

$$a(\nabla, \quad \tau_*)y_t = b(\nabla, \quad \tau_*)u_t + v_t \tag{3.7}$$

where the polynomials $a(\lambda)$ and $b(\lambda)$ have the usual expressions :

$$a(\lambda, \tau_*) = 1 + \lambda a_1 + \ldots + \lambda^n a_n,$$

$$b(\lambda, \tau_*) = \lambda b_1 \ldots + \lambda_n b_n.$$

Here $\{v_t\}$ is the white noise with independent values. Some of the coefficients of the equation (3.7) may be unknown and their set determines the parameter vector (τ_*) which is estimated from the observations of control and output signals. Attempts will be made to obtain such estimates with the help of a suitable forecasting model. In the present case, the output can be expressed in the augmented form as

$$y_t = [1 - a(\nabla, \tau_*)]\, y_t + b(\nabla, \tau_*)u_t + v_t.$$

Hence the forecasting model can be written as

$$\hat{y}_t(\tau) = [1 - a(\nabla, \tau)]\, y_t + b(\nabla, \tau)u_t. \tag{3.8}$$

If the row-vector Φ_{t-1} and the scalar quantity ϕ_{t-1} are given by the relations

$$\phi_{t-1} = [1 - a(\nabla, 0)]\, y_t + b(\nabla, 0)u_t,$$

$$[1 - a(\nabla, \tau)y_t + b(\nabla, \tau)u_t = \Phi_{t-1}\tau + \phi_{t-1}, \tag{3.9}$$

then the forecasting equation takes a specially simple form as

$$\hat{y}_t(\tau) = \Phi_{t-1}\tau + \phi_{t-1}, \tag{3.10}$$

i.e., it will be linear with respect to the parameter τ. The functional (3.6) can be expressed as

$$J_t(\tau) = (\tau - \tau_*)^* \,(\mathbf{M}\Phi^*_{t-1}\Phi_{t-1})\,(\tau - \tau_*) + \mathbf{M}\,|\,v_t\,|^2. \tag{3.11}$$

For the stationary version (v^∞, y^∞ are stationary processes) the above relationship does not depend on t. If the matrix $\mathbf{M}\Phi^*_{t-1}\,\Phi_{t-1}$ is nonsingular, then the minimum of the functional (3.11) is unique and coincides with the known value of the parameter (τ_*) of the object (3.7).

Replacement of the functional (3.6) to be minimized by its empirical analogue will lead to the least square method (LSM) :

$$\bar{J}_t(\tau) = t^{-1} \sum_{s=1}^{t} |\,\varepsilon_s(\tau)\,|^2 = t^{-1} \sum_{s=1}^{t} |\,y_s - \Phi_{s-1}\tau - \phi_{s-1}\,|^2 .$$

b) Regressive model for a correlated disturbance. The presence of a correlated disturbance is a more realistic situation. To account for this the object will be expressed by the equation

$$a(\nabla, \tau_*)y_t = b(\nabla, \tau_*)u_t + c(\nabla, \tau_*)v_t, \tag{3.12}$$

where $a(\lambda, \mathcal{T}_*)$ and $b(\lambda, \tau_*)$ are the same polynomials as in (3.7); $C(\lambda, \tau_*) = 1 + \lambda C_1 + \ldots\ldots + \lambda^n C_n$ are the coefficients of the polynomial $C(\lambda, \tau_*)$ and they may be unknown like the coefficients of the polynomials $a(\lambda, \tau_*)$ and $b(\lambda, \tau_*)$. The set of unknown coefficients will be again designated through τ_*.

Since the disturbances v_t are not observable, then the expression of the output of the object (3.12) in the augmentated form is more complex. It is expressed by assuming that the polynomial $C(\lambda, \tau)$ is stable for all $\tau \in \mathcal{T}$.

$\bar{y}_t$ and $\bar{u}_t$ will be now introduced. They are given by the relations

$$c(\nabla, \tau_*)\bar{y}_t = y_t,\; c(\nabla, \tau_*)\bar{u}_t = u_t. \tag{3.13}$$

As it follows from (3.13), $\bar{y}_t$ and $\bar{u}_t$ are the outputs of the linear stable filter of transfer function $H(\lambda) = [c(\lambda, \tau_*)]^{-1}$. The inputs to this transfer function are the corresponding processes y^∞, u^∞. The outputs of this filter, *i.e.*, $\bar{y}_t$ and $\bar{u}_t$, are also determined by given initial conditions. But we shall neglect such a situation in the perspective of the stability of the polynomial $C(\lambda, \tau_*)$ and the assumption that the instant (t) under consideration is such that the transient response of the filter can be neglected. Putting (3.13) in (3.12), we have

$$a(\nabla, \tau_*)\tilde{y} = b(\nabla, \tau_*)\tilde{u}_t + v_t + f_t,$$

where f_t satisfies the equation $c(\nabla, \tau_*)f_t = 0$ (hence $f_t \to 0$ when $t \to \infty$ and is determined by the transient responses of the filter (3.13). Neglecting f_t we find out

$$a(\nabla, \tau_*)\tilde{y}_t = b(\nabla, \tau_*)\tilde{u}_t + v_t$$

and consequently,

$$\tilde{y}_t = [1 - a(\nabla, \tau_*)]\,\tilde{y}_t + b(\nabla, \tau_*)\tilde{u}_t + v_t.$$

The augmented form of y_t will be determined from here after consideration of the fact that by virtue of (3.13) we have

$\bar{y}_t = y_t[1 - c(\nabla, \tau_*)]\,\tilde{y}_t$. Then

$$y_t = [c(\nabla, \tau_*) - a(\nabla, \tau_*)]\,\tilde{y}_t + b(\nabla, \tau_*)\tilde{u}_t + v_t$$

and the forecasting model can be taken in the form

$$\hat{y}_t(\tau) = [c(\nabla, \tau) - a(\nabla, \tau)]\,\tilde{y}_t + b(\nabla, \tau)\tilde{u}_t. \tag{3.14}$$

Though $\tilde{y}_t$ and $\tilde{u}_t$ are determined from (3.13), pre-formulation of the observation process and its formation are difficult as because the filter in (3.13) depends on the unknown parameter τ_* in (3.13). Hence, $\tilde{y}_t$ and $\tilde{u}_t$ in (3.14) signify the filter outputs when the filter transfer function is $[c(\lambda, \tau)]^{-1}$, *i.e.*, additional relations are to be supplemented to the model in (314)

$$c(\nabla, \tau)\tilde{y}_t = y_t, \; c(\nabla, \tau)\tilde{u}_t = u_t. \tag{3.15}$$

The usual forecasting model (3.14) and (3.15) can be rewritten in the following recursive form

$$c(\nabla, \tau)\hat{y}_t(\tau) = [c(\nabla, \tau) - a(\nabla, \tau)]\,y_t + b(\nabla, \tau)u_t. \tag{3.16}$$

It is obtained by operating the equation (3.14) with $c(\nabla, \tau)$ and considering the relationship of (3.15). When $c(\lambda, \tau) \equiv 1$ the model of (3.8) is arrived at.

If the polynomial $c(\lambda, \tau)$ contains unknown coefficients, then the model (3.14) will be nonlinear with respect to parameter τ.

Replacement of the filters (3.13) by (3.15) is natural, but the above mentioned method of formulating the forecasting model is done heuristically. Another natural (at the same time heuristic) method of formulation of the forecasting model will be considered. The equation (3.12) is written in the form

$$y_t = \{[1 - a(\nabla, \tau_*)]\,yt + b(\nabla, \tau_*)u_t + [c(\nabla, \tau_*) - 1]\,v_t\} + v_t.$$

The expression in the second bracket is determined by the pre-formulation of the control process and hence the expression can be considered to be the augmented one for y_t. The model can be chosen in the form

$$\hat{y}_t(\tau) = [1 - a(\nabla, \tau)]\,y_t + b(\nabla, \tau)u_t + [c(\nabla, t) - 1]\,v_t.$$

However, this type of forecasting models can only be used in cases where the previous values of the disturbance variables are known. (It is sometimes possible to observe the previous values of the disturbances in identification problems.) If v^{t-1} is known at the instant t, then construction of the estimate $\hat{v}^{t-1}$ may be tried and it can be used along with v^{t-1}. The estimate $\hat{v}^{t-1}$ can be determined with the help of the 'forecasting' model

$$c(\nabla, \tau)\hat{v}_t(\tau) = a(\nabla, \tau)y_t - b(\nabla, \tau)u_t. \tag{3.17}$$

The meaning of the above equation is well-understood. The forecast $y_t(\tau)$ will be then

$$\hat{y}_t(\tau) = y_t - \hat{v}_t(\tau). \tag{3.18}$$

The system of equations (3.17) and (3.18) gives the method of determining the forecast $\hat{y}_t(\tau)$ from the observed values y^{t-1} and u^{t-1}. If $c(\lambda, \tau) \equiv 1$, then the equation (3.18) becomes the same as (3.8) (by virtue of (3.17)) and the equation (3.17) then determines the forecast of the disturbance at the future instant.

Then the functional (3.6) has the following expression

$$J_t(\tau) = \mathbf{M} \mid \varepsilon_t(\tau) \mid^2 = \mathbf{M} \mid \hat{v}_t(\tau) \mid^2 . \tag{3.19}$$

The method of formulating the estimates from the condition of minimization of the functional (3.19) (to be precise from the empirical formula responding to (3.19)) is known as the extended method of the least squares.

It may be observed that the forecasting models (3.17), (3.18) and (3.16) practically coincide. In fact, because of (3.17), the equation (3.16) can be written as

$$c(\nabla, \tau)\hat{y}_t(\tau) = c(\nabla, \tau)y_t - c(\nabla, \tau)\hat{v}_t.$$

The relationship in (3.18) results from here within the transient regime.

The mathematical model (3.12) of the control object is rather a wide one. When $b(\lambda, \tau_*) \equiv 0$, its various forms are well-known in the probability theory and mathematical statistics as: autoregressive model (for $c(\lambda, \tau_*) \equiv 1$), sliding mean model (for $a(\lambda, \tau_*) \equiv 1$) and ARMA model (in the general case). The equation (3.12) is sometimes known as an ARMAC model, when $b(\lambda, \tau_*) \not\equiv 0$.

c) The generalised expression for a linear forecasting model. The above forecasting models were linear with respect to observed values. A more generalised linear forecasting model may be taken in the form

$$z_{t+1}(\tau) = F(\tau)z_t(\tau) + G(\tau)\begin{pmatrix} y_t \\ u_t \end{pmatrix},$$

$$\hat{y}_t(\tau) = H(\tau)z_t(\tau), \tag{3.20}$$

where $F(\tau)$, $G(\tau)$ and $H(\tau)$ are the matrices of proper dimensions determined on the set $\mathcal{T}$. During the formulation of the gradient algorithms the following matrix is required to be introduced.

$$\hat{V}_t(\tau) = \text{grad}_\tau \hat{y}_t(\tau) = -\text{grad}_\tau \varepsilon_t(\tau). \tag{3.21}$$

The above matrix satisfies the linear system of equations (because of (3.20)). Introducing the 'extended' state vector of the model

$$x_t(\tau) = \text{col}\ (z_t(\tau), \hat{V}_t(\tau)), \tag{3.22}$$

the forecasting model can be expressed as

$$x_{t+1}(\tau) = A(\tau)x_t(\tau) + B(\tau)\begin{pmatrix} y_t \\ u_t \end{pmatrix},$$

$$\begin{pmatrix} \hat{y}_t(\tau) \\ \mathrm{col}\ \hat{V}_t(\tau) \end{pmatrix} = C(\tau)x_t(\tau). \qquad (3.23)$$

Here col $\hat{V}_t(\tau)$ signifies a vector composed of the elements from the matrix $\hat{V}_t(\tau)$. (The method of composition is not essential, but it is fixed once for all).

The model of (3.23) shows how the forecasting $(\bar{y}_{t+1}(\tau))$ of the output $\hat{y}_{t+1}$ at a future instant is carried out from the observed data y^t and u^t, and the gradient of the forecasting error with respect to τ.

Usually a stable forecasting model is desirable (boundedness of y^∞ and u^∞ must bring forth boundedness of $\hat{y}^\infty$). In that case the matrix $F(\tau)$ in the system (3.20) must be stable and then the matrix $A(\tau)$ in the system (3.23) will be stable.

5.3.3 *Recursive Schemes for Estimation*

The fundamental idea behind the recursive identification methods lies in the use of gradient and pseudogradient methods for minimization of the functional of forecasting error. Thus, for the stochastic model of the object, the set of observed data y^t and u^t is interpreted as a realization of random quantities, for which the ensemble operation *i.e.*, averaging along the ensemble, is determined. Then the identification objective is the minimization of the mean square error of the forecast (or some function of these values). This means that the following is required to be minimized.

$$J(\tau) = \frac{1}{2}\mathbf{M}\,|\,\varepsilon_t(\tau)\,|^2 = \frac{1}{2}\mathbf{M}\,|\,y_t - \hat{y}_t(\tau)\,|^2. \qquad (3.24)$$

Here the observation scheme is assumed, for simplicity's sake, to be stationary and $J(\tau)$ is independent of t*. If $\hat{y}(\tau)$ depends smoothly on τ, then it follows that the extremal value of the parameter is to be searched out from the solutions of the equation

$$\mathrm{grad}\, J(\tau) = -\mathbf{M}\hat{V}_t(\tau)\varepsilon_t(\tau) = 0. \qquad (3.25)$$

The above equation is known as the regressive equation in *mathematical statistics*. $V_t(\tau)\varepsilon_t(\tau)$ is known as the *stochastic gradient of the functional* $J(\tau)$. To minimize $J(\tau)$, the gradient method can be used stochastically :

$$\tau_t = \tau_{t-1} + \gamma_t \hat{V}_t(\tau_{t-1})\varepsilon_t(\tau_{t-1}). \qquad (3.26)$$

Here γ_t are scalar or matrix quantities determining the magnitude of the step of an algorithm and chosen by suitable means. In mathematical statistics, the algorithm of (3.26) is known after Robbins-Monro.

The possibility of computation of the values V_t and ε_t for current values of the parameter τ, is a supposition in the use of the procedure laid in (3.26). This, generally speaking, demands storage of all pre-formulation of observations y^t and u^t, as because the forecasting models 'operate' when the parameter τ is fixed (vide, for example, (3.23)). To obtain the recursive procedure, it is necessary to change the method of formation of $V_t(\tau_{t-1})$ and $\varepsilon(\tau_{t-1})$ by permitting the possibility of their recomputation in the forecasting models.

---

*On the other hand minimization of the forecast is realized at the above mentioned instant, or a more complex function of forecasting error is given, such as, $J(\tau) = \overline{\lim_{t\to\infty}}\ \mathbf{M}\,|\,\varepsilon_t(\tau)\,|^2$.

a) Recursive estimation method based in linear forecasting model. We introduce the recursive estimation algorithm resulting from the forecasting model of (3.23):

$$\tau_{t+1}=\tau_t+\gamma_t\hat{V}_t\varepsilon_t,\ \varepsilon_t=y_t-\hat{y}_t,$$

$$x_{t+1}=A(\tau_t)x_t+B(\tau_t)\begin{pmatrix}y_t\\u_t\end{pmatrix},\tag{3.27}$$

$$\begin{pmatrix}\hat{y}_t\\ \mathrm{col}\,\hat{V}_t\end{pmatrix}=C(\tau_t)x_t.$$

The quantities $\hat{y}_t$ and $\hat{V}_t$ in the algorithm of (3.27) do not already coincide with the quantities $\hat{y}_t(\tau_{t-1})$ and $\hat{V}_t(\tau_{t-1})$. They are determined by the algorithm itself after the initial conditions and the nature of the coefficients γ_t are known. The dependence of the matrices A, B and C on τ are assumed to be known. If the matrices A, B and C are determined only on the convex set $\mathcal{T}\subseteq R^p$, then the first equation in (3.27) can be replaced by

$$\tau_t=P_{\mathcal{T}}[\tau_{t-1}+\gamma_t\hat{V}_t\varepsilon_t],\tag{3.28}$$

where $P_{\mathcal{T}}$ is the projection on $\mathcal{T}$.

The choice of the positive numbers or the matrices γ_t in the procedure (3.23), results in various algorithms. The numbers γ_t in the Robbins-Monro procedure do not depend on the observation process and satisfy the conditions below

$$\gamma_t\geq 0,\quad \sum_{t=0}^{\infty}\gamma_t=\infty,\quad \sum_{t=0}^{\infty}\gamma_t^2<\infty.\tag{3.29}$$

If we choose $\gamma_t=\tilde{\gamma}_tR_t^{-1}$, where the numbers $\tilde{\gamma}_t$ possess the properties of (3.29) and

$$R_t=t^{-1}\sum_{s=l}^{t}\hat{V}_s^*\hat{V}_s,\tag{3.30}$$

then we get the stochastic version of the Gauss-Newton algorithm, etc.

b) *Recursive algorithm for the least square method.* The algorithm is obtained on the basis of the forecasting model (3.8). In the present case $\mathrm{grad}_\tau\hat{y}_t^*(\tau)=\Phi_{t-1}^*$, where the matrix Φ_t is determined by the relations (3.11) and the LSM algorithm can be presented in the form

$$\tau_{t+1}=\tau_t+\gamma_t\Phi_t^*(y_{t+1}-\Phi_t\tau_t-\phi_t),$$

$$\gamma_t^{-1}=\gamma\sum_{s=1}^{t}\Phi_s^*\Phi_s.\tag{3.31}$$

Using the matrix identity, it is possible to rewrite the relationship for the matrix coefficients γ_t in the recursive form as

$$\gamma_t=\gamma_{t-1}-\gamma_{t-1}\Phi_t^*(1+\Phi_t\gamma_{t-1}\Phi_t^*)^{-1}\Phi_t\gamma_{t-1}.\tag{3.32}$$

The relations in (3.31) and (3.32) represent the Kalman-Bucy filter for state estimation $x_{t+1}=x_t=\tau$, based on 'observations' $y_{t+1}=\Phi_t\tau+\phi_t+v_{t+1}$ and are known as the *recursive LSM*.

c) Recursive modification of the extended least square method. The algorithm is obtained from the forecasting model of (3.17) and (3.18). The function $\hat{V}_t(\tau)$ then takes the form

$$\hat{V}_t(\tau) = \text{grad}_t \hat{y}_t(\tau) = -\text{grad}_t \hat{v}_t(\tau).$$

Differentiation of both sides of the equation (3.17) with respect to τ, gives us the equation for $\hat{V}_t(\tau)$:

$$c(\nabla, \tau)\hat{V}_t(\tau) + [\,\text{grad}_\tau c(\nabla, \tau)]\,\hat{v}_t(\tau)$$

$$= \text{grad}_\tau[a(\nabla, \tau)]\,y_t - [\,\text{grad}_\tau b(\nabla, \tau)]\,u_t. \qquad (3.33)$$

Here $\text{grad}_\tau c(\lambda, \tau)$, $\text{grad}_\tau a(\lambda, \tau)$ and $\text{grad}_\tau b(\lambda, \tau)$ are the polynomials of λ with vector coefficients of the scalar polynomials of $c(\lambda, \tau)$, $a(\lambda, \tau)$ and $b(\lambda, \tau)$ with respect to τ. Considering the relations (3.17), (3.18) and (3.33) we write the algorithm of the extended least square method as

$$\tau_{t+1} = \tau_t - \gamma_t \tau_t \hat{v}_t \hat{V}_t, \quad \hat{y}_t = y_t - \hat{v}_t,$$

$$c(\nabla, \tau_{t-1})\hat{v}_t = a(\nabla, \tau_{t-1})y_t - b(\nabla, \tau_{t-1})u_t, \qquad (3.34)$$

$$c(\nabla, \tau_{t-1})\hat{V}_t = -[\text{grad}_\tau c(\nabla, \tau)]\,\hat{v}_t +$$

$$+[\text{grad}_\tau a(\nabla, \tau)]\,y_t - [\text{grad}_\tau b(\nabla, \tau)]\,u_t,$$

where γ_t are the non-negative numbers or matrices, whose choice is still not indefinite. It may be observed that since the polynomials $a(\lambda, \tau)$, $b(\lambda, \tau)$ and $c(\lambda, \tau)$ are linearly dependent on parameter $\tau(\tau_*$ coincides with the set of unknown coefficients of these polynomials), the polynomials grad $\tau a(\lambda, \tau)$, grad $\tau b(\lambda, \tau)$, and grad $\tau c(\lambda, \tau)$, possess coefficients independent of τ.

It is understood that the basis of the algorithm (3.34) (it usually consists of establishing the foundation of the estimates τ_t) makes a special supposition of the choice of coefficients γ_t and sometimes of the initial conditions τ_0, $\hat{v}^0_{-n}$, V^0_{-n} in the algorithm (3.34).

§ 5.4 Identification of a Linear Control Object in the Presence of Correlated Noise

The approach mentioned in section 5.3 for estimation of the parameters of a dynamic object will be adopted, as an illustration, to the solution of the problem of identification of the control object expressed by the equation (3.12). In accordance with the recommendations in section 5.3.3, for which the algorithm in (3.34) may be made use of, and then only the basis of the estimates $\{\tau_t\}$ is to be established, superimposing additional limits, if necessary, on identifiability of the object and the set $\mathcal{T}$ (its indefiniteness). In reality a rigid conclusion about the properties of the estimates $\{\tau_t\}$ is obtained by overcoming considerable difficulties. The studies carried out below, to some extent characterize the difference between the solution obtained empirically (it is sufficient to use the algorithm in (3.34) based on real and sufficiently generalised considerations) and that from the formal means (an accurate assessment of the achievable estimation algorithm (3.34) is required).*

5.4.1 *Uniqueness of the Minimum of the Forecasting Performance Criterion*

The algorithm (3.34) is essentially a stochastically gradient with respect to the functional

$$J(\tau) = \overline{\lim_{t\to\infty}}\ J_t(\tau) = \overline{\lim_{t\to\infty}}\ \mathbf{M}\,|\,\hat{v}_t(\tau)\,|^2. \tag{4.1}$$

The functional determines the bounded mean square values of the forecast (vide (3.19)). Let the control u^∞ for the object (3.12) be formulated with the help of the stabilizing regulator

$$\alpha(\nabla)u_t = \beta(\nabla)y_t, \tag{4.2}$$

whose coefficients are known, $\alpha(0) = 1$. It may be recalled that the stabilizability of the regulator (4.2) signifies stability of the polynomial $g(\lambda, \tau_*)$ of the control system (3.12) and (4.2) :

$$g(\lambda, \tau_*) = a(\lambda, \tau_*)\alpha(\lambda) - b(\lambda, \tau_*)\beta(\lambda). \tag{4.3}$$

Then the processes $\hat{v}^\infty$, y^∞ and u^∞, which can be determined by the system (3.17), (3.12) and (4.2) for $t \to \infty$, will converge to a stationary processes and the functional of (4.1) is expressed through their spectral density. By elementary calculations, we get the formula

$$J(\tau) = \frac{\sigma_v^2}{2\pi i}\oint\left[\frac{g(\lambda, \tau)c(\lambda, \tau_*)}{g(\lambda, \tau_*)c(\lambda, \tau)}\right]\times\left[\frac{g(\lambda, \tau)c(\lambda, \tau_*)}{g(\lambda, \tau_*)c(\lambda, \tau)}\right]^\nabla \frac{d\lambda}{\lambda}, \tag{4.4}$$

where

$$g(\lambda, \tau) = a(\lambda, \tau)\alpha(\lambda) - b(\lambda, \tau)\beta(\lambda), \tag{4.5}$$

$$\sigma_v^2 = \mathbf{M}v_t^2. \tag{4.6}$$

Let

$$\tau_* = \text{col}\,(a_{i1}, \ldots, a_{ip}, b_{j1}, \ldots, b_{jr}, c_1, \ldots, c_n), \tag{4.7}$$

i.e., all the coefficients of the polynomial $c(\lambda, \tau_*)$ are not known except c_o and may be some of the coefficients of the polynomials $a(\lambda, \tau_*)$ and $b(\lambda, \tau_*)$. The vector τ is also assumed to have an analogous structure and the following suppositions are made based on their possible values with respect to the set :

(1) $\deg c(\lambda, \tau) = \deg c(\lambda, \tau_*) = n$, $\deg g(\lambda, \tau) = \deg g(\lambda, \tau_*) = l$, where polynomial $g(\lambda, \tau)$ is given by the formula (4.5) and $l = p + r$;

(2) $c(\lambda, \tau) \neq 0$ for $|\,\lambda\,| \leq 1$, $c(0, \tau) = a(0, \tau) = 1$;

(3) let $\lambda_1(\tau), \ldots, \lambda_n(\tau), \mu_1, \ldots, \mu_{p+r}$ be the roots of the corresponding polynomials

*It is possible that this difference is due to improper applications during the proof of the consistency of the estimates $\{\tau_t\}$: the conditions determined appear to be sufficient. Recently there is a trend to weaken the limits of the formal assertions by way of modelling the identification process in a digital computer and treating the required properties of the estimates through the programme.

$c_*(\lambda, \tau) = \lambda^n c(\lambda^{-1}, \tau)$, $g_*(\lambda, \tau_*) = \lambda^l g(\lambda^{-1}, \tau_*)$ all different and $H(\lambda, \tau) = c(\lambda, \tau) \times \text{grad}_\tau[g(\lambda, \tau)\, c^{-1}(\lambda, \tau)]$ be the polynomial vector of dimension $(n+p+r)$. Then

$$\det |H(\lambda_1, \tau), \ \ldots, \ H(\lambda_n, \tau), \ H(\mu_1, \tau), \ \ldots, \ H(\mu_{p+r}, \tau)| \neq 0; \tag{4.8}$$

(4) $|c(\lambda, \tau| + |g(\lambda, \tau)| \not\equiv 0(\forall\lambda)$.

Lemma 5.4.1 *Let the open and bounded set satisfy the above mentioned conditions (1) to (4) and* $\tau_* \in \mathcal{T}$. *Then the functional (4.4) assumes the least value when* $\tau = \tau_*$, *where*

$$\text{grad}\, J(\tau) \neq 0, \ \text{if}\ \tau \neq \tau_*. \tag{4.9}$$

5.4.2 *Modification of the Estimation Algorithm*

Substantiation of the convergence of the estimates obtained by the algorithm (3.34) encounters difficulties. The dynamic operator $c(\nabla, \tau_{t-1})$ in the algorithm is non-stationary and in spite of the stability of the polynomial $c(\lambda, \tau_{t-1})$ for all τ_{t-1}, the system of (3.34) may generate divergent estimates because of possible 'parametric resonance'.*

Hence the algorithm of (3.34) will be somewhat altered without permitting frequent changes in the estimates τ_{t-1}. The following estimation algorithm will be studied.

Let $t_1 = 1$, t_2, ..., be the increasing sequence of instants of time, $\lim_{k\to\infty} t_k = \infty$. The sequences $\hat{v}^\infty$, $\hat{V}^\infty$ and τ^∞ will be determined with the help of recursive relations

$$c(\nabla, \tau_{t-1})\hat{v}_t = a(\nabla, \tau_{t-1})y_t - b(\nabla, \tau_{t-1})u_t,$$

$$c(\nabla, \tau_{t-1})\hat{V}_t = -[\text{grad}_\tau c(\nabla, \tau)]\,\hat{v}_t$$

$$+[\text{grad}_\tau a(\nabla, \tau)]\,y_t - [\text{grad}_\tau b(\nabla, \tau)]\,u_t; \tag{4.10}$$

$$\tau_{t_{k+1}} = P_{\mathcal{T}}\left[\tau_{t_k} - \gamma_k \hat{v}_{t_{k+1}} \hat{V}_{t_{k+1}}\right], \tag{4.11}$$

$$\tau_t = \tau_{t_k}, \ t_k \le t < t_{k+1}.$$

y_t and u_t in the algorithm (4.11) are determined by the relations (3.12) and (4.2). γ_k is the sequence of non-negative numbers and $P_{\mathcal{T}}$ is the projector on the set $\mathcal{T}$, which is assumed to be convex compact.

The identification procedure of (4.10) and (4.11) with given initial values $\hat{v}^0_{-n}$, $\hat{V}^0_{-n}$ and $\tau_0 \in \mathcal{T}$ and the time sequence (t_1^∞) is determined completely. The initial conditions may be chosen as deterministic or random quantities. In either case, we assume that

$$\sum_{j=-n}^{0} \mathbf{M}\hat{v}_j^4 < \infty,$$

$$\sum_{j=-n}^{0} \mathbf{M}\,|\hat{V}_j|^4 < \infty, \quad \mathbf{M}\,|\tau_0|^2 < \infty. \tag{4.12}$$

---

*The appearance of the 'parametric resonance' in linear nonstationary systems like $a(\nabla, t)y_t = 0$, with the variables y_t increasing with time without any limit, is well-known. It is true even for the 'frozen' coefficients when the equation $a(\nabla, t)y_t = 0$ is asymptotically stable (the polynomial $a(\lambda, t)$ is stable for all t).

The correction of the estimate τ_t is carried out at the instant t_k and for the rest of the period t it does not change. As mentioned before, the estimates τ_t are to be changed quite rarely, so that the nonstationary relations in (4.10) remained stable.

5.4.3 *Consistency of the Estimates of the Identification Algorithm*

The artificial slowing down the rate of change of estimates in algorithm (4.11), in comparison with the algorithm (3.34), leads to the fact that the sequences $\hat{v}^{\infty}$ and $\hat{V}^{\infty}$ determinable by relations (4.10) are close to stationary sequences, for which the functional (4.1) has the form of (4.4).

The procedure in (4.11) (if no attention is given on $P_{\mathcal{T}}$) is stochastically antigradient with respect to the above functional. This means that in the sequence of estimates τ_t, the functional $\mathbf{M}J(\tau_t)$ is the Liapunov function, $\mathbf{M}J(\tau_t) \to \min$ for $t \to \infty$ and because of the lemma 5.4.1 it leads to $\tau_t \to \tau_*$. Such a fundamental series of considerations can be used while establishing the consistency of the identification algorithm (4.10) and (4.11). The above properties are true for some suppositions about the properties of the set $\mathcal{T}$ and disturbance v^{∞}, and parameters γ_t of algorithm (4.11). We reconsider the sequential expression of these conditions.

Let us designate the stationary processes through $\{\bar{y}_t\}$, $\{\bar{u}_t\}$, $\{\bar{v}_t(\tau)\}$ and $\{\bar{V}_t(\tau)\}$ determinable by the relations

$$
\begin{aligned}
&a(\nabla, \tau_*)\bar{y}_t = b(\nabla, \tau_*)\bar{u}_t + c(\nabla, \tau_*)v_t,\\
&\alpha(\nabla)\bar{u}_t = \beta(\nabla)\bar{y}_t,\\
&c(\nabla, \tau)\bar{v}_t(\tau) = a(\nabla, \tau)\bar{y}_t - b(\nabla, \tau)\bar{u}_t, \qquad (4.13)\\
&c(\nabla, \tau)\bar{V}_t(\tau) = -[\mathrm{grad}_\tau c(\nabla, \tau)]\bar{v}_t(\tau)\\
&+[\mathrm{grad}_\tau a(\nabla, \tau)]\bar{y}_t - [\mathrm{grad}_\tau b(\nabla, \tau)]\bar{u}_t.
\end{aligned}
$$

Lemma 5.4.2 *Suppose the following conditions are satisfied :*

1) *the white noise* v^{∞} *is centered and*

$$\mathbf{M}v_t = \sigma_v^2 > 0, \quad \mathbf{M}\,|\,v_t\,|^4 \le \sigma^4 < \infty; \qquad (4.14)$$

2) *the set* $\mathcal{T}$ *is bounded,* $\tau_* \in \mathcal{T}$ *and for any* $\tau \in \mathcal{T}$, *the polynomial* $c(\lambda, \tau)$ *is stable;*

3) *for the sequence* $\{t_k\}$ *in* (4.11), *we have*

$$\lim_{k\to\infty} (t_k - t_{k-1}) = \infty.$$

Then the following assertions are true :

1) *for any* τ, *the functions* $\bar{v}_t(\tau)$ *and* $\bar{V}_t(\tau)$ *determined by the equations (4.13) have the following valid inequalities :*

$$\sup_t \mathbf{M}\,|\,\bar{v}_t(\tau)\,|^2 < \infty, \quad \sup_t \mathbf{M}\,|\,\bar{V}_t(\tau)\,|^2 < \infty, \qquad (4.15)$$

$$\sup_t \mathbf{M}\,|\,\bar{v}_t(\tau)\,|^2 |\,\bar{V}_t(\tau)\,|^2 < \infty; \qquad (4.16)$$

2) *the functions* $\hat{v}_t$ *and* $\hat{V}_t$ *determinable by the relations (4.10) satisfy the inequalities*

$$\mathbf{M}\,|\,\bar{v}_{t_k}(\tau_{t_k}) - \hat{v}_{t_k}\,|^2 \leq C\rho^{t_k}, \tag{4.17}$$

$$\mathbf{M}\,|\,\bar{V}_{t_k}(\tau_{t_k}) - \hat{V}_{t_k}\,|^2 \leq C\rho^{t_k}, \tag{4.18}$$

$$\sup_k \mathbf{M}\,|\,\hat{v}_{t_k}\,|^2 |\,\hat{V}_{t_k}\,|^2 < \infty, \tag{4.19}$$

where C and ρ *are positive constants,* $\rho < 1$.

The inequalities (4.17) and (4.18) ensure necessary proximity of the processes $\hat{v}^\infty$ and $\hat{V}^\infty$ to the stationary processes $\bar{v}^\infty(\tau_t)$ and V $\bar{V}^\infty(\tau_t)$. It is observed that

$$\mathbf{M}\{|\,\bar{v}_t(\tau_s)\,|^2 |\,\tau_s\} = J(\tau_s),\ \mathbf{M}\{\bar{v}_t(\tau_s)\bar{V}_t(\tau_s)\,|\,\tau_s\} = \text{grad}\,J(\tau_s).$$

These permit the use of pseudogradient properties of the procedure (4.11).

Theorem 5.4.1 *Suppose the following conditions are satisfied under the conditions of the lemmas 5.4.1 and 5.4.2 :*

1) the set $\mathcal{T}$ *is convex and the projector* $P_{\mathcal{T}}$ *on this set is known ;*

2) the functional (4.4) on the set $\mathcal{T}$ *satisfies the inequality*

$$|\,J(\tau'') - J(\tau') - (\tau'' - \tau')\,\text{grad}\,J(\tau)\,| \leq C\,|\,\tau'' - \tau'\,|^2 \tag{4.20}$$

where C is a constant, $C > 0$;

3) there exists $\gamma_0 > 0$ *such that for any* $\tau \in \mathcal{T}$ *and* $\gamma \in (0, \gamma_0)$, *the following inequality is satisfied for stationary processes* $\{\bar{v}_t(\tau), \bar{V}_t(\tau)\}$ *resulting from the relations in (4.13)*

$$[\text{grad}\,J(\tau)]^*\,\mathbf{M}\{P_{\mathcal{T}}[\tau - \gamma\bar{v}_t(\tau)\bar{V}_t(\tau)] - \tau\,|\,\tau\}$$

$$\leq -(\gamma/2)\,|\,\text{grad}\,J(\tau)\,|^2; \tag{4.21}$$

4) the numbers γ_t *in the algorithm (4.11) possess the properties :*

$$\gamma_t \geq 0, \quad \sum_{t=0}^{\infty} \gamma_t = \infty, \quad \sum_{t=0}^{\infty} \gamma_t^2 < \infty.$$

Then for the estimates $\{\tau_t\}$ *which are obtained by the identification procedure of (4.10) and (4.11), independent of choice of initial conditions, and which satisfy (4.12), the limits below are valid with probability 1 :*

$$\lim_{t\to\infty} \tau_t = \tau_*, \quad \lim_{t\to\infty} J(\tau_t) = J(\tau_*),$$

$$\lim_{t\to\infty} \mathbf{M}J(\tau_t) = J(\tau_*).$$

5.4.4 *Identification of Linear Systems with Known Spectral Density of Noise*

If all the coefficients of the polynomial $c(\lambda, \tau_*)$ are known, *i.e.*,

$$c(\lambda, \tau_*) = c(\lambda), \tag{4.22}$$

then the identification scheme becomes quite simple. We shall demonstrate this. The equation (3.12) may be then written as

$$z_t = \Phi_{t-1}\tau_* + c(\nabla)v_t. \tag{4.23}$$

where $z_t = y_t - \phi_{t-1}$ and the functions ϕ_{t-1} and Φ_{t-1} are determined by the relations in (3.9).

It is convenient to introduce $\tilde{y}_t$ and $\tilde{u}_t$ to be determined by the stable filters of (3.13), *i.e.*,

$$c(\nabla)\tilde{y}_t = y_t, \quad c(\nabla)\tilde{u}_t = u_t. \tag{4.24}$$

The equation (4.23) uses these variables within the transient response and can be expressed as

$$y_t = [c(\nabla) - a(\nabla, \tau_*)]\,\tilde{y}_t + b(\nabla, \tau_*)\tilde{u}_t + v_t. \tag{4.25}$$

The functions $\tilde{\phi}_{t-1}$ and $\tilde{\Phi}_{t-1}$ are introduced by means of the relations

$$\tilde{\phi}_{t-1} = [c(\nabla) - a(\nabla, 0)]\,\tilde{y}_t + b(\nabla, 0)\tilde{u}_t, \text{ and}$$

$$\tilde{\Phi}_{t-1}\tau_* = [a(\nabla, 0) - a(\nabla, \tau_*)]\,\tilde{y}_t + [b(\nabla, \tau_*) - b(\nabla, 0)]\,\tilde{u}_t. \tag{4.26}$$

We can then write

$$y_t = \tilde{\Phi}_{t-1}\tau_* + \tilde{\phi}_{t-1} + v_t. \tag{4.27}$$

Hence the forecasting model can be expressed as

$$\hat{y}_t(\tau) = \tilde{\Phi}_{t-1}\tau + \tilde{\phi}_{t-1}. \tag{4.28}$$

Thus, the model of (3.10) that results in the LSM is obtained. The recursive procedure of (3.31) is written in the form

$$\tau_{t+1} = \tau_t + \gamma_t \tilde{\Phi}_t^*(y_{t+1} - \tilde{\Phi}_t \tau_t - \tilde{\phi}_t), \tag{4.29}$$

where γ_t is chosen as non-negative numbers. The following assertion can be made with respect to the procedure in (4.29).

Theorem 5.4.2 *Suppose the following conditions are satisfied*:

1) the numbers γ_t *satisfy the condition* (4) *in theorem* 5.4.1;

2) the following inequality is satisfied

$$p > n - k, \tag{4.30}$$

where n and p are the orders of the equations in (3.12), *and consequently k is the control delay;*[*]

3) the disturbance v^∞ *is bounded as*

$$|v_t| \le C_v. \tag{4.31}$$

Then the constrained equality $\lim\limits_{t \to \infty} \tau_t = \tau_*$ *is true while determining the estimates* τ_t *formed by the identification algorithm* (4.29) *with probability 1 and in the least square sense.*

The conditions in the theorem give additional boundary conditions which can be slackened. For example, the boundary on the noise can be replaced by presupposition of a boundary of fourth moments. The proof will then be more complicated.

---

[*]P is the greatest index (other than zero) of coefficient of the polynomials $\alpha(\lambda)$ and $\beta(\lambda)$. n is the greatest index (other than zero) of the coefficient of the polynomials $a(\lambda, \tau_*)$ and $b(\lambda, \tau_*)$, k is the least index (other than zero) of the coefficient of the polynomial $b(\lambda, \tau_*)$.

§ 5.5 Identification of Control Objects Using Test Signals

The identification problem is presented in a complex form where in the feedback control the current resutls of identification are made use of. The main difficulty lies in the little latitude in varying the control signal formulated on the basis of feedback theory. Identification in the presence of additive white noise is well-studied. Modification of the least square method, as shown in section 5.4, gives good estimates for correlated disturbances. But in this case the current values of estimation were assumed not to be made use of. (Here a stabilizing regulator without parameter tuning was considered.)

The identification problem becomes more complicated if the disturbances on the object do not possess useful statistical characteristics like stationarity, centricity, etc. A satisfactory solution to such identification problems is possible if special measures are taken to 'augment' the control signal. A more realistic approach in augmenting the control signal is the addition (to it) of a 'test' signal of sufficient intensity and spectral characteristics. If the noise in the system with a test signal is independent then the identification problem may be expected to be solved. This is established in the following section. The test signal used is a white noise. The parametric identification for the object with scalar input and output is based on the general assumptions regarding the feedback character, so that it is applicable to adaptive control design (vide section 6.3).

5.5.1 *Statement of the Identification Problem*

Let the control object be expressed as

$$a(\lambda, \tau)y_t = b(\nabla, \tau)u_t + v_t. \tag{5.1}$$

Here the polynomials $a(\lambda, \tau)$ and $b(\lambda, \tau)$ have the following forms

$$a(\lambda, \tau) = 1 + \lambda a_1 + \quad \dots \quad + \lambda^n a_n, \; b(\lambda, \tau) = \lambda^k b_k + \quad \dots \quad + \lambda^m b_m, \tag{5.2}$$

where k is the control delay, $1 \le k \le m$, and τ is a vector of unknown control coefficients (5.1) such that

$$\tau = \text{col}\,(a_1, \quad \dots, \quad a_n, b_k, \quad \dots, \quad b_m) \tag{5.3}$$

we assume that $\tau \in \mathcal{T}$, where $\mathcal{T}$ is a bounded close set of possible values of the vector (5.3).

The identification problem is to obtain the estimates $\{\tau_t\}$ of the vector (5.3) with property

$$\lim_{t \to \infty} \quad \tau_t = \tau. \tag{5.4}$$

The estimate τ_t is formed at the instant t as a function of sets of control variables u^t and output variables y^t. It is understood that this function must not depend on the unknown value of τ that determines the definite control object, but it can be found out by the properties the set $\mathcal{T}$.

If the noise and/or the control variables in (5.1) appear to be random quantities, then the convergence in (5.4) is to be understood with probability 1.

It is also supposed that

$$u_t = \overline{u}_t + \overline{w}_t, \tag{5.5.}$$

where $\overline{u}_t$ are the proper control signals formed, say, with the help of feedback signal; $\overline{w}_t$ is the specially introduced signals, the aggregate of which $\overline{w}^\infty = (\overline{w}_0, \overline{w}_1 \dots)$ will be known as the test signal. The properties of the sequences $\overline{u}^{\infty}$ and $\overline{w}^{\infty}$, which ensure the possibility of obtaining consistent estimates $\{\tau_t\}$, will be specified later.

5.5.2 *Introduction of the Estimation Parameter*

A parameter θ will be introduced along with the parameter in (5.3) with the same dimension, such as,

$$\theta = \text{col}\,(\theta_0, \quad \ldots, \quad \theta_{p-1}),$$

$$p = m + n - k + 1, \tag{5.6}$$

where the components θ_t are determined from the linear control systems

$$a(\nabla, \tau)\theta_t = b_{k+t}, \quad t = 0, 1, \quad \ldots, \quad p-1. \tag{5.7}$$

The determinant of this system is equal to 1 and hence the system can be solved uniquely with respect to the values $\theta_t = \theta_t(\tau)$. A deeper relationship between the vectors τ and θ is shown by the assertion to follow.

Lemma 5.5.1 *Suppose the pair of polynomials $\{a(\lambda), b(\lambda)\}$ is controllable (vide definition 3.4.1). Then the following vectors are linearly independent.*

$$\tilde{\theta}_t = \text{col}\,(\theta_t, \theta_{t-1}, \ldots, \theta_{t-n+1}),$$

$$t = m - k, \quad \ldots, \quad p-1,$$

Hence the quantities θ_t are determined (for $t \geq 0$) from the system (5.7) and $\theta_t = 0$ (when $t < 0$).

It follows from the lemma of 5.5.1 that the reflection $\theta(\tau) : \mathcal{T} \rightarrow \theta$, determined by the system of (5.7), is uniquely inversible, if the polynomials $\{a(\lambda, \tau), b(\lambda, \tau)\}$ are controllable for all $\tau \in \mathcal{T}$, *i.e.*, there exists an inverted reflection $\tau = \tau(\theta)$. Actually, the last n relations in the system of (5.7) can be written as

$$\tilde{\theta}_{m+n-k} + |\, \tilde{\theta}_{m+n-k-1}, \quad \ldots, \quad \tilde{\theta}_{m-k} \,|\, \text{col}\,(a_1, \quad \ldots, \quad a_n) = 0.$$

In view of the linear independence of the vectors in the lemma, this system uniquely determines the parameters $a_1, \ldots, a_n$. The parameters $b_k \ldots, b_m$ are then found out from the first $(m - k + l)$ equations of the system in (5.7). The above gives the evidence of continuity of the function $\tau(\theta)$ determined in the set $\Theta = \theta(\mathcal{T}.)$ Hence within the conditions of the lemma 5.5.1, it is sufficient to obtain consistent estimates $\theta(t)$ of the vector θ. The estimates τ_t will then be found out from the system of (5.7), in which the components $\theta_j(t)$ of the vector $\theta(t)$ are also presented along with quantities θ_j, *i.e.*, $\tau_t = \tau[\theta(t)]$.

5.5.3 *Estimation Algorithm*

We assume that the set Θ of vectors θ, which can be determined by the system of (5.7), when $\tau \in \mathcal{T}$, can be expressed as

$$\Theta = \prod_{j=0}^{p-1} \Theta_j, \quad p = m + n - k + 1, \tag{5.8}$$

where Θ_j are the bounded closed intervals, *i.e.*, Θ is a 'parallelopiped' in the Euclidean space $\mathbf{R}^p$. Let the projector on Θ_j be represented by P_j.*

The identification algorithm is taken in the form

$$\theta_j(t+1) = P_j\left[\theta_j(t) + \gamma_t\left(y_{pt+k+j} - \sum_{s=0}^{j} \theta_s(t) u_{pt-s+j}\right)\overline{w}_{pt}\right], \tag{5.9}$$

where γ_t are some non-negative quantities, $j = 0, 1, ..., p-1$; $t = 0, 1, ...$ and $\overline{w}_t$ are the test signals (vide (5.5)). The procedure of (5.9) assumes that the initial set $\theta(0) = \text{col}\,(\theta_0(0), \ldots, \theta_{p-1}(0))$ is given and it is an arbitrary vector from the set Θ.

5.5.4 *Consistency of the Estimates*

The assertion on the consistency of the estimates obtained by the algorithm of (5.9) will be formulated as follows.

Theorem 5.5.1 *Suppose the following conditions are satisfied :*

1) for all $\tau \in \mathcal{T}$, the polynomials $\{a(\lambda, \tau), b(\lambda, \tau)\}$ are the controllable pair;

2) if $\theta_j(\tau)$, $j = 0,1, \ldots\ldots, p-1$, are the functions determined by the system of (5.7), then the set $\Theta_j = \theta_j(\mathcal{T})$ generated by them will be closed bounded intervals ;

3) the test signal $\overline{w}^\infty$ is determined by the formulae

$$\overline{w}_t = 0 \text{ when } t \neq ps, \quad p = n + m - k + 1, \tag{5.10}$$

s is an arbitrary natural number

$$\overline{w}_{pt} = \sqrt{\eta_t R_t \delta_t^{-1}}\, w_t, \tag{5.11}$$

where w^∞ is a random process with independent values and

$$\mathbf{M}w_t = 0, \quad \mathbf{M}\,|\,w_t\,|^2 = \sigma_w^2 > 0, \; |\,w_t\,| \leq C_w < \infty, \tag{5.12}$$

$\{\delta_t, \eta_t\}$ are the sequences of non-negative numbers such that

$$\sum_{t=0}^{\infty} \eta_t = \infty, \quad \sum_{t=0}^{\infty} \eta_t^2 < \infty, \quad \sum_{t=0}^{\infty} \eta_t \delta_t < \infty, \quad \lim_{t\to\infty} \eta_t \delta_t^{-1} = 0, \tag{5.13}$$

$$R_t = \rho + \sum_{s=0}^{t} |\,y_{pt-s}\,|^2 + \sum_{s=0}^{t} |\,\overline{u}_{pt-s}\,|^2 ; \tag{5.14}$$

4) the noise v^∞ is independent of the test signal w^∞ and satisfies the inequality

$$\sup_t \mathbf{M}\,|\,v_t\,|^2 < \infty; \tag{5.15}$$

5) the specific control is formed by the following law

$$\overline{u}_0 = U_0(y_0),\; \overline{u}_t = U_t(\overline{u}^{t-1}, y^t), \quad t = 1, 2, ..,$$

where $U^\infty(\cdot)$ is the control strategy ensuring fulfilment of the following inequalities with probability 1

$$|\,\mathbf{M}\{\overline{u}_{pt+j}\overline{w}_{pt}\,|\,F^t\}\,| \leq C_1 \mathbf{M}\{|\,\overline{w}_{pt}\,|^2 |\,F^t\}, \tag{5.16}$$

$$\mathbf{M}\{|\,\overline{u}_{pt+j}\,|^2 |\,F^t\} \leq C_2 R_t, \tag{5.17}$$

---

* $P_j\theta_j = \theta_j$, if $\theta_j \in \Theta_j$, otherwise $P_j\theta_j$ is the end of the interval Θ_j approaching θ_j.

where $j = 0, \quad 1, \ldots, p-1, t = 1, 2, \ldots, \quad ; C_1, \quad C_2$ are some positive constants to be determined; R_t is a function (5.14) and F^t is the σ–algebra of the random quantities y^{pt+k-1} and u^{pt-1}.

Then the algorithm of (5.9) will give strongly consistent estimates $\theta(t)$, for any arbitrary initial condition $\theta(0) \in \Theta$ and

$$\gamma_t = \delta_t R_t^{-1}. \tag{5.18}$$

Then the bounded equality of (5.4) is fulfilled with probability 1 for the estimates $\tau_t = \tau[\theta(t)]$. Here $\tau(\theta)$ is determined by the system of (5.7).

It may be noted that if $\sup_t R_t < \infty$, then from (5.11) we have $\lim_{t \to \infty} \overline{w}_t = 0$ with probability 1, *i.e.*, the test signal decays with time. This property of the identification algorithm permits design of adaptive constrained optimal control strategies (vide section 6.3).

The noise v^∞ can be determined. Then the condition (4) of the theorem signifies a uniform limit (in the time domain) on the noise signals.

§ 5.P Proofs of Lemmas and Theorems

5.P.1 *Proof of the Theorem 5.1.1*

Let $Z_{t,s}$ signify the quadratic matrix $(n \times n)$ determinable by the conditions (for $t < s$)

$$Z_{t,s} = (F_t^* - \Phi_t^* K_t^*) Z_{t+1,s}, \; Z_{s,s} = I_n. \tag{P.1}$$

The solution of the equation (1.7) can be written as

$Z_t = Z_{t,t_*+1}$. The obvious properties of the matrix function follow (P.1) :

$$Z_{t+1,t_*+1} = Z_{t+1,t_*} Z_{t_*,t_*+1}, \tag{P.2}$$

$$Z_{t,t+1} = (F_t^* - \Phi_t^* K_t^*) Z_{t+1,t+1} = F_t^* - \Phi_t^* K_t^*.$$

Using the same notations, the estimate of (1.25) is written as

$$\hat{x}_{t+1} = \sum_{s=0}^{t} Z_{s+1,t+1}^* K_s y_s,$$

$$\hat{x}_t = \sum_{s=0}^{t-1} Z_{s+1,t}^* K_s y_s.$$

Computation of these equalities and use of the relationship in (P.2) give

$$\hat{x}_{t+1} - x_{t+1} = K_t y_t + \sum_{s=0}^{t-1} (Z_{s+1,t+1}^* - Z_{s+1,t}^*) K_s y_s$$

$$= K_t y_t + (Z_{t,t+1}^* - I_n) \sum_{s=0}^{t-1} Z_{s+1,t}^* K_s y_s = K_t y_t$$

$$+ (Z_{t,t+1}^* - I_n)\hat{x}_t = K_t y_t + (F_t - K_t \Phi_t - I_n)\hat{x}_t.$$

The result coincides with (1.27). The choice of the initial estimate $\hat{x}_0$ is to be made from the condition in (1.28). The theorem 5.1.1 is proved.

5.P.2 *Proof of the Lemma 5.1.1*

From (1.29) we get

$$\mathrm{M}(x_{t+1}-\hat{x}_{t+1})(x_{t+1}-\hat{x}_{t+1})^{*}=G_{t+1}(Z_0^{t+1},h_0^t),$$

i.e., the problem of mean square estimate is equivalent to the matrix problem of optimization of the object (1.7) for the performance criterion

$$J_t[h_0^t] = \mathrm{Sp}\, G_{t+1}(Z_0^{t+1},h_0^t). \tag{P.3}$$

It can be easily shown that this is equivalent to the linear quadratic matrix problem of (1.7) and (1.11) when $r_t = t$. If in reality it is not so, then matrices $\bar{h}_o^t$ are found out such that

$$\mathrm{Sp}\, G_{t+1}(\bar{Z}_0^{t+1},\bar{h}_0^t) < \mathrm{Sp}\, G_{t+1}(Z_0^{t+1},h_0^t). \tag{P.4}$$

Here $\bar{Z}_0^{t+1}$ is the solution of the system (1.7) for a choice of matrices $h_t = \bar{h}_t$, where h_0^t is the sequence of matrices determinable by the theorem 5.1.1.

For any n-vector a we get from the theorem 5.1.1

$$a^{*}G_{t+1}(Z_0^{t+1},h_0^t)a \le a^{*}G_{t+1}(\bar{Z}_0^{t+1},\bar{h}_0^t)a. \tag{P.5}$$

Choosing basis vectors of the space $\mathbf{R}^n$ as the a vectors and adding (P.5) we have

$$\mathrm{Sp}\, G_{t+1}(Z_0^{t+1},h_0^t) \le \mathrm{Sp}\, G_{t+1}(\bar{Z}_0^{t+1},\bar{h}_0^t).$$

It contradicts (P.4). The above equivalence of matrix optimization problems is established. From here and using the theorem 5.1.1 the optimality of estimates of Kalman-Bucy filter follows.

In case of Gaussian v_t the linearity of optimal estimates $\hat{x}_0^t$ is well-known. The lemma 5.1.1 is proved.

5.P.3 *Proof of the Lemma 5.1.2*

From (1.8), the Schwarz inequality for any n-vector $a \neq 0$ is obtained as

$$|a^{*}(x_{t+1}-\hat{x}_{t+1})|^2 \le a^{*}G_{t+1}a\sum_{s=0}^{t+1} v_s^{*}N_s^{-1}v_s.$$

Hence we can find out using (1.33)

$$|a^{*}(x_{t+1}-\hat{x}_{t+1})|^2 \le a^{*}G_{t+1}(Z_0^{t+1},h_0^t)ar_{t+1}.$$

$$\text{But }\sup_{|a|\le 1}\frac{a^{*}G_{t+1}(Z_0^{t+1},h_0^t)a}{|a|^2}=\Lambda_{t+1}$$

where Λ_{t+1} is the greatest eigenvalue of the matrix G_{t+1}, as because

$$\sup_{v_0^t\in V}|x_{t+1}-\hat{x}_{t+1}|^2 \le r_{t+1}\Lambda_{t+1}. \tag{P.6}$$

We hall show that the inequality of (P.6) is an exact one. Let a_{t+1} be the normalised eigen vector of matrix G_{t+1} for the eigen value of Λ_{t+1}. We shall determine the sequence v_0^{t+1} by the formulae

$$v_s = \alpha_{t+1}N_s\begin{Vmatrix} Z_s \\ -h_{s-1}^{*}\end{Vmatrix}a_{t+1},$$

$s=0,\ 1,\ \ldots,\ t+1,\ h_{-1}=0.$

For a choice of $\alpha_{t+1}=\sqrt{r_{t+1}\Lambda_{t+1}^{-1}}$ for v_0^{t+1} we have

$$\sum_{s=0}^{t+1} v_s^* N_s^{-1} v_s = r_{t+1},$$

i.e., $v_0^{t+1}\in V^{t+1}$. The relationship of (1.8) will then have the expression as

$$x_{t+1}-\hat{x}_{t+1}=a_{t+1}G_{t+1}a_{t+1}=\sqrt{r_{t+1}\Lambda_{t+1}a_{t+1}}.$$

This means that $|x_{t+1}-\hat{x}_{t+1}|^2=r_{t+1}\Lambda_{t+1}$, and this is what we wanted to prove. Thus, the optimal estimate problem becomes equivalent to the problem of optimization of the quantity $\Lambda_{t+1}=\Lambda_{t+1}(z_0^{t+1},h_0^t)$ for the constraints as in (1.7). The latter is equivalent to the linear quadratic matrix problem. Actually it can be argued otherwise, *i.e.*, existence of matrices $\tilde{h}_0^t$ can be assumed, for which the greatest eigenvalue of the matrix $\tilde{G}_{t+1}=G_{t+1}(\tilde{Z}_0^{t+1},\tilde{h}_0^t)$ is less than the similar eigenvalue of the matrix $G_{t+1}(Z_0^{t+1},h_0^t)$ obtained from the solution of the quadratic linear problem. But then $a^*G_{t+1}a\le a^*\hat{G}_{t+1}a$. Choosing $a=a_{t+1}$, where a_{t+1} is the eigen vector of the matrix G_{t+1} for the maximum eigen value, we find out $\Lambda_{t+1}\le a_{t+1}^*\cdot\tilde{G}_{t+1}a_{t+1}\le\tilde{\Lambda}_{t+1}$. It contradicts the assumption made. The reference to the theorem 5.1.1 completes the proof of the lemma 5.1.2.

5.P.4 *Proof of the Lemma 5.1.3*

We shall make use of the following simple assertion.

Lemma 5.P.1 [177] *For arbitrary initial conditions det* τ_0, γ_0, *det* $\gamma_0\neq 0$, *the relations* (1.37) *are equivalent to equalities*

$$\gamma_{t+1}^{-1}=\gamma_0^{-1}+\sum_{k=1}^{t+1}\Phi_k^*R^{-1}\Phi_k,$$

$$\tau_{t+1}=\gamma_{t+1}\gamma_0^{-1}\tau_0+\gamma_{t+1}\sum_{k=1}^{t+1}\Phi_k^*R^{-1}y_k. \tag{P.7}$$

Making use of (1.36) and (P.7), we have

$$\tau_{t+1}=\tau+\gamma_{t+1}\sum_{k=1}^{t+1}\Phi_k^*R^{-1}\bar{v}_k+\gamma_{t+1}\sum_{k=1}^{t+1}\Phi_k^*R^{-1}\tilde{v}_k+\gamma_{t+1}\gamma_0^{-1}(\tau_0-\tau). \tag{P.8}$$

Under the conditions of the lemma and by virtue of the theorem 2.P.2, we have

$$\lim_{t\to\infty}\ t^{-1}\sum_{k=0}^{t}\Phi_k^*R^{-1}\tilde{v}_k=0$$

and $\sup_t t\gamma_t<\infty$. Hence the estimation error $\tau_t-\tau$ is determined by the quality $\gamma_t\sum_{k=0}^{t}\Phi_k^*\bar{v}_k$.

We shall examine the observation scheme alongwith (1.36) :

$$\bar{y}_t=\Phi_t\tau+\bar{v}_t. \tag{P.9}$$

The recursive relations of (1.37) are also valid for the above scheme. Only difference is that y_t in (1.37) is to be replaced by $\bar{y}_t$.

Let the following set be the uncertainty set for the observation scheme of (P.9)

$$Q_{t+1}=\left\{\tau,\quad v^t: |\tau|^2+\sum_{k=1}^{t+1}\bar{v}_k^2\le C_{\bar{v}}^2(t+1)\right\}. \tag{P.10}$$

Then because of the lemma 5.1.2, the procedure of (1.37) is optimal in the minimax sense and the estimate of (1.35) gives

$$|\tau-\tau_t|^2\le C_{\bar{v}}^2(t+1)\Lambda_{t+1}, \tag{P.11}$$

where Λ_{t+1} is the greatest eigen value of the matrix γ_{t+1}. But, from (P.7), we have

$$\frac{1}{t+1}\gamma_{t+1}^{-1}=\frac{1}{t+1}\gamma_0^{-1}+\frac{1}{t+1}\sum_{k=1}^{t+1}\Phi_k^*R^{-1}\Phi_k.$$

Hence

$$(t+1)\gamma_{t+1}=\left(\frac{1}{t+1}\gamma_0^{-1}+\frac{1}{t+1}\sum_{k=0}^{t}\Phi_k^*R^{-1}\Phi_k\right)^{-1}$$

and because of (1.40), we have

$$\overline{\lim_{t\to\infty}}\quad (t+1)\Lambda_{t+1}\le\lambda_0^{-1}.$$

Hence the inequality of (1.41) follows from (P.11). The constraints (1.38) on the disturbances $\{\bar{v}_k\}$ and (P.10) do not have any essential difference when $t\to\infty$. Thus, the inequality of (1.41) is valid for the constraints in (1.38). The lemma 5.1.3 is proved.

5.P.5 *Proof of the Theorem 5.2.1*

Making use of the Bayesian rule for combined density of random quantities x^{t+1}, u^t and τ^{t+1} we have

$$p(x^{t+1},\quad u^t,\quad \tau^{t+1})$$

$$=p(\tau_{t+1}\,|\,x^{t+1},\quad u^t,\quad \tau^t)p(x_{t+1}\,|\,x^t,\quad u^t,\tau^t)$$

$$\times p(u_t\,|\,x^t,\quad u^{t-1},\quad \tau^t)p(x^t,\quad u^{t-1},\quad \tau^t). \tag{P.12}$$

But because of the assumptions made about the independence of random quantities v_t, w_t and e_t and because of (2.3), (2.1) and (2.5), we have

$$p(\tau_{t+1}\,|\,x^{t+1},u^t,\tau^t)=p_w(\tau_{t+1}-F_t\tau_t\,|\,t+1),$$

$$p(x_{t+1}\,|\,x^t,u^t,\tau^t)=p_v(x_{t+1}-A_t(\tau_t)x_t$$

$$-B_t(\tau_t)u_t\,|\,t+1)=p_v(y_t-\Phi_t\tau_t\,|\,t+1),$$

$$p(u_t\,|\,x^t,u^{t-1},\tau^t)=p_e(u_t-U_t(u^{t-1},x^t)\,|\,t). \tag{P.13}$$

Integrating both sides of the equality of (P.12) with respect to τ^t and considering the dependence in (P.13), we find out

$$p(x^{t+1},u^t,\tau_{t+1})=p_e(u_t-U_t)\int p_w(\tau_{t+1}-F_t\tau_t)$$

$$\times p_v(y_t-\Phi_t\tau_t)p(x^t,u^{t-1},\tau_t)d\tau_t, \tag{P.14}$$

where for brevity's sake, the dependence of the density on t, is not shown. In accordance with the definition of the *a posteriori* density of random quantities τ_{t+1} (vide section 2.4), we have

$$p_{\varpi+1}(\tau \mid x^{t+1}, u^t) = C_{t+1} p(x^{t+1}, u^t, \tau),$$

where $C_{t+1} = C_{t+1}(x^{t+1}, u^t)$ is a quantity to be determined by the condition of normalization, *i.e.*,

$$\int p_{\varpi+1}(\tau \mid x^{t+1}, \quad u^t) d\tau = 1. \tag{P.15}$$

Here we have considered that the chosen control strategy (vide (2.5)) does not depend on τ. The formula of (P.14) results in a recursive relationship

$$p_{\varpi+1}(\tau \mid x^{t+1}, \quad u^t)$$

$$= \frac{C_{t+1}}{C_t} \int p_w(\tau - F_t \tau') p_v(y_t - \Phi_t \tau') p_\varpi(\tau' \mid x^t, \quad u^{t-1}) d\tau'.$$

The quantity $C_{t+1}\, C_t^{-1}$ does not depend on τ and is determined from the normalization condition of (P.15). Thus the following expression

$$p_{\varpi+1}(\tau \mid x^{t+1}, \quad u^t)$$

$$= \frac{\int p_w(\tau - F_t \tau') p_v(y_t - \Phi_t \tau') p_\varpi(\tau' \mid x^t, \quad u^{t-1}) d\tau'}{\int \int p_w(\tau - F_t \tau') p_v(y_t - \Phi_t \tau') p_\varpi(\tau' \mid x^t, \quad u^{t-1}) d\tau' d\tau} \tag{P.16}$$

is the initial recursive formula for later consideration in case of *a posteriori* densities of random quantities τ_t.

Since the right side in (P.16) determines Gaussian density (because of (1.53) and for $t = 0$), then it follows from (P.16) that for any t the *a posteriori* density $p_\varpi(\cdot \mid x^t, u^{t-1})$ is a Gaussian. We designate the mean and the covariance of this density through τ_{t+1} and P_{t+1} respectively. We observe that the following relations permit obvious integration of (P.16) with respect to τ'.

$$p_w(\tau - F_t \tau') \sim N(\tau - F_t \tau', \quad R_w),$$

$$p_v(y_t - \Phi_t \tau') \sim N(y_t - \Phi_t \tau', \quad R_v),$$

$$p_\varpi(\tau') \sim N(\hat{\tau}_t, \quad P_t).$$

Actually, separating out the exponential indices of the terms dependent on τ', we get

$$\exp \frac{1}{2} \{ -(\tau')^* \, [\Phi_t^* R_v^{-1}(t+1) \Phi_t + F_t^* R_w^{-1}(t+1) F_t + P_t] \tau'$$

$$-2(\tau')^* \, [\Phi_t^* R_v^{-1}(t+1) y_t + F_t^* R_w^{-1}(t+1) \tau + P_t^{-1} \hat{\tau}_t] + \quad \ldots \},$$

where the dots indicate values not dependent on τ or τ'. Separating out the complete square in τ', we have

$$\exp \frac{1}{2} \{ - \mid L_t^{-1/2} \tau' - L_t^{1/2} \, [\Phi_t^* R_v^{-1}(t+1) y_t + F_t^* R_w^{-1}(t+1) \tau +$$

$$+P_t^{-1}\hat{\tau}_t]\,|^2+|\,L_t^{1/2}\,[\Phi_t^* R_v^{-1}(t+1)y_t+F_t^* R_w^{-1}(t+1)\tau$$

$$+P_t^{-1}\hat{\tau}_t]\,|^2-\tau^* R_w^{-1}(t+1)\tau+\quad\ldots\}$$

where for brevity's sake we have used

$$L_t=[P_t^{-1}+\Phi_t^* R_v^{-1}(t+1)\Phi_t+F_t^* R_w^{-1}(t+1)F_t]^{-1}. \tag{P.17}$$

Integration with respect to τ' leads to a function in which the dependence on τ can be expressed as

$$\exp\frac{1}{2}\{-\tau^*[R_w^{-1}-R_w^{-1}F_t L_t F_t^* R_w^{-1}]\tau$$

$$+2\tau^* R_w^{-1}F_t^* L_t^*[\Phi_t^* R_v^{-1}y_t+P_t^{-1}\hat{\tau}_t]+\quad\ldots\}.$$

This function determines the right-hand side of the formula (P.16) within the normalized terms dependent on x^{t+1} and u^t. Since as per definition $p_{\tau_{t+1}}(\tau\,|\,x^{t+1},u^t)\sim N(\hat{\tau}_{t+1}\,P_{t+1})$, then the comparison of the terms (for τ) in the indices of the left and right-hand sides of the formula (P.16) leads to the relations

$$P_{t+1}^{-1}=R_w^{-1}(t+1)-R_w^{-1}(t+1)F_t L_t F_t^* R_w^{-1}(t+1),$$

$$\hat{\tau}_{t+1}=P_{t+1}R_w^{-1}(t+1)F_t L_t[\Phi_t^* R_v^{-1}(t+1)y_t+P_t^{-1}\hat{\tau}_t]. \tag{P.18}$$

These differ from the relations (2.14) to (2.16) only by identity transformations. This will be shown now

$$P_{t+1}=[R_w^{-1}-R_w^{-1}F_t L_t F_t^* R_w^{-1}]^{-1}$$

$$=R_w-F_t(F_t^* R_w F_t-L_t^{-1})^{-1}F_t^*$$

$$=R_w+F_t(P_t^{-1}+\Phi_t^* R_v^{-1}\Phi_t)^{-1}F_t^*-R_w$$

$$=F_t\left[P_t-P_t\Phi_t^*(R_v+\Phi_t P_t\Phi_t^*)^{-1}\Phi_t P_t\right]F_t,$$

and it coincides with (1.57). Using the obvious equalities and considering the relationship found out for P_{t+1}, we get from the second equality of (P.18)

$$\hat{\tau}_{t+1}=F_t L_t(\Phi_t^* R_v^{-1}y_t+P_t^{-1}\hat{\tau}_t)+F_t P_t[I_p$$

$$-(\Phi_t^* R_v^{-1}\Phi_t+P_t^{-1})L_t]\,(\Phi_t^* R_v^{-1}y_t+P_t^{-1}\hat{\tau}_t)-F_t P_t\Phi_t^*(R_v^{-1}$$

$$+\Phi_t P_t\Phi_t^*)^{-1}\Phi_t P_t[I_p-(\Phi_t^* R_v^{-1}\Phi_t+P_t^{-1})L_t]\,(\Phi_t^* R_v^{-1}y_t$$

$$+P_t^{-1}\hat{\tau}_t)=F_t P_t\Phi_t^* R_v^{-1}y_t+F_t F_t\hat{\tau}_t-F_t P_t\Phi_t^*(R_v+$$

$$+\Phi_t P_t\Phi_t^*)^{-1}\Phi_t P_t(\Phi_t^* R_v^{-1}y_t+P_t^{-1}\hat{\tau}_t)=F_t\hat{\tau}_t$$

$$+F_t P_t\Phi_t^*(R_v+\Phi_t P_t\Phi_t^*)^{-1}(y_t-\Phi_t\hat{\tau}_t).$$

This coincides with (2.14). The choice of initial conditions (2.17) is evident. The formula of (2.18) follows during determination of covariance of the Gaussian distribution of the random quantities τ_{t+1} during observation of y_t and Φ_t in accordance with (2.9). The established optimality of statistics (2.14) to (2.16) invites (2.19). The theorem is proved.

5.P.6 *Proof of the Theorem 5.2.2*

We can consider $H_0 > 0$. when the general constraint is absent, because if $H_0 \geq 0$, then for a choice of α from the condition $\alpha I_p \geq H_0$, the solution $\{\tilde{H}_t\}$, determinable by the initial condition $\tilde{H}_0 = \alpha i_p$, satisfies the inequality $\tilde{H}_t \geq H_t^*$ (for all t). Using the matrix identity in (2.24) (for $H_0 > 0$) and considering (2.28), we have

$$H_{t+1}^{-1} = (F^*)^{-1} H_t^{-1} F^{-1} + (F^*)^{-1} \Phi_t R_t^{-1} \Phi_t^* F^{-1}.$$

$$\text{So } H_t^{-1} = (F^*)^{-t} H_0^{-1} F^{-t} + \sum_{s=1}^{t} (F^*)^{-s} \Phi_{t-s} R_{t-s}^{-1} \phi_{t-s}^* F^{-s}. \tag{P.19}$$

It will be shown that for any p-vector a the fulfilment of $\lim_{t\to\infty} a^* H_t a = 0$ is ensured. It is obvious when a belongs to the root subspace of the matrix F^{-1}, which provides eigenvalue outside the unit circle. Let a belong to the root subspace for the eigenvalue λ of the matrix F^{-1} on the unit circle. If λ corresponds to the simple elementary divisors, then $F^{-1}a = \lambda a$ and $|\lambda| = 1$. We find out the following from (P.19) and the equality $\lim_{t\to\infty} a^* H_t a = 0$ follows from (2.25) by virtue of $\sup_t |R_t| < \infty$:

$$a^* H_t^{-1} a = a^* H_0^{-1} a + \sum_{s=1}^{t} a^* \Phi_{t-s} R_{t-s}^{-1} \Phi_{t-s}^* a. \tag{P.20}$$

Let λ correspond to the multiple divisors. Without the generalised constraint, we may assume that a corresponds to the subspace with cyclic basis $a_0, \quad a_1, \quad \ldots, \quad a_{k-1}$:

$$F^{-1} a_s = \lambda a_s + a_{s-1}, \quad s = 1, \quad \ldots, \quad k-1, \quad F^{-1} a_0 = \lambda a_0, \quad |\lambda| = 1$$

i.e., the Jordanian of dimension $k \times k$ corresponds to the respective part of the matrix F. It is well-known [29] that for any vector of arbitrary reference a_s, $s \neq 0$, we have

$$F^{-t} a_s = \lambda^t a_s + t\lambda^{t-1} a_{s-1} + \quad \ldots$$

$$+ \frac{t(t-1) \quad \ldots \quad (t-s+1)}{s!} \lambda^{t-s} a_0$$

$$= \frac{t^s}{s!} \left[\lambda^{t-s} a_0 + O\left(\frac{1}{t}\right) \right].$$

Hence $\lim_{t\to\infty} a_s^* (F^*)^{-t} H_0^{-1} F^{-t} a_s = \infty$. It leads to the equality $\lim_{t\to\infty} a^* H_t a = 0$ (because of (P.19)). If $s = \infty$, then for the vector $a = a_o$, the condition in (P.20) is fulfilled and again we arrive at the same equality. Thus, the relationship $\lim_{t\to\infty} a^* H_t a = 0$ is established for

---

*The proof is established in the same way as the assertion in the theorem 3.2.1.

any reference vector a_s, and so for any vector from the corresponding root subspace. It establishes (2.26). From (2.24), we get $F-\tilde{K}_t\Phi_t = H_{t+1}(H_tF^*)^{-1} = H_{t+1}(F^*)^{-t}H_t^{-1}$. We express (2.27) with the consideration of the latter formula as : $G_{t+1}^0 = H_{t+1}(F^*)^{-1}H_t^{-1}G_t^0H_t^{-1}F^{-1}H_{t+1}$, and so $G_t^0 = H_t(F^*)^{-t}H_0^{-1}G_0^0H_0F^{-t}H_t$. Because of (P.19), we have $(F^*)^{-t}H_0^{-1}F^{-t} \le H_t^{-1}$ and hence $G_t^0 \le | G_0^0 \| H_0^{-1} | H_t(F^*)^{-t}H_0^{-1}F^{-t}H_t$ $\le | G_0^0 | H_0^{-1} | H_t$. Now the relationship (2.29) results from (2.26). The theorem is proved.

5.P.7 *Proof of the Theorem 5.2.3*

We can put $Q = \varepsilon I_p$ without any generalised constraint. Here ε is a small positive parameter. Actually, replacing Q by εI_p, $Q \le \varepsilon I_p$ will give an equation. The solution of the latter majorizes the solution of the equation in (2.31) (for an identical initial condition). Hence it is sufficient to show that the solution of the equation given below possesses the property $\lim_{t\to\infty} | P_t(\varepsilon) | \to 0$ when $\varepsilon \to 0$:

$$P_{t+1}(\varepsilon) = f[P_t(\varepsilon), \varepsilon], \tag{P.21}$$

$$f(P, \varepsilon) = FPF^* - FP\Phi^*(R + \Phi P\Phi^*)^{-1}\Phi PF^* + \varepsilon I_p.$$

We examine the following equation side by side with (P.21)

$$P_{t+1} = f(P_t, \varepsilon_t). \tag{P.22}$$

Here the function $f(P, \varepsilon)$ has been already determined. It can be shown that a choice of a monotonously decaying sequence of positive numbers ε_t is possible, such that $\lim_{t\to\infty} \varepsilon_t = 0$ and the solutions $\{P_t\}$ of the equation (P.22) possesses the property $\lim_{t\to\infty} P_t = 0$ for any choice of $P_0 \ge 0$. According to the lemma 3.P.2

$$f(P, \varepsilon) \le \phi(P, K, \varepsilon), \tag{P.23}$$

$$\phi(P, K, \varepsilon) = (F - K\Phi)P(F - K\Phi)^* + KRK^* + \varepsilon I_p,$$

where K is an arbitrary matrix of appropriate dimension. Again the equality in (P.23) is obtained for

$$K = K(\varepsilon) = FP\Phi^*(\Phi P\Phi^* + R)^{-1}.$$

Let the sequence of matrices $\tilde{P}_t$ be determined by the conditions

$$\tilde{P}_{t+1} = \phi(\tilde{P}_t, K_t, \varepsilon_t), \quad \tilde{P}_0 = P_0,$$

$$K_t = FP_t(0)\Phi^*[\Phi P_t(0)\Phi^* + R]^{-1},$$

where $\{P_t(0)\}$ is the solution of the equation (P.21) for $\varepsilon = 0$. Using (P.23), it can be easily shown that the numbers $\{\varepsilon_t\}$ can be chosen in such a manner that $\lim_{t\to\infty} \tilde{P}_t = 0$, and so $\lim_{t\to\infty} P_t = 0 (P_t \le \tilde{P}_t)$. We put $\tilde{P}_t = P_t(0) + G_t$ and $P_0(0) = P_0$. Then $G_{t=1} = V_t G_t V_t^*$ $+ \varepsilon_t I_p$, $G_0 = 0$, and $V_t = F - K_t\Phi$. Using the formula

$F - \tilde{K}_t \Phi = H_{t+1}(F^*)^{-1} H_t^{-1}$, we get

$V_t = P_{t+1}(0)(F^*)^{-1} P_t^{-1}(0)$, and hence

$$G_{t+1} = \varepsilon_t I_p + \varepsilon_{t-1} v_t V_t^* + \quad \dots$$

$$+\varepsilon_1 V_t \quad \dots \quad V_2 V_2^* \quad \dots \quad V_t^* + \varepsilon_0 V_t \quad \dots \quad V_1 V_1^* \quad \dots \quad V_t^*$$

$$= \varepsilon_t I_p + \varepsilon_{t-1} P_{t+1}(0)(F^*)^{-1} P_t^{-2}(0)$$

$$\times F^{-1} P_{t+1}(0) + \quad \dots \quad + \varepsilon_0 P_{t+1}(0)(F^*)^{-t} P_1(0) F^{-t} P_{t+1}(0).$$

We choose ε_t such that $\varepsilon P_{t+1}^{-1}(0)_t \le \delta_t I_p$, $\sum_{t=0}^{\infty} \delta_t < \infty$, $\varepsilon_{t+1} \le \varepsilon_t$ and $\lim_{t\to\infty} \varepsilon_t = 0$. Considering $(F^*)^{-k} P_{t+1-k}^{-1}(0) F^{-k} \le P_{t+1}^{-1}(0)$, and because of the inequality $H_{t+1}^{-1} \ge (F^*)^{-1} H_t^{-1} F^{-1}$, we get $G_{t+1} \le \varepsilon_t I_p + \sum_{k=0}^{t-1} \delta_k P_{k+1}(0)$. From the theorem 5.2.2, we get $\lim_{t\to\infty} P_t(0) = 0$, and so $\lim_{t\to\infty} G_t = 0$. Consequently, $\lim_{t\to\infty} \tilde{P}_t = \lim_{t\to\infty} P_t(0) + \lim_{t\to\infty} G_t = 0$, since $P_t \le \tilde{P}_t$ and $\lim_{t\to\infty} P_t = 0$. The montonity of the function $f(P, \varepsilon)$, namely, $P_1 \le P_2$ and $\varepsilon \le \varepsilon_2 \Rightarrow f(P_1, \varepsilon_1) \le f(P_2, \varepsilon_2)$, we get

$$P(\varepsilon_1) = f[P(\varepsilon_1), \varepsilon_1] \le f[P(\varepsilon_0), \varepsilon_0] \le f(P_0, \varepsilon_0) = P_1.$$

By induction $P(\varepsilon_t) \le P_t$ for all t and since $\lim_{t\to\infty} P_t = 0$, then $\lim_{t\to\infty} P(\varepsilon_t) = 0$. Because of the monotonous dependence of $P(\varepsilon)$ on ε, we get $P(\varepsilon) \to 0$ for $\varepsilon \to 0$. It proves the theorem 5.2.3 for $Q = \varepsilon I_p$, and so for any arbitrary matrix Q.

5.P.8 *Proof of the Theorem 5.2.4*

We note that if $P_0^{\infty}(\varepsilon)$ is the solution of the equation $P_{t+1}(\varepsilon) = f_t[P_t(\varepsilon), \varepsilon]$, $f_t(P, \varepsilon) = FPF^* - FP\Phi_t^*(\Phi_t P \Phi_t^* + R)^{-1} \Phi_t P F^* + Q$, then for any s, $0 \le s \le k-1$, (k is the number from the condition of the theorem), the sequence $\{P_{tk} + s(\varepsilon)\}, t = 0, 1, \dots$, is a solution of the Riccati equation as given below :

$$P_{(t+1)k+s}(\varepsilon) = \overline{F} P_{tk+s}(\varepsilon) \overline{F}^* + R_2(\varepsilon) - [\overline{F} P_{tk+s}(\varepsilon) \overline{\Phi}_t^*$$

$$+R_{3t}(\varepsilon)] \quad \left[\overline{\Phi}_t P_{tk+s}(\varepsilon) \overline{\Phi}_t^* + R_{1t}(\varepsilon)\right]^{-1}$$

$$\times \left[\overline{\Phi}_t P_{tk+s}(\varepsilon) \overline{F}^* + R_{3t}^*(\varepsilon)\right]. \qquad \text{(P.24)}$$

We shall show how the equation (P.24) is obtained for $s = 0$ (the case of $s \ne 0$ is completely analogous).

The estimation scheme given below is examined for independent noises $\{v_t, w_t\}$ which satisfy $Mw_tw_t^* = \varepsilon I$, $Mv_tv_t^* = R$ and $Mv_tw_t^* = 0$.

$$\tau_{t+1} = F\tau_t + w_{t+1}, \quad y_t = \Phi_t\tau_t + v_t, \tag{P.25}$$

we rewrite the system (P.25) in the form of $\xi_{t+1} = \overline{F}\xi_t + \eta_t$ and $z_t = \overline{\Phi}_t\xi_t + \zeta_t$, where

$$\xi_t = x_{tk}, \quad z_t = \text{col}\,(y_{tk+k-1}, \quad \ldots, y_{tk}),$$

$$\eta_t = \sum_{j=0}^{k-1} F^j w_{tk+k-j},$$

$$\zeta_t = \left\| \begin{matrix} \Phi_{tk+k-1} \sum_{j=0}^{k-1} F^j w_{tk+k-1-j} + v_{tk+k-1} \\ \cdot \\ \cdot \\ \cdot \\ \Phi_{tk+1} w_{tk+1} + v_{tk+1} \\ v_{tk} \end{matrix} \right\| .$$

The covariance matrix v_{tk} of the optimal error of the estimate $\hat{\xi}_t$, for the quantity ξ_t, from the observation z_0^{t-1} will satisfy the equation (P.24) for $\overline{F} = F^k$, $\overline{\Phi}_t = \text{col}(\Phi_{tk+k-1}F^{k-1}, \ldots, \Phi_{tk+1}F, \Phi_{tk})$, $R_{1t}(\varepsilon) = \mathbf{M}\zeta_t\zeta_t^*$, $R_2(\varepsilon) = \mathbf{M}\eta_t\eta_t^*$, and $R_{3t}(\varepsilon) = \mathbf{M}\eta_t\zeta_t^*$. But this covariance matrix will coincide with the matrix $P_{tk}(\varepsilon) = \mathbf{M}(\tau_{tk} - \hat{\tau}_{tk})(\tau_{tk} - \hat{\tau}_{tk})^*$ (at the instant t), which satisfies (P.24) for $s = 0$. We use the identity below which can be easily verified :

$$APA^* - (APB^* + S)(BPB^* + R)^{-1}(BPA^* + S^*) + Q$$

$$= (A - SP^{-1}B)P(A - SP^{-1}B)^* + Q - SR^{-1}S^*$$

$$-(A - SR^{-1}B)PB^*(BPB^* + R)^{-1}BP(A - SR^{-1}B)^*$$

Let $\tilde{P}_t(\varepsilon) = P_{tk}(\varepsilon)$, $\tilde{F}_t(\varepsilon) = \overline{F}R_{3t}(\varepsilon)R_{1t}(\varepsilon)\overline{\Phi}_t$ and $\tilde{R}_{2t}(\varepsilon) = R_2(\varepsilon) - R_{3t}(\varepsilon)R_{1t}^{-1}(\varepsilon)R_{3t}^*(\varepsilon)$. Then the equation (P.24) can be rewritten as

$$\tilde{P}_{t+1}(\varepsilon) = \tilde{F}_t(\varepsilon)\tilde{P}_t(\varepsilon)\tilde{F}_t^*(\varepsilon) + \tilde{R}_{2t}(\varepsilon)$$

$$-\tilde{F}_t(\varepsilon)\tilde{P}_t(\varepsilon)\overline{\Phi}_t^*\left[\overline{\Phi}_t\overline{P}_t(\varepsilon)\overline{\Phi}_t^* + R_{1t}(\varepsilon)\right]^{-1}\overline{\Phi}_t\tilde{\Phi}_t(\varepsilon)\tilde{F}_t^*(\varepsilon). \tag{P.26}$$

We shall later need the following inequality, which is valid for $B^*R^{-1}B \geq B_1^*R_1B_1$,

$$APA^* - APB^*(BPB^* + R)^{-1} \times BPA^* \leq APA^*$$

$$-APB_1^*(B_1PB_1^* + R_1)^{-1}B_1PA^*.$$

This inequality can be easily shown if we consider the validity of the identity for R > 0, *i.e.*,

$$APA^* - APB^*(BHB^* + R)^{-1}BPA^* = A(P^{-1} + B^*R^{-1}B)^{-1}A^*.$$

It is observed that for $\beta > 0$ and for all t, $\overline{\Phi}_t^* R_{1t}^{-1} \overline{\Phi}_t \geq \beta I$ because of the uniform observation condition (2.40) for the matrices (F, Φ_0^∞) and since $\overline{\lim}_{t \to \infty} |\Phi_t| < \infty$. From (P.26) it can be found out that $\tilde{P}_t \leq \hat{P}_t(\varepsilon)$, where the matrices $\hat{P}_t(\varepsilon)$ are determined by the equation (for $\Phi = \sqrt{\beta} I$)

$$\hat{P}_{t+1}(\varepsilon) = \tilde{F}_t(\varepsilon)\hat{P}_t(\varepsilon)\tilde{F}_t^*(\varepsilon) + \tilde{R}_{2t}(\varepsilon)$$

$$-\tilde{F}_t(\varepsilon)\hat{P}_t(\varepsilon)\Phi^*[\Phi\hat{P}_t(\varepsilon)\Phi^* + I]^{-1}\Phi\hat{P}_t(\varepsilon)\tilde{F}_t^*(\varepsilon).$$

Here I is the identity matrix of appropriate dimension. The matrices $\overline{P}_t(\varepsilon)$, $t = 0, 1, \ldots,$ will be determined by the conditions

$$\overline{P}_0(\varepsilon) = \hat{P}(\varepsilon),$$

$$\overline{P}_1(\varepsilon) = [\tilde{F}_0(\varepsilon) - N_0(\varepsilon)\Phi]\overline{P}_0(\varepsilon)\,[\tilde{F}_0(\varepsilon) - N_0(\varepsilon)\Phi]^*$$

$$+N_0(\varepsilon)N_0^*(\varepsilon) + \tilde{R}_{20}(\varepsilon),$$

where the matrix $N_0(\varepsilon)$ can be found out from the relations

$$\tilde{F}_0(\varepsilon) - N_0(\varepsilon)\Phi = \overline{F} - K_0(\varepsilon)\Phi,$$

$$K_0(\varepsilon) = \tilde{F}\overline{P}_0(\varepsilon)\Phi^*[\Phi\overline{P}_0(\varepsilon)\Phi^* + I]^{-1},$$

i.e. $N_0(\varepsilon) = K_0(\varepsilon) - R_{30}(\varepsilon)R_{10}^{-1}(\varepsilon)\overline{\Phi}_0\Phi^{-1}$.

Because of the inequality (P.23), we have then $\hat{P}_1(\varepsilon) \leq \overline{P}_1(\varepsilon)$. Let there be $\overline{P}_k(\varepsilon)$, $k = 0, \ldots, t$, and $\overline{P}_k(\varepsilon) \geq \hat{P}_k(\varepsilon)$. We determine

$$\overline{P}_{t+1}(\varepsilon) = [\tilde{F}_t(\varepsilon) - N_t(\varepsilon)\Phi]\overline{P}_t(\varepsilon)\,[\tilde{F}_t(\varepsilon) - N_t(\varepsilon)\Phi]^*$$

$$+N_t(\varepsilon)N_t^*(\varepsilon) + \tilde{R}_{2t}(\varepsilon), \tag{P.27}$$

$$N_t(\varepsilon) = K_t(\varepsilon) - R_{3t}(\varepsilon)R_{1t}^{-1}(\varepsilon)\overline{\Phi}_t\Phi^{-1},$$

$$K_t(\varepsilon) = \overline{F}P_t(\varepsilon)\Phi^*[\Phi\overline{P}_t(\varepsilon)\Phi^* + I]^{-1}.$$

It can be shown analogously that $\overline{P}_{t+1}(\varepsilon) \leq \hat{P}_{t+1}(\varepsilon)$. The equation of (P.27) can be presented in the form

$$\overline{P}_{t+1}(\varepsilon) = [\overline{F}_t - K_t(\varepsilon)\Phi]\overline{P}_t(\varepsilon)\,[\overline{F}_t - K_t(\varepsilon)\Phi]^* + \overline{Q}_t(\varepsilon)$$

$$+K_t(\varepsilon)K_t^*(\varepsilon) + D_t(\varepsilon) = \overline{F}\overline{P}_t(\varepsilon)\overline{F}^* + D_t(\varepsilon)$$

$$-\overline{F}\overline{P}_t(\varepsilon)\Phi^*[\Phi\overline{P}_t(\varepsilon)\Phi^* + I]^{-1}\Phi\overline{P}_t(\varepsilon)\overline{F}^* + \overline{Q}_t(\varepsilon),$$

Here $\overline{Q}_t(\varepsilon)$ is a sequence of symmetric matrices, for which $\overline{Q}_t(\varepsilon) \leq \rho(\varepsilon)I$ within the conditions of the theorem. Here $\rho(\varepsilon)$ is a positive function such that $\rho(\varepsilon) \to 0$ when

$\varepsilon \to 0$. According to the theorem 5.2.3, $\overline{\lim}_{t\to\infty} |\overline{P}_t(\varepsilon)|$ can be made as small as necessary by the choice of a small $\varepsilon > 0$.

But $\overline{\lim}_{t\to\infty} |\tilde{P}_t(\varepsilon)| \le \overline{\lim}_{t\to\infty} |\hat{P}_t(\varepsilon)| \le \overline{\lim}_{t\to\infty} |\overline{P}_t(\varepsilon)|$.

Also $\overline{\lim}_{t\to\infty} |P_{tk}(\varepsilon)| \to 0$.

The smallness of $\overline{\lim}_{t\to\infty} |P_{tk+s}(\varepsilon)|$ for $s = 1, \ldots, k-1$ is established in the same manner.

The theorem 5.2.4 is proved.

5.P.9 *Proof of the Lemma 5.4.1*

Let the F.R.F. $h(\lambda, \tau)$ be given by

$$g(\lambda, \tau)c^{-1}(\lambda, \tau) = 1 + \lambda h(\lambda, \tau). \tag{P.28}$$

The formula (P.28) gives a stable F.R.F. $h(\lambda, \tau)$ of the argument λ. The functional of (4.4) is rewritten using (P.28):

$$J(\tau) = \frac{\sigma_v^2}{2\pi i} \oint [R(\lambda) + \Pi(\lambda)h(\lambda)]$$

$$\times [R(\lambda^{-1}) + \Pi(\lambda^{-1})h(\lambda^{-1})] \frac{d\lambda}{\lambda}, \tag{P.29}$$

where $\Pi(\lambda) = g(\lambda, \tau_*)c^{-1}(\lambda, \tau_*)$ and $R(\lambda) = \lambda^{-1}\Pi(\lambda)$. Carrying out separation of the function $R(\lambda) = [c(\lambda, \tau_*) - g(\lambda, \tau_*)].\lambda^{-1}g^{-1}(\lambda, \tau_*) - \lambda^{-1}$, we can easily find out (vide, for example, proof of the theorem 3.3.2) that the minimum of the functional (P.29) will be attained (in a class of arbitrary stable F.R.F. $h(\lambda)$) when the transfer function is

$$h(\lambda) = \frac{g(\lambda, \tau_*) - c(\lambda, \tau_*)}{\lambda c(\lambda, \tau_*)} = -\frac{1}{\lambda} - \frac{1}{\lambda}\,\frac{g(\lambda, \tau_*)}{c(\lambda, \tau_*)}.$$

By virtue of (P.28), it now follows that $c(\lambda, \tau)g^{-1}(\lambda, \tau) = c(\lambda, \tau_*)g^{-1}(\lambda, \tau_*)$. It proves the first assertion of the lemma.

Considering the notations in the condition (3), we get from (4.4)

$$\text{grad}\, J(\tau) = \frac{\sigma_v^2}{\pi i} \oint \frac{c(\lambda, \tau_*)}{g(\lambda, \tau_*)}\,\frac{c_*(\lambda, \tau_*)}{g_*(\lambda, \tau_*)} \times$$

$$\times \frac{g_*(\lambda, \tau)H(\lambda, \tau)}{c_*(\lambda, \tau)c_*(\lambda, \tau)}\,\frac{d\lambda}{\lambda}. \tag{P.30}$$

The expression under integral in (P.30) possesses singularities inside the unit circle only at the roots of the polynomials $g_*(\lambda, \tau_*)$ and $c_*(\lambda, \tau)$ and for $\lambda = 0$, because of the stability of the polynomials $g(\lambda, \tau_*)$ and $c(\lambda, \tau)$. It results from the condition (2) that $H(0, \tau) = 0$ and so the equality grad $J(\tau) = 0$ leads to the relations (by virtue of the conditions (3) of the lemma)

$$\frac{c(\lambda_i, \tau_*)c_*(\lambda_j, \tau_*)g_*(\lambda_j, \tau)}{g(\lambda_j, \tau_*)g_*(\lambda_j, \tau_*)c(\lambda_j, \tau)c'_*(\lambda_j, \tau)} = 0,$$

$$j=1, \quad \ldots, \quad n, \tag{P.31}$$

$$\frac{c(\mu_i, \tau_*)c_*(\mu_j, \tau_*)g_*(\mu_j, \tau)}{g(\mu_j, \tau_*)c_*(\mu_j, \tau)c(\mu_j, \tau)g'_*(\mu_j, \tau_*)}=0, \quad j=1, \quad \ldots, \quad p+r,$$

where g'_* and c'_* designate the derivatives of the respective polynomials with respect to λ. Since the polynomials $c(\lambda, \tau)$ and $g(\lambda, \tau_*)$ are stable, the conditions $|\lambda_j| \leq 1$, $|\mu_j| \leq 1$, $c(\lambda_j, \tau) \neq 0$ and $g(\mu_j, \tau)=0$ are fulfilled, and so from (P.31) the following equalities are obtained

$$\frac{c_*(\lambda_j, \tau_*)g_*(\lambda_j, \tau)}{g_*(\lambda_j, \tau_*)}=0,$$

$$\text{and } \frac{c_*(\mu_k, \tau_*)g_*(\mu_k, \tau)}{c_*(\mu_k, \tau)}=0, \tag{P.32}$$

$$j=1, \quad \ldots, \quad n \text{ and } k=1, \quad \ldots, \quad p+r.$$

$c_*(\lambda_j, \tau_*)=0, \quad j=1, \quad \ldots, \quad n$, and

$g_*(\mu_K, \tau)=0$, $k=1, \ldots, p+r$ are found out from (P.32) considering the conditions (1) to (4) of the lemma. Then, because of the condition (1), the validity of the identity (in terms of λ) can be shown :

$$c(\lambda, \tau)=c(\lambda, \tau_*), \quad \text{and } g(\lambda, \tau)=g(\lambda, \tau_*). \tag{P.33}$$

By virtue of (4.7), the identity (P.33) signifies satisfaction of (4.9).

5.P.10 *Proof of the Lemma 5.4.2*

The relationship below follows equations in (4.13)

$$g(\nabla, \tau_*)c(\nabla, \tau)\overline{v}_t(\tau)=g(\nabla, \tau)c(\nabla, \tau_*)v_t. \tag{P.34}$$

The transfer function $h(\lambda, \tau)$ for this equation is a stable F.R.F. So the following expression is true for the transfer function

$$h(\lambda, \tau)=\frac{g(\lambda, \tau)c(\lambda, \tau_*)}{g(\lambda, \tau_*)c(\lambda, \tau)}=\sum_{k=0}^{\infty} h_k(\tau)\lambda^k. \tag{P.35}$$

Here the series in (P.35) converges uniformly when $|\lambda| \leq 1$. Consequently, for $\overline{v}_t(\tau)$ the expression $\overline{v}_t(\tau)=\sum_{k=0}^{\infty} h_k(\tau)v_{t-k}$, is valid. So considering (4.14), we get

$$\mathbf{M}|\overline{v}_t(\tau)|^2=\sigma_v^2 \sum_{k=0}^{\infty} |h_k(\tau)|^2, \quad \mathbf{M}|\overline{v}_t(\tau)|^4 \leq \sigma^4 \sum_{k=0}^{\infty} |h_k(\tau)|^4$$

$$+\sigma_v^4\left[\sum_{k=0}^{\infty} |h_k(\tau)|^2\right] < \infty.$$

Analogous considerations for the vector function $\overline{V}_t(\tau)$ are true. Thus, for vector coefficients $H_K(\tau)$, we have

$$\overline{V}_t(\tau)=\sum_{k=0}^{\infty} H_k(\tau)v_{t-k}. \tag{P.36}$$

We find out the following from (P.36)

$$\mathbf{M}\,|\,\overline{V}_t(\tau)\,|^2 = \sigma_v^2 \sum_{k=0}^{\infty} |\,H_k(\tau)\,|^2,$$

$$\mathbf{M}\,|\,\overline{V}_t(\tau)\,|^4 \le \sigma^4 \sum_{k=0}^{\infty} |\,H_k(\tau)\,|^4 + \sigma_v^4\left[\sum_{k=0}^{\infty} |\,H_k(\tau)\,|^2\right]^2 < \infty.$$

Thus, $2\mathbf{M}\,|\,\overline{v}_t(\tau)\,|^2|\,\overline{V}_t(\tau)\,|^2 \le \mathbf{M}\,|\,\overline{v}_t(\tau)\,|^4 + \mathbf{M}\,|\,\overline{V}_t(\tau)\,|^4 < \infty.$

The first assertion is proved.

If $\{\overline{y}_t, \overline{u}_t\}$ are the stationary processes determined by the pair of equations (4.13) and $\{y_t, u_t\}$ are arbitrary sequences that satisfy the equations (3.12) and (4.2), then for random quantities $\Delta y_t = y_t - \overline{y}_t$ and $\Delta u_t = u_t - \overline{u}_t$, the equalitites $a(\nabla, \tau_*)\Delta y_t = b(\nabla, \tau_*)\Delta u_t$ and $\alpha(\nabla)\Delta u_t = \beta(\nabla)\Delta y_t$ will be satisfied. Hence, $g(\nabla, \tau_*)\Delta y_t = 0$ and $g(\nabla, \tau_*)\Delta u_t = 0$. Because of the stability of the polynomial $g(\lambda, \tau_*)$, we can obtain the estimates using standard procedures :

$$\mathbf{M}(|\,y_t - \overline{y}_t\,|^2 + |\,u_t - \overline{u}_t\,|^2) \le C\rho^t\mathbf{M}(|\,y_0 - \overline{y}_0\,|^2 + |\,u_0 - \overline{u}_0\,|^2),$$

and (P.37)

$$\mathbf{M}(|\,y_t - \overline{y}_t\,|^4 + |\,u_t - \overline{u}_t\,|^4) \le C\rho^t\mathbf{M}(|\,y_0 - \overline{y}_0\,|^4 + |\,u_0 - \overline{u}_0\,|^4).$$

Here C and ρ are positive constants $\rho < 1$.

Using (P.37) we have from (4.10) and (4.13)

$$c(\nabla, \tau_{t-1})\,[\overline{v}(\tau_{t-1}) - \hat{v}_t] = a(\nabla, \tau_{t-1})\,[\overline{y}_t - y_t]$$

$$- b(\nabla, \tau_{t-1})\,[\overline{u}_t - u_t].$$

Since $\tau_{t-1} \in \mathcal{T}$ and $\mathcal{T}$ is compact, the coefficients of the polynomials $a(\lambda, \tau_{t-1})$ and $b(\lambda, \tau_{t-1})$ are bounded by some constants to be determined. Hence, the function

$$\nu_t = a(\nabla, \tau_{t-1})\,[\overline{y}_t - y_t] - b(\nabla, \tau_{t-1})\,[\overline{u}_t - u_t]$$

will satisfy the inequalities

$$\mathbf{M}\nu_t^2 \le C\rho^t\mathbf{M}\nu_0^2, \quad \mathbf{M}\nu_t^4 \le C\rho^t\mathbf{M}\nu_0^4 \tag{P.39}$$

where C and ρ are positive constants, $\rho < 1$. Assuming that $t_k \le t \le t_{k=1}$, we rewrite the equation (P.38) in the standard form as

$$x_{t+1} = A\left(\tau_{t_k}\right)x_t + b\nu_{t+1}. \tag{P.40}$$

The vector

$$x_t = \operatorname{col}\left(\overline{v}\left(\tau_{t_k}\right) - \hat{v}_{t_k}, \quad \ldots, \quad \overline{v}\left(\tau_{t_k}\right) - \hat{v}_{t_k+n}\right).$$

Since the polynomial $c(\lambda, \tau_{tk})$ is stable, the matrix $A(\tau_{tk})$ is stable, and so for any $s > 0$, the following inequality is true

$$\left|A^s\left(\tau_{t_k}\right)\right| \le C_k\rho_k^s. \tag{P.41}$$

where C_k and ρ_k are positive ($\rho_k \le \rho < 1$) depending on τ_{tk}. Iteration of the equation (P.30) and use of the inequality (P.41) gives us

$$|x_{t_{k+1}}| \le C_k\left[\rho^{t_{k+1}-t_k} | x_{t_k} | + \sum_{s=0}^{t_{k+1}-t_k-1} \rho^s | v_{t-s} |\right]. \tag{P.42}$$

The compactness of $\mathcal{T}$ gives $\sup_k C_k < \infty$, and since $\lim_{k\to\infty} (t_{k+1} - t_k) = \infty$, then for sufficiently large values of k, we have $C_k \rho^{t_{k+1}-t_k-1} \le \rho/2 < 1$. Hence from (P.42) the following inequality will result when alongwith it (P.39) is also taken into account. Thus,

$$\mathbf{M} | x_{t_{k+1}} |^2 \le \rho^2 \mathbf{M} | x_{t_k} |^2 + C\rho^{t_k} \frac{1-\rho^{t_{k+1}-t_k}}{1-\rho} (t_{k+1} - t_k).$$

Iteration of the above inequality and consideration of the expression for the vector x_t gives us the validity of the inequality in (4.17), where C and ρ are constants ($\rho < 1$). (The constant C, generally speaking, depends on the choice of initial conditions in the equations for $\hat{v}_t$ and $\overline{v}_t(\tau)$.)

The inequality of (4.18) can be established in a completely analogous manner. The inequality of (4.19) is a corollary of the inequalities in (4.17) and (4.18). The lemma is proved.

5.P.11 *Proof of the Theorem 5.4.1*

Making use of the condition (2) of the theorem for $t' = t_k$ and $t'' = t_{k+1}$, we have

$$J\left(\tau_{t_{k+1}}\right) \le J\left(\tau_{t_k}\right) + \left[\text{grad}\, J\left(\tau_{t_k}\right)\right]^* \left(\tau_{t_{k+1}} - \tau_{t_k}\right) +$$

$$+ C\, | \tau_{t_{k+1}} - \tau_{t_k} |^2 . \tag{P.43}$$

Because of the convexity of the set $\mathcal{T}$, we have from (4.11)

$$| \tau_{t_{k+1}} - \tau_{t_k} |^2 \le \gamma_k^2 | \hat{v}_{t_{k+1}} |^2 \left| \hat{V}_{t_{k+1}} \right|^2 .$$

Considering the condition (4) of the theorem and the inequality (4.19) we ascertain that

$$\sum_{k=0}^{\infty} \mathbf{M} | \tau_{t_{k+1}} - \tau_{t_k} |^2 < \infty. \tag{P.44}$$

Considering the inequality $| P_{\mathcal{T}}\tau'' - P_{\mathcal{T}}\tau' | \le | \tau'' - \tau' |$, which is valid for any τ'' and τ' (because of the convexity of $\mathcal{T}$), we evaluate the second term of the right-hand side of (P.43) as

$$\left[\text{grad}\, J\left(\tau_{t_k}\right)\right]^* \left(\tau_{t_{k+1}} - \tau_{t_k}\right) \le \left[\text{grad}\, J\left(\tau_{t_k}\right)\right]^*$$

$$\times \left\{P_{\mathcal{T}}\left[\tau_{t_k} - \gamma_k \overline{v}_{t_{k+1}}\left(\tau_{t_{k+1}}\right) \overline{V}_{t_{k+1}}\left(\tau_{t_{k+1}}\right)\right] - \tau_{t_k}\right\}$$

$$+ \gamma_k \,|\, \text{grad}\, J\left(\tau_{t_k}\right) \left| \overline{v}_{t_{k+1}}\right)\left(\tau_{t_{k+1}}\right)$$

$$\times \overline{V}_{t_{k+1}}\left(\tau_{t_{k+1}}\right) - \hat{v}_{t_{k+1}} \hat{V}_{t_{k+1}} | .$$

From the inequality of (4.21) (for $\gamma_k \le \gamma_0$), because of the following equalities (following (4.1) and (4.13))

$$\mathbf{M} \mid \bar{v}_t(\tau) \mid^2 = J(\tau),$$

$$\text{and } 2\mathbf{M}\bar{v}_t(\tau)\bar{V}_t(\tau) = \text{grad } J(\tau),$$

we have, in view of the stationarity of the processes $\bar{v}^{\infty}(\tau)$ and $\bar{V}^{\infty}(\tau)$:

$$\mathbf{M}\left[\text{grad } J\left(\tau_{t_k}\right)\right]^* \left\{P_r\left[\tau_{t_k} - \gamma_k \bar{v}_t\left(\tau_{t_k}\right)\bar{V}_t\left(\tau_{t_k}\right)\right] - \tau_{t_k}\right\}$$

$$\le -\frac{\gamma_k}{2} \mid \text{grad } J\left(\tau_{t_k}\right) \mid^2.$$

The inequality below follows the lemma 5.4.2

$$\mathbf{M}\left\{\bar{v}_{t_{k+1}}\left(\tau_{t_{k+1}}\right)\bar{V}_{t_{k+1}}\left(\tau_{t_{k+1}}\right) - \hat{v}_{t_{k+1}}\hat{V}_{t_{k+1}} \mid \tau_{t_k}\right\} \le C\rho^{t_k}$$

where C, ρ are to be determined ($\rho < 1$).

Hence

$$\mathbf{M}\left\{\left[\text{grad } J\left(\tau_{t_k}\right)\right]^* \left(\tau_{t_{k+1}} - \tau_{t_k}\right) \mid \tau_{t_k}\right\}$$

$$\le (\gamma_k/2) \mid \text{grad } J\left(\tau_{t_k}\right) \mid^2 + C\gamma_k\rho^{t_k}.$$

Computation of the conditional mathematical expectation of both parts of the inequality in (P.43) and assorting the estimates obtained above, we find out

$$\mathbf{M}\left\{J\left(\tau_{t_{k+1}}\right) \mid \tau_{t_k}\right\} \le J\left(\tau_{t_k}\right) - (\gamma_k/2) \mid \text{grad } J\left(\tau_{t_k}\right) \mid^2$$

$$+ C\mathbf{M} \mid \tau_{t_{k+1}} - \tau_{t_k} \mid^2 + C\gamma_k\rho^{t_k}. \tag{P.45}$$

The theorem 2A.1 can be made use of alongwith (P.44) to observe that the following expression exists in the mean square sense with probability 1.

$$\lim_{t \to \infty} J(\tau_t) = J_*. \tag{P.46}$$

The inequality below also follows (P.45)

$$\sum_{k=0}^{\infty} \gamma_k \mathbf{M} \mid \text{grad } J\left(\tau_{t_k}\right) \mid^2 < \infty.$$

From the condition (4) of the theorem the mean square equality $\lim_{k \to \infty} \text{grad } J(\tau_{t_k}) = 0$ exists on some subsequence with probability 1. We designate the subsequence again by $\{t_k\}$. This means (as it results from the lemma (5.4.1)) that $\tau_{tk} \to \tau_*$ with probability 1. Hence we get from (P.46) $J_* = J(\tau_*)$ and the convergence $\tau_t \to \tau_*$ exists with probability 1. The theorem is proved.

5.P.12 ***Proof of the Theorem 5.4.2***

Considering (4.27) and (4.28), we get from (4.29)

$$\mid \tau_{t+1} - \tau \mid^2 = \mid \tau_t - \tau \mid^2 + 2\gamma_t \bar{\Phi}_t(\tau_t - \tau)\left[-\bar{\Phi}_t(\tau_t - \tau) + v_{t+1}\right]$$

$$+ \gamma_t^2 \mid \bar{\Phi}_t \mid^2 \left[\Phi_t(\tau - \tau_t) + v_{t+1}\right]^2.$$

Carrying out conditional averaging, we get

$$\mathbf{M}(|\tau_{t+1}-\tau|^2| y^t, u^t) = |\tau_t-\tau|^2 - 2\gamma_t |\Phi_t(\tau_t-\tau)|^2$$

$$+\gamma_t^2 |\tilde{\Phi}_t|^2 (|\Phi_t(\tau-\tau_t)|^2 + \sigma_v^2).$$

From (4.31) we have $\sup_t |\tilde{\Phi}_t| \leq C_\Phi < \infty$, where C_Φ is some nonrandom constant. We find out,

$$\mathbf{M}\{|\tau_{t+1}-\tau|^2| y^t, u^t\} \leq (1+\gamma_t^2 C_\Phi^4)|\tau_t-\tau|^2$$

$$-2\gamma_t |\hat{\Phi}_t(\tau_t-\tau)|^2 + \gamma_t^2 C_\Phi^2 \sigma_v^2. \qquad \text{(P.47)}$$

Because of the condition (1) of the theorem, we may make use of the theorem 2.A.1, and so the following limit exists with probability 1

$$\lim_{t\to\infty} |\tau_t-\tau|^2 = \Delta_*. \qquad \text{(P.48)}$$

Here $\mathbf{M}|\tau_{tl}-\tau_*|^2 \to \mathbf{M}\Delta_*$ for $t\to\infty$, and

$$\sum_{t=0}^{\infty} \gamma_t \mathbf{M}|\tilde{\Phi}_t(\tau_t-\tau)|^2 < \infty. \qquad \text{(P.49)}$$

It will be shown that under the conditions of the theorem, the inequality below is fulfilied with some $\rho > 0$

$$\mathbf{M}|\tilde{\Phi}_t(\tau-\tau_t)|^2 \geq \rho \mathbf{M}|\tau-\tau_t|^2. \qquad \text{(P.50)}$$

Then from (P.49) and $\sum \gamma_t = \infty$, it follows that $\tau_{t_k} \to \tau$ for the subsequence $\{t_k\}$, $t_k\to\infty$. This means that $\Delta_* = 0$ is (P.48), *i.e.*, $\tau_t \to \tau$ with probability 1. It is not difficult to show the convergence of $\tau_t \to \tau$ from (P.47) in the mean square sense. Thus, to complete the proof of the theorem, it is sufficient to establish the inequality of (P.50).

If we introduce the vector

$$x_t = \mathrm{col}\,(\bar{y}_t, \ \ldots, \ \bar{y}_{t-k-p+1}, \bar{u}_{t-k}, \ \ldots, \ \bar{u}_{t-k-p+1}),$$

then consideration of the relations below

$$a(\nabla, \tau_*)\bar{y}_t = b(\nabla, \tau_*)\bar{u}_t + v_t, \quad \alpha(\nabla)\bar{u}_t = \beta(\nabla)\bar{y}_t \qquad \text{(P.51)}$$

gives (for x_t) the following equation in the usual form

$$x_{t+1} = Ax_t + Bv_{t+1}.$$

The condition (2) of the theorem provides [175] complete controllability of the pair (A,B), and so [6, 177], we have

$$\mathbf{M}\{\tilde{\Phi}^*_{t+1}\tilde{\Phi}_{t+1}| x^{t-n+1}\} \geq \rho I, \qquad \text{(P.52)}$$

where ρ is a positive number and I is an identity matrix of appropriate dimension. Because of the conditions (1) and (3) of the theorem, the inequality of (P.50) follows [6] the inequality of (P.52). It completes the proof of the theorem.

5.P.13 *Proof of the Lemma 5.5.1*

The expression in (5.7) is written in the usual form of equation.

$$\tilde{\theta}_{t+1} = A\tilde{\theta}_t + \bar{b}_{t+k+1}e, \qquad \text{(P.53)}$$

where $\overline{b}_s = b_s$, for $s \le m$, $\overline{b}_s = 0$ for $s > m$,

$$A = \begin{Vmatrix} -a_1 & -a_2 & \dots & -a_{n-1} & -a_n \\ 1 & 0 & \dots & 0 & 0 \\ \cdot & \cdot & \cdot & \cdot & \cdot \\ \cdot & \cdot & \cdot & \cdot & \cdot \\ \cdot & \cdot & \cdot & \cdot & \cdot \\ 0 & 0 & \dots & 1 & 0 \end{Vmatrix}, \quad e = \begin{Vmatrix} 1 \\ 0 \\ \cdot \\ \cdot \\ \cdot \\ 0 \end{Vmatrix}, \tag{P.54}$$

$\tilde{\theta}_t = \mathrm{col}\,(\theta_t, \;\dots,\; \theta_{t-n+1})$.

From (P.53) we get

$$\tilde{\theta}_t = A^t\tilde{\theta}_0 + \sum_{s=0}^{t-1} A^s e \overline{b}_{t-s+k} = \sum_{s=0}^{t} A^s \overline{b}_{t-s+k} e.$$

Partially $\tilde{\theta}_{m-k} = De$, where

$$D = \sum_{s=0}^{m-k} A^s b_{m-s}. \tag{P.55}$$

Analogously

$$\tilde{\theta}_{m-k+s} = DA^s e, \quad s = 1, \quad \dots, \quad n. \tag{P.56}$$

Let the vectors $\tilde{\theta}_{m-K}, \dots, \tilde{\theta}_{m+n-k+1}$ be linearly dependent. Then there exists an n-vector $d \neq 0$, such that $d^*\tilde{\theta}_{m-k+s} = 0$, $s = 0, 1, \dots, n-1$. Or from (P.56), we have

$$(D^*d, \quad A^s e) = 0, \quad S = 0, \quad 1, \quad \dots, \quad n-1. \tag{P.57}$$

Since the vectors e, Ae, ..., $A^{n-1}e$ (as is evident from (P.54)) are linearly independent, then from (P.57), we have the equality $D^*d = 0$. Let e_j be the eigen vectors of the matrix A*, A* $e_j = \lambda_j e_j$, j = 1, ..., n (for simplicity's sake it is assumed that the eigen values λ_j do not have multiple divisors). Assuming $d = \sum_{j=1}^{n} d_j e_j$, we find out

$$0 = D^*d = \sum_{j=1}^{n} d_j D^* e_j$$

$$= \sum_{j=1}^{n} d_j(\lambda_j^{m-k} b_k + \quad \dots \quad + b_m) e_j.$$

Because of linear independence of the vectors e_j, the above means that

$$d_j(\lambda_j^{m-k} b_k + \quad \dots \quad + b_m) = 0, \quad j = 1, \quad \dots, \quad n. \tag{P.58}$$

If $\lambda_j = 0$, then $d_j = 0$, since the matrix A^* can have null eigen value only when $a_n = 0$. Then because of the conditions in the lemma, we have $b_m \neq 0$. If $\lambda_j \neq 0$, then it is not difficult to show that $a(\lambda^{-1}) = 0$. Again because of the conditions of the lemma $b(\lambda_j^{-1}) \neq 0$ and from (P.58) we have $d_j = 0$. Thus, $d = 0$ and the vectors $\tilde{\theta}_{m-k}, \dots, \tilde{\theta}_{m+n-k-1}$ can not be linearly dependent.

5.P.14 *Proof of the Theorem 5.5.1*

For $\Delta\theta_j(t) = \theta_j(t) - \theta_j$, we have from the relations in (5.9),

$$|\,\Delta\theta_j(t+1)\,|^2 \le |\,\Delta\theta_j(t)\,|^2 + 2\gamma_t \Delta\theta_j(t) \overline{w}_{pt} [y_{pt+k+j}$$

$$-\sum_{s=0}^{j}\theta_s(t)u_{pt+j-s}] + \gamma_t^2\overline{w}_{pt}^2\Big[y_{pt+k+j} - \sum_{s=0}^{j}\theta_s(t)u_{pt+j-s}\Big]^2.$$

Eliminating the variables $y_{pt+k+j}, ..., y_{pt+k}$ from the equation $y_{pt+k+j} = [1 - a(\nabla, \tau)]\cdot$

$$\times y_{pt+k+j} + b(\nabla, \tau)u_{pt+k+j} + v_{pt+k+j}$$

(by virtue of the equation (5.1) and taking into account (5.7), we have

$$y_{pt+k+j} = f_j(y_{pt+k-n}^{pt+k-1},\ u_{pt-m}^{pt-1})\times$$

$$+\sum_{s=0}^{j}\theta_s u_{pt+j-s} + \overline{f}_j(v_{pt+k}^{pt+k+j}), \tag{P.60}$$

$j = 0, 1, ..., p-1$, where $f_j(\cdot)$ and $\overline{f}_j(\cdot)$ are the linear functions of corresponding arguments. Putting (P.60) in (P.59), and considering the compactness of the set Θ and the property in (5.10), we have

$$|\Delta\theta_j(t+1)|^2 \le |\Delta\theta_j(t)|^2 + 2\gamma_t\Delta\theta_j(t)\overline{w}_{pt}[f_j(y_{pt+k-n}^{pt+k-1},\ u_{pt-m}^{pt-1})$$

$$+\overline{f}_j(v_{pt+k}^{pt+k+j}) - \sum_{s=0}^{j}\Delta\theta_s(t)u_{pt+j-s}] + C_1\gamma_t^2\overline{w}_{pt}^2[\sum_{j=1}^{n}|y_{pt+k-j}|^2$$

$$+1 + \sum_{j=1}^{m}|u_{pt-j}|^2 + \sum_{j=0}^{p-1}(|v_{pt+k+j}|^2 + |\overline{u}_{pt+j}|^2)] \tag{P.61}$$

where C_1 is a constant to be determined. Because of (5.12), (5.11), (5.16) and the condition (4) of the theorem, we have

$$\mathbf{M}\{[f_j(y_{pt+k-n}^{pt+k-1},\ u_{pt-m}^{pt-1}) + f_j(v_{pt+k}^{pt+k+j})]\overline{w}_{pt}\mid F^t\} = 0$$

and

$$|\mathbf{M}\{u_{pt+j-s}\overline{w}_{pt}\mid F^t\}| = |\mathbf{M}\{|\overline{w}_{pt}|^2| F^t\} + \mathbf{M}\{\overline{u}_{pt+j-s}\overline{w}_{pt}\mid F^t\}|$$

$$\le C\mathbf{M}\{|\overline{w}_{pt}|^2| F^t\} = C\sigma_w^2\eta_t R_t\delta_t^{-1}$$

where C is a constant $C > 0$. Introducing the conditional mean (for condition F^t) in (P.61) and considering (5.17), (5.18) and (5.14), we get

$$\mathbf{M}\{|\Delta\theta_j(t+1)|^2| F^t\} \le |\Delta\theta_j(t)|^2(1 - 2\sigma_w^2\eta_t)$$

$$+C_3\eta_t|\Delta\theta_j(t)|\sum_{s=0}^{i-1}|\Delta\theta_s(t)| + C_4\delta_t\eta_t + C_5\eta_t^2 \tag{P.62}$$

where C_3, C_4, C_5 are the constants to be determined. In particular, when $j = 0$, we have

$$\mathbf{M}\{|\Delta\theta_0(t+1)|^2| F^t\} \le |\Delta\theta_0(t)|^2$$

$$\times(1 - 2\sigma_w^2 n_t) + C_4\eta_t\delta_t + C_5\eta_t^2.$$

By virtue of (5.13) and the theorem 2.A.1, we can find out from the above result that the $\lim_{t\to\infty}|\Delta\theta_0(t)| = 0$ exists with probability 1. Moreover,

$$\sum_{t=0}^{\infty}\mathbf{M}|\Delta\theta_0(t)|^2\eta_t < \infty$$

can be found out. We can ascertain, by induction and from (P.62), the existence of the limits $\lim_{t \to \infty} \Delta\theta_j(t) = 0$ (with probability 1).

Here $\sum_{t=0}^{\infty} \mathbf{M} \mid \Delta\theta_j(t) \mid^2 \eta_t < \infty$.

This establishes a strong consistently of the estimates $\theta_j(t)$, $j = 0, \ldots\ldots, p\text{-}1$. Also by virtue of the continuous invertibility of the reflection $\tau = \tau(\theta)$, obtained from the system in (5.7), the estimates $\tau_t = \tau[\theta(t)]$ will also be strongly consistent. Thus the result is established.

6

Adaptive Control of Stochastic Systems

In this chapter we shall consider design of adaptive regulators for bounded optimal control of a class of linear control objects in the presence of a random noise. The performance criterion for control is quadratic and may consist of mean operation of the ensemble of the realizations of the controllable process. Design of the adaptive regulator is done mainly through the identification approach using the results of chapter 5 on estimation of unknown control object parameters and those of chapter 3 on design of optimal regulators for linear stochastic control objects having completely known parameters.

§ 6.1 Dual Control

Unobservable (unmeasureable) parameters characterizing the control objects and/or disturbances sometimes exist side by side with the state variables of the control system. These parameters may be constant or time variant. In such a case, the control is said to be realized under *a priori* uncertainty conditions. Such problems are the topics of the adaptive control theory.

However, if the unknown parameters are random and their distributions are known, then for a sufficiently generalized control problem (vide, for example, Sections 2.2 and 2.3), the unknown parameters can be attributed to the control object states (in such a case we talk of state space expansion for the control object). The only important aspect is then that the evolution (in time) of the random parameters satisfies the conditions of causality (about which reference has been made in Section 1.2*), and the relations determining the evolution parameters are known to be within the random disturbance. Expansion of the state vector related to inclusion of random parameters in it is not essential from the abstract point of view. Under these conditions we arrive at the ordinary,

*In stochastic systems, the expression for causality often contains the Markov property of inputs and outputs of the processes.

stochastic control problem for 'sufficient' determinacy conditions.*. During finite time period control we can make use of the Bayesian control strategy (vide Section 2.4). Such a strategy may be realized in some cases of unspecified time period control.

The particular case of the above situation i.e., when the random parameters do not vary with time is important. For sufficiently generalized conditions and unbounded observation time, the *a posteriori* distributions, corresponding to these parameters, possess a tendency of localization around the parameter realization for the control process. It gives the procedure for construction of *a posteriori* distributions, which can be termed as an 'identification' technique.

Generally speaking, the control object identification may be realized by simpler methods unrelated to construction of *a posteriori* distributions. For consistency of the estimates, their use in the feedback loop also leads to a bounded optimal control problem. Here, degeneration of the optimization problem is distinctly visible and is evident from the fact that the performance criterion is independent of control system transients.

The role of transient response is quite important for applications. Hence among the bounded optimal control strategies, an attempt to separate out the 'best', in the transient response sense, is quite realistic. It may be possible to introduce, in different ways, functionals which characterize the intensity of the transient responses. Minimization of such functions in a class of bounded optimal control problems permit us to find out the 'global optimal' control strategy.

However, design of the latter invites considerable difficulties because of the nonlinear character of the optimization problem. Hence it is usual to formulate nearly optimal (from the point of view of consideration of the transient responses) control strategies. In such formulation, the concept of duality of control will be helpful : 'the control signals must be to a certain extent learning type but to a certain extent directing type'. Here the underlined meaning is that the control is then given by a new function that ensures system identification to a certain degree. In the Bayesian approach to finite time period control similar problems did not arise. The only control function was the performance criterion minimization in a class of admissible controls. The 'bifurcation' of control functions is possible only if the optimization problem permits nonunique control strategy. Such a possibility has been discussed above for bounded optimal control problems.

The interpretation of the control system in the light of dual control was done by A.A. Fel'dbaum. It gives the possibility of a fresh look at the control theory in totality. It stimulates study of real-life important control problems related to deterministic as well as 'purely stochastic' systems.

It can be intuitively understood that the quicker the completion of the indentification (adaptation) process, the better is the transient response of the control system. The best, in the mean square sense, are the Bayesian estimates,** and the control strategy based on Bayesian estimtes are the best. Strictly speaking, such a concept has not been substantiated, although the simulation results (in cases where the difficulties related to

*Similar expansion of state vector of linear control objects with unknown coefficients can 'change' into nonlinear objects. It considerably complicates the solution of the optimal control problem.

**Here Bayesian estimates are those which have been obtained from *a posteriori* distribution. Mean value of the *a posteriori* distribution may be mentioned as one such estimate.

computation of *a posteriori* distributions can be overcome) usually testify the usefulness of the Bayesian estimates from the point of view of the quality of transient responses. Formulation of strategies which optimize transient responses has not been well-studied.*

In this section an approach to adaptive control based on the idea of dual control is being discussed. The possibility of adaptive control synthesis using Bayesian estimates has also been studied. The generalised approach to formulation of adaptive strategy (identification technique) is given. Here the estimates possessing the property of consistency have been considered along with the Bayesian estimates. Such estimates are based on the recursive procedure of LSM and other modifications of the method of stochastic approximation. This is studied in greater details in Section 6.2.

6.1.1 *Bayesian Approach to Adaptive Control Problems*

(*a*) *Formulation of the problem.* Let the states of the control object and its outputs be expressed by the equations,

$$x_0 = v_0', \; x_{t+1} = X_{t+1}(x_t, u^t, \tau_t) + v'_{t+1}, \tag{1.1}$$

$$y_t = Y_t(x_t, \tau_t) + v''t. \tag{1.2}$$

Here $t = 0, 1 \ldots.., u_t \in \mathbf{R}^m, y_t \in \mathbf{R}^l, \tau_t \in \mathbf{R}^p$,

and $X_t(\cdot)$, $Y_t(\cdot)$ are the given functions of their own arguments, $v_t = \text{col}(v'_t, v_t'')$ are the noise present in the observation channel of the system and acting on it. It is the time variant vector parameter. It is assumed that the evolution of this parameter is from the equations below :

$$\tau_0 = e_0, \; \tau_{t+1} = T_{t+1}(\tau_t) + e_{(t+1)}, \; t = 0, 1, \ldots \tag{1.3}$$

where $T_t(\cdot)$ are given functions, e_t is random component of the drifting parameter τ_t.

The control objective is to minimize the functional below in a class $\mathbf{U}^p$ of realizable control strategies :

$$J[U_0^\infty(\cdot)] = \overline{\lim_{t\to\infty}}\, t^{-1} \sum_{s=1}^{t} \mathbf{M}q_{s-1}(x_s, u_{s-1}, \tau_s) \tag{1.4}$$

The weighting functions in (1.4) are assumed to be non-negative. All the sequences $U_0^\infty(\cdot)$ of functions $U_t(.)$ with values in $\mathbf{R}^m$ are assumed to constitute $\mathbf{U}^r$. These make the control u^∞ according to the rule :

$$u_0 = U_0(y_0) + w_0, \; u_t = U_t(u^{t-1}, y^t) + w_t \tag{1.5}$$

where w^∞ are the disturbances randomizing the control strategy. Again, the quantities $M(|y_t|^2 + |u_t|^2)$ exist and the following inequality is satisfied,

$$\overline{\lim_{t\to\infty}}\, t^{-1} \sum_{s=1}^{t} \mathbf{M}\,(|x_s|^2 + |u_s|^2) < \infty. \tag{1.6}$$

Let us assume that the random quantities $\{v'_t, v_t'', e_t, w_t\}$ are independent in totality and the probability distributions corresponding to them have the densities denoted by

*It is possible that alongwith the quality of the transient responses, we are required to consider the sequential problem of 'global optimization' and complexity of the realizable control strategies.

$p_v(\cdot)=p_v(\cdot\mid t)$, $p_e(\cdot)=p_e(\cdot\mid t)$, and $p_w(\cdot)=p_w(\cdot\mid t)$ respectively. These densities are assumed to be known at all instants. The time dependence may not be mentioned if the context of the instant of time under consideration is clear.

(*b*) *Extension of the state space.* The variable τ_t in the system of (1.1) and (1.2) is interpreted to be unknown parameter. However, consideration of the relations in (1.3) permits us to cross over to an 'extended' object which does not contain 'unkown parameters'. With this in mind we denote,

$$z_t = \text{col}\,(x_t,\ \tau_t). \tag{1.7}$$

It is a vector for which the following conditions are satisfied (by virtue of (1.1) and (1.3))

$$z_0 = \text{col}\,(v'_0,\ e_0),\ z_{t+1} = Z_{t+1}(z_t,\ u^t) + \text{col}\,(v'_{t+1},\ e_{t+1}). \tag{1.8}$$

Here $t = 0, 1, \ldots.$ and

$$Z_{t+1}(z_t,\ u^t) = \text{col}(X_{t+1}(x_t,\ u^t,\ \tau_t),\ T_{t+1}(\tau_t)). \tag{1.9}$$

The 'extended object' equation of (1.8) supplements the output equation

$$y_t = \tilde{Y}_t(z_t) + v'_t,\ \tilde{Y}_t = Y_t(x_t,\ \tau_t). \tag{1.10}$$

The control problem of (1.8), (1.10 and (1.7) does not already contain unknown parameters but has the form of a usual stochastic control problem with incomplete observation of state vectors, which can be evaluated from the given observations y^t and u^t.

(*c*) *Optimality of the realizable control strategy and Bellman's equation.* It will be shown that the control strategy that minimizes the functional (1.4) may be linked with Bellman-Liapunov function in the same manner as it was done in paragraph 3.4.1. Let $\{V_t\}$ be the sequence of functions $V_t = V_t(y^t, u^{t-1})$ determined for all values of the arguments. For all t, the function below will be introduced,

$$W_t(y^t,\ u^t) = \int [V_{t+1}(y^{t+1},\ u^t) + q_t(x_{t+1},\ u_t,\ \tau_{t+1}]\, p_{v'_{t+1}}[y_{t+1}$$

$$-Y_{t-1}(x_{t+1},\ \tau_{t+1})]\, p(x_{t+1},\ \tau_{t+1} \mid y^t,\ u^t)dy_{t+1}dx_{t+1}d\tau_{t+1} \tag{1.11}$$

where $\rho(x_{t+1}, \tau_{t+1} \mid y^t, u^t)$, is the *a posteriori* density function of the random quantities (x_{t+1}, τ_{t+1}) for observations y^t, u^t. Let the strategy $\tilde{U}^\infty(\cdot)$ be determined as per following rules,

$$\tilde{U}_t = \tilde{U}_t(u^{t-1},\ y^t) = \arg\min_{v_t} W_t(y^t,\ u^t). \tag{1.12}$$

Theorem 6.1.1 *We assume that the following conditions are satisfied for the strategy $\tilde{U}^\infty(\cdot)$ determined by the relations (1.11) and (1.12) :*

(1) $\tilde{U}^\infty(\cdot) \in \mathbf{U}^P$,

(2) for the random process (y^∞, u^∞) determinable by the control system of (1.1) - (1.3) and (1.5), the following equality holds when $U_t(\cdot) = \tilde{U}_t(\cdot)$,

$$W_t(y^t,\ u^t) - V_t(y^t,\ u^{t-1}) = r_t(y^t,\ u^{t-1}), \tag{1.13}$$

where $r_t(\cdot)$ is the sequence of functions such that with the strategy $\tilde{U}\infty(\cdot)$, the magnitude of $J_\bullet$ takes the minimum value in $\mathbf{U}^P$, where,

$$J_* = \lim_{t\to\infty} t^{-1} \sum_{s=1}^{t} \mathbf{M} r_t, \tag{1.14}$$

(3) the following inequality is true for the random process (y^∞, u^∞) resulting from the control system in (1.1) to (1.3) and (1.5). Here the strategy $U^\infty(\cdot) \in \mathbf{U}^p$ is arbitrary.*

$$\lim_{t\to\infty} \mathbf{M} V_t(y^t, u^{t-1}) < \infty. \tag{1.15}$$

Then the strategy $\tilde{U}^\infty(\cdot)$ is optimal in $\mathbf{U}^p$ and

$$\min_{U^\infty(\cdot)\in \mathbf{U}^p} J[U^\infty(\cdot)] = J_*. \tag{1.16}$$

The Bellman's equation, in this case, results from (1.13) and (1.12) :

$$V_t(y^t, u^{t-1}) = \arg\min_{u_t \in \mathbf{R}^m} W_t(y^t, u^t) + r_t \tag{1.17}$$

and the theorem 6.1.1 postulates the existence of the solution $V_t(y^t, u^{t-1})$ of this equation satisfying the 'boundary' condition of (1.15). The formulae of (1.12) and (1.11) show how the optimal control strategy is to be synthesized, if the Bellman-Liapunov function $V_t(\cdot)$ is known. In addition, this synthesis pre-supposes a known *a posteriori* density function $\rho(x_{t+1}, \tau_{t+1} \mid y^t, u^t)$ vide (1.11)).

6.1.2 *Adaptive Version of the Gaussian Linear Quadratic Control Problems with Observable Vector States*

The possibility of a Bayesian approach to a sufficiently generalised linear-quadratic control problem will be shown.

(*a*) *Formulation of the problem.* Let the control object be expressed as,

$$x_{t+1} = A(\tau)x_t + Bu_t + v_{t+1}, \tag{1.18}$$

$$\text{where } A(\tau) = A_0 + \sum_{j=1}^{l} A_j \tau^{(j)} \tag{1.19}$$

is a coefficient matrix ($n \times n$ diamension) dependent on l - parameter vector,

$$\tau = \mathrm{col}(\tau^{1)}, \ldots, \tau^{(l)}). \tag{1.20}$$

The above is a Gaussian random quantity with a density function $p_\tau(\cdot) \sim N(\tau_0, R_\tau)$. It is assumed that the disturbances v_t are also Gaussian random quantities $p_{v_t}(\cdot) \sim N(0, R_v)$, which do not have any autocorrelation and correlation with the vector τ. The matrices A_j (j = 0,1, ... l), B, R_τ and R_v of dimensions $n \times n$, $n \times m$, $l \times l$ and $n \times m$ respectively are time invariant and assumed to be known.

We assume that the state vector x_t is known for all t and formally put $y_t = x_t$. The performance criterion is chosen as :

$$J[U^\infty(\cdot)] = \lim_{t\to\infty} t^{-1} \sum_{s=1}^{t} \mathbf{M} \begin{pmatrix} x_s \\ u_s \end{pmatrix}^* N \begin{pmatrix} x_s \\ u_s \end{pmatrix} \tag{1.21}$$

---

*If the functions $V_t(\cdot)$ are non-negative, then the fulfilment of the inequality (1.15) is sufficient only for strategy $\tilde{U}^\infty(\cdot)$.

with a non-negative matrix

$$N = \left\| \begin{matrix} Q & S \\ S^* & R \end{matrix} \right\|. \tag{1.22}$$

The functional of (1.21) and (1.22) is defined on the set U^p of realizable control strategies. The latter includes all the regular strategies (vide para 1.2.2), which ensure fulfilment of the inequality (1.6) independent of the choice of the initial states of the control object (1.18).

(*b*) *Bellman's equation.* For synthesis of optimal control strategy in ^{p}, the Bellman-Liapunov function is formed as follows,

$$V_t(x) = x^* H_t x. \tag{1.23}$$

Attempt will be made to choose the symmetric non-negative matrix H_t that satisfies the Bellman's equation. The function (1.11) in the present case has the expression,

$$W(x^t, u^t) = [A(\tau_t)x_t + Bu_t]^* H_{t+1}[A(\tau_t)x_t + Bu_t]$$

$$+\mathrm{Sp} H_{t+1}^{1/2} R_v H_{t+1}^{1/2} + x_t^* L_t H_{t+1} x_t + \begin{pmatrix} x_t \\ u_t \end{pmatrix}^* N \begin{pmatrix} x_t \\ u_t \end{pmatrix} \tag{1.24}$$

where the following notations have been used,

$$L_t H_{t+1} = \sum_{j=1}^{l} \sum_{k=1}^{l} P_t^{(jk)} A_j^* H_{t+1} A_k, \tag{1.25}$$

$$\tau_t^{(j)} = \mathbf{M}\,(\tau^{(j)} \mid y^t,\, u^t), \tag{1.26}$$

$$P_t^{(jk)} = \mathbf{M}\{(\tau^{(j)} - \tau_t^{(j)})(\tau^{(k)} - \tau_t^{(k)})^* \mid y^t,\, u^t\}. \tag{1.27}$$

According to the theorem 5.2:1 *a posteriori* density $p_t(\cdot \mid y^t, u^t)$ of random quantities τ is Gaussian. Its vector τ_t and matrix P_t (which can be determined by the formulae (1.26) and (1.27)) are sufficiently statistical. These statistics satisfy the recursive relations written in a more generalised fashion as in Section 5.2.1. It follows naturally that matrix P_t does not depend on control signal u_t. Thus, the matrix (1.25) does not depend on u_t. The minimization of the function (1.24) with respect to u_t is realized in a simple manner. Minimum is achieved when,

$$u_t = -(B^*H_{t+1}B + R)^{-1}\,[B^*H_{t+1}A(\tau_t) + S^*]x_t \tag{1.28}$$

and equal to ,

$$\min_{u_t} W_t(x^t, u^t) = x_t^*[A^*(\tau_t)H_{t+1}A(\tau_t) + L_t H_{t+1} + Q$$

$$-(B^*H_{t+1}A(\tau_t) + S^*)^*\,(B^*H_{t+1}B + R)^{-1}\,(B^*H_{t+1}A(\tau_t) + S^*)]x_t$$

$$+\mathrm{Sp}\, H_{t+1}^{1/2} R_v H_{t+1}^{1/2}. \tag{1.29}$$

It follows from (1.29) that the function in (1.23) will satisfy the Bellman's equation (1.17), if the values of r_t and matrix H_t satisfy the following correlations,

$$r_t = \mathrm{Sp} H_{t+1}^{1/2} R_v H_{t+1}^{1/2} + x_t^* L_t H_{t+1} x_t, \tag{1.30}$$

$$H_t = A^*(\tau_t)H_{t+1}A(\tau_t) + Q + [B^*H_{t+1}A(\tau_t) + S^*]^* K(H_{t+1},\, \tau_t), \tag{1.31}$$

$$K(H,\, \tau) = -(B^*HB + R)^{-1}\,[B^*HA(\tau) + S^*]. \tag{1.32}$$

Thus, the relations (1.28), (1.31), (1.32) and (1.25) to (1.27) determine the control strategy which will be realizable and optimal in U^p, if the following conditions are fulfilled :

1. Matrices H_t which satisfy the equation (1.31) are uniformly bounded along t and non-negative, and the following equation is asymptotically stable.

$$\tilde{x}_{t+1} = [A(\tau_t) + BK(H_{t+1}, \tau_t)]\tilde{x}_t. \tag{1.33}$$

2. The following quantity exists and it can be determined by feedback (1.28) in a control strategy. It takes the least value in U^p

$$J_* = \lim_{t\to\infty} t^{-1} \sum_{s=1}^{t} [\mathrm{SpM}H_{s+1}^{1/2} R_v H_{s+1}^{1/2} + \mathrm{M}x_s^* L_s H_{s+1} x_s]. \tag{1.34}$$

The condition 1 ensures realizability of the control strategy under consideration and fulfilment of the condition in (1.15) for the function (1.23). The condition 2 ensures adaptability of the theorem 6.1.1. The above two conditions are, in a general case, difficult to be verified. Moreover, even when they are fulfilled, finding out the matrix H_t from equation (1.31) is difficult.

(c) *Particular case.* A special form of the linear quadratic adaptive control problem will be considered, where the above difficulties are not so significant. To be precise, let the matrix (1.19) be represented by

$$A(\tau) = A_0 + BC\tau^* D \tag{1.35}$$

where the $n \times m$ matrix B is given in equation (1.18), C and D are m-vector and $l \times n$ matrix* respectively. Moreover, we assume that in the performance criterion of (1.21) and (1.22), the matrices S and R are null matrices, *i.e.*,

$$J[U^\infty(\cdot)] = \lim_{t\to\infty} t^{-1} \sum_{s=1}^{t} \mathrm{M}x_s^* Q x_s. \tag{1.36}$$

Taking the above properties into account, the expression for equation (1.31) is,

$$H_t = A_0^* H_{t+1} A_0 + Q - A_0^* H_{t+1} B (B^* H_{t+1} B)^{-1} B^* H_{t+1} A_0, \tag{1.37}$$

i.e., its coefficients do not depend on statistics of τ_t. Assuming that the Lure' equation,

$$H = A_0^* H A_0 + Q - A_0^* HB(B^* HB)^{-1} B^* HA_0 \tag{1.38}$$

has a solution $H \geq 0$, for which $B^*HB > 0$ and the matrix $A_0 - B(B^*HB)^{-1}B^*HA_0$ is stable, we can choose $H_t \equiv H$. Then the feedback (1.28) is expressed as

$$u_t = -(B^*HB)^{-1} B^* HA(\tau_t) x_t. \tag{1.39}$$

Considering (1.17), (1.30), (1.31) and (1.25), it is easy to calculate for $H_t \equiv H$ that

$$(t+1)^{-1} \sum_{s=0}^{t} \mathrm{M}x_s^* Q x_s = (t+1)^{-1} \mathrm{M}(x_0^* H x_0 - x_{t+1}^* H x_{t+1})$$

$$+ \mathrm{Sp}H^{1/2} R_v H^{1/2} + (t+1)^{-1} \mathrm{M} \sum_{s=0}^{t} C^* B^* HBC x_s^* D^* P_s D x_s,$$

where the matrix P_t is determined by the formula of (1.27). Hence we get the estimates,

---

*The following equation leads, as a particular case, to the equation (1.18) and (1.35). In terms of 'input-output' variables it is expressed as : $a(\nabla, \tau) y_t = u_{t-1} + v_t$, $a(\lambda, \tau) = 1 + \lambda a_1 + \ldots + \lambda^n a_n$, and $\tau = \mathrm{col}\,(a_{i_1}, \ldots, a_{i_p})$.

$$\lim_{t\to\infty} t^{-1}\sum_{s=0}^{t} \mathrm{M}x_s^* Q x_s \le \mathrm{Sp}\, H^{1/2} R H^{1/2}$$

$$+C^* B^* H B C \lim_{t\to\infty} t^{-1}\sum_{s=0}^{t} \mathrm{M}x_s^* D^* P_s D x_s. \tag{1.40}$$

It will be shown below that the matrices P_t decay monotonously, where $P_t \ge P_{t+1}$. Assuming for sufficiently large t, that the following is satisfied, i.e.,

$$Q > C^*B^*HBCD^*P_tD. \tag{1.41}$$

We conclude from (1.40) that the feedback of (1.39) results in realizable (which ensures fulfilment of inequality in (1.6)) control strategy. If,

$$\lim_{t\to\infty} P_t = 0, \tag{1.42}$$

then the above control strategy is optimal in $\mathbf{U}^p$. Actually, if we examine the class $\mathbf{U}^a$ of all admissible control strategies $U^\infty(\cdot)$, functions $U_t = U_t(u^{t-1}, y^t, \tau)$, in which the inequality of (1.6) is ensured (but the functions can depend on parameter τ), then it can be seen that the feedback of (1.39) realizes the optimal control strategy in $\mathbf{U}^p$ for $\tau_t \equiv \tau$. Then the functional (1.36) takes the value $\mathrm{Sp}\, H^{1/2} R_v H^{1/2}$. Since $\mathbf{U}^r \subseteq \mathbf{U}^a$, then from (1.40) and (1.42) the optimality of the control strategy (resulting from feedback of (1.39)) follows in $\mathbf{U}^p$ (and even in $\mathbf{U}^a$).

d) Recursive relations for statistics of a posteriori distribution and their characteristics. In the present case the effectiveness of the optimal control strategy is supported by the possibility of recursive computation of statistics in (1.26) and (1.27). To express the recursive formulae, we write the equation of (1.18) as

$$y_t = \Phi_t \tau + v_{t+1}, \tag{1.43}$$

Here $\quad y_t = x_{t+1} - A_0 x_t - B u_t,\ \Phi_t = \| A_1 x_t, \ldots, A_l x_t \|$. (1.44)

The equation (1.43) represents the particular case of the equation (2.9) in chapter 5. Hence the recursive relations of the theorem 5.2.1 can be written as,

$$\tau_{t+1} = \tau_t + K_t[x_{t+1} - A(\tau_t)x_t - Bu_t], \tag{1.45}$$

$$P_{t+1} = P_t + K_t \Phi_t P_t, \tag{1.46}$$

$$K_t = -P_t \Phi_t^* (R_v + \Phi_t P_t \Phi_t^*)^{-1}, \tag{1.47}$$

$t = 0, 1, \ldots$, and the initial conditions are $\tau_0, P_o = R_\tau$. The procedure of (1.45) to (1.47) represents the recursive form of the least square algorithm. The characteristics of the estimates obtained through this method have been well-studied. The least square estimates in the non-recursive form are computed by the formulae :

$$P_{t+1} = \left(R_\tau + \sum_{s=0}^{t} \Phi_s^* R_v^{-1} \Phi_s \right)^{-1}, \tag{1.48}$$

$$\tau_{t+1} = \tau + P_{t+1} R_\tau^{-1}(\tau_0 - \tau) + P_{t+1} \sum_{s=0}^{t} \Phi_s^* R_v^{-1} v_{s+1}. \tag{1.49}$$

The above formulae have been derived from (1.45) to (1.47)*. It results from (1.48) that, if the information matrix,

---

*Because of (1.48), the inequality $P_{t+1} \le P_t$ is evident for all t.

$$I_t = (t+1)^{-1} \sum_{s=0}^{t} \Phi_s^* R_v^{-1} \Phi_s \tag{1.50}$$

is limitingly non-degenerated, such as,

$$\lim_{t\to\infty} I_t > 0 \tag{1.51}$$

then the following limit equalities are satisfied with a probability of 1,

$$\lim_{t\to\infty} P_t = 0, \ \lim_{t\to\infty} \tau_t = \tau, \tag{1.52}$$

i.e., the statistics $\{\tau_t\}$ are strongly consistent estimates of the unknown parameter τ.

e) The limiting form of Riccati equation. Re-examination of the equation (1.31) gives that the fulfilment of the inequality (1.51) for $t \to \infty$, transforms it into Lure' equation :

$$H = A^*(\tau)HA(\tau) + Q -$$
$$-[B^*HA(\tau)+S^*]^*(B^*HB+R)^{-1}[B^*HA(\tau)+S^*]. \tag{1.53}$$

If $R>0$ and for every vector $\tau \varepsilon R^1$, the matrix pair $\{A(\tau), B\}$ is stabilizable, and the matrix pair $\{A(\tau) - BR^{-1}S^*, [Q - SR^{-1}S^*]^{1/2}\}$ is detectable, then the equation (1.53) has a unique non-negative solution $H = H(\tau)$ (this follows from the theorem 3.22) and the matrix $A + BK(H, \tau)$ is stable. This means that the limiting non-degeneration of the information matrix (1.50) ensures asymptotic stability of equation (1.33). Moreover, the matrix $H(\tau)$determines the feedback that degenerates the optimal strategy in a class U^p introduced above during the discussion on limiting equality (1.42). Thus,

$$u_t = K[H(\tau), \tau]x_t. \tag{1.54}$$

Then,

$$\min_{U^{\infty}(\cdot)\in U^a} J[U^{\infty}(\cdot)] = \mathrm{MSp} H^{1/2}(\tau) R_v H^{1/2}(\tau), \tag{1.55}$$

and since $U^p \subseteq U^a$, then for $H_t \to H(T)$, $t\to\infty$, the value (1.34) takes the least value in U^p, which is equal to (1.55). Thus, the control strategy degenerated from the relations in (1.28), (1.25), (1.31), (1.32) and (1.45) to (1.47) is optimal in U^p.

f) Identification approach to formulation of optimal control strategy. The realization of the control strategy under consideration is difficult even when the limiting equalities of (1.52) are available. The reason can be attributed to the necessity of computing the matrices H_t which satisfy the equation (1.31) and (1.32). Hence the control strategy degenerated from the feedback, *i.e.*,

$$u_t = K(H_t, \tau_t)x_t, \tag{1.56}$$

will be studied. Here the matrix $K(H, \tau)$ is determined by the formula (1.32), the estimates τ_t are computed from the formulae (1.45) to (1.47) and the matrices H_t are found out from the equation,

$$H_{t+1} = A^*(\tau_t)H_t A(\tau_t) + Q + [B^*H_t A(\tau_t) + S^*]^* K(H_t, \tau_t). \tag{1.57}$$

For the limiting nondegenerative matrix (1.50), the Riccati equation of (1.57) can be considered to be a recursive procedure for determination of the required solution $H = H(\tau)$ of the Lur'e equation (1.53). When the corresponding conditions of the theorem 3.2.2 are satisfied, then the convergence $H_t \to H(\tau)$, $t\to\infty$ may be present for an arbitrary choice of the initial matrix $H_0 \geq 0$.

The feedback below is equally useful :

$$u_t = K[H(\tau_t), \tau_t]x_t. \tag{1.58}$$

Here $H(\tau)$ is the solution of the Lur'e equation (1.53). If the limiting equalities of (1.52) are present, then we can again consider the optimal control strategy. But its formulation will be more complicated because of the necessity of searching out the solution to the Lur'e equation at every instant of control.

The feedback of (1.58) more or less clearly characterizes the essence of the identification technique as applied to adaptive control : in the optimal feedback (vide (1.54)), both the unknown parameter τ and its estimate are made use of. The estimate is found out at corresponding instants of time (in the present case the estimate is computed with the help of a recursive procedure for the least square method and in a general case the estimates can be determined by other methods). Then the strong consistency of the estimates $\{\tau_t\}$ (in the present case the limiting non-degeneration of the matrix (1.50) is sufficient) ensures realizability and optimality of the control strategy thus obtained in $\mathbf{U}^r$ (as well as in $\mathbf{U}^p$).

g) Consistency of the estimates. Various methods of synthesis of realizable control strategies result from the above analysis. In every case, establishment of the optimality of the strategy is based on the consistency of the estimates $\{\tau_t\}$. Establishment of the consistency of the estimates may demand additional assumptions about the control system. The situation becomes easier if the synthesized feedback signals are randomized by additive test signals. Thus, the feedback of (1.58) may be used along with the feedback shown below :

$$u_t = K[H(\tau_t), \tau_t]x_t + w_t. \tag{1.59}$$

Here the test signal w^∞ is composed of unknown random quantities ω_t satisfying the conditions,

$$\mathbf{M}w_t = 0, \ \mathbf{M} \,|\, w_t \,|^2 > C \ln t, \ \lim_{t\to\infty} \mathbf{M} \,|\, w_t \,|^4 = 0, \tag{1.60}$$

where C is a constant $C > 0$. The presence of a very quickly decaying test signal in the feedback loop (vide inequality of (1.60)) permits establishment of equality (1.43), and since the test signal w^∞ decays with time to zero, then the control, as specified, by the feedback of (1.59), will be limiting optimal.

We shall limit our attention to these general observations about the possibility of synthesis of adaptive limiting optimal control. Rigid assertions demand more agreement between the properties of strongly consistent estimates (convergence $\tau_t \to \tau$ with probability 1) and the mean operation in the performance criterion (1.21). Precise results of synthesis of adaptive limiting optimal control (within the framework of identification approach) will be presented in Section 6.2.

h) Loss estimation in adaptation in identification approach. It was earlier observed that the performance criterion of (1.21) and (1.22) does not depend on the transient responses and so the optimal control, when solvable, possesses various solutions from the point of view of transient responses. It is interesting to understand the difference of the limiting optimal controls in respect of transient responses. Keeping this in mind we shall recall the equation (1.18) (vide (1.35)). The following discrete performance criterion (vide section 3.2.4b) will be made use of to take the transient responses into account,

$$J[U^\infty(\cdot), \gamma^2] = (1-\gamma^2) \sum_{t=0}^{\infty} \gamma^{2t} \mathbf{M} x_t^* Q x_t. \tag{1.61}$$

Here Q is a positive matrix and γ is a parameter ($O < \gamma < 1$). It can be shown by the methods presented in Section 3.2.4, that for a known parameter τ the admissible control strategy which minimizes the functional (1.61) is determined by the feedback

$$u_t = -(B^*HB)^{-1}B^*HA(\tau)x_t. \tag{1.62}$$

Then the functional (1.61) takes a value as,

$$\min_{U^\infty(\cdot)\in U^a} J[U^\infty(\cdot),\gamma] = (1-\gamma^2)x_0^*Hx_0 + \gamma^2 \mathrm{Sp} H^{1/2}R_v H^{1/2}. \tag{1.63}$$

We assume that there exists a sequence of estimates $\{\hat{\tau}_t\}$ having the characteristic of consistency $\lim_{t\to\infty}\hat{\tau}_t = \tau$. The control corresponding to the identification approach is realized with the help of feedback

$$u_t = -(B^*HB)^{-1}B^*HA(\hat{\tau}_t)x_t. \tag{1.64}$$

Computation of the functional (1.61) in such a control is not difficult. Actually, considering (1.18), (1.38) and (1.35), we find out

$$\mathrm{M}\{x_{t+1}^*Qx_{t+1} \mid x^t\} = \mathrm{Sp}H^{1/2}R_vH^{1/2} + x_t^*C^*B^*HBCD^*\hat{P}_tDx_t,$$

where

$$\hat{P}_t = \mathrm{M}\{(\tau-\hat{\tau}_t)(\tau-\hat{\tau}_t)^* \mid x^t\}. \tag{1.65}$$

Hence by virtue of (1.61)

$$J[U^\infty(\cdot),\gamma] = (1-\gamma^2)x_0^*Qx_0 + \gamma^2\,\mathrm{Sp}\,\mathrm{H}^{1/2}R_vH^{1/2}$$

$$+(1-\gamma^2)C^*B^*HBC\sum_{t=0}^{\infty}\gamma^{2t}\mathrm{M}x_t^*D^*\hat{P}_tDx_t. \tag{1.66}$$

Comparison of (1.66) with (1.63) shows that the last item in the right-hand side of (1.66) may be interpreted as an unavoidable loss 'in adaptation' : it is the control cost due to uncertainty of control object parameters. This item 'worsens the quality of the transient responses of control systems with respect to control under conditions of sufficient determinacy. It is well-known that $\hat{P}_t \geq P_t$, where the matrix P_t is determined by the formula (1.27). Evidently it can be assumed that the last item in the right-hand side of (1.66) takes the least value, if the estimates of the parameter τ are found out through algorithm in (1.45) to (1.47). In such a case, the Bayesian control strategy will be the best of all strategies (from the point of view of transient responses), which can be synthesized within the frame work of identification approach using feedbacks of (1.64). In general, the assumption made is needed as a foundation, since the third item in (1.66) still depends on the response $\{x_t\}$ that can be determined by the control strategy used.

6.1.3 *Bayesian Control Strategy*

Determination of the optimal strategy in U^p with the help of the theorem 6.1.1 is apparently possible only in special cases (one of them has been examined in section 6.1.2). Here we shall express the general form for synthesis of realizable strategies. It leads to limiting optimal control in a series of cases.

Let $\mathrm{U}^a = \mathrm{U}^a(\tau^\infty)$ be the set of all admissible strategies $U^\infty(\cdot)$ and the functions $U_t = U_t(u^{t-1}, y^t, \tau_t)$ ensuring the inequality (1.6) cannot depend on parameter τ_t. The problem of minimization of the functional (1.4) in a class $\mathbf{U}^a(\tau^\infty)$ is sometimes easier than that in $\mathbf{U}^p$. Here the applied methods of control theory may be the conditions of complete determinacy of control object parameters. Without discussing the practical methods of determining the optimal control strategy in $\mathbf{U}^a(\tau^\infty)$, we assume that such a stragety $\tilde{U}^\infty(\cdot)$ has been found out, *i.e.*, the constituents of its function $\tilde{U}_t = \tilde{U}_t(u^{t-1}, y^t, \tau_t)$ are known for all possible values of the arguments. Let $p(\tau_t \mid y^t, u^{t-1})$ denote *a posteriori* density function of the random quantity τ_t for all observations (y^t, u^{t-1}).* The following functions are compared with the functions :

$$\hat{U}_t(u^{t-1}, y^t) = \int \hat{U}(u^{t-1}, y^t, \tau_t)p(\tau_t \mid y^t, u^{t-1})d\tau_t. \tag{1.67}$$

The above functions do not depend on the parameter τ_t. If this gives a realizable strategy $\tilde{U}^\infty(\cdot)$ (*i.e.*, the fulfilment of the inequality in (1.6) is ensured), then it will be called Bayesian control strategy.

It will now be shown, for example, how the Bayesian control strategy is realized and its optimal properties are established.

(*a*) *Non-stationary minimum - phase object for a special performance criterion.* Let the control object be expressed through the scalar equation

$$a(\nabla, \tau_t)y_t = u_{t-1} + v_t \tag{1.68}$$

where, $a(\lambda, \tau_t) = 1 + \lambda a_1(t) + \ldots + \lambda^n a_n(t), a_j(t)$ are the variable (with time) parameters of the control object,

$$\tau_t = \mathrm{col}\left(a_{i_1}(t), \ldots, a_{i_p}(t)\right), \tag{1.69}$$

and $\{v_t\}$ are the disturbances composed of independent random quantities with the following properties

$$\mathbf{M}v_t = 0, \ \mathbf{M}v_t^2 = \sigma_v^2 > 0. \tag{1.70}$$

The performance criterion is taken in the form,

$$J[U^\infty(\cdot)] = \overline{\lim_{t\to\infty}}\, t^{-1} \sum_{s=1}^{t} \mathbf{M}y_s^2. \tag{1.71}$$

We shall examine the problem of minimization of (1.71) in $\mathbf{U}^p$ - the realizable control strategies.

The optimal control strategy is easily found out in the admissible class of control strategies, $\mathbf{U}^a = \mathbf{U}^a(\tau^\infty)$. It is determined by the feedback $u_t = \tilde{a}(\nabla, \tau_t)y_t$, where $\tilde{a}(\lambda, \tau_t) = \lambda^{-1}[a(\lambda, \tau_t) - 1]$. The coefficients of this feedback linearly depend on parameter τ_t. Hence the Bayesian control strategy, when it exists, is determined (vide (1.67)) by the feedback,

$$u_t = \tilde{a}(\nabla, \tilde{\tau}_t)y_t, \tag{1.72}$$

where,
$$\tilde{\tau}_t = \int \tau_t p(\tau_t \mid y^t, u^{t-1})d\tau_t. \tag{1.73}$$

---

*It will be shown in section 6.1.4 that the expression $p(\tau_t \mid y^t, u^{t-1})$ does not depend on the choice of the control strategy. Hence there is no need to recall the control strategy which gave rise to this function.

Here it is assumed that the parameter τ_t is a random quantity for all values, of t, and its time evolution is determined by the equations like (1.3). Thus, for the formulation of a Bayesian control strategy, it is sufficient to know, in this case, only the mean value $\hat{\tau}_t$ of the *a posteriori* distribution of the random quantities τ_t. The feedback (1.72) has a structure that appears during the identification process : in the optimal feedback, the estimates of the unknown parameters are used along with the parameters at any instant of time.

To establish the realizability of the strategy degenerated by the feedback (1.72), additional presuppositions are required. We shall examine two particular cases.

1. We assume that in (1.3) the initial condition $e_0=\tau_0$ is the value determined and $T_{t+1}(\tau)=\tau, t=0,1,\ldots, i.e.,\quad \tau_t=\tau_0+w_t$. Let $p_w(\cdot\mid t)$ be the densities of independent random quantities, and w_t have the properties,

$$\int \omega p_\omega(\omega\mid t)d\omega=0,\ \int \omega\omega^* p_w(\omega\mid t)d\omega=\operatorname{diag}(\rho_1,\ \ldots,\ \rho_p).$$

Then for the statistics of (1.73) we have $\hat{\tau}_t=\tau_0$ and for a feedback of (1.72) we can find out from (1.68)

$$\mathbf{M}\mid y_t\mid^2=\sum_{k=1}^{p}\rho_k\mathbf{M}\mid y_{t-i_k}\mid^2+\sigma_v^2.$$

It is not difficult to conclude from here that the feedback of (1.72) degenerates the realizable strategy which is optimal in $\mathbf{U}^p$ when

$$\sum_{k=1}^{p}\rho_k<1.$$

Then,

$$\min_{U^\infty(\cdot)\in \mathbf{U}^p} J[U^\infty(\cdot)]=\sigma_v^2\left(1-\sum_{k=1}^{p}\rho_k\right)^{-1}=$$

$$=\min_{U^\infty(\cdot)\in \mathbf{U}^a} J[U^\infty(\cdot)]\left(1-\sum_{k=1}^{p}\rho_k\right)^{-1}.$$

The above formula shows how the quality of control worsens under the conditions of uncertainty of randomly varying coefficients of the object.

2. We then assume that in the equations of (1.3) $\tau_{t+1}(\tau)=\tau$ and $w_t\equiv 0$, and e_0 is the random quantity with density $p_e(\cdot\mid 0)\sim N(\bar{\tau},\ R_\tau)$. The disturbances v_t are taken as independent Gaussian values, $v_t\sim N(0,\ \sigma_v^2)$. Then the *a posteriori* density $p(\tau_0\mid y_t,\ u^{t-1})$ is Gaussian, and the vectors in (1.73) are found out from the recursive correlations. To express the latter, the equation in (1.68) (for $\tau_t\equiv\tau_0=\tau$) is presented in the form :

$$y_t=\Phi_{t-1}^*\tau+\phi_{t-1}+u_{t-1}+v_t,$$

where, $\Phi_t=\operatorname{col}\left(-y_{t-i_1},\ \ldots,\ -y_{t-i_p}\right),\ \phi_{t-1}=-\tilde{a}(\nabla,\ 0)y_t.$

Then the recursive correlations of the theorem 5.2.1 can be adopted. The limiting nondegeneration of information matrix (1.50) is ensured, as in section 6.1.2, by the consistency of the estimates $\{\hat{\tau}_t\}$. In the same way, the existence of the Bayesian strategy and its optimality in the spaces $\mathbf{U}^r$ and $\mathbf{U}^a$ are also established. To provide the consistency of the estimates $\{\hat{\tau}_t\}$, the recommendations of section 6.1.2g may be considered.

(b) Bayesian strategy for an object of general nature with scalar input-output. Let the control object be expressed as,

$$a(\nabla, \tau)y_t = b(\nabla, \tau)u_t + v_t, \tag{1.74}$$

where, $a(\lambda, \tau) = 1 + \lambda a_1 + \ \dots\ + \lambda^n a_n$, $b(\lambda, \tau) = \lambda b_1 + \ \dots\ + \lambda^n b_n$,

$$\tau = \text{col}\left(a_{i_1}, \dots a_{i_p}, \dots b_{j_r}\right). \tag{1.75}$$

The performance criterion has the expression,

$$J[U^\infty(\cdot)] = \overline{\lim_{t\to\infty}}\, t^{-1}\mathbf{M} \sum_{s=1}^{t} \begin{pmatrix} y_s \\ u_s \end{pmatrix}^* \mathrm{N} \begin{pmatrix} y_s \\ u_s \end{pmatrix}. \tag{1.76}$$

Here the matrix N is of 2×2 dimension and non-negative. The noise v^∞ with independent values possesses the properties of (1.70).

In chapter 3, it has been established that the optimal control strategy in $\mathbf{U}^a$ has an expression of the feedback,

$$u_t = -\alpha(\nabla, \tau)ut + \beta(\nabla, \tau)y_t. \tag{1.77}$$

Its coefficients depend on the parameter τ in a known manner. Let $\tau \in \mathcal{T}$. Here the set of uncertainty $\mathcal{T}$ is such that for any $\tilde{\tau} \in \mathcal{T}$ the polynomial pair $\{a(\lambda, \tilde{\tau}), b(\lambda, \tilde{\tau})\}$ is stabilizable and $a(\lambda, \tilde{\tau}) \neq 0$ for $|\lambda| = 1$. In accordance with the results in section 3.4.4, the polynomial coefficients α_k, β_k are determined on the set $\mathcal{T}$, where $\alpha(\lambda, \tau) = \lambda\alpha_1(\tau) + \dots + \lambda^m\alpha_m(\tau)$, $\beta(\lambda, \tau) = \beta_0(\tau) + \dots + \lambda^m\beta_m(\tau)$. They can also be found out with the help of the methods mentioned in section 3.4.4.

If the parameter τ is a random quantity with its values in $\mathcal{T}$, then the Bayesian strategy, in conformity with the formulae (1.67) and (1.77), is determined by the non-stationary feedback

$$u_t = -\alpha_t(\nabla)u_t + \beta_t + (\nabla)y_t, \tag{1.78}$$

where, $\alpha_t(\lambda) = \lambda\alpha_{1,t} + \ \dots\ + \lambda^m\alpha_{m,t}$, $\beta_t(\lambda) = \beta_{0,t} + \ \dots\ + \lambda^m\beta_{m,t}$,

$$\alpha_{k,t} = \int \alpha_k(\tau)p(\tau \mid y^t, u^{t-1})d\tau,\ \beta_{k,t} = \int \beta_k(\tau)p(\tau \mid y^t, u^{t-1})d\tau. \tag{1.79}$$

The coefficients $\alpha_k(\tau)$, $\beta_k(\tau)$ are usually rational functions of parameter τ and so the relations in (1.79) and (1.78) may be used, in principle, for formulation of the optimal control strategy in $\mathbf{U}^p$. Establishment of the realizability and optimality corresponding to the feedback (1.78) of the control strategy also requires additional presuppositions. In the present case presupposition about the Gaussian nature of the vector τ is not acceptable : the polynomials $\alpha(\lambda, \tau)$, $\beta(\lambda, \tau)$ are not defined for all τ and the formulae (1.79) may not make any sense. Hence we assume that the *a priori* distribution of random quantities τ has the expression of a truncated Gaussian distribution, *i.e.*, the random quantity τ has a density,

$$p(\tau \mid 0) = \frac{I_{\mathcal{T}}(\tau)\exp\left[-(1/2)(\tau - \bar{\tau})^* R_\tau^{-1}(\tau - \bar{\tau})\right]}{\int I_{\mathcal{T}}(\tau)\exp\left[-(1/2)(\tau - \bar{\tau})^* R_\tau^{-1}(\tau - \bar{\tau})\right] d\tau}. \tag{1.80}$$

Here $I_{\mathcal{T}}(\tau)$ is the characteristic function of the set $\mathcal{T}$. Consequently the density $P(\tau \mid 0)$ is determined by thre e 'parameters' : vector $\bar{\tau}$, positive matrix R_τ and indicator $I_{\mathcal{T}}(\tau)$ of the set $\mathcal{T}$. For $\mathcal{T}=\mathbf{R}^{p+r}$ we have $I_{\mathcal{T}}(\tau)\equiv 1$ and we arrive at the Gaussian distribution. If the noise v_t is Gaussian, $p_v(\cdot)\sim N(0, R_v)$, then it can be shown (it follows from the relations obtained in section 6.1.4) that the *a posteriori* distribution $p(\tau \mid y^t, \quad u^{t-1})$ can be written as

$$p(\tau \mid y^t, u^{t-1}) = \frac{I_{\mathcal{T}}(\tau)\exp\left[-(1/2)(\tau-\hat{\tau}_t)^* p_t^{-1}(\tau-\hat{\tau}_t)\right]}{\int I_{\mathcal{T}}(\tau)\exp\left[-(1/2)(\tau-\hat{\tau}_t)^* P_t^{-1}(\tau-\hat{\tau}_t)\right]d\tau} \tag{1.81}$$

and it is determined by the statistics, *i.e.*, vector $\hat{\tau}_t=\hat{\tau}_t(y^t, \quad u^{t-1})$ and positive matrix $P_t=P_t(y^t, \quad u^{t-1})$. These statistics satisfy the recursive relations. To express the latter of the equaiton (1.74), we present it in the form,

$$y_t=\Phi^*_{t-1}\tau+\phi_{t-1}+v_t, \tag{1.82}$$

where, $\Phi_{t-1}=\operatorname{col}\left(-y_{t-i_1}, \ldots, -y_{t-i_p}, \ldots, u_{t-j_r}\right)$,

$\phi_{t-1}=[1-a(\nabla, 0)]y_t+b(\nabla, 0)u_t$.

The recursive formulae will then have the same form as in theorem 5.2.1 (for $\tau_t\equiv\tau$). As in section 6.1.2, the limiting nondegeneration of the information matrix (1.50) ensures consistency of the estimates $\{\hat{\tau}_t\}$ and in the same manner the existence of the Bayesian strategy and its optimality in U^P are also established.

(*c*) *Scalar object with measurement delay.* We examine the simplest form of the equation (1.68) $y_t+a_1y_{t-1}=b_ku_{t-k}+v_t$. With some approximation it represents the mathematical model of the technology of cellulose-paper production. The disturbances v_t are assumed to be Gaussian, uncorrelated and they satisfy the conditions in (1.70). The amplification b_k, $b_k\neq 0$, and the control delay k, $k>1$, and the time constant a_1 are assumed to be known. The control objective is the requirement of the least value of the functional $J[U^\infty(\cdot)]=\overline{\lim_{t\to\infty}}\mathbf{M}\mid y_t-y_*\mid^2$, where y_* is the nominal value of the output.

For a known time constant a_1, the optimal strategy is given by the linear feedback

$$u_t=b_k^{-1}[-(-a_1)^k y_t+y_*]. \tag{1.83}$$

Then $\quad \min\limits_{U^\infty(\cdot)\in U^\infty} J[U^\infty(\cdot)]=\sigma_v^2\sum\limits_{l=0}^{k-1}(-a_1)^l,$

i.e., the least value of the performance criterion depends on the unknown parameter a_1.

In accordance with (1.67) and (1.83), the Bayesian strategy has the form of a feedback, such as,

$$u_t=b_k^{-1}[-a_1(t)y_t+y_*], \tag{1.84}$$

where the gain $a_1(t)$ is computed as

$$a_1(t)=\int(-a_1)^k p(a_1 \mid y^t, u^{t-1})da_1.$$

If $a_1=\tau$ is a Gaussian random quantity $\tau\sim N(\tau_0, R_\tau)$, then $p(a_1\mid y^t, u^{t-1})\sim N(\tau, P_t)$ and the statistics τ_t and P_t are computed following the recursive procedure of the least square methods :

$$\tau_{t+1}=\tau_t+K_t(y_{t+1}-\tau_t y_t-b_k u_{t-k+1}),$$

$$P_{t+1}=\frac{\sigma_v^2 p_t}{\sigma_v^2+p_t y_t^2},\ K_t=\frac{p_t y_t}{\sigma_v^2+p_t y_t^2},\tag{1.85}$$

Here $P_0=R_\tau$. The coefficient $a_1(\tau)$ is then computed without any specific difficulty.

It is not difficult to show the consistency of the estimates in the present case using relations in (1.85). It stands for the realizability and optimality of Bayesian control strategy (1.84) in $\mathbf{U}^p$.

We observe that the adaptive strategy for the same performance criterion as in the present example can be formed without computing the coefficients of the feedback $a_1(t)$. Truly the identification approach leads to the feedback

$$u_t=b_k^{-1}[-(-\tau_t)^k y_t+y_*],\tag{1.86}$$

which, by virtue of the consistency of the estimates $\{\tau_t\}$, gives rise to optimal control strategy in $\mathbf{U}^p$. The Bayesian strategy coincides with the 'identification' strategy only for $k=1$. From the point of view of computation, the feedback of (1.86) is preferable to (1.84). Moreover, for the condition of consistency of estimates $\{\tau_t\}$ the feedback of (1.86) remains optimal even when the disturbances do not remain Gaussian and do not possess a density function.

From the point of view of the transient response (vide section 6.1.2h), the possible preference of the Bayesian can not be substantiated for this simple example.

6.1.4 *Recursive Relations for a Posteriori Distributions*

We once more consider the control problem (1.1) to (1.6). It is observed that for the formulation of the Bayesian control strategies, the possibility recursive evaluation of *a posteriori* distributions of control object states and parameters is a reality. A few of the properties of recursive relations for *a posteriori* distributions of random control object parameters are discussed below.

(*a*) *Recursive relations for a posteriori distributions of states and control object parameters.* We assume that the control strategy $[U^\infty(\cdot)]\in\mathbf{U}^p$ is fixed and consider the following relations which follow (1.8) to (1.10) and (1.5) (because of independence of the random quantities $\{v'_t, v''_t, e_t, w_t\}$,

$$p(z_{t+1}\mid z^t, y^t, u^t)=p_{v'}[x_{t+1}-X_{t+1}(x_t, u^t, \tau_t)]\,p_e[\tau_{t+1}-T_{t+1}(\tau_t)],$$

$$p(u_t\mid z^t, y^t, u^{t-1})=p_w[u_t-U_t(u^{t-1}, y^t)],$$

$$p(y_t\mid z^t, y^{t-1}, u^{t-1})=p_{v''}[y_t-Y_t(x_t, \tau_t)].$$

Using the Bayesian formulae of section 2.4 for *a posteriori* density,

$$p_t(x_t, \tau_t)=p(z_t\mid y^t, u^{t-1})=p(x_t, \tau_t\mid y^t, u^{t-1}),\tag{1.87}$$

we can easily find out the required recursive relations

$$p_{t+1}(x, \tau)=\frac{p_{v''}(y_{t+1}-Y_{t+1})\int p_{v'}(x-X_{t+1})p_e(\tau-T_{t+1})p_t(x', \tau')dx'd\tau'}{\int p_{v''}(y_{t+1}-Y_{t+1})p_{v'}(x-X_{t+1})p_e(\tau-T_{t+1})p_t(x', \tau')dx'd\tau'dxd\tau}\tag{1.88}$$

where $t = 0, 1, \ldots, p_0(x', \tau') = p_v(x' \mid 0) p_e(\tau' \mid 0)$
and the arguments of the functions are dropped, for brevity's sake, *i.e.*,

$$X_{t+1} = X_{t+1}(x', u', \tau'),\ T_{t+1} = T_{t+1}(\tau'),\ Y_{t+1} = Y_{t+1}(x, \tau).$$

The relations in (1.88) determine the *a posteriori* density $p_t(x, \tau)$ of the random quantity x_t, τ_t (for observations y^t, u^t) as the function of the variables x, τ, y^t and u^t. The expression for this function (vide (1.88)) is independent of the choice of the control strategy $U^\infty(\cdot) \in \mathbf{U}^p$.

Integrating the density $p_t(x, \tau)$ with respect of τ or x, we get the *a posteriori* density $p_x(\cdot \mid y^t, u^{t-1}), p_\tau(\cdot \mid y^t, u^{t-1})$ of the random quantity x_t, τ_t. Thus,

$$p_{x_t}(x \mid y^t, u^{t-1}) = \int p_t(x, \tau) d\tau,\ p_{\tau_t}(\tau \mid y^t, u^{t-1}) = \int p_t(x, \tau) dx. \tag{1.89}$$

The *a posteriori* density $p(x_{t+1}, \tau_{t+1} \mid y^t, u^t)$ of the random quantity (x_{t+1}, τ_{t+1}) is computed through $p(x_t, \tau_t \mid y^t, u^t)$ (when the observations are y^t, u^t) as per following formula

$$p(x_{t+1}, \tau_{t+1} \mid y^t, u^t) = \int p_{v^t}[x_{t+1} - X_{t+1}(x_t, u^t, \tau_t)] p_e[\tau_{t+1}$$

$$-T_{t+1}(\tau_t)]^1 p(x_t, \tau_t \mid y^t, u^{t-1}) dx_t d\tau_t. \tag{1.90}$$

The performance criterion (1.4) has the following expression in terms of the density,

$$J[U^\infty(\cdot)] = \overline{\lim_{t \to \infty}}\, t^{-1} \sum_{s=1}^{t} \mathbf{M} \int q_{s-1}(x_s, \tau_s u_{s-1})$$

$$\times p(x_s \tau_s \mid y^{s-1}, u^{s-1}) dx_s d\tau_s. \tag{1.91}$$

(*b*) *Bayesian strategies for a time invariant parameter.* The analysis of the relations for the *a posteriori* distributions will be continued for the special case, when,

$$T_t(\tau) \equiv \tau,\ e_t \equiv 0. \tag{1.92}$$

The control strategy $U^\infty(\cdot) \in \mathbf{U}^p$ is assumed to be arbitrary but fixed.

In accordance with (1.92) and (1.3), the parameter τ_t takes a random value at the initial stage $t = 0$, which is $\tau = \tau_0 = e_0$ and then it retains its value unchanged. Exactly in such a situation the possibility of interpretation of a Bayesian strategy (in a class $\mathbf{U}^p$) as an adaptive control strategy is considered. If the *a posteriori* density $p(\tau \mid y^t, u^t)$ of the random parameter τ is localised at the point τ_0 with time (and this fact does not depend on distribution of the random quantity e_0), then the unknown value of the parameter τ_0 is restored (control object is identified). Hence it is possible to synthesize the limiting optimal control with respect to the functional which depends on the value of the parameter τ_0.

(*c*) *Recursive relations for a posteriori densities.* The conditions of (1.92) simplify the recursive relations of (1.88). For simplicity in writing, we bring for the disturbance w^∞ (randomizing the control strategy) to the output y^∞ and determine the following from (1.88) :

$$p(\tau \mid y^t) = p(\tau \mid y^{t-1}) \frac{\int p_{v''}(y_t - Y_t) p(x_t \mid y^{t-1}, \tau) dx_t}{\int\int p_{v''}(y_t - Y_t) p(x_t \mid y^{t-1}) p(\tau \mid y^{t-1}) dx_t d\tau}. \tag{1.93}$$

The procedure in (1.93) is examined for the initial condition $p(\tau \mid y_0) = p_*(\tau \mid 0)$.

It depends on the conditioned density $p(x_t \mid y^{t-1}, \ \tau)$ and can be recursively computed as,

$$p(x_{t+1} \mid y^t, \tau) = \frac{\int p_{v'}(x_{t+1} - X_{t+1}) p_{v''}(y_t - Y_t) p(x_t \mid y^{t-1}, \tau) dx_t}{\int p_{v''}(y_t - Y_t) p(x_t \mid y^{t-1}. \tau) dx_t}. \tag{1.94}$$

The initial condition is

$$p(x_1 \mid y_0, \tau) = \frac{\int p v'[x_1 - X_1(x_0, u_0, \tau)] \, p_{v''}[y_0 - Y_0(x_0, \tau)] \, p_v(x_0) dx_0}{\int p_{v''}[y_0 - Y_0(x_0, \tau)] \, p_v(x_0) dx_0}. \tag{1.95}$$

(*d*) *Inequality of information.* We shall examine the sensitivity of the procedure (1.93) in choosing the initial condition, namely the density $p_*(\tau \mid 0)$. Let $\tilde{p}(\tau \mid y^t)$ denote the function to be determined by the procedure of (1.93) with the initial condition $\tilde{p}(\tau \mid y_0) = \tilde{p}_*(\tau \mid 0)$, which is different from $p_*(\tau \mid 0)$. (The observations remain the same).

It may be observed that the independence of the conditional density $p(x_{t+1} \mid y_t, \ \tau)$, from the choice of the function $\tilde{p}(\tau \mid y_0)$, follows from the formulae (1.93) to (1.95). It appears that the conditional density (also known as the transient density) given by

$$\tilde{p}(y_t \mid y^{t-1}) = \int\int p_{v''}(y_t - Y_t) p(x_t \mid y^{t-1}, \tau) \tilde{p}(\tau \mid y^{t-1}) dx_t d\tau \tag{1.96}$$

becomes the same as the transient density below (when $t \to \infty$) independent of the difference between the initial densities $\tilde{p}(\tau \mid y_o)$ and $p(\tau \mid y_0)$:

$$p(y_t \mid y^{t-1}) = \int\int p_{v''}(y_t - Y_t) p(x_t \mid y^{t-1}, \tau) p(\tau \mid y^{t-1}) dx_t d\tau. \tag{1.97}$$

As a measure of the proximity of the densities $p(y_t \mid y^{t-1})$ and $\tilde{p}(y_t \mid y^{t-1})$, we choose

$$I_t(p; \tilde{p}) = \int \left[\ln \frac{p(y_t \mid y^{t-1})}{\tilde{p}(y_t \mid y^{t-1})} \right] p(y_t \mid y^{t-1}) dy_t. \tag{1.98}$$

It is known as the Kul'bak - Leybler information number. Since the logarithm is a convex function, then due to the Yensen inequality we can find

$$\int \left[\ln \frac{\tilde{p}(y_t \mid y^{t-1})}{p(y_t \mid y^{t-1})} \right] p(y_t \mid y^{t-1}) dy_t \le$$

$$\le \ln \int \frac{\tilde{p}(y_t \mid y^{t-1})}{p(y_t \mid y^{t-1})} p(y_t \mid y^{t-1}) dy_t = \ln \int \tilde{p}(y_t \mid y^{t-1}) dy_t = 0,$$

i.e., if the integral in (1.98) exists, then, $I_t(p;\bar{p})\geq 0$, but $I_t(p;\bar{p})=0$ only when $p(y_t\mid y^{t-1})=\bar{p}(y_t\mid y^{t-1})$ for almost all $y_t\in\mathbf{R}^l$.

Theorem 6.1.2 *We assume that t for densities* $p_t(\tau\mid 0)$ and $\bar{p}_t(\tau\mid 0)$ *in the procedure of (1.93) the following inequality is satisfied:*

$$h_0=\int\left[\ln\frac{p_t(\tau\mid 0)}{\bar{p}_t(\tau\mid 0)}\right]p(\tau\mid y^t)d\tau<\infty. \tag{1.99}$$

Then for the random variables

$$h_t=\int\left[\ln\frac{p(\tau\mid y^t)}{\bar{p}(\tau\mid y^t)}\right]p(\tau\mid y^t)d\tau. \tag{1.100}$$

there exist a limit $h_\infty=\lim\limits_{t\to\infty}h_t$ *in the square sense and with probability 1 for all t, and for the variables (1.98) the following equality holds*

$$\mathbf{M}I_{t+1}(p;\bar{p})=\mathbf{M}h_t-\mathbf{M}h_{t+1}. \tag{1.101}$$

The corollary. The following relations hold for a probability 1 and in the mean square sense

$$\lim_{t\to\infty}I_t(p;\bar{p})=0,\quad \sum_{t=0}^{\infty}\mathbf{M}I_t(p;\bar{p})<h_0. \tag{1.102}$$

(*e*) *Interpretation of the technique for computation of a posteriori probabilities.* The theorem 6.1.2 and its corollary give a different angle on the procedure (1.93). We assume, for example, that optimization of the control objective for the control object (1.1) (where $\tau_t\equiv\tau_0$ is a known parameter) is to be ensured. Then the synthesis of the realizable control strategy will be done in the following manner. We consider that a small neighbourhood $\mathcal{T}_0$ of the point τ_0 is chosen. The density $p(\cdot\mid y_0)$ of the random variableτ* is located on this neighbourhood, for example $p(\tau\mid y_0)$ responds to the uniform distribution on $\mathcal{T}_0$. It is understood that the density $p(\tau\mid y_0)$ is unknown (to be precise, the set $\mathcal{T}_0$ of its localisation is known). It follows from the formula of (1.93) that all the densities $p(\tau\mid y^t)$ will also be localized on the set $\mathcal{T}_0$. We then introduce the random quantity τ_t with a conditional density function $p(\tau\mid y^t)$. Then the process $\{y_{t+1},\ \ p(\tau\mid y^t)\}$ can be seen as a process with a transient function $p(y_{t+1}\mid y^t,\ \ \tau_t)$ which depends on the parameter τ_t. The latter is subject to 'small' disturbances (because of the above mentioned $\tau_t\in\mathcal{T}_0$ for all t). It is evident that $\mathbf{M}\{p(y_{t+1}\mid y^t,\ \ \tau_t)\mid y^t\}=p(y_{t+1}\mid y^t)$. The right-hand side is given by the formula of (1.97) and can be asymptotically evaluated (for $t\to\infty$) using theorem 6.1.2 with the help of the sequence $\{\bar{p}(y_{t+1}\mid y^t)\}$. The latter can be obtained by the procedure laid in (1.93) and (1.96) for any arbitrary choice of the density $\bar{p}(\tau\mid y_0)$. (It is only required that the inequality of (1.99) is satisfied and for this condition it is sufficnet, for example, that we choose for density $\bar{p}(\tau\mid y_0)$ that relates the uniform dis-

*To be precise, $p(\tau\mid y_0)\neq 0$ only when $\tau\in\mathcal{T}_0$.

tribution in some compact $\mathcal{T} \supseteq \mathcal{T}_0$.) Since the neighbourhood $\mathcal{T}_0$ of the point τ_0 can be chosen as small as possible, then we can really expect that the control strategy $U^\infty(\cdot)$ ensuring the control objective for the process with 'disturbance' and having a transient function $p(y_{t+1} \mid y^t, \tau_t)$, will provide the control objective, with a corresponding degree of accuracy, for the process with a transient function $p(y_{t+1} \mid y^t, \tau_0)$, where the parameter τ_0 is unknown.

If the set of $\mathcal{T}$ consists of finite points or countable number of points, such as, $\mathcal{T} = \{\tau^{(1)}, \tau^{(2)}, \ldots\}$, then the above relations for the density will enter in the corresponding relations for probabilities. It will be required to examine the probabilisitc sequences $\{\tilde{p}_t(k)\}$ instead of the functions $\{\tilde{p}(\tau \mid y^t)\}$. Here $\tilde{p}_t(k) = \tilde{P}\{\tau = \tau^{(k)} \mid y^t\}$ is the *a posteriori* probability of that, for which the parameter τ takes the value $\tau^{(k)}$. Then,

$$\tilde{p}(y_{t+1} \mid y^t) = \sum_k p_k(y_{t+1} \mid y^t)\tilde{p}_t(k)$$

where $p_k(y_{t+1} \mid y_t)$ is the transient function related to the value $\tau^{(k)}$ of the parameter τ. Choosing $\tilde{p}_0(k) = 1$, if $\tau^{(k)} = \tau_0$, we get $\tilde{p}_t(k) \equiv 1$ for all t and $\tilde{p}(y_{t+1} \mid y^t) = p_k(y_{t+1} \mid y^t)$. Since $\tilde{p}(y_{t+1} \mid y^t) \to p_k(y_{t+1} \mid y^t)$ and using the theorem 6.1.2, we find that the observation of $\tilde{p}(y_{t+1} \mid y^t)$, generally speaking, permits regeneration of index k and this helps in finding out the unknown value $\tau^{(k)}$ of the parameter τ. It is possible to realize the bounded-optimal control for the functional, whose least value depends on τ, by disposing of the value $\tau^{(k)}$.

(*f*) *Observation on identifiability of the control object.* Whether the density $p(\tau \mid y^t)$ converges to a δ-function in a concentrated 'true' parameter τ_0, remains as an open question.

It is not difficult to establish that for an arbitrary set $\mathcal{T}_0$ of positive Lebegian measure, the non-negative random quantities $\xi_t(\mathcal{T}_0) = \int_{\mathcal{T}_0} p(\tau \mid y^t)d\tau$ will be martingale, *i.e.* $\mathbf{M}\{\xi_{t+1}(\mathcal{T}_0) \mid y^t\} = \xi_t(\mathcal{T}_0)$, and so in view of theorem 2.A.1 the following relationship exists with probability 1

$$\lim_{t\to\infty} \xi_t(\mathcal{T}_0) = \xi_*(\mathcal{T}_0). \tag{1.103}$$

The random quantities $\hat{\tau}_t = \mathbf{M}(\tau \mid y^t) = \int \tau p(\tau \mid y^t)d\tau$ will be introduced with the presupposition that the incident $\Omega_{\tau_0} = \left\{\lim_{t\to\infty} \hat{\tau}_t = \tau_0\right\}$ possesses a positive probability. Let $\tau = \tau(\omega)$ be the random quantity with the density function $p_*(\cdot \mid 0)$ and $\mathcal{T}_{\tau_0} = \tau(\Omega_{\tau_0})$ be the set of the values of this random quantity related to the elementary incidents from Ω_{τ_0}. Let $\mathcal{T}_0$ be an arbitrary set in $\mathcal{T}$ of positive Lebegian measure, such that $\mathcal{T}_0 \supseteq \mathcal{T}_{\tau_0}$. Then because the event Ω_{τ_0} belongs to σ-algebra of the random quantity y^∞ that can be determined,

$$\xi_*(\mathcal{T}_0) = \lim_{t\to\infty} \mathbf{M}\left\{I_{\mathcal{T}_0}(\tau) \mid y^t\right\} = \mathbf{M}\left\{I_{\mathcal{T}_0}(\tau) \mid y^\infty\right\}$$

$$\geq \mathbf{M}\left\{I_{\mathcal{T}_{\tau_0}}(\tau) \mid y^\infty\right\} \geq \mathbf{M}\left\{I_{\Omega_{\tau_0}} \mid y^\infty\right\} = I_{\Omega_{\tau_0}}.$$

Thus, almost for all elementary events from Ω_{τ_0} the following is satisfied if, $p_*(\tau_0 \mid 0) > 0$,

$$\lim_{t\to\infty} \int_{\mathcal{T}_0} p(\tau \mid y^t)d\tau = 1. \tag{1.104}$$

If $\mathcal{T}_0$ is a single point set, $\mathcal{T}_0 = \tau_0$, then (1.104) signifies that the event $\{\tau_t \to \tau_0, t \to \infty\}$ draws, with probability 1, the event $\{p(\tau \mid y^t) \to \delta(\tau - \tau_0), t \to \infty\}$, where $\delta(\tau)$ is the δ-function. This means that there is a localization of the density $p(\tau \mid y^t)$ at the point τ_o. If $\mathcal{T}_0$ possesses a positive Lebegian measure, then the equality of (1.104) is true for $\mathcal{T}_0 = \mathcal{T}_0$.. This means that the density $p(\tau \mid y^t)$ is localised on the set $\mathcal{T}_0$ but not necessarily at the point τ_0 (although the 'mean values' τ_t of the densities $p(\tau \mid y^t)$ converge to τ_0 as per assumption made).

The behaviour of $p(\tau \mid y^t)$ substantially depends on the choice of the control strategy $U^\infty(\cdot)$. Actually, let the equation (1.1) possess the expression of $y_{t+1} = \tau_0[y_t - u_t] + v_{t+1}$ for $x_t = y_t$ and $\tau_t \equiv \tau_0$. The transient function of the process $\{y_t\}$ is easily determined from :

$$p(y_{t+1} \mid y^t, \tau_0) = p_{v'}[y_{t+1} - \tau_0(y_t - u_t)].$$

If the control strategy is determined by the equality $u_t = y_t$, then, according to the Bayesian formulation, the relationship $p(\tau \mid y^{t+1}) = p(\tau \mid y^t) = p_e(\tau \mid 0)$ is fulfilled, *i.e.*, the localization of the densities $p(\tau \mid y^t)$ at the point τ_0 (or at any other point τ) does not take place. Randomization of the control allows us to avoid similar degeneration.

(*g*) *Example : Computer simulation of the bicyclist-robot.* To illustrate the workability of the formulation of the adaptive control strategy as discussed above, we present the results of computer simulation of the control system described by the equation

$$y_t + a_1 y_{t-1} + a_2 y_{t-2} = b_1 u_{t-1} + b_2 u_{t-1} + b_2 u_{t-2} + v_t. \tag{1.105}$$

The equation of a double-wheel bicycle in an elementary form is described by the equation (1.105). For typical parameters of the bicycle (bicycle speed is constant and equal to 7 m/s, the sampling period of control is l s, the distance between the two axes of the wheels is 1.1 m, coordinates of the centre of gravity of the system are 0.6 m and 1.1 m, the initial coordinate is at the point of tangency of the rear wheel with the support surface), its equation is given by

$$y_t - 16.7 y_{t-1} + y_t = 46,9 u_{t-1} + 23,4 u_{t-2} + v_t \tag{1.106}$$

(all the parameters are expressed in the SI units). It is required to construct an automat (of a bicyclist-robot) that is capable of keeping the bicylce in the vertical position under the conditions that the data, enumerated above, for a concrete model of a bicycle is not known, and consequently, the coefficients a_1, a_2, b_1, b_2 in the equation (1.105) are also not known. The bicyclist robot or the controller must 'correct itself' with any bicycle whose parameter vector τ is contained in a set $\mathcal{T} \subseteq \mathbf{R}^4$. Here

$$\tau = \text{col}\,(a_1,\ a_2,\ b_1,\ b_2). \tag{1.107}$$

The disturbance acting on the bicycle was assumed to be a Gaussian random quantity with the parameters $\mathbf{M}v_t = 0$, $\mathbf{M}v_t^2 = 0.01$ and it was simulated with the help of a standard program.

The regulator expressing the action of the bicyclist-robot was taken in the form

$$\alpha(\nabla,\ \tau_t)u_t = \beta(\nabla,\ \tau_t)y_t, \qquad (1.108)$$

and the polynomials $\alpha(\lambda,\ \tau)$ and $\beta(\lambda,\ \tau)$ and of the varibale λ were determined from the polynomials $a(\lambda)$ and $b(\lambda)$ using the conditions,

$$(1+\lambda a_1+\lambda^2 a_2)\alpha(\lambda,\ \tau)-(\lambda b_1+\lambda^2 b_2)\beta(\lambda,\ \tau) = 1,$$

$$\alpha(0,\ \tau) = 1. \qquad (1.109)$$

The initial conditions of the control system were $y_{-1} = 8.72.\times 10^{-3}$, $y_{-2} = 0$, $u_{-1} = 0$, and $u_{-2} = 0$. The estimates of the unknown parameters (1.07) were computed as per algorithm (1.45) to (1.47). Here τ_0=col (-20.0; 1.2; 50.0; 27.0). It is the 20% throw of the true coefficients of the equation (1.106). $R_\tau = 400I_4$, where I_4 is the 4×4 unit matrix. Then in case of degeneration of the coefficient matrix of the equation (1.109), the value obtained at the previous step is retained. Simulation according to equations (1.106) and (1.107) yielded so far $|\ y_t\ |< 0.52$. It corresponded to the bicycle frame deviation within a limit of 30°. The bicycle at the limiting stage was considered to be 'falling', and then it came back to the initial position, the simulation process continued. The variation in y_t is shown in Fig. 3 (perfectly linear variation) for 50 iterations (intervals) of control. At the second and twelfth iterations the bicycle frame deviated from the vertical plane by more than 30°. The bicycle 'fell down'. At this instant sufficiently good estimates of the control parameters (1.105) were obtained and this could be utilised to stop the bicycle from 'falling'. (The experiments were conducted for 100 intervals, but the first 50 of them are only presented in Fig.3.) The thin line in this figure shows the variation of the values $|\ \tau - \tau_t\ |$ with time, where $\tau =$ col (16.7; 1.0; 46.9; 23.4).

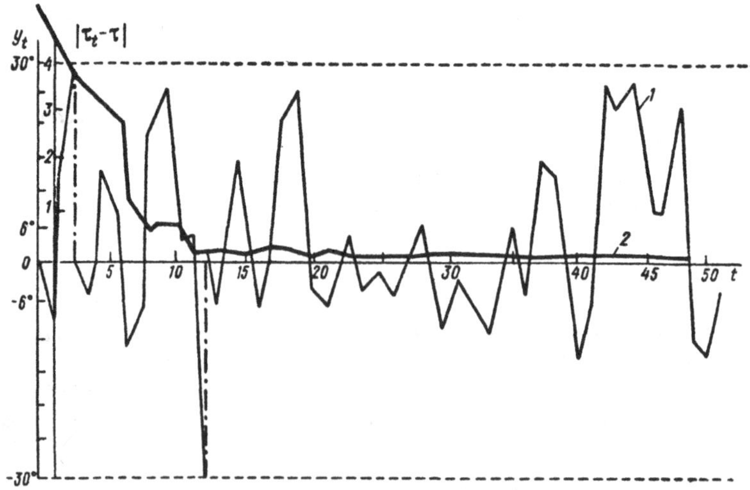

Fig. 3. The transient response of the bycyclist robot system.

The above data apparently show the effectiveness of the estimate algorithm in (1.45) to (1.47).

§ 6.2 Initial Synthesis of Adaptive Control Strategy in the Presence of the Correlated Noise

In the presence of the correlated disturbance on the control object the Bayesian approach becomes complicated. The technique of adaptive control synthesis given below uses mainly the identification method based on consistent estimates of unknown parameters. The stochastic approximation method for modification of recursive procedures has been used as the estimation (adaptation) algorithm. During the control system analysis with adaptable parameters the maximum difficulty was caued by the possible instability of the system in the transient state. The synthesis of adaptive control strategy virtually leads to the identification problem sufficiently studied in chapter 5, if the control system stability is guaranteed beforehand (for example, if it is known that the control object is stable). The possible instability of the control system may lead to divergence of the estimates formed by the adaptor.

6.2.1 *Adaptive Optimal Control for a Performance Criterion Dependent on Events*

(*a*) *Formulation of the problem.* Let the control object be expressed through 'input-output' variables through the equation,

$$a(\nabla, \tau)y_t = b(\nabla, \tau)u_t + c(\nabla, \tau)v_t. \tag{2.1}$$

Here y_t is the output variable, $y_t \in R^l$, u_t and v_t are the control and noise variables, $u_t \in R^m$ and $v_t \in R^l$, and $a(\lambda, \tau), b(\lambda, \tau), c(\lambda, \tau)$ are polynomials, such that,

$$a(\lambda, \tau) = I_l + \lambda a_1 + \ldots + \lambda^n a_n,\ b(\lambda, \tau) = \lambda b_1 + \ldots + \lambda^n b_n,$$

$$c(\lambda, \tau) = I_l + \lambda c_1 + \ldots + \lambda^n c_n. \tag{2.2}$$

The coefficients in these expressions for the polynomials are the matrices of appropriate dimensions. They are continuous functions of the parameter τ in the set $\mathcal{T} \subseteq \mathbf{R}^p$. The disturbance v^∞ is assumed to be a random process with unknown values and properties as below,

$$\mathbf{M}v_t = 0,\ \mathbf{M}v_t v_t^* = R_v > 0,\ \mathbf{M}\,|\,v_t\,|^4 < C_v < \infty. \tag{2.3}$$

It is assumed that the class of realizable control strategies have been made up of sequences $U^\infty(\cdot)$ of functions $U_t = U_t(u^{t-1},\ \ y^t)$ which determine the control u^∞ as per rule below,

$$u_t = U_t(u^{t-1}, y^t) + w_t. \tag{2.4}$$

Here w_t is the randomizing control (test) signal. Its characteristic will be specified later. It is understood that the strategies from $\mathbf{U}^p$ ensure fulfilment of the inequality* below, independent of the choice of the initial conditions of the control and object (2.1),

$$\overline{\lim_{t\to\infty}}\, t^{-1} \sum_{s=1}^{t} (|\,y_s\,|^2 + |\,u_s\,|^2) < \infty. \tag{2.5}$$

---

*We recall that the correlations between the random values are understood to be with probability 1.

The performance criterion is taken as

$$J[U^\infty(\cdot), \tau, v^\infty, \omega^\infty] = \overline{\lim_{t\to\infty}}\ t^{-1}\sum_{s=1}^{t} q(x_s, u_s), \tag{2.6}$$

where $x_t = \mathrm{col}\,(y_t, \ \ldots, \ y_{t-n+1}, u_{t-1}, \ \ldots, \ u_{t-n+1})$ (2.7)

and $q(x, u)$ is a nonnegative quadratic form of variables x and u :

$$q(x, u) = \begin{pmatrix} x \\ u \end{pmatrix}^* N \begin{pmatrix} x \\ u \end{pmatrix}, \ N = \begin{Vmatrix} Q & S \\ S^* & R \end{Vmatrix} \geq 0. \tag{2.8}$$

It has been shown above that when the control strategy $U^\infty(\cdot) \in \mathbf{U}^p$ is fixed, by virtue of (2.1) and (2.4), the functional in (2.6) can be seen as a random quantity dependent on parameter τ and disturbances v^∞, w^∞.

The control objective is to ensure inequality,

$$J[U^\infty(\cdot), \tau, v^\infty, w^\infty] \leq r^{-1}(\tau, v^\infty, w^\infty). \tag{2.9}$$

Here $r(\cdot)$ is the level of control quality given as a function of corresponding arguments. In accordance with the definition in 4.1.1, the strategy $U^\infty(\cdot) \in \mathbf{U}^p$ ensuring fulfilment of the inequality (2.9) for all $\tau\in\mathcal{T}$, will be known as adaptive in a class of uncertainty $\mathcal{T}$.

The optimizing control objective with the quality level $r(\cdot)$ in it chosen as the greatest, will be of interest for us later on. To be precise, let $\mathbf{U}^a = \mathbf{U}^a(\tau)$ be the set of admissible strategies $U^\infty(\cdot)$. The functions $U_t = U_t(u^{t-1}, \ y^t, \ \tau)$ in them may depend on the parameter τ. The control u^∞ is formed for $U^\infty(\cdot) \in \mathbf{U}^a(\tau)$ as per rule $u_t = U_t(u^{t-1}, y^t, \tau) + w_t$ and it ensures fulfilment of the relations,

$$\overline{\lim_{t\to\infty}}\, t^{-1}\sum_{s=1}^{t}(|\, y_s\,|^2 + |\, u_s\,|^2) < \infty, \ \lim_{t\to\infty} t^{-1}(|\, y_t\,|^2 + |\, u_t\,|^2) = 0,$$

$$\sup_t \mathbf{M}(|\, y_t\,|^2 + |\, u_t\,|^2) < \infty. \tag{2.10}$$

Let us denote :

$$r^{-1}(\tau, v^\infty, w^\infty) = \inf_{U^\infty(\cdot)\in \mathbf{U}^a} J[U^\infty(\cdot), \tau, v^\infty, w^\infty]. \tag{2.11}$$

It is known from the results of the para 3.4.3 that for sufficiently general conditions, the lower boundary in (2.11) is obtained on the admissible strategy that formulates the limiting optimal control. The adaptive optimal control consists of a choice (synthesis) of the strategy $U^\infty(\cdot) \in \mathbf{U}^p$ that ensures the control objective in (2.9) and (2.11).

(b) Linear stationary control strategies. In the set of admissible strategies, the linear stationary strategies degenerated by the feedback, as given below, have special values

$$\alpha(\nabla, \tau)u_t = \beta(\nabla, \tau)y_t. \tag{2.12}$$

Here $\alpha(\lambda, \tau)$, $\beta(\lambda, \ \tau)$ are the polynomials of λ, whose matrix coefficients may depend on parameter τ and $\alpha(0, \tau) = I_m$.

Lemma 6.2.1 *Let us assume that the disturbance v^∞ on the control object (2.1) have independent values and the properties as given in (2.3).*

Then the control strategy as given by the feedback of (2.12) will be admissible, if the characteristic polynomial of the control system (2.1) and (2.2), as given below, is stable :

$$g(\lambda, \tau) = \det \begin{Vmatrix} a(\lambda, \tau) & -b(\lambda, \tau) \\ \beta(\lambda, \tau) & -\alpha(\lambda, \tau) \end{Vmatrix}. \tag{2.13}$$

Then the following relations hold

$$\lim_{t\to\infty} t^{-1}(|y_t|^2 + |u_t|^2) = 0, \ \sup_t \mathbf{M}(|y_t|^2 + |u_t|^2) < \infty, \tag{2.14}$$

$$\lim_{t\to\infty} t^{-1} \sum_{s=1}^{t} \mathbf{M}(|y_s|^4 + |u_s|^4) < \infty, \tag{2.15}$$

$$\lim_{t\to\infty} t^{-1} \sum_{s=1}^{t} q(x_s, u_s) = \lim_{t\to\infty} t^{-1} \sum_{s=1}^{t} \mathbf{M}q(x_s, u_s). \tag{2.16}$$

From (2.16) it follows that the limits lead to degenerated random quantities which are, with probability 1, constants. Moreover, it follows from (2.14) and (2.15) that the stabilizing feedbacks *i.e.*, with a stable characteristic polynomial of (2.13)) give admissible control strategies.

If the linear stationary control strategy is randomized by the additive test signal which decays quite fast as $t \to \infty$, then the control strategy has the same characteristics as in the corresponding regular strategy. The precise assertion will be formulated.

Lemma 6.2.2 *Let the control strategy with the stable polynomial (2.13) and the noise* (v^∞) *with independent values and properties as in (2.3), be generated by the feedback as,*

$$\alpha(\nabla, \tau)u_t = \beta(\nabla, \tau)y_t + w_t. \tag{2.17}$$

Then : (1) if the test signal w^∞ *has a limiting matrix of covariances and it satisfies, with probability 1, the limiting equality given below, then for the control system of (2.1) and (2.17), the inequality of (2.5) is valid :*

$$\lim_{t\to\infty} t^{-1} \sum_{s=0}^{t-1} |w_s|^2 = 0, \tag{2.18}$$

(2) if the random quantities w_t *satisfy the condition*

$$\lim_{t\to\infty} t^{-1} \sum_{s=0}^{t-1} \mathbf{M}|w_s|^2 = 0 \tag{2.19}$$

then for the control system of (2.1) and (2.17), the relationship in (2.16) is valid and

$$\lim_{t\to\infty} t^{-1}\mathbf{M}(|y_t|^2 + |u_t|^2) = 0,$$

$$\overline{\lim_{t\to\infty}}\, t^{-1} \sum_{s=0}^{t-1} \mathbf{M}(|y_s|^2 + |u_s|^2) < \infty, \tag{2.20}$$

(3) if the random quantities w_t *satisfy at the same time the condition**

$$\lim_{t\to\infty} t^{-1} w_t = 0 \tag{2.21}$$

and the relationship (2.19), then (2.14) is satisfied.

---

*The equality (2.21) is deliberately made valid if the random quantities w_t are independent and possess bounded fourth moments (vide the proof of the equality (A.17)). If $\lim_{t\to\infty} \mathbf{M}|w_t|^4 = 0$, then obviously (2.19) is satisfied.

Thus with the conditions of the lemma 6.2.2 the equalities of (2.19) and (2.21) ensure admissibility of control strategy given by the feedback of (2.17).

(c) Existence of stabilizing regulators. We recall that the regulator of (2.12) is known to be stabilizing, if the polynomial (2.13) is stable. From the observations on lemma 3.4.2, it results that the stabilizability of the pair of polynomials $\{a(\lambda), b(\lambda)\}$ is the necessary condition for existence of a stabilizing regulator. It wil be shown that this property of the polynomials $\{a(\lambda), b(\lambda)\}$ is sufficient.

Let the control object (2.1) be expressed in a standard form of state vector (2.7) :

$$x_{t+1} = A(\tau)x_t + B(\tau)u_t + C_c(\nabla, \tau)v_{t+1}, \tag{2.22}$$

$$A = \left\| \begin{matrix} -a_1 & \dots & -a_{n-1} & -a_n & b_2 & \dots & b_{n-1} & b_n \\ I_l & \dots & 0 & 0 & 0 & \dots & 0 & 0 \\ \cdot & & & & & & & \\ \cdot & & & & & & & \\ 0 & \dots & I_l & 0 & 0 & \dots & 0 & 0 \\ 0 & \dots & 0 & 0 & 0 & \dots & 0 & 0 \\ 0 & \dots & 0 & 0 & I_m & \dots & 0 & 0 \\ \cdot & & \cdot & \cdot & \cdot & & \cdot & \cdot \\ \cdot & & \cdot & \cdot & \cdot & & \cdot & \cdot \\ 0 & \dots & 0 & 0 & 0 & \dots & I_m & 0 \end{matrix} \right\|,$$

$$B = \left\| \begin{matrix} b_1 \\ 0 \\ \cdot \\ \cdot \\ \cdot \\ 0 \\ I_m \\ 0 \\ \cdot \\ \cdot \\ \cdot \\ 0 \end{matrix} \right\|, \quad C = \left\| \begin{matrix} I_l \\ 0 \\ \cdot \\ \cdot \\ \cdot \\ 0 \\ 0 \\ 0 \\ \cdot \\ \cdot \\ \cdot \\ 0 \end{matrix} \right\|. \tag{2.23}$$

Lemma 6.2.3 *The following equivalence of the inequalities holds, because of (2.23), for any complex* λ *:*

$$\det\| \, |\lambda|^{2n} \, [a(\lambda^{-1})a^*(\lambda^{-1}) + b(\lambda^{-1})b^*(\lambda^{-1})]\| \neq 0 \Leftrightarrow$$

$$\Leftrightarrow \det\|(A - \lambda I)(A^* - \lambda^* I) + BB^*\| \neq 0. \tag{2.24}$$

It results from the lemma 6.2.3 that the stabilizability (controllability) of the pair of polynomials $\{a(\lambda), \ b(\lambda)\}$ is equivalent to the corresponding stabilizability (controllability) of the matrix pair $\{A, B\}$. Since a matrix D exists for the stabilizable pair $\{A, B\}$, such that $A + BD$ is a stable matrix, then the feedback $u_t = D\,x_t$ takes the expression of (2.12), because of (2.17) and it is stabilizing for the control object (2.1). Thus the stabilizability, in the sense of the definition in 3.4.1, of the pair of polynomials $\{a(\lambda), \ b(\lambda)\}$ appears to be equivalent to the existence of the stabilizing regulator of (2.12).

(*d*) *Identification approach to the control problem*: Let us assume that the test signal satisfies the conditions of the lemma 6.2.2, for all $\tau \in \mathcal{T}$ the pair $\{A(\tau) \quad B(\tau)\}$ is stabilizable, the frequency condition (A.2) of the chapter 3 is fulfilled for the quadratic form (2.8), and the polynomial det $c(\lambda, \tau)$ is stable. Then it can be asserted from the theorem 3.4.5 and the lemma 6.2.2 that an optimal control strategy exists in the class $\mathbf{U}^a$, which can be determined by the feedback (2.17), and whose polynomials $\alpha(\lambda, \tau)$ and $\beta(\lambda, \tau)$ can be determined in terms of the control coefficient (2.1) following the methods mentioned in Section 3.4.

The realizable strategy is searched out in the form of a feedback as is done in the identification approach, *i.e.*,

$$\alpha(\nabla, \tau_t)u_t = \beta(\nabla, \tau)y_t + w_t, \tag{2.25}$$

where $\{\tau_t\}$ are the estimates of the unknown parameter τ to be chosen by a suitable method. If the estimates τ_t are found as per law,

$$\tau_{t+1} = T_{t+1}(\tau^t, y^t, u^t), \tag{2.26}$$

where $T_t(\cdot)$ are the functions of its own arguments, then the relations in (2.25) and (2.26) determine the unasserting control strategy. If the equality,

$$\lim_{t\to\infty} \tau_t = \tau \tag{2.27}$$

is satisfied with probability 1, for all values of the parameter $\tau \in \mathcal{T}$ which determines the correspondence between the inputs and control states (2.1), then the aforesaid strategy will be known as identifying.

Lemma 6.2.4. *Let the coefficients of the polynomials* $\alpha(\lambda, \tau)$ and $\beta(\lambda, \tau)$ *related to the optimal regulator be continuous with respect to* τ *in the set* $\mathcal{T}$. *Let the noise* v^∞ *take independent values and satisfy the conditions in (2.3) and the test signal* w^∞ *also assume independent values and satisfy the conditions,*

$$\mathbf{M}w_t = 0, \ \lim_{t\to\infty} \mathbf{M}\,|\,w_t\,|^4 = 0. \tag{2.28}$$

Then (1) if the control strategy of (2.25) and (2.26) is realizable and identifying, then the following inequality is satisfied with probability 1 :

$$\lim_{t\to\infty} t^{-1} \sum_{s=0}^{t-1} q(s_s, u_s) = J[\hat{U}^\infty(\cdot), \tau]. \tag{2.29}$$

Here $J[\hat{U}^\infty(\cdot), \tau]$ *are the values of the functional (2.6) related to the optimal regulator (2.12).**

(2) if, in addition, the set $\mathcal{T}$ *is compact and the control strategy of (2.25) and (2.26) ensures fulfilment of the relations*

$$\tau_t \in \mathcal{T}, t^{-1} \sum_{s=0}^{t-1} (|\,x_s\,|^2 + |\,u_s\,|^2) \le \upsilon, \tag{2.30}$$

where υ *is a non-negative random quantity with a finite mean value* $\mathbf{M}\upsilon < \infty$, *then the following equality is satisfied*

---

*In conformity with the lemma 6.2.1, the value $J[\hat{U}^\infty(\cdot), \tau]$ is determined.

$$\lim_{t\to\infty} t^{-1}\sum_{s=0}^{t-1} \mathbf{M}_q(x_s, u_s) = J[\hat{U}^\infty(\cdot), \tau]. \tag{2.31}$$

It follows from the above assertion that the realizable and at the same time identifying control strategy is adaptive (in the class $\mathcal{T}$ in relation to the optimizing control objective of (2.9) and (2.11). The adaptive strategy may not then satisfy the conditions in (2.10) and hence it is not necessarily the admissible one.

For 'unconditional' performance criteria of the type (1.21) related to the optimality of the strategy, a more constrained condition (2.30) is required to be fulfilled. The first condition in (2.30) is usually ensured by introducing the projection (on set $\mathcal{T}$) operator. The set $\mathcal{T}$ must be a convex one. The second condition in (2.30) usually assumes the existence of the summable majorants for the sequences of time averaged squares of the state and control variables.

Thus the problem of ensuring the event dependent control objective of (2.9) and (2.11) requires less constraints on identifying strategy (sufficient to ensure fulfilment of the inequality (2.5)) than for the case with control objective of (2.9) and (1.21) (where the condition (2.30) is also to be ensured).

As it is known, establishment of the properties of the estimate consistency to be obtained by identification algorithm, usually presupposes the boundedness (in some sense) of the system phase variables. If the current estimates of the unknown parameters are used for formation of control signals (as is found in the identification approach (vide (2.25)), then substantiation of the estimate consistency may invite difficulties due to the requirement of the constraint on the control system phase variables. When the uncertainty of the control object parameters is appreciable then the latter condition may pose itself as a separate problem.

A useful assertion about control strategy realizability will be now presented.

Lemma 6.2.5 *The control strategy of (2.25) and (2.26) will be realizable, if for any small value $\varepsilon > 0$ it ensures fulfilment of the relations*

$$\tau_t \in \mathcal{T}, \ \lim_{t\to\infty}(\tau_{t+1} - \tau_t) = 0, \tag{2.32}$$

$$\lim_{t\to\infty} t^{-1}\sum_{s=0}^{t-1} |\eta_s|^2 \le \left(\lim_{t\to\infty} t^{-1}\sum_{s=0}^{t-1} |y_s|^2 + \lim_{t\to\infty} t^{-1}\sum_{s=0}^{t-1} |u_s|^2\right) \tag{2.33}$$

where the random quantities η_t are determined by the formulae

$$\eta_t = -[a(\nabla, \tau) - a(\nabla, \tau_t)]\,y_t + [b(\nabla, \tau) - b(\nabla, \tau_t)]\,u_t$$

$$+c\,[(\nabla, \tau) - c(\nabla, \tau_t)]\,v_t. \tag{2.34}$$

We observe that the equality in (2.32) does not signify the convergence of the estimates $\{\tau_t\}$. It is valid for a broad class of estimate algorithms obtained in the structures of the method of the recursive objective inequalities and of the method of stochastic approximation. The lemma 6.2.5, is based on the use of the 'freezing' Liapunov functions.

(*e*) *Identification algorithm based on stochastic approximation method.* We asume that the coefficients of the equation (2.1) linearly depend on the parameter τ:

$$a_j(\tau) = a_j^{(0)} + a_j^{(1)}\tau^* a_j^{(2)},\ b_j(\tau) = b_j^{(0)} + b_j^{(1)}\tau^* b_j^{(2)},$$

$$C_j(\tau) = C_j^{(0)} + C_j^{(1)}\tau^* C_j^{(2)},\ j = 1, 2, \ldots, n. \tag{2.35}$$

Here $a_j^{(0)}$ and $c_j^{(0)}$ are $l \times l$ square matrices, $b_j^{(0)}$ are $l \times$ m right matrices, $a_j^{(1)}$, $b_j^{(1)}$ and $c_j^{(1)}$ are l-vectors, $a_j^{(2)}$ and $c_j^{(2)}$ are $p \times 1$ right matrices, $b_j^{(2)}$ are $p \times m$ matrices. The matrices with top indices in (2.35) are assumed to be known.

The equation (2.1) can be expressed (taking into account (2.35)) as,

$$y_t = \Phi_{t-1}\tau + \phi_{t-1} + [c(\nabla, \tau) - I_l]\,(v_t - \tilde{v}_t) + v_t, \tag{2.36}$$

where,
$$\phi_{t-1} = -\sum_{j=1}^{n} a_j^{(0)} y_{t-j} + \sum_{j=1}^{n} b_j^{(0)} u_{t-j} + \sum_{j=1}^{n} c_j^{(0)} \tilde{v}_{t-1}, \tag{2.37}$$

$$\Phi_{t-1}\tau = -\sum_{j=1}^{n} a_j^{(1)}\tau^* a_j^{(2)} y_{t-j} + \sum_{j=1}^{n} b_j^{(1)}\tau^* b_j^{(2)} + \sum_{j=1}^{n} c_j^{(1)}\tau^* c_j^{(2)} \tilde{v}_{t-1}$$

and $\{\tilde{v}_t\}$ is an arbitrary sequence of l-vectors. $\tilde{v}_t$ will be chosen further in the form of functions of observable values y^t and u^{t-1}, given by the recursive relations

$$\tilde{v}_t = y_t - \phi_{t-1} - \Phi_{t-1}[\tau_{t-1} + \gamma_{t-1}\Phi^*_{t-1}$$

$$\times (y_t - \phi_{t-1} - \Phi_{t-1}\tau_{t-1})], \tag{2.38}$$

$$\gamma_t^{-1} = \gamma_{t-1}^{-1} + \mathrm{Sp}\,\Phi_{t-1}\Phi^*_{t-1} + 1, \quad \gamma_0 = 1, \tag{2.39}$$

$$\tau_t = P_{\mathcal{T}}[\tau_{t-1} + \gamma_{t-1}\Phi^*_{t-1}\tilde{v}_t], \quad t = 1, 2, \quad \ldots, \tag{2.40}$$

and will be interpreted as estimates of unobservable disturbances v_t. The relations shown include the algorithm (2.40) to get the estimates τ_t. The algorithm is a modification of the recursive least square method. This modification consists of application of the projector $P_{\mathcal{T}}$ on the set $\mathcal{T}$ (the set $\mathcal{T}$ is then assumed to be convex, so that the operation $P_{\mathcal{T}}$ while comparing any point from $\mathbf{R}^p$ with a point from $\mathcal{T}$ and approaching it, could be determined for the whole of $\mathbf{R}^p$) and of the random positive coefficients γ_t given by the recursive formula (2.39).*

The algorithm of (2.39) and (2.40) coincides in essence with the broad LSM algorithm (vide section 5.3). It only differs in fixation of the choice of the values γ_t. The algorithm assumes that the initial estimate $\tau_0 \in \mathcal{T}$ is given.

Theorem 6.2.1 *We assume that the following conditions are fulfilled:*

1) the parameter τ in the equation (2.1) belongs to the convex compact $\mathcal{T}$,

2) the dependence of the polynomial coefficients (2.2) on τ is determined by the formulae (2.35),

3) for all $\tau \in \mathcal{T}$ the pair of polynomials $\{a(\lambda, \tau), b(\lambda, \tau)\}$ is stabilizable, det $b(\lambda, \tau) \neq 0$ for $|\lambda| = 1$. *For the transfer function $W(\lambda, \tau) = a^{-1}(\lambda, \tau)\, b(\lambda, \tau)$ the frequency condition of section 3.A is satisfied, det $c(\lambda, \tau) \neq 0$* for $|\lambda| \le 1$ and for $|\lambda| = 1$ *the matrix polynomial $c(\lambda, \tau)$ satisfies the inequality,*

---

*In the standard Robbin-Monro stochastic approximation method, the coefficients γ_t are chosen by the deterministic values which satisfy the conditions $\gamma_t \geq 0$,

$$\sum_{t=1}^{\infty} \gamma_t = \infty, \quad \sum_{t=1}^{\infty} \gamma_t^2 < \infty.$$

$$\operatorname{Re} c(\lambda, \tau) = \frac{1}{2}[c^{\nabla}(\lambda, \tau) + c(\lambda, \tau)] \geq \rho I_l \tag{2.41}$$

where $p > 0$,

4) the noise and test signals $\{v_t, \quad \omega_t\}$ *are independent in totality and satisfy the conditions in (2.3) and (2.28).*

Then independent of the choice of the initial conditions the following inequality is satisfied for the control system of (2.1), (2.25) and (2.38) to (2.40) :

$$\overline{\lim_{t\to\infty}}\, t^{-1} \sum_{s=0}^{t-1} (|y_s|^2 + |u_s|^2 + |\tilde{v}_s|^2) < \infty. \tag{2.42}$$

In addition, let the following conditions be also satisfied :

5) for all complex λ *the following inequality** *is satisfied :*

$$\det \| |\lambda|^{2q} [a(\lambda^{-1}, \tau)a^*(\lambda^{-1}, \tau) + b(\lambda^{-1}, \tau)b^*(\lambda^{-1}, \quad \tau)] +$$

$$+ |\lambda|^{2n} c(\lambda^{-1}, \tau)c^*(\lambda^{-1}, \tau)\| \neq 0; \tag{2.43}$$

6) the columns $\Phi_t^{(j)}$ *of the matrix* Φ_t^* *from the expression (2.36) has a form* $\Phi_t^{(j)} = Q^{(j)}\tilde{z}_t$,

where $\tilde{z}_t = \operatorname{col}(y_t, \ldots, y_{t-n+1}, u_t, \ldots, u_{t-n+1}, \tilde{v}_t, \ldots, \tilde{v}_{t-n+1})$ *and* $Q^{(j)}$ *are constant matrices such that,*

$$\sum_j^{(j)} Q^{(j)}[Q^{(j)}]^* > o;$$

7) for the test signal the following inequality is valid with a constant $C_w > 0$ *:*

$$\mathrm{M} w_t w_t^* \geq C_w (\mathrm{In}\, t)^{-1} I_m \tag{2.44}$$

Then the control strategy, determined by the relations (2.25) and (2.38) to (2.40), is realizable, identifying and adaptive in the class $\mathcal{T}$ in respect of the control objective (2.9) and (2.11).

It may be noted that the conditions (1), (3) and (4) of the theorem ensure a continuous dependence of the optimal regulator (2.25) on the parameter τ.

6.2.2 *Direct Method of Adaptive Control Formulation*

Identification approach does not happen to be the only possible method for adaptive control formulation. For formation of the adaptive strategy optimizing the functional in (2.6) to (2.8), it is possible to make use of the direct method, when a set of admissible control strategies is parametrized with the help of a vector parameter. (It differs, in general, from the unknown parameter of the control object). The performance criterion is given as a function of this parameter and it permits the formulation of the realizable control strategy using a stochastic gradient of the performance criterion. One such parameter, as given below, is the set of coefficients of optimal regulator. The dependence of these coefficients on unknown control object parameters is not made use of.

(*a*) *Description of the direct method of performance criterion minimization.* We examine the adaptive control problem of the control object (2.1) for optimization of the control objective (2.9) and (2.11) with the performance criterion (2.6) to (2.8). But, for

---

*Here $q = \max(n, r)$ and r is the degree of the polynomials $\alpha(\lambda, \quad \tau)$, $\beta(\lambda, \quad \tau)$ in (2.25).

the simplicity of the statement, we assume that the control, noise and output variables of the control and output variables are scalar quantities ($l = 1$) and $x_t = y_t$ in the functional (2.6). Let the stabilizing feedback signal be fixed as,

$$\tilde{\alpha}(\nabla, \theta)u_t = \tilde{\beta}(\nabla, \theta)y_t, \tag{2.45}$$

and let it be determined by the collection of coefficients,

$$\theta = \text{col}\,(\alpha_0, \alpha_1, \ldots, \alpha_\gamma, \beta_0, \ldots, \beta_r), \ \alpha_0 \neq 0, \tag{2.46}$$

of the polynomials $\tilde{\alpha}(\lambda, \theta)$, $\tilde{\beta}(\lambda, \theta)$, $\theta \in \Theta$. Here Θ is the set of possible values of vector parameter θ. The set Θ is assumed to be convex and the projector P_Θ on this set is assumed to be known. As it is already known, the feedback (2.45) may be optimal in the class $\mathbf{U}^a$ of admissible control strategies for the special choice $\theta = \theta(\tau)$ of the coefficients.

For a stabilizing feedback, i.e., for a stable polynomial,

$$g(\lambda, \theta, \tau) = a(\lambda, \tau)\,\tilde{\alpha}(\lambda, \quad \theta) - b(\lambda, \tau).\tilde{\beta}(\lambda, \theta), \tag{2.47}$$

the functional (2.6) to (2.8) is a determined quantity (because of the lemma 6.2.1) not depending on the choice of the initial conditions in the control system of (2.1) and (2.45). It may be constrained by the consideration of the stationary processes $\hat{y}^\infty$, $\hat{u}^\infty$ generated by the control system. Then the quantity $\mathbf{M}q(\hat{y}_t, \quad \hat{u}_t)$ does not depend on t and we find out from (2.16) :

$$J[U^\infty(\cdot), \tau, \theta] = \mathbf{M}q\,(\hat{y}_t, \hat{u}_t). \tag{2.48}$$

For a fixed τ, the random quantity $\hat{y}_t$, $\hat{u}_t$ is uniquely determined by the stabilizing loop of (2.45), $\hat{y}_t = y_t(\theta)$ and $\hat{u}_t = u_t(\theta)$, *i.e.*, the functional (2.48) becomes a functional of the function θ

$$\mathbf{M}q\,(\hat{y}_t, \hat{u}_t) = \tilde{J}\,(\theta). \tag{2.49}$$

It takes the least value for $\theta = \theta(\tau)$ when $\tilde{\alpha}[\lambda, \theta(\tau)] = \alpha(\lambda, \tau)$, $\tilde{\beta}[\lambda, \theta(\tau)] = \beta(\lambda, \tau)$. Hence for the formation of the realizable strategy we make use of the feedback,

$$\tilde{\alpha}(\nabla, \theta_t)u_t = \tilde{\beta}(\nabla, \theta_t)y_t. \tag{2.50}$$

The set of coefficients θ_t in it is determined by the recursive procedure of the stochastic gradient of the functional (2.49). To describe this procedure we observe that,

$$\text{grad}_\theta \tilde{J}\,(\theta) = 2\mathbf{M}\left[\frac{\partial q\,(\hat{y}_t, \hat{u}_t)}{\partial \hat{y}_t}\phi_t(\theta) + \frac{\partial q\,(\hat{y}_t, \hat{u}_t)}{\partial \hat{u}_t}\psi_t(\theta)\right], \tag{2.51}$$

$$\text{where } \phi_t(\theta) = \text{grad}_\theta\, y_t(\theta), \text{ and } \psi_t(\theta) = \text{grad}_\theta\, u_t(\theta) \tag{2.52}$$

are the respective gradients with respect to θ. It is not difficult to show that when the control system linearity in (2.1) and (2.45) is taken into account, the vectors $\phi_t(\theta)$, and $\psi_t(\theta)$ satisfy the linear equations,

$$g(\nabla, \tau, \theta)\phi_t(\theta) = b(\nabla, \tau)\Phi_t(\theta),$$

$$\text{and } g(\nabla, \tau, \theta)\psi_t(\theta) = a(\nabla, \tau)\Phi_t(\theta). \tag{2.53}$$

Here the polynomial $g(\lambda, \tau, \theta)$ is determined by the formula (2.47) and

$$\Phi_t(\theta) = \text{ col}\,(-u_t(\theta), \ldots, -u_{t-r}(\theta), y_t(\theta), \ldots, y_{t-r}(\theta). \tag{2.54}$$

The relations in (2.51) to (2.54) do not permit computation of stochastic gradient of the functional $\bar{J}(\theta)$ when the parameter τ is unknown. However, they show how the synthesis of realizable control strategy may be tried. Precisely, it may be assumed that the algorithm of (2.26), for achieving the estimates of the unknown parameter τ, is available. Then the algorithm for computation of the estimates θ_t may be taken in the form,

$$\theta_{t+1} = P_{\Theta}[\theta_t - \gamma_t([Qy_t + Su_t]\phi_t + [Sy_t + Ru_t]\psi_t)]. \tag{2.55}$$

Here Q, S, R are the elements of the matrix (2.8) and vectors ϕ_t and ψ_t are determined by the recursive relations,

$$g(\nabla, \tau_t, \theta_t)\phi_t = b(\nabla, \tau_t)\Phi_t,$$

$$\text{and } g(\nabla, \tau_t, \theta_t)\psi_t = a(\nabla, \tau_t)\Phi_t, \tag{2.56}$$

$$\text{where } \Phi_t = \text{col}\,(-u_t, \quad \ldots, \quad -u_{t-r}, \quad y_t, \quad \ldots, \quad y_{t-r}). \tag{2.57}$$

After the initial conditions and the quantities $\{\gamma_t\}$ are given, the relations (2.50), (2.26), (2.55) to (2.57) give non-asserting control strategy. In order that this strategy is realizable, the inequality (2.5) is to be ensured. The proof of its optimality is related to the establishment of the property $\text{grad}_\theta \bar{J}(\theta) \neq 0$ outside the set of the points of the global minimum of the functional $\bar{J}(\theta)$.

This is the general scheme for synthesis of adaptive strategy by direct minimization of the performance criterion. Any one of the algorithms of the stochastic approximation method may be chosen as the procedure in (2.26).

The substantiation of the described scheme for direct minimization in the general case is difficult. The performance criterion may have multiextrema and the non-stationary feedback of (2.50), (2.55) to (2.57) does not give guarantee to fulfilment of the inequality (2.5). Hence a simplified adaptive control problem, where the above substantiation is possible, is discussed below :

(*b*) *Direct method in the limiting optimal tracking problem for a minimum phase control object.* Let the control object be given by

$$a(\nabla, \tau)y_t = b(\nabla, \tau)u_t + e_t \tag{2.58}$$

and the disturbance e_t be formulated by the filter

$$d(\nabla, \tau)e_t = c(\nabla)v_t \tag{2.59}$$

where $\{v_t\}$ is the sequence of independent random quantities with the properties as in (2.3). It is assumed that for all $\tau \in \mathcal{T}$ the following conditions are fulfilled,

$$a(0, \tau) = 1, \quad b(\lambda, \tau) = \lambda^k b_+(\lambda, \tau), \quad k \geq 1,$$

$$b_+(\lambda, \tau) \neq 0, \quad |\lambda| \leq 1, \tag{2.60}$$

i.e., the minimum-phase control objects with a fixed value of control delay k are being discussed :

$$d(\lambda, \tau) \neq 0, \quad c(\lambda) \neq 0 \text{ for } |\lambda| \leq 1, \text{ and } c(0) = 1. \tag{2.61}$$

This signifies the stability of the formulating filter (2.59) in terms of input and output variables. The polynomial $C(\lambda)$ is then independent of the parameter τ, *i.e.*, it is assumed to the known. The performance criterion is expressed as,

$$J[U^\infty(\cdot), \tau, v^\infty] = \overline{\lim_{t \to \infty}}\; t^{-1} \sum_{s=0}^{t-1} |y_s - y_{s*}|^2. \tag{2.62}$$

Here $\{y_{t*}\}$ is the trackable (programmed) stationary sequence of random quantities independent of random quantities $\{v_t\}$ and possessing bounded second moments with observable realizations of at least at the k-th instant forward.*

If the equation (2.58) is rewritten as,

$$d(\nabla, \tau)a(\nabla, \tau)y_t = d(\nabla, \tau)b_+(\nabla, \tau)u_{t-k} + c(\nabla)v_t, \tag{2.63}$$

then it is not difficult to formulate the optimal feedback in $\mathbf{U}^a$. In fact let $F(\lambda, \tau)$, $G(\lambda, \tau)$ be the polynomials satisfying the relationship.

$$d(\lambda, \tau)a(\lambda, \tau)F(\lambda) - c(\lambda) = \lambda^k G(\lambda). \tag{2.64}$$

The relationship (2.64) uniquely determines the polynomials $F(\lambda) = F(\lambda, \tau)$, $G(\lambda) = G(\lambda, \tau)$ if it is required to satisfy the condition $\deg F(\lambda) < k$. Applying the operation, $F(\nabla, \tau)$ to the equation (2.63) and considering (2.64) we get the relationship :

$$c(\nabla)\,[y_{t+k} - F(\nabla)v_{t+k} - y_{(t+k)*}]$$

$$= F\,(\nabla, \tau)d(\nabla, \tau)b_+\,(\nabla, \tau)u_t$$

$$-G(\nabla, \tau)y_t - c(\nabla)y_{(t+k)*}. \tag{2.65}$$

It is not difficult to establish the optimality of the feedback from the relationship (2.65) and the independence of the random quantities v_t^{t+k} from the random quantities y^t, u^t, y_*^{t+k},

$$F\,(\nabla, \tau)d(\nabla, \tau)b_+\,\,(\nabla, \tau)u_t$$

$$= G(\nabla, \tau)y_t + c(\nabla)y_{(t+k)*}. \tag{2.66}$$

The functional (2.62) takes the following value in it

$$\min_{U^\infty(\cdot) \in U^a} J[U^\infty(\cdot), \tau, v^\infty] = \sigma_v^2 \sum_{j=0}^{k-1} |F_j(\tau)|^2. \tag{2.67}$$

Here $F_j(\tau)$ are the coefficients of the polynomial $F(\lambda, \tau)$. The equation (2.66) can be written in terms of the notations used in (2.46) and (2.57) as

$$\Phi_t^*\theta = c(\nabla)y_{(t+k)*},$$

or introducing the process ϕ_t through the relationship

$$c(\nabla)\phi_t = \Phi_t \tag{2.68}$$

and 'neglecting' the stable polynomial $c(\lambda)$** it can be expressed as

$$\phi_t^*\theta = -y_{(t+k)*}. \tag{2.69}$$

The expression of (2.65), because of (2.69) and after the polynomial $c(\lambda)$ is neglected, can be simplified as :

---

*This means that for every instant t, the values y_*^{t+k} enter the sensor (observable) and may be used for formulating the control.

**Here and henceforth the independence of the functional (2.62) from the transient states of the control system is assumed. Due to this reason any process $\{f_t\}$ satisfying the equation $c(\nabla)f_t = 0$ with stable polynomial $c(\lambda)$ can be identified with the forces $f_t \equiv 0$.

$$y_t = -\phi^*_{t-k}\theta + F(\nabla, \tau)v_t. \qquad (2.70)$$

It is not difficult to compute the characteristic polynomial (2.47) for the regulator (2.66):

$$g(\lambda, \tau, \theta) = b_+(\lambda, \tau)c(\lambda). \qquad (2.71)$$

The first of the relations (2.53) takes the form of (2.68) when the stable polynomial $b_+(\lambda, \tau)$ is neglected. The algorithm of (2.55) can be rewritten as :

$$\theta_{t+1+k} = P_{\Theta}\ \left[\theta_t - \gamma_t\phi_t\left(y_{t+k} - y_{(t+k)*}\right)\right]. \qquad (2.72)$$

after the above is taken into account. When the initial estimates θ_o^{k-1} and the values $\{\gamma_t\}$ are given, the procedure in (2.72), (2.68) and (2.57) determines the adaptation algorithm. The coefficients of the feedback

$$\phi_t^*\theta_t = -y_{(t+k)*} \qquad (2.73)$$

are 'tuned' using the algorithm when the feedback is optimal (vide (2.69)) for $\theta_t \equiv \theta(\tau)$.

Theorem 6.2.2 *We assume that in the optimal tracking problem of (2.58) and (2.59) with the performance criterion (2.62) the following conditions are satisfied :*

1) the disturbances $\{v_t\}$ are independent in totality and possess the following properties

$$\mathbf{M}v_t = 0,\ \mathbf{M}\,|\,v_t\,|^2 = \sigma_v^2 = \lim_{t\to\infty} t^{-1}\sum_{s=0}^{t-1}|\,v_s\,|^2 > 0; \qquad (2.74)$$

2) the trackable sequence $\{y_{t*}\}$ has second moments and

$$\sup_t \mathbf{M}\,|\,y_{t*}\,|^2 < \infty,\ \lim_{t\to\infty} t^{-1}\sum_{s=0}^{t-1}|\,y_{s*}\,|^2 < \infty. \qquad (2.75)$$

Here the values y_*^{t+k} are known for all t ;

3) for any $\tau \in \mathcal{T}$, the conditions in (2.60) and (2.61) are satisfied and the polynomial $e(\lambda)$ does not depend on τ ;

4) Θ is the arbitrary convex set containing the set $\theta(\mathcal{T}) = \{\theta : \theta = \theta(\tau),\ \tau \in \mathcal{T}\}$ and such that $\bar{\alpha}(0, \theta) \neq 0$ for $\theta \in \Theta$;

5) the sequence $\{\gamma_t\}$ is determined by the recursive relations

$$\gamma_{t+1}^{-1} = \gamma_t^{-1} + |\,\phi_t\,|^2,\ \gamma_0 = 1. \qquad (2.76)$$

Then the control strategy determined by the relations (2.73), (2.72), (2.68) and (2.57) is adaptive in the class $\mathcal{T}$ in respect of the control objective

$$\lim_{t\to\infty} t^{-1}\sum_{s=0}^{t-1}|\,y_s - y_{s*}\,|^2 \le \sigma_v^2\sum_{j=0}^{k-1}|\,F_j(\tau)\,|^2 = J_*. \qquad (2.77)$$

We note that the condition (4) of the theorem ensures solvability of the equation (2.73) with respect to u_t.

In case the polynomial $c(\lambda)$ in the filter (2.59) depends on τ, $c(\lambda) = c(\lambda, \tau)$, the use of the above formulation is not possible, but with additional assumptions about the polynomial $c(\lambda, \tau)$, it can be modified basically. The regulator of (2.69) will then be expressed, to achieve this objective, as

$$\Phi_t^* \theta = -c(\nabla, \tau) y_{(t+k)\cdot}. \tag{2.78}$$

Introducing the 'extended' vectors,

$$\tilde{\phi}_t = \mathrm{col}\,(\Phi_t, y_{(t+k-1)\cdot}, \ldots, y_{(t+k-p)\cdot}), \tilde{\theta} = \mathrm{col}\,(\theta, c_1, \ \ldots \ c_p). \tag{2.79}$$

Here p is the degree of the polynomial $c(\lambda, \tau)$. The equation (2.78) is rewritten in the form $\tilde{\phi}_t^* \tilde{\theta} = -y_{(t+k)}$. (analogous to (2.69)). The adaptive version of the regulator equation will be taken as

$$\tilde{\phi}_t \tilde{\theta}_t = -y_{(t+k)\cdot\cdot}, \tag{2.80}$$

and its parameter tuning algorithm (adaptation algorithm) will be in the form

$$\tilde{\theta}_{t+1+k} = P_{\tilde{\Theta}}\left[\tilde{\theta}_t - \gamma_t \tilde{\phi}_t\left(y_{t+k} - y_{(t+k)\cdot}\right)\right]. \tag{2.81}$$

It is also analogous to (2.72).

Here $\tilde{\Theta}$ is a convex set of possible values of the parameter $\tilde{\theta}$ and $P_{\tilde{\Theta}}$ is the projector on this $\tilde{\Theta}$. The result about adaptability of the corresponding control strategy will be carried out.

Theorem 6.2.3 *We assume that for the polynomial $c(\lambda) = c(\lambda, \tau)$ the following condition is valid,*

$$\mathrm{Re}\, c(\lambda, \tau) \geq \rho > 0, \ \tau \in \mathcal{T}, |\lambda| = 1, \tag{2.82}$$

and the condition of the theorem 6.2.2 be satisfied, in which the set Θ and the vectors θ, ϕ_t have been replaced by $\tilde{\Theta}$, $\tilde{\theta}, \tilde{\phi}_t$ respectively.

Then the control strategy determinable by the relations (2.80), (2.81) and (2.79) is adaptive in the class $\mathcal{T}$ in respect of the control objective (2.77).

In the theorems 6.2.2 and 6.2.3, it is not asserted that the corresponding control strategies are identifying.

6.2.3 *Adaptive Optimization of the Unconditional Performance Criterion*

The adaptive strategy that satisfies (2.30) is difficult to be formulated since the turning procedure of (2.26) does not give guarantee of uniformity (in realizations) of convergence of the estimates $\{\tau_t\}$. This complicates the establishment of the inequality in (2.30) which may be brought in the identifying control strategy related to (2.31).

A method of formulation of control strategy, which is identifying and at the same time it ensures fulfilment of the inequality of (2.30), is given below. This formulation, along with the 'optimizing' regulator, makes use of the additional 'stabilizing' regulator and the adaptive strategy is realized through 'switching' of these regulators. The presence of the stabilizing regulator gives possibility of establishing the equality (2.30), and at the same time the stabilizing regulator 'operates' for a finite period practically for each control process realization. It provides the optimal properties of the control strategy obtained.

(*a*) *Formulation of the problem.* We examine, again the adaptive control problem of the control object (2.1) whose coefficients linearly depend on $\tau \in \mathcal{T}$. The performance criterion is taken in the form

$$J[U^\infty(\cdot), \tau] = \lim_{t \to \infty} t^{-1} \sum_{s=0}^{t-1} \mathrm{M} q(x_s, u_s). \tag{2.83}$$

Here $q(x, u)$ is a quadratic non-negative function (2.8) and the vector x_t is given by the formulae (2.7). Unlike the functional in (2.6), the functional in (2.83) does not depend on the events as a result of the consideration of the ensemble operation of the unconditional averaging. Hence it is known as unconditional. The functional (2.83) is defined on the set $\mathbf{U}^p$ of realizable strategies $U^\infty(\cdot)$ which formulate control law according to (2.4) and ensure fulfilment of the inequality,

$$\lim_{t\to\infty} t^{-1} \sum_{s=0}^{t-1} \mathbf{M}(|y_s|^2 + |u_s|^2) < \infty. \tag{2.84}$$

The adaptive optimization control problem is to formulate the realizable control strategy that satisfies the inequality,

$$J[U^\infty(\cdot), \tau] \le \inf_{\tilde{U}^\infty(\cdot) \in \mathbf{U}^a(\tau)} J[\tilde{U}^\infty(\cdot), \tau], \tag{2.85}$$

where $\mathbf{U}^a(\tau)$ is the set of admissible strategies $U^\infty(\cdot)$ which formulate control law according to (2.4) and ensure fulfilment of the relations

$$\sup_t \mathbf{M}(|y_t|^2 + |u_t^2|) < \infty. \tag{2.86}$$

But these strategies may depend on the parameter τ. Experience from the solution of the adaptive optimization control problems shows that under altogether general conditions, the set $\mathbf{U}^p$ is not empty and it contains control strategy which is optimal in all the classes $\mathbf{U}^a(\tau)$, $\tau \in \mathcal{T}$. This strategy forms the adaptive control.

(*b*) *The properties of the realizable strategy in the presence of a stabilizing regulator.* We asume that the regulator,

$$\alpha^c(\nabla)u_t = \beta^c(\nabla)u_t + w_t, \tag{2.87}$$

is known. It stabilizes the control object (2.1) for all $\tau \in \mathcal{T}'$, where $\mathcal{T}'$ is the bounded open (not necessarily convex) set of possible values of the parameter τ. Let $\mathcal{T}$ be the arbitrary convex compact set containing $\mathcal{T}'$, $\mathcal{T}' \subseteq \mathcal{T}$. Let $P_{\mathcal{T}}$ be the projector on $\mathcal{T}$. The estimates $\{\tau_t\}$ of the known parameter τ will be formed as per algorithm (2.38) to (2.40) where the matrix θ_t and vector ϕ_t are determined by the relations in (2.37) as functions of variables u^t and y^t. Under the conditions of the theorem 6.2.1 the control strategy generated by the regulator (2.87) is identifying (vide (2.27)). Apart from this, the strategy possesses additional useful properties.

Lemma 6.2.6 *We asume that the random quantity,*

$$\mu = \sup_t t^{-1} \sum_{s=0}^{t-1} |v_s|^2 + \sup_t t^{-1} \sum_{s=0}^{t-1} |w_s|^2 \tag{2.88}$$

can be additive, $\mathbf{M}\mu < \infty$.

Then the random quanitity

$$\upsilon = \sup_t t^{-1} \sum_{s=0}^{t-1} (|y_s|^2 + |u_s|^2) \tag{2.89}$$

will be additive for the control system of (2.1) *and* (2.87).

(*c*) *Formulation of the adaptive control strategy in the presence of a stabilizing regulator.* We assume that there are two regulators (2.25) and (2.87). The first one depends on the parameters $\{\tau_t\}$ which can be framed up. The second stabilizes the control object (2.1) for all $\tau \in \mathcal{T}'$ we write the method of formulating, the optimal control based on use of these regulators by turn.

Let the determined constant C_z and the initial estimate $\tau_0 \in \mathcal{T}'$ be given. The control signal is formed by the regulator (2.25) up to the instant $t_1 \geq t_0 = 0$. The estimates τ_t in it are given by the algorithm (2.38) to (2.40). The instant t_l is determined as the first instant of breakeven if we have from one of the conditions

$$\tau_t \in \mathcal{T}', t^{-1} \sum_{s=0}^{t-1} | z_s |^2 \leq C_z, \qquad (2.90)$$

where $$z_t = \text{col}\,(y_t, \ldots, \ y_{t-q}, u_t, \ldots, u_{t-q}) \qquad (2.91)$$

and q is an estimate from top for the indices of non-zero co-efficients of the equations (2.1), (2.25) and (2.87). For $t > t_1$, the control signal is formed by the stabilizing regulator (2.87) up to the first post t_l instant of time t_2 and then both the conditions in (2.90) appear to be satisfied. If, such an instant t_2 is reached, then for $t > t_2$, there will be a changeover to the regulator (2.25). It can be used up to the instant t_3 when at least one of the conditions in (2.90) is broken, etc.

If the constant C_z is chosen not so large, then the switching with the stabilizing regulator to the optimal may not take place.

Theorem 6.2.4. *Let the random quantity (2.89) be additive and the conditions of the theorem 6.2.1 be satisfied. Let $\mathcal{T}$ be the arbitrary convex compact containing $\mathcal{T}'$.*

Then for the choice of a sufficiently large value of C_z^*, the formulation procedure described determines the adaptive control strategy in the class $\mathcal{T}'$ and with respect to the control objective (2.85). Then the general 'operational' time of the stabilizing regulator is finite with probability 1.

(d) Design of the stabilizing regulator. The assumption of the presence of the regulator (2.87) stabilizing the control object (2.1) for all $\tau \in \mathcal{T}'$ is sufficiently restricted. A method of designing the feedback is suggested below. The feedback appears to be stationary and stabilizing except for some (random and definite) time of 'adaptation'. This design is based on the method of recursive objective inequalities (vide section 4.3.4). The important feature of the method is its possibility to form the 'piecewise - stationary' feedback whose general number of changing the coefficients permits the estimation from top independent of the noise realization. The demerit of the method is its requirements of removal of important constraints on disturbances.

---

*The following may be chosen, for example, as C_z :

$$C_z = (4q+1)\max\,(J_0, J_c),$$

$$J_0 = \sup_{\tau \in \mathcal{T}} \frac{1}{2\pi i} \oint \text{Sp}\left[W_0(\lambda, \tau) c(\lambda, \tau) R_v c^{\nabla}(\lambda, \tau) W_0^{\nabla}(\lambda, \tau)\right] \frac{d\lambda}{\lambda},$$

$$J_c = \sup_{\tau \in \mathcal{T}} \frac{1}{2\pi i} \oint \text{Sp}\left[W_c(\lambda, \tau) c(\lambda, \tau) R_v c^{\nabla}(\lambda, \tau) W_c^{\nabla}(\lambda, \tau)\right] \frac{d\lambda}{\lambda}.$$

Here $W_0(\lambda, \tau)$ and $W_c(\lambda, \tau)$ are the transfer functions (vide formula (4.59) chapter 3 corresponding to control systems of (2.1), (2.12) and (2.1), (2.87) respectively. The constants J_0, J_c are computed with respect to stationary processes of the optimal and stabilizing regulators, when $w_t = 0$.

Thus, let us assume that the disturbance v^∞ in equation (2.1) is bounded. If the set $\mathcal{T}' = \mathcal{T}$(of possible values of the vector τ) is compact, then for all t and with some constant C_v independent of $\tau \in \mathcal{T}$ the inequality holds,

$$|c(\nabla, \tau)v_t| \le C_v. \tag{2.92}$$

Assuming the dependence of the polynomial $a(\lambda, \tau)$, $b(\lambda, \tau)$ on τ to be linear, we arrive at the problem of adaptive stabilization as examined in section 4.3.3.

The feedback* thus formulated, namely,

$$\tilde{\alpha}(\nabla, \tilde{\tau}_t)u_t = \tilde{\beta}(\nabla, \tilde{\tau}_t)y_t + w_t \tag{2.93}$$

uses the 'stripe' technique (precisely 'extended stripe') as the adaptation algorithm. With the conditions of the theorem 4.3.3, the following limit is attained in a finite time.

$$\lim_{t\to\infty} \tilde{\tau}_t = \tilde{\tau}_\infty. \tag{2.94}$$

Here it is not necessary that $\tilde{\tau}_\infty = \tau$ and the regulator (2.93) is stabilizing for the control object (2.1) when $\tilde{\tau}_t \equiv \tilde{\tau}_\infty$. For the 'extended stripe' algorithm it is difficult to indicate the estimate from top for the number of corrections of the estimates $\{\tilde{\tau}_t\}$. Hence a different version of the 'stripe' technique will be used here. It may be mentioned as the 'modified stripe' technique.

The formulae (2.37) determine the matrices $\tilde{\theta}_{t-1}$ and the vectors $\tilde{\phi}_{t-1}$ for $\tilde{v}_t \equiv 0$. With these notations, the control object (2.1) is written as :

$$y_t = \tilde{\Phi}_{t-1}\tau + \tilde{\phi}_{t-1} + c(\nabla, \tau)v_t. \tag{2.95}$$

The 'modified band' algorithm formulates the sequence of estimates $\{\tilde{\tau}_t\}$ according to the formulae,

$$\tilde{\tau}_{t+1} = P_{\mathcal{T}}\left[\tilde{\tau}_t + \theta_t \frac{\tilde{\Phi}_t \eta_{t+1}}{|\tilde{\Phi}_t \tilde{\Phi}_t^*|}\right], \tag{2.96}$$

where, $\quad \eta_{t+1} = y_{t+1} - \tilde{\Phi}_t\tilde{\tau}_t - \tilde{\phi}_t,$

$$\theta_t = \begin{cases} 0, \text{ if } |\eta_{t+1}| < 2C_v + \varepsilon\sqrt{|\tilde{\Phi}_t\tilde{\Phi}_t^*|}, \\ 1, \text{ if } |\eta_t| \ge 2C_v + \varepsilon\sqrt{|\tilde{\Phi}_t\tilde{\Phi}_t^* t|}, \end{cases} \tag{2.97}$$

$P_{\mathcal{T}}$ is the projector on the convex set $\mathcal{T}$ of possible values of parameter τ and ε is a parameter of the algorithm. (For $\varepsilon = 0$ the basic 'stripe' algorithm is reached - vide section 4.3.4.)

Lemma 6.2.7 *For any $\varepsilon > 0$ the algorithm of (2.96) and (2.97) finitely converges independent of the choice of the initial estimate $\tau_0 \in \mathcal{T}$. Then the inequality below holds*

$$\sum_{t=0}^{\infty} \theta_t \le |\tau - \tau_0|^2 \varepsilon^2. \tag{2.98}$$

The inequality (2.98) shows that the number of corrections (variations) of estimates is limited from the top by the quantity independent of realization of the noise (although

---

*For $\tilde{\tau}_t \equiv \tau$ this feedback is stabilizing for the control object of (2.1) and the order of the difference equation does not excel the number q.

the time taken to achieve its own limit $\tilde{\tau}_\infty$ (vide (2.94)) by the sequence $\{\tau_t\}$, uniform in respect of the realizations of the noise, is not bounded). It permits the use of the control system in (2.93), (2.96) and (2.97) alongwith the stabilizing regulator (2.87) for synthesis of the adaptive strategy in theorem 6.2.4. The control strategy is then realized by the following method. Let the constant $C_z > 0$ and the initial estimate $\tau_0 \in \mathcal{T}$ be given. For $t > t_0 = 0$ the control signal is formulated by the regulator of (2.25). The estimates τ_t in it are formed by the algorithm (2.38) to (2.40) up to the instant t_1 when for the first time at least one of the conditions in (2.90) is broken. Here z_t is the vector in (2.91). The control is formulated by the regulator (2.93) from the instant t_1. The estimates $\{\tilde{\tau}_t\}$ in it are computed as per algorithm (2.96) and (2.97). The computation of the estimates $\{\tau_t\}$, in accordance with the algorithm (2.38) to (2.40), is carried out at the same time. During the first post t_2 instant t_3 when both the conditions in (2.90) appear to be fulfilled, the control 'goes over' to the regulator (2.25), the estimate τ_{t3} is 'frozen'. During the next transition to the regulator (2.93) (at the instant t_4 when one of the conditions in (2.90) is broken) the above estimate is used as the initial value for the algorithm (2.96) and (2.97). The above method of 'switching' of the regulators is further carried on. Thus, the estimates $\{\tau_t\}$ are obtained by the algorithm (2.38) to (2.40) at all instants of time. Then, as the estimates $\{\tilde{\tau}_t\}$ are formulated only during the operation of the regulator (2.93), the initial estimate for each transition is taken from the value obtained at the end of the previous interval of its operation.

Theorem 6.2.5 *Under the conditions of the theorem 6.2.1, we assume that for all* $\tau \in \mathcal{T}$, *the pair of polynomials* $\{a(\lambda, \tau), b(\lambda, \tau)\}$ *is stabilizable, the test signal and the noise* $\{vt\}$ *are bounded so that the inequality (2.92) is satisfied.*

Then the control strategy which is being realized by the above method of regulator ((2.25) and (2.93)) 'switching' is admissible and adaptive in the class $\mathcal{T}$ *with respect to the control objective (2.85) and (2.83), when the constant* $C_z > 0$ *is sufficiently large and the number* $\varepsilon > 0$ *is sufficiently small.*

If the noise v^∞ is not bounded, the design of the stabilizing regulator may be done with the help of the method given in section 5.5. In order to make the identification process finitely convergent, the estimates $\{\tau_t\}$ obtained from the conditions from the theorem 5.5.1 (vide page 393) may be used in the following form

Let $\tau_{t1} = \hat{\tau}_{t1}$ at any instant t_1. Then $\tau_t = \hat{\tau}_{t1}$ up to the instant t_2 when for the first time we have $|\hat{\tau}_{t_2} - \hat{\tau}_{t_1}| \geq \varepsilon$ and then $\tau_{t_2} = \hat{\tau}_{t_2}$. Here ε is a positive parameter. Its choice affects the identification accuracy. The choice of the sufficiently small $\varepsilon > 0$ ensures achievement of a stabilizing regulator for the control object (2.1) in finite time.

§ 6.3 Design of the Adaptive Minimax Control

Design of the adaptive minimax control strategy based on computation of information sets has been already initiated in section 4.2. From the point of view of computation, this method of design is acknowledged to be complex since formation of information sets potentially require large (practically infinitely large) memory of the computers. It is shown in the present section that the use of the test signals in the form of a white noise (that decays to the zero intensity) permits us to continue, and substantiate the recursive scheme of adaptive strategy synthesis so that the control quality does not go worse than the worst obtained by minimax strategy (given by stationary feedback under the conditions of complete determinacy of control object parameters). This scheme is based on the identification algorithms given in section 5.5.

6.3.1 *Statement of the Problem*

We recall the problem formulation for minimax control. The control object is taken with scalar 'input - output' and its expression is given by the equation

$$a(\nabla, \tau)y_t = b(\nabla, \tau)u_t + v_t. \tag{3.1}$$

The polynomials $a(\lambda, \tau)$ and $b(\lambda, \tau)$ in it are determined to within the vector parameter τ:

$$a(\lambda, \tau) = 1 + \lambda a_1 + \dots + \lambda^n a_n,$$

$$b(\lambda, \tau) = \lambda^k b_k + \dots + \lambda^m b_m,\ k \geq 1, \tag{3.2}$$

$$\tau = \text{col}\,(a_1, \dots, a_n, b_k, \dots, b_m). \tag{3.3}$$

The only thing known about the parameter τ is the switching $\tau \in \mathcal{T}$, where $\mathcal{T}$ is the convex eucledian space $\mathbf{R}^{n+m-k+1}$. The disturbance v^∞ belongs to the class $\mathbf{V}_\rho$ of sequences which can be determined by the condition for a given $\rho > 0$

$$|v_t| \leq \rho. \tag{3.4}$$

The realizable control strategy class $\mathbf{U}^p$ has been formed by the sequences $U^\infty(\cdot)$ of functions $U_t(\cdot)$ formulating the control u^∞ as per following laws,

$$u_t = U_t(u^{t-1}, y^t, w^t), t = 1, 2, \dots. \tag{3.5}$$

They ensure the fulfillment of the inequality :

$$\overline{\lim_{t\to\infty}}\,(|y_t| + |u_t|) < \infty \tag{3.6}$$

independent of the choice of the initial conditions of the control object (3.1). The functional below is defined in the set $\mathbf{U}^p$,

$$J[U^\infty(\cdot), \tau, v^\infty, w^\infty] = \overline{\lim_{t\to\infty}}(|y_t| + R\,|u_t|), \tag{3.7}$$

and it characterizes the control quality. R is the non-negative coefficient. To formulate the optimizing control objective we introduce the class $\mathbf{U}^a(\tau)$ of the admissible control strategies given by the stabilizing regulators,

$$\alpha(\nabla, \tau)u_t = \beta(\nabla, \tau)y_t + e_t. \tag{3.8}$$

The coefficients of these regulators may depend on the parameter τ, $\lim_{t\to\infty} e_t = 0$.

The control objective is to ensure the following inequality for all $\tau \in \mathcal{T}$ by choosing the strategy $U^\infty(\cdot) \in \mathbf{U}^p$,

$$J[U^\infty(\cdot), \tau, v^\infty, w^\infty]$$

$$\leq \inf_{U^\infty(\cdot)\in \mathbf{U}^a(\tau)} \sup_{v^\infty \in \mathbf{V}_\rho} \overline{\lim_{t\to\infty}}\,(|y_t| + R\,|u_t|). \tag{3.9}$$

If such a strategy exists, then it is known as adaptive in the class $\mathcal{T}\times \mathbf{V}_\rho$ in respect of the control objective (3.9).

Under the conditions of the theorem 3.5.1 for $R = 0$, $e_t = 0$, the lower limit in (3.9) is achieved on the feedback of the form (3.8). Its coefficients may be found out by the methods mentioned in the theorem 3.5.1. The optimal synthesis problem in $\mathbf{U}^a(\tau)$ will have also a solution for the control strategy when $R > 0$. But at the present moment nothing is proposed, in general, about the effectiveness of the computation methods for the polynomials $\alpha(\lambda, \tau)$ and $\beta(\lambda, \tau)$, which determine the optimal feedback. Nevertheless it will be later assumed that the dependence of $\alpha(\lambda, \tau)$ and $\beta(\lambda, \tau)$ on τ (on the set $\mathcal{T}$) is known and continuous.

6.3.2 *Formulation of the Adaptive Control Strategy*

The realizable control strategy will be formulated using feedback and following the identification approach as,

$$\alpha(\nabla, \tau_s)u_t = \beta(\nabla, \tau_s)y_t + \tilde{\omega}_t, \; s = \left[\left[\frac{t}{n+m-k+1}\right]\right], \tag{3.10}$$

where [[ ]] is the whole part of the corresponding number, $\alpha(\lambda, \tau)$ and $\beta(\lambda, \tau)$ are the polynomials of the optimal regulator (3.8) in $\mathbf{U}^a(\tau)$, and $\tilde{w}_t$ is the excitation of the form,

$$\tilde{w}_t = \alpha(\nabla, \tau_s)\,[x_t w_t], \tag{3.11}$$

where w^∞ is the test signal. The gains x_t of the test signals and the estimates $\tau_t = \tau(\theta_t)$ are the same as in the theorem 5.5.1. For $t = (n+m-k+1)l$, $(n+m-k+1)l+1$, ..., $(n+m-k)\times l+n+m-k$ (l is the arbitrary natural number), the feedback coefficients (3.10) do not change. For the control strategy formulated by this method it can be shown that the conditions of the theorem 5.5.1 are fulfilled and hence, the strategy is realizable and identifying. Hence it is adaptive. The final result will be formulated.

Theorem 6.3.1. *We assume that in addition to the conditions (1) to (3) of the theorem the following conditions are satisfied :*

4) *the coefficients of the polynomials* $\alpha(\lambda, \tau)$ and $\beta(\lambda, \tau)$ *are continuous in the space* $\mathcal{T}$ *and the degrees of those polynomials are bounded by the numbers* l_1 *and* l_2 *;*

5) *the control is formulated in accordance with (3.10) and (3.11) where the values* x_t *and* $\tau_1 = \tau(\theta_t)$ *are also the same as in the theorem 5.5.1.*

Then $\lim_{t\to\infty} \tau_t = \tau$ *and the control strategy described above is adaptive in the class* $\mathcal{T}\times \mathbf{V}_\rho$ *in respect of the control objective (3.9).*

We note that for $R = 0$ the condition (4) of the theorem is satisfied if $b(\lambda, \tau) \neq 0$ when $|\lambda| = 1$.

§ 6.P Proofs of the Lemmas and the Theorems

6.P.1 *Proof of the Theorem 6.1.1*

For each strategy $U^\infty(\cdot) \in \mathbf{U}^p$ the following is satisfied by virtue of (1.12) and (1.13) :

$$W_t(y^t, u^t) - V_t(y^t, u^{t-1}) \geq r_t(y^t, u^{t-1}),$$

or taking into account of (1.11)

$$\mathbf{M}q_t(x_{t+1}, u_t, \tau_{t+1}) \geq \mathbf{M}V_{t+1}(y^{t+1}, u^t)$$

$$-\mathbf{M}V_t(y^t, u^{t-1}) + \mathbf{M}r_t(y^t, u^{t-1}). \tag{P.1}$$

Summing up the inequalities obtained and using (1.4) and (1.15), we have,

$$J[U^\infty(\cdot)] \geq \overline{\lim_{t\to\infty}}\, t^{-1} \sum_{s=1}^{t} \mathbf{M}r_t(y^t, u^{t-1}). \tag{P.2}$$

Since the quantity (1.14) takes that the least value (in $\mathbf{U}^p$) in the strategy $\tilde{U}^\infty(\cdot)$ which can be determined from (1.12), then from (P.1) the optimatety of the control strategy $\tilde{U}^\infty(\cdot)$ in $\mathbf{U}^p$ follows.

6.P.2 *Proof of the Theorem 6.1.2*

Starting from (1.100) we find out

$$\mathbf{M}(h_{t+1} \mid y^t) = \int\int\left[\ln\frac{p(\tau \mid y^{t+1})}{\bar{p}(\tau \mid y^{t+1})}\right] p(\tau \mid y^{t+1}) d\tau\, p(y_{t+1} \mid y^t) dy_{t+1}. \tag{P.3}$$

Because of (1.93) we have

$$\ln\frac{p(\tau \mid y^{t+1})}{\bar{p}(\tau \mid y^{t+1})} = \ln\frac{p(\tau \mid y^t)}{\bar{p}(\tau \mid y^t)} + \ln\frac{\bar{p}(y_{t+1} \mid y^t)}{p(y_{t+1} \mid y^t)},$$

where the independence of the functions in (1.94) from the choice of the initial density $p_*(\cdot \mid 0)$ has been taken into account. Moreover, the notations of (1.96) and (1.97) have been used. We continue the computation in (P.3) as follows :

$$\mathbf{M}\{h_{t+1} \mid y^t\} = \int\int\left[\ln\frac{p(\tau \mid y^t)}{\bar{p}(\tau \mid y^t)}\right] p(\tau \mid y^{t+1}) d\tau\, p(y_{t+1} \mid y^t) dy_{t+1}$$

$$+\int\int\left[\ln\frac{\bar{p}(y_{t+1} \mid y^t)}{p(y_{t+1} \mid y^t)}\right] p(\tau \mid y^{t+1}) d\tau\, p(y_{t+1} \mid y^t) dy_{t+1}$$

$$= \int\left[\ln\frac{p(\tau \mid y^t)}{\bar{p}(\tau \mid y^t)}\right] p(\tau \mid y^t) d\tau$$

$$+\int\left[\ln\frac{\bar{p}(y_{t+1} \mid y^t)}{p(y_{t+1} \mid y^t)}\right] p(y_{t+1} \mid y^t) dy_{t+1} = h_t - I_{t-1}(p; \bar{p}). \tag{P.4}$$

Since $I_{t+1}(p{:}\bar{p}) \geq 0$, then from (P.4) follows the poly martingality of the sequence of non-negative quantities h_t. In accordance with the theorem 2.A.1, the convergence of $h_t \to h,\quad t \to \infty$, to some finite random quantity h_∞ takes place with probability 1. Unconditional averaging of (P4) is carried out and it results in (1.102). The theorem is proved.

6.P.3 Proof of the Lemma 6.2.1

Let r represent the greatest index, other than zero, of the coefficients α_j and β_j of the equation (2.12) and let $q = n + r$. Because of (2.1) and (2.12), we have, the vector,

$$z_t = \mathrm{col}\,(y_t, y_{t-1}, \ \ldots \ y_{t-q}, u_t, \ \ldots, \ u_{t-q}), \tag{P.5}$$

which satisfies the equation

$$z_{t+1} = Dz_t + Ec(\nabla, \tau)v_{t+1}. \tag{P.6}$$

The above matrices D and E with appropriate dimensions can be written by the standard method. They are the functions of the coefficients of the equations (2.1) and (2.12). Because of the stability of the polynomial (2.13), the matrix D is stable, *i.e.*, there exist positive numbers C and ρ, $\rho < 1$ (may be dependent on τ) such that for all $t \geq 0$, the inequalities below are satisfied

$$|D^t| \leq C\rho^t. \tag{P.7}$$

We find out from (P.6)

$$z_{t+1} = D^t z_0 + \sum_{s=0}^{t-1} D^s Ec(\nabla, \tau)v_{t-s}. \tag{P.8}$$

Let Γ be an arbitrary non-negative matrix of dimensions $(q + 1)(l + m) \times (q + l)(l + m)$. The quadratic form $J_t = t^{-1} \sum_{s=1}^{t} z_s^* \Gamma z_s$ will be now examined. Introducing the following notations

$$J_t^{(1)} = t^{-1} \sum_{s=1}^{1} (D^s z_0)^* \Gamma D^s z_0,$$

$$J_t^{(2)} = t^{-1} \sum_{s=1}^{t} \left[\sum_{j=0}^{s} D^j Ec(\nabla)v_{s-j} \right]^* \Gamma \left[\sum_{j=0}^{s} D^j Ec(\nabla)v_{s-j} \right] \tag{P.9}$$

we may write for any

$$(1 - \varepsilon^{-1})J_t^{(1)} + (1 - \varepsilon)J_t^{(2)} \leq J_t \leq (1 + \varepsilon^{-1})J_t^{(1)} + (1 + \varepsilon)J_t^{(2)}. \tag{P.10}$$

Because of (P.10) we have, with probability 1,

$$\lim_{t \to \infty} J_t^{(1)} = 0 \tag{P.11}$$

and the following exists in the mean square sense,

$$\lim_{t \to \infty} J_t^{(2)} = J_*^{(2)}. \tag{P.12}$$

The value of $J_*^{(2)}$ will be computed. To do this $J_t^{(2)}$ is transformed to $J_t^{(2)} = t^{-1} \sum_{s=1}^{t} \mu_s$, where

$$\mu_t = \sum_{k=1}^{t} \sum_{k'=1}^{t} \sum_{j=0}^{n} \sum_{j'=0}^{n} c_j v_{k-j} \gamma(t, k, k') c_{j'} v_{k'-j'} \tag{P.13}$$

and $\gamma(t, k, k') = (D^{t-k}E)^* \Gamma D^{t-k'} E$. Because of (P.7) the inequality, such as, $|\gamma(t, k, k')| \leq C_\gamma \rho^{2t-k-k'}$, $k \leq t, k' \leq t$, are satisfied, where C_γ, ρ, $\rho < 1$ are some positive constants.

Considering the independence and the properties (2.3) of the random values v_t and carrying out simple but cumbersome computations for the function $R(t, \bar{t}) = \mathbf{M}(\mu_t - \mathbf{M}\mu_t)^* (\mu_t - \mathbf{M}\mu_t)$, we get the estimate $|R(t, \bar{t})| \leq C_R \rho^{2|t-\bar{t}|}$, where $C_R > 0$. This means that the conditions of the theorem 2.A.3 are fulfilled for the random quantity $(\mu_t - \mathbf{M}\mu_t)$ (and for $\rho = 0$). Hence,

$$\lim_{t\to\infty} t^{-1}\sum_{s=0}^{t-1}(\mu_s - \mathbf{M}\mu_s) = 0. \tag{P.14}$$

Carrying out the obvious transformations for μ_t, we get,

$$\mathbf{M}\mu_t = \sum_{k=1}^{t}\sum_{k'=1}^{t} \mathbf{M}(c(\nabla)v_k)^*\gamma(t,k,k')c(\nabla)v_k'$$

$$= Sp\sum_{k=1}^{t}\sum_{k'=1}^{t}\gamma^{1/2}(t,k,k')\,\mathbf{M}c(\nabla)v_k'(c(\nabla)v_k)^*\gamma^{1/2}(t,k,k')$$

$$= Sp\sum_{k=1}^{t}\sum_{k'=1}^{t}\gamma^{1/2}(t,k,k')\frac{1}{2\pi i}\oint c(\lambda)R_v c^{\nabla}(\lambda)\lambda^{k'-k}\frac{d\lambda}{\lambda}$$

$$\times\gamma^{1/2}(t,k,k') = \mathrm{Sp}\,R_v^{1/2}\frac{1}{2\pi i}\oint c^{\nabla}(\lambda)\sum_{k=1}^{t}\sum_{k'=1}^{t}\gamma(t,k,k')$$

$$\times c(\lambda)\lambda^{k-k'}\frac{d\lambda}{\lambda}R_v^{1/2} = Sp R_v^{1/2}\frac{1}{2\pi i}\oint c^{\nabla}(\lambda)\sum_{k=1}^{t}(\lambda^{-k}D^{t-k}E)^*$$

$$\times\Gamma\sum_{k'=1}^{t}\lambda^{-k'}D^{t-k'}Ec(\lambda)\frac{d\lambda}{\lambda}R_v^{1/2}$$

$$= \mathrm{Sp}R_v^{1/2}\frac{1}{2\pi i}\oint c^{\nabla}(\lambda)E^*(\lambda^{-1}I - D^t)^*(\lambda^{-1}I - D^*)^{-1}\Gamma(\lambda^{-1}I$$

$$-D)^{-1}(\lambda^{-1}I - D^t)Ec(\lambda)\frac{d\lambda}{\lambda}R_v^{1/2}. \tag{P.15}$$

The relations $\mathrm{Sp}\{a^*b\} = \mathrm{Sp}\{ba^*\}$, which are valid for right matrices a, b of identical dimension, have been made use of alongwith the formulae below

$$\mathbf{M}c(\nabla)v_s(c(\nabla)v_{s'})^* = \frac{1}{2\pi i}\oint c(\lambda)R_v c^{\nabla}(\lambda)\lambda^{-s+s'-1}\frac{d\lambda}{\lambda}$$

and $$\sum_{s=1}^{t}\lambda^{-s}D^{t-s} = (\lambda^{-1}I - D^t)(I - \lambda D)^{-1}.$$

It results from (P.15) that

$$\lim\mu_t = \mathrm{Sp}R_v^{1/2}\frac{1}{2\pi i}\oint c^{\nabla}(\lambda)\,[I - \lambda D)^{-1}E]^{\nabla}$$

$$\times\Gamma(I - \lambda D)^{-1}Ec(\lambda)\frac{d\lambda}{\lambda}R_v^{1/2}.$$

Because of (P.14) and (P.9) we have

$$\lim_{t\to\infty} J_t^{(2)} = \lim_{t\to\infty}\mathbf{M}J_t^{(2)} = \lim_{t\to\infty}\mathbf{M}\mu_t.$$

Because of (P.11) and (P.10) we get the formula :

$$\lim_{t\to\infty} t^{-1}\sum_{s=0}^{t-1} z_s^*\Gamma z_s = \lim_{t\to\infty} t^{-1}\sum_{s=0}^{t-1}\mathbf{M}z_s^*\Gamma z_s$$

$$= \mathrm{Sp}R_v^{1/2}\frac{1}{2\pi i}\oint[(I - \lambda D)^{-1}Ec(\lambda)]^{\nabla}\Gamma[(I - \lambda D)^{-1}Ec(\lambda)]\frac{d\lambda}{\lambda}R_v^{1/2}. \tag{P.16}$$

The inequality of (2.5) results from (P.16). It is not difficult to verify the equality $E^*(\lambda I - D^*)^{-1}, \Gamma(\lambda^{-1}I - D)^{-1}E = \bar{q}[W_1(\lambda), W_2(\lambda)]$, by choosing the matrix Γ in (P.16), from the condition $z^*_t\Gamma z_t = q(x_t, u_t)$. Here $W_1(\lambda)$ and $W_2(\lambda)$ are the control system ((2.1) and (2.12)) transfer functions determined by the formulae (3.52) and (3.55) of chapter 3. Because of the lemma 3.4.1, the right-hand side in (2.16) coincides with $\overline{\lim_{t\to\infty}} t^{-1} \sum_{s=1}^{t} \mathrm{M}q(x_s, u_s)$.

It shows the equality of (2.16).

Further for an arbitrary non-negative matrix Γ we have, after consideration of (P.8) and (P.7),

$$z_t^*\Gamma z_t \le 2(D^t z_0)^*\Gamma D^t z_0 + 2\sum_{s=0}^{t-1}\sum_{s'=0}^{t-1} c^*(\nabla)v_{t-s}E^*(D^*)^s$$

$$\times\Gamma D^s E c(\nabla)v_{t-s} \le C_1 + C_2\left(\sum_{s=0}^{t-1}\rho^s \mid v_{t-s}\mid\right)^2$$

$$\le C_1 + C_3\left(\sum_{s=0}^{t-1}\rho^s + \sum_{s=0}^{t-1}\rho^s \mid v_{t-s}\mid^2\right).$$

Here Cj are the constants determined. Consequently,

$$\overline{\lim_{t\to\infty}} t^{-1} z_t^*\Gamma z_t \le C_3 \overline{\lim_{t\to\infty}} t^{-1} \sum_{s=0}^{t}\rho^{t-s} \mid v_s\mid^2 . \tag{P.17}$$

It can be easily established for the function $V_t = \sum_{s=1}^{t} s^{-1}(\mid v_s\mid^2 - \mathrm{Sp}R_v)$ that $\mathrm{M}(V_{t+1}^2 \mid v^t) \le V_t^2 + (t+1)^{-2}\,\mathrm{M}[\mid v_{t+1}\mid^2 - \mathrm{Sp}R_v]^2$. Since the random quantities $\{v_t\}$ satisfy the conditions (2.3), then because of the theorem 2.A.1 it can be asserted that $\lim_{t\to\infty} V_t^2$ exists with probability 1 and it is a finite random quantity. Hence the following equality is fulfilled

$$\lim_{t\to\infty} t^{-1} \mid v_t\mid^2 = 0, \tag{P.18}$$

when (P.18) is considered in (P.17), we can establish the equality $\lim_{t\to\infty} t^{-1} z_t^*\Gamma z_t = 0$. With a special choice of Γ we get the equality in (2.14).

Lastly, we find out from (P.8)

$$\mid z_t\mid^4 = (\mid D^t z_0\mid^2 + 2(D^t z_0)^* \sum_{s=0}^{t-1} D^s E c(\nabla)v_{t-s}$$

$$+\left|\sum_{s=0}^{t-1} D^s E c(\nabla)v_{t-s}\right|^2)^2.$$

Carrying out averaging of the equality obtained and considering the properties (2.3) of the random quantity, we can verify the uniformity (along t) of the boundedness of

$M|z_t|^4$. The inequality (2.15) is established by the same process. The proof of the lemma 6.2.1 is thus completed.

6.P.4 ***Proof of the Lemma 6.2.2***

Let $y_t^{(0)}$ and $u_t^{(0)}$ denote the output and control variables of the control system (2.1) and (2.12). Let $\tilde{y}_t = y_t - y_t^{(0)}$, $\tilde{u}_t = u_t - u_t^{(0)}$. Then considering (2.1) and (2.17) we find out $a(\nabla, \tau)\tilde{y}_t = b(\nabla, \tau)\tilde{u}_t$, and $\alpha(\nabla, \tau)\tilde{u}_t = \beta(\nabla, \tau)\tilde{y}_t + w_t$.

Turning to vector $\tilde{z}_t =$ col $(\tilde{y}, \tilde{y}_{t-1}, \ldots, \tilde{y}_{t-q}, \tilde{u}_t, \ldots, \tilde{u}_{t-q})$ we get the equation

$$z_{t+1} = D\tilde{z}_t + Ew_{t+1}. \tag{P.19}$$

For the matrix D the inequality of (P.7) is valid. From (P.19) the inequality $|\tilde{z}_t|^2 \leq C_1 + C_2 \sum_{s=0}^{t-1} \rho^s |w_{t-s}|^2$ can be introduced, where C_1 and C_2 are constants. Hence,

$$t^{-1}\sum_{s=0}^{t-1} |\tilde{z}_s|^2 \leq C_1 + C_2 t^{-1} \sum_{s=0}^{t-1} |w_s|^2 \sum_{j=s}^{t} \rho^{j-s} \leq C_1 + C_3 t^{-1} \sum_{s=0}^{t-1} |w_s|^2. \tag{P.20}$$

Consequently, $\lim_{t\to\infty} t^{-1} \sum_{s=0}^{t-1} |\tilde{z}_s|^2 = 0$ (because of (2.18)). Since for an arbitrary non-negative matrix Γ and a positive number ε, the inequalities below are valid,

$$(1-\varepsilon)[z_t^{(0)}]^*\Gamma z_t^{(0)} + (1-\varepsilon^{-1})\tilde{z}_t^*\Gamma\tilde{z}_t \leq z_t^*\Gamma z_t$$

$$\leq (1+\varepsilon)[z_t^{(0)}]^*\Gamma z_t^{(0)} + (1+\varepsilon^{-1})\tilde{z}_t^*\Gamma\tilde{z}_t,$$

then taking into account (P.20) and (2.18) and with an arbitrary choice of ε, we have

$$\lim_{t\to\infty} t^{-1} \sum_{s=0}^{t-1} z_s^*\Gamma z_s = \lim_{t\to\infty} t^{-1} \sum_{s=0}^{t-1} [z_s^{(0)}]^*\Gamma z_s^{(0)}.$$

Here, $z_t^{(0)}$ is the vector (P.5) for the control system of (2.1) and (2.12) and z_t is the analogous vector for the control system of (2.1) and (2.17).

Next we consider the lemma 6.2.1 and obtain the inequality (2.5) and equality (2.16). When (2.21) is satisfied, we can find out in a similar manner

$$t^{-1}|\tilde{z}_t|^2 \leq t^{-1}C_1 + t^{-1}C_2 \sum_{s=0}^{t-1} \rho^s |\omega_{t-s}|^2.$$

From here by computations we find out the equality $\lim_{t\to\infty} t^{-1}|\tilde{z}_t|^2 = 0$. The equality of (2.14) is then elementarily established by taking the lemma 6.2.1 into account.

6.P.5 ***Proof of the Lemma 6.2.3***

The inequality in the right-hand side of equivalence relationship (2.24) means that there is no vector $f \in R^{n(l+m)}$, $f \neq 0$, such that the equalities

$$A^*f = \lambda^*f, \; B^*f = 0 \tag{P.21]}$$

are simultaneously satisfied. We assume that such a vector $f =$ col $(p_1, \ldots, p_n, r_1, \ldots, r_n)$ exists. The first equality of (P.21) (vide (2.83)) is equivalent to the relations :

$$-a_1^* p_1 + p_2 = \lambda^* p \uparrow {}^{\text{'}}02, \ldots, -a_{n-1}^* p_1 + p_n = \lambda^* p_{n-1},$$

$$-a_n^* p_1 = \lambda^* p_1,$$

$$b_2^* p_1 + r_2 = \lambda^* r_1, \ldots, b_{n-1}^* p_1 + r_n = \lambda^* r_{n-1}, b_n^* p_1 = \lambda^* r_n.$$

Taking the second formula of (P.21) into account we get, by virtue of (2.23), the equality

$r_1 = b_1^* p_1$. We re-write these relations in the form,

$$(\lambda^n)^* a^*(\lambda^{-1}) p_1 = 0, \ (\lambda^n)^* b^*(\lambda^{-1}) p_1 = 0,$$

$$i.e.\ \det \| \, |\lambda|^{2n} [a(\lambda^{-1}) a^*(\lambda^{-1}) + b(\lambda^{-1}) b^*(\lambda^{-1})] \| = 0. \tag{P.22}$$

Carrying out the reasoning in the reverse order we introduce (P.21) from (P.22). This establishes the equivalence of the inequality in (2.24).

6.P.6 *Proof of the Lemma 6.2.4*

We rewrite the equation (2.25) in the form,

$$\alpha(\nabla, \tau) u_t = \beta(\nabla, \tau) y_t + \tilde{w}_t, \tag{P.23}$$

$$\text{where} \qquad \tilde{\omega}_t = \omega_t + x_t^* [\xi(\tau) - \xi(\tau_t)]. \tag{P.24}$$

The vector z_t is determined by the formula (P.5) and ,

$$\xi(\tau) = \text{col}(-\alpha_0(\tau), \ \ldots, \ -\alpha_q(\tau), \beta_0(\tau), \ \ldots, \ \beta_q(\tau)) \tag{P.25}$$

is the set of the coefficients of the optimal regulator (2.17) $\xi(\tau_t)$ is an analogous set for the regulator (2.25). In (P.25) it is assumed that $\alpha_j(\tau) = 0$ and $\beta_j(\tau) = 0$ for $j > r$, $\alpha_0(\tau) = I_m$. The inequality of (2.5) is equivalent to $\overline{\lim_{t \to \infty}}\, t^{-1} \sum_{s=0}^{t-1} |z_s|^2 < \infty$.

Hence taking (2.27) into account we can verify the validity of the equality (2.18) for the random quantity of (P.24). Since the optimal regulator (2.17) is stabilizing, then by virtue of the lemma 6.2.2., we have the relationship in (2.29) fulfilled.

Let the condition (2.30) be satisfied. Than we can indicate an additive random quantity $\tilde{\upsilon}$ for the functions z_t, such that for all t the inequalities below are satisfied,

$$t^{-1} \sum_{s=0}^{t-1} |z_s|^2 \le \tilde{\upsilon}, \ \mathbf{M}\tilde{\upsilon} < \infty. \tag{P.26}$$

Moreover, it can be shown (by taking any $\varepsilon > 0$) that the incident Ω_ε and the instant t_* are such that for all $t \ge t_*$ the inequalities below are satisfied

$$|\xi(\tau) - \xi(\tau_t)| < \varepsilon \text{ on } \Omega_\varepsilon, \ \int_{\Omega \backslash \Omega_\varepsilon} \tilde{\upsilon} dP < \varepsilon. \tag{P.27}$$

The first inequality of (P.27) is the consequence of the strong consistency of the estimates τ_t and continuity of the function $\xi(\tau)$ on $\mathcal{T}$. Since the probability of the even Ω_ε can be chosen as close as possible to unity (because of the choice of sufficiently large t_*), then it completes the fulfilment of the second inequality in (P.27). Let $I_{\Omega\varepsilon}$ be the indicator of the even Ω_ε. Then with the help of the obvious estimates we get

$$t^{-1}\sum_{s=0}^{t-1}\mathbf{M}\,|\,\tilde{w}_s\,|^2 \le t^{-1}\sum_{s=0}^{t-1}\mathbf{M}\,|\,w_s\,|^2 + t^{-1}\sum_{s=0}^{t-1}\mathbf{M}\,|\,z_s\,|^2|\,\xi(\tau)-\xi(\tau_s)\,|^2$$

$$\le t^{-1}\sum_{s=0}^{t-1}\mathbf{M}\,|\,w_s\,|^2 + t^{-1}\sum_{s=0}^{t-1}\{\mathbf{M}\,|\,z_s\,|^2|\,\xi(\tau)-\xi(\tau_s)\,|^2 I_{\Omega_\varepsilon}$$

$$+\mathbf{M}\,|\,z_s\,|^2|\,\xi(\tau)-\xi(\tau_s)\,|^2\left(1-I_{\Omega_\varepsilon}\right)\}$$

$$\le t^{-1}\sum_{s=0}^{t-1}\mathbf{M}\,|\,w_s\,|^2 + \varepsilon\mathbf{M}\tilde{\upsilon} + \varepsilon C_{\mathcal{T}}$$

where $\quad C_{\mathcal{T}} = \sup_{\tau',\tau''\in\mathcal{T}} |\,\xi(\tau')-\xi(\tau'')\,|^2.$

The first term in the right-hand side of the inequality satisfies (2.19) (because of (2.28)). Hence,

$$\lim_{t\to\infty} t^{-1}\sum_{s=0}^{t-1}\mathbf{M}\,|\,\tilde{w}_s\,|^2 \le \varepsilon(\mathbf{M}\tilde{\upsilon}+C_{\mathcal{T}}).$$

Because of the arbitrariness of ε, it can be shown that the inequality (2.19) is fulfilled for the random quantity $\tilde{w}_t$. Hence by virtue of the lemma 6.2.2, the formula of (2.29) holds. The lemma 6.2.4 is proved.

6.P.7 ***Proof of the Lemma 6.2.5***

We rewrite the control system of (2.1) and (2.25) in the form,

$$a(\nabla,\tau_t)y_t = b(\nabla,\tau_t)u_t + c(\nabla,\tau_t)v_t + \eta_t, \tag{P.28}$$

and $\quad \alpha(\nabla,\tau_t)u_t = \beta(\nabla\,\tau_t)y_t + w_t$

Because of (P.28) the vector,

$$z_t = \mathrm{col}\,(y_{t-1},\ \ldots\ y_{t-q-1}, u_{t-1},\ \ldots,\ u_{t-q-1}), \tag{P.29}$$

satisfies the equation,

$$z_{t+1} = D(\tau_t)z_t + E(\tau_t)f_t. \tag{P.30}$$

Here q = max (n, r) and the matrices D and E are expressed in the standard form as,

$$f_t = \mathrm{col}(\eta_t + c(\nabla,\tau_t)v_t, w_t). \tag{P.31}$$

From the method for formulation of the polynomials $\alpha(\lambda,\tau_t)$ and $\beta(\lambda,\tau_t)$, it results that the matrix $D(\tau_t)$ is stable. Hence, the formula below determines a positive matrix

$$H(\tau_t) = \sum_{s=0}^{\infty}[D^*(\tau_t)]^k D^k(\tau_t).$$

For the Liapunov function $V_t = z_t^* H(\tau_{t-1})z_t$ (vide (P.30)) we have :

$$V_{t+1} = V_t + z_t^*[H(\tau_t)-H(\tau_{t-1})]\,z_t - |\,z_t\,|^2 +$$

$$+(E(\tau_t)f_t)^* H(\tau_t)E(\tau_t)f_t + 2z_t^* D^*(\tau_t)H(\tau_t)E(\tau_t)f_t. \tag{P.32}$$

Considering (2.32) and the continuity of $H(\tau)$ in the compact set we have the following relations fulfilled

$$\lim_{t\to\infty}|\,H(\tau_t)-H(\tau_{t-1})\,| = 0,$$

$$\sup_t |\,D(\tau_t)\,| < \infty,\ \sup_t |\,H(\tau_t)\,| < \infty. \tag{P.33}$$

The last inequality ensures fulfilment of the following inequalities in view of (P.31), (2.34), (P.29), (2.3) and (2.28) :

$$\overline{\lim_{t\to\infty}} t^{-1}\sum_{s=0}^{t-1} |f_s|^2 \le C_1 + \varepsilon C_2 \overline{\lim_{t\to\infty}} t^{-1}\sum_{s=0}^{t-1} |z_s|^2 . \tag{P.34}$$

Here C_1 and C_2 are finite random quantities. Taking (P.33) and (P.34) into account, we have, from (P.32),

$$\lim_{t\to\infty} t^{-1}\sum_{s=0}^{t-1} \gamma_s |z_s|^2 \le C_3 + \varepsilon C_4 \overline{\lim_{t\to\infty}} \sum_{s=0}^{t-1} |z_s|^2 .$$

Here C_3 and C_4 are finite random quantities and $\overline{\lim_{t\to\infty}} \gamma_t > 0$. Since the number ε can be chosen as small as possible, then from the inequality obtained we have the inequality, $\overline{\lim_{t\to\infty}} t^{-1}\sum_{s=0}^{t-1} |z_s|^2 < \infty$, which is equivalent to (2.5). The lemma 6.2.5 is proved.

6.P.8 ***Proof of the Theorem 6.2.1***

The proof will be given preceded by a few auxiliary assertions.

Lemma 6.P.1 *Let F_t be the* $\sigma-$algebra *formed by the random quantities v_t and w^t. Under the conditions of the theorem the random quantities satisfy the inequality*

$$\mathbf{M}\{|\tau_{t+1}-\tau|^2 | F^t\} \le |\tau_t - \tau|^2 - 2\gamma_t \mathbf{M}\{(c(\nabla,\tau)\Delta v_{t+1})^* \nabla v_{t+1} | F^t\}$$

$$+2\gamma_t^2\sigma_v^2 \mathrm{Sp}\Phi_t\Phi_t^* - \gamma_t^2(\Phi_t\Phi_t^* L_t \bar{v}_{t+1})^* v_{t+1}, \tag{P.35}$$

$$\text{where,}\quad \Delta v_{t+1} = v_{t+1} - \bar{v}_{t+1},\ L_t = (I_t - \gamma_t\Phi_t\Phi_t^*)^{-1}. \tag{P.36}$$

Proof Because of (2.40) and (2.39) and the inequality

$$|P_{\mathcal{T}}\tau' - \tau| \le |\tau' - \tau|,\ \tau \in \mathcal{T},$$

we have

$$|\tau_{t+1} - \tau|^2 \le |\tau_t - \tau|^2 + 2\gamma_t[\Phi_t(\tau_t - \tau)]^* \bar{v}_{t+1} + \gamma_t^2 |\Phi_t^* \bar{v}_{t+1}|^2 . \tag{P.37}$$

With the help of the formulae (2.36) and (2.38) the following formula is easily introduced

$$\Phi_t(\tau_t - \tau) = c(\nabla,\tau)\Delta v_{t+1} - \gamma_t\Phi_t\Phi_t^* L_t \bar{v}_{t+1}. \tag{P.38}$$

Exercise of (P.38) the inequality of (P.37) can be rewritten as

$$|\tau_{t+1} - \tau|^2 \le |\tau_t - \tau|^2 - 2\gamma_t(c(\nabla,\tau)\Delta v_{t+1})^* \Delta v_{t+1}$$

$$+2\gamma_t(c(\nabla,\tau)\Delta v_{t+1})^* v_{t+1} - \gamma_t^2(\Phi_t\Phi_t^* L_t \bar{v}_{t+1})^* \bar{v}_{t+1}. \tag{P.39}$$

Here we have used the relations, $\Phi_t\Phi_t^* L_t = \Phi_t L_t \Phi_t^* \ge \Phi_t\Phi_t^*$ which can be obtained easily from (P.36). Carrying out the conditional mean in the inequality (P.39) and considering the fact that the following relationship is fulfilled because of (2.2), (P.36), (2.3) and (2.1), we get (2.35) :

$$\mathbf{M}\{(c(\nabla,\tau)\Delta v_{t+1})^* v_{t+1} | F^t\} = \mathbf{M}\{v_{t+1} - \bar{v}_{t+1})^* v_{t+1} | F_t\}$$

$$= \sigma_v^2 - \mathbf{M}\{\bar{v}_{t+1}^* v_{t+1} | F^t\} = \mathrm{Sp}(R_v - R_v^{1/2} L_t^{-1} R_v^{1/2})$$

$$= \gamma_t \mathrm{Sp} R_v \Phi_t\Phi_t^* \le \gamma_t \sigma_v^2 \mathrm{Sp}\Phi_t\Phi_t^*.$$

Lemma 6.P.2 *There exists a random quantity R such that* $\mathbf{M}R<\infty$, *for fulfilment of the frequency condition of (2.41), and then for all t the following inequality is satisfied*

$$\rho_t = 2\sum_{k=1}^{t}([c(\nabla,\tau)-\rho I_l]\Delta v_k)^*\Delta v_k + R > 0. \tag{P.40}$$

For scalar quantities Δv_k, the proof of the lemma can be found out, for example, in ([177], lemma 2.P.2). The proof is similar* for vector Δv_k.

Lemma 6.P.3 *For the quantities*

$$\Gamma_t = |\tau_t - \tau|^2 + \rho_t\gamma_t \tag{P.41}$$

with probability 1 and in the mean square sense there exists

$$\lim_{t\to\infty}\Gamma_t = \Gamma_*,\ \mathbf{M}\Gamma_* < \infty \tag{P.42}$$

and in addition the inequalities below are satisfied

$$\sum_{t=0}^{\infty}\mathbf{M}\gamma_t|\Delta v_{t+1}|^2<\infty,\ \sum_{t=0}^{\infty}\mathbf{M}\gamma_t^2|\Phi_t^*\bar{v}_{t+1}|^2<\infty. \tag{P.43}$$

Proof By virtue of (P.35) and (P.41) we find out

$$\mathbf{M}\{\Gamma_{t+1}|F^t\} \le \Gamma_t - 2\rho\gamma_t\mathbf{M}\{|\Delta v_{t+1}|^2|F^t\}$$

$$+2\gamma_t^2\sigma_v^2\mathrm{Sp}\,\Phi_t\Phi_t^* - \gamma_t^2\mathbf{M}\{(\Phi_t\Phi_t^*L_t\bar{v}_{t+1})^*\bar{v}_{t+1}|F^t\}$$

$$\le \Gamma_t + 2\gamma_t^2\sigma_v^2\,\mathrm{Sp}\Phi_t\Phi_t^*. \tag{P.44}$$

But because of (2.39), we have,

$$\gamma_t^2\mathrm{Sp}\Phi_t\Phi_t^* \le \gamma_t\gamma_{t-1}\mathrm{Sp}\Phi_t\Phi_t^* \le \gamma_t\gamma_{t-1}(\gamma_t^{-1}-\gamma_{t-1}^{-1}) = \gamma_{t-1}-\gamma_t,$$

Hence $\sum_{t=1}^{\infty}\gamma_t^2\mathrm{Sp}\Phi_t\Phi_t^* \le \gamma_0 = 1.$

So (P.42) exists with probability 1 because of the theorem 2.A.1. The mean square convergence in (P.42) easily results from the second inequality of (P.44). Carrying out unconditional mean in the first inequality of (P.44) and taking into account of the obvious estimate $L_t \ge I_1$ we get the inequalities of (P.43).

Lemma 6.P.4 *Under the conditions of the theorem the following inequalities are satisfied*

$$\sum_{t=1}^{\infty}\mathbf{M}|\tau_{t+1}-\tau_t|^2<\infty,\ \sum_{t=0}^{\infty}\mathbf{M}\gamma_t|\Phi_t(\tau-\tau_t)|^2<\infty. \tag{P.45}$$

Proof From (2.40), the estimate $|\tau_{t+1}-\tau_t| \le \gamma_t|\Phi_t^*\bar{v}_{t+1}|$, leading to the first inequality in (P.45), is evident by virtue of (P.43). Further, from (P.38) and taking into account $\gamma_t S_p\Phi_t\Phi_t^* \le 1$ we have $|\Phi_t(\tau-\tau_t)|^2 \le 2|c(\nabla,\tau)\Delta v_{t+1}|^2 + 2\gamma_t^2|\Phi_t^*\bar{v}_{t+1}|^2$.

---

*For a scalar polynominal $c(\lambda,\tau)$, the stability results (2.41). The analystical function $c(\lambda,\tau)$ does not have specificities for $|\lambda|\le 1$ and hence $\min_{|\lambda|\le 1}\mathrm{Re}\,c(\lambda,\tau) = \min_{|\lambda|\le 1}\mathrm{Re}\,c(\lambda,\tau) \ge \rho$. Hence $\min_{|\lambda|\le 1}|c(\lambda,\tau)| \ge \rho > 0$, i.e., the polynominal $c(\lambda,\tau)$ does not have roots when $|\lambda|\le 1$.

Hence consideration of (P.43) leads to the second inequality in (P.45).

Corollary. The limiting equalities along with (2.32) are satisfied as

$$\lim_{t\to\infty}\gamma_t\sum_{s=0}^{t-1}|v_{s+1}-\bar{v}_{s+1}|^2=0,\ \lim_{t\to\infty}\gamma_t\sum_{s=0}^{t-1}|\Phi_s(\tau_s-\tau)|^2=0. \tag{P.46}$$

The equality (2.32) follows directly from (P.45). The following inequalities follow from (P.43) and (P.45) :

$$\sum_{t=0}^{\infty}\gamma_t|\Delta v_{t+1}|^2<\infty,\ \sum_{t=0}^{\infty}\gamma_t|\Phi_t(\tau-\tau_t)|^2<\infty.$$

Application of the Kronecker lemma (vide, for example, ([177], lemma 2.P.3)) leads to the equalities in (P.46).

Lemma 6.P.5 *Under the conditions of the theorem the inequality of (2.42) is satisfied.*

Proof The equation (2.1) can be written in the same form as that of the first equation of (P.28) where the quantity η_t is determined by the formula (2.34), and using the notations in (2.37) it can be presented as

$$\eta_t=\Phi_{t-1}(\tau-\tau_{t-1})+\Phi_{t-1}(\tau_{t-1}-\tau_t)$$

$$+[c(\nabla,\tau)-c(\nabla,\tau_t)]\,(v_t-\bar{v}_t). \tag{P.47}$$

By virtue of (P.45) and (P.43), and inequalities $\gamma_t\le\gamma_{t-1}\le 1$ and $\gamma_t\mathrm{Sp}\Phi_t\Phi_t^*\le 1$, the random quantity of (P.47) satisfies the inequality

$$\gamma_t\sum_{s=0}^{t-1}|\eta_{s+1}|^2=\varepsilon$$

for ε as small as possible and for all sufficiently large t. Considering (2.33) we have,

$$\gamma_t=\left(t+\mathrm{Sp}\sum_{s=0}^{t-1}\phi_s\Phi_s^*\right)^{-1}$$

and consequently for sufficiently large t the inequality below is satisfied.

$$t^{-1}\sum_{s=0}^{t-1}|\eta_{s+1}|\le\varepsilon\left(t^{-1}\sum_{s=0}^{t-1}\mathrm{Sp}\Phi_s\Phi_s^*+1\right). \tag{P.48}$$

From the notations in (2.37) we have :

$$\mathrm{Sp}\Phi_t\Phi_t^*\le C_1\left[\sum_{s=t-n}^{t}(|y_s|^2+|u_s|^2+|\bar{v}_s|^2)\right]. \tag{P.49}$$

Here C_1, is a constant, $C_1>0$. Hence the inequality (P.48) can be strengthened as

$$\overline{\lim_{t\to\infty}}\,t^{-1}\sum_{s=0}^{t-1}|\eta_{s+1}|^2\le C_2+\varepsilon C_3(\overline{\lim_{t\to\infty}}\,t^{-1}\sum_{s=0}^{t-1}|y_s|^2$$

$$+\overline{\lim_{t\to\infty}}\,t^{-1}\sum_{s=0}^{t-1}|u_s|^2+\overline{\lim_{t\to\infty}}\,t^{-1}\sum_{s=0}^{t-1}|\bar{v}_s|^2). \tag{P.50}$$

From the obvious inequality

$$t\gamma_t t^{-1}\sum_{s=1}^{t}|\bar{v}_s|^2\le 2t\gamma_t t^{-1}\sum_{s=1}^{t}|v_s|^2+2\gamma_t\sum_{s=1}^{t}|v_s-\bar{v}_s|^2$$

and from (2.3), (P.46) and (2.39) we find out :

$$\overline{\lim_{t\to\infty}}\,t^{-1}\sum_{s=1}^{t}|\bar{v}_s|^2\le 2\sigma_v^2.$$

Considering the last inequality of (P.50) we arrive at (2.33) and hence by virtue of the lemma 6.2.5, the inequalities (2.5) and (2.42) are obtained.

Lemma 6.P.6 *Under the conditions of the theorem and for any natural s the following inequality holds*

$$\sum_{t=s}^{\infty} \mathbf{M}\gamma_{t-s} \,|\, \Phi_t(\tau_{t-s}-\tau) \,|^2 < \infty. \tag{P.51}$$

Proof Let us first show that

$$\sum_{t=s}^{\infty} \mathbf{M}(\gamma_{t-s}-\gamma_t) \,|\, \Phi_t(\tau_t-\tau) \,|^2 < \infty. \tag{P.52}$$

From (2.39) we have

$$\gamma_{t-s}-\gamma_t = \gamma_{t-s}\left(s + \sum_{s'=t-s+1}^{t} \mathrm{Sp}\Phi_{s'}\Phi_{s'}^{*}\right)\gamma_t. \tag{P.53}$$

We write the control system of (2.1) and (2.25) in the form

$$a(\nabla, \tau_t)y_t = b(\nabla, \tau_t)u_t + [c(\nabla, \tau) - 1]\tilde{v}_t + \eta_t^{(1)} + v_t,$$

$$\alpha(\nabla, \tau_t)u_t = \beta(\nabla, \tau_t)y_t + w_t, \tag{P.54}$$

$$\tilde{v}_t = \eta_t^{(2)} + v_t,$$

where by virtue of (P.47), (P.46) and (P.43) the random quantities

$$\eta_t^{(1)} = \eta_t + [c(\nabla, \tau_t) - 1]\,(v_t - \tilde{v}_t),\ \eta_t^{(2)} = v_t - \tilde{v}_t \tag{P.55}$$

satisfy the inequality

$$\sum_{t=0}^{\infty} \mathbf{M}\gamma_t[\,|\,\eta_t^{(1)}\,|^2 + |\,\eta_t^{(2)}\,|^2] < \infty, \tag{P.56}$$

i.e., it is 'small' for $t \to \infty$. The vector function

$$z_t = \mathrm{col}(y_{t-1}, \ldots\, y_{t-1}, u_{t-1}, \quad \ldots,$$

$$u_{t-q-1}, \tilde{v}_{t-1}, \quad \ldots, \quad \tilde{v}_{t-q-1}) \tag{P.57}$$

satisfies the equation (vide (P.54))

$$z_{t+1} = D(\tau_t)z_t + E(\tau_t)\begin{Vmatrix} v_t \\ w_t \end{Vmatrix} + F(\tau_t)\begin{Vmatrix} \eta_t^{(1)} \\ \eta_t^{(2)} \end{Vmatrix}, \tag{P.58}$$

$$D = \left\| \begin{array}{cccccccccccc}
-a_1 & \cdots & -a_{q-1} & -a_q & b_1 & \cdots & b_{q-1} & b_q & c_1 & \cdots & c_{n-1} & c_n \\
I_l & \cdots & 0 & 0 & 0 & \cdots & 0 & 0 & 0 & \cdots & 0 & 0 \\
\vdots & \vdots & \vdots & \vdots & \vdots & \vdots & \vdots & \vdots & \vdots & \vdots & \vdots & \vdots \\
0 & \cdots & I_l & 0 & 0 & \cdots & 0 & 0 & 0 & \cdots & 0 & 0 \\
\beta_1 - \beta_0 a_1 & \cdots & & \beta_q - \beta_0 a_q & -a_1 + \beta_0 b_1 & \cdots & -a_q + \beta_0 b_q & b_q & \beta_0 c_1 & \cdots & \beta_0 & c_n \\
0 & \cdots & 0 & 0 & I_m & \cdots & 0 & 0 & 0 & \cdots & 0 & 0 \\
\vdots & \vdots & \vdots & \vdots & \vdots & \vdots & \vdots & \vdots & \vdots & \vdots & \vdots & \vdots \\
0 & \cdots & 0 & 0 & 0 & \cdots & I_m & 0 & 0 & \cdots & 0 & 0 \\
0 & \cdots & 0 & 0 & 0 & \cdots & 0 & 0 & 0 & \cdots & 0 & 0 \\
0 & \cdots & 0 & 0 & 0 & \cdots & 0 & 0 & I_l & \cdots & 0 & 0 \\
\vdots & \vdots & \vdots & \vdots & \vdots & \vdots & \vdots & \vdots & \vdots & \vdots & \vdots & \vdots \\
0 & \cdots & 0 & 0 & 0 & \cdots & 0 & 0 & 0 & \cdots & I_l & 0
\end{array} \right\|,$$

$$E^* = \left\| \begin{array}{cccccccccccc}
I_l & \cdots & 0 & 0 & \beta_0^* & \cdots & 0 & 0 & I_l & \cdots & 0 & 0 \\
0 & \cdots & 0 & 0 & I_m & \cdots & 0 & 0 & 0 & \cdots & 0 & 0
\end{array} \right\|,$$

$$F^* = \left\| \begin{array}{cccccccccccc}
I_l & \cdots & 0 & 0 & \beta_0^* & \cdots & 0 & 0 & 0 & \cdots & 0 & 0 \\
0 & \cdots & 0 & 0 & 0 & \cdots & 0 & 0 & I_l & \cdots & 0 & 0
\end{array} \right\|,$$

$q = \max(n, r)$ and it is assumed that a_j, b_j and c_j are null matrices for $j > n$, α_j and β_j are the null matrices for $j > r$.

From (P.58) we determine

$$\sum_{s'=t-s+1}^{t} |z_{s'}|^2 \le C_1\{|z_{t-s}|^2 + \sum_{s'=t-1}^{t-1} (|v_{s'}|^2 + |\omega_{s'}|^2)$$

$$+ \sum_{s'=t-s}^{t} (|\eta_{s'}^{(1)}|^2 + |\eta_{s'}^{(2)}|^2)\},$$

where C_1 is a constant. Taking (P.56) and (P.53) into account, we get

$$(\gamma_{t-s} - \gamma_t)|\Phi_t(\tau_t - \tau)|^2 \le C_1\gamma_{t-s}\gamma_t|\Phi_t(\tau_t - \tau)|^2 [s + |z_{t-s}|^2$$

$$+ \sum_{s'=t-1}^{t} (|\eta_{s'}^{(1)}|^2 + |\eta_{s'}^{(2)}|^2 + |v_{s'}|^2 + |\omega_{s'}|^2)| \le C_1\gamma_t|\Phi_t(\tau_t - \tau)|^2$$

$$+C_2(t-s)^{-2}\left|\sum_{s'=t-s}^{t} (|v_{s'}|^2 + |\omega_{s'}|^2)\right|^2$$

$$+C_1\gamma_t^2|\Phi_t|^2|\tau_t - \tau|^2|\Phi_t(\tau_t - \tau)|^2 \le C_3\gamma_t|\Phi_t(\tau_t - \tau)|^2$$

$$C_2(t-s)^{-2}\left|\sum_{s'=t-s}^{t} (|v_{s'}|^2 + |\omega_{s'}|^2)\right|^2,$$

where C_j are positive constants. Here the inequalities below obtained directly from the formula (2.39) and the set $\mathcal{T}$compactness have been taken into account :

$$\gamma_{t-s}|\Phi_{t-s}|^2 \le 1,$$

$$\gamma_t \le t^{-1}, \text{ and } \sup_{\tau', \tau'' \in \mathcal{T}} |\tau' - \tau''| < \infty.$$

Since the random quantities v_t and w_t possess bounded fourth moments, then from the inequality obtained, we get

$$\sum_{t=s}^{\infty} \mathbf{M}\gamma_{t-s}|\Phi_t(\tau_t - \tau)|^2 < \infty$$

and so considering (P.45) we arrive at (P.52). By virtue of (P.52) and (P.54) we get the inequality of (P.51) :

$$\sum_{t=s}^{\infty} \mathbf{M}\gamma_{t-s}|\Phi_t(\tau_{t-s} - \tau)|^2 \le 2\sum_{t=s}^{\infty} \mathbf{M}\gamma_{t-s}|\Phi_t(\tau_t - \tau)|^2$$

$$+2\sum_{t=s}^{\infty} \mathbf{M}\gamma_{t-s}|\Phi_t|^2|\tau_{t-s} - \tau_t|^2 \le 2\sum_{t=s}^{\infty} \mathbf{M}\gamma_{t-s}|\Phi_t(\tau_t - \tau)|^2$$

$$+2\sum_{t=s}^{\infty} \mathbf{M}\gamma_t|\Phi_t|^2|\tau_{t-s} - \tau_t|^2 \le 2\sum_{t=s}^{\infty} \mathbf{M}\gamma_{t-s}|\Phi_t(\tau - \tau_t)|^2$$

$$+2\sum_{t=s}^{\infty} \mathbf{M}|\tau_{t-s} - \tau_t|^2 < \infty.$$

Lemma 6.P.7 *Under the conditions of the theorem the pair of matrices* $\{D(\tau), E(\tau)\}$, *which can be determined by the formulae in (P.59), is controllable.*

Proof The controllability of the pair (D,E) is equivalent to (vide definition 3.4.1) the absence of y of the matrix D^* of the eigenvector f:

$$D^*f = \lambda f \tag{P.60}$$

for which

$$E^*f = 0. \tag{P.61}$$

We assume that there exists a non-zero vector

$$f = \mathrm{col}(p_1, \ \ldots, \ p_q, \ r_1, \ \ldots \ r_q, \ s_1, \ \ldots, \ s_n) \tag{P.62}$$

that satisfies (P.60) and (P.61). Because of (P.59) we get

$$r_1 = 0, \ p_1 + s_1 = 0. \tag{P.63}$$

Taking (P.63) into account, we write (P.60) as

$$-a_1^* p_1 + p_2 = \lambda p_1, \quad \ldots, \ -a_{q-1}^* p_1 + p_q = \lambda p_{q-1}, \ -a_q^* p_1 = \lambda p_q,$$

$$b_1^* p_1 + r_2 = 0, \quad \ldots, b_{q-1}^* p_1 + r_q = \lambda r_{q-1}, \ b_q^* p_1 = \lambda r_q,$$

$$c_1^* p_1 + s_2 = -\lambda p_1, \quad \ldots, c_{n-1}^* p_1 + s_n = \lambda s_{n-1}, \ c_{n-1}^* p_1 = \lambda s_n,$$

$$[\lambda^q I_l + \lambda^{q-1} a_1 + \ \ldots \ + \lambda^{q-n} a_n] p_1 = 0,$$

$$[\lambda^{q-1} b_1 + \ \ldots \ + \lambda^{q-n} b_n] p_1 = 0, \tag{P.64}$$

$$[\lambda^n I_l + \ \ldots \ + c_n] p_1 = 0.$$

By virtue of (2.43) we get from (P.64), $p_1 = 0$, and consequently $f = 0$. Thus, the pair (D, E) is controllable.

Lemma 6.P.8 *Under the conditions of the theorem, the matrices* Φ_t, *can be expressed as* $\Phi_t = \Phi'_t + \Phi''_t$, *where the matrices* Φ'_t and Φ''_t *satisfy the inequalities :*

$$\sum_{t=x}^{\infty} \mathrm{M}\gamma_t \, \mathrm{Sp}\Phi''_t(\Phi''_t)^* < \infty, \tag{P.65}$$

$$\mathrm{M}\{(\Phi'_t)^*\Phi'_t \mid F^{t-\kappa}\} \geq C_\Phi \ln(t-\kappa)^{-1} I \tag{P.66}$$

where $\kappa = 2q + n$, C_Φ *is a positive random quantity and I is the unit matrix of a proper dimension.*

Proof Let N be a natural number and $\nu = 3Nn$. For $t = \nu, \ \nu+1, \ \ldots \nu + N - 1$, we examine the equations

$$z'_{t+1} = D(\tau_\nu) z'_t + E(\tau_t) \begin{Vmatrix} v_t \\ w_t \end{Vmatrix}, \tag{P.67}$$

$$z''_t = [D(\tau_t) - D(\tau_\nu)] z''_t + F(\tau_t) \begin{Vmatrix} \eta_t^{(1)} \\ \eta_t^{(2)} \end{Vmatrix}. \tag{P.68}$$

Evidently, $z_t = z'_t + z''_t$, where z_t is the solution of the equation (P.58). From (P.68) the estimate $\sum_{t=0}^{\infty} \gamma_t |z''_t|^2 < \infty$ can be easily established for z''_t when we consider the compactness of the set $\mathcal{T}$, the inequality in (P.56) and the equality $\lim_{N\to\infty} [D(\tau_t) - D(\tau_v)] = 0$ which is fulfilled with probability 1 (vide (2.32)). Hence the equality below holds

$$\sum_{t=0}^{\infty} \mathbf{M}\gamma_t |z_t''|^2 < \infty. \tag{P.69}$$

We find out from (P.67)

$$z'_{\upsilon+\kappa} = D^{\kappa}(\tau_\upsilon) z_v' + \sum_{s=0}^{\kappa-1} D^s(\tau_\upsilon) E(\tau_{\upsilon+\kappa-s-1}) \begin{Vmatrix} v_{\upsilon+\kappa-s-1} \\ w_{\upsilon+\kappa-s-1} \end{Vmatrix}.$$

Hence by virtue of the independence in totality of the random quantities v_t and w_t, and the last condition of the theorem, the inequality below is satisfied :

$$\mathbf{M}\{z'_{\upsilon+\kappa}(z'_\upsilon + \kappa)^* | F^\upsilon\} = D^\kappa(\tau_\upsilon) z'_\upsilon (z'_\upsilon)^* [D^\kappa(\tau_\upsilon)]^*$$

$$+ \sum_{s=0}^{\kappa-1} D^s(\tau_\upsilon) \mathbf{M}\{E(\tau_{\upsilon+x-s-1}) \begin{Vmatrix} R_v & 0 \\ 0 & R_{w(t)} \end{Vmatrix} \times$$

$$\times E^*(\tau_{\upsilon+\kappa-s-1}) | F^\upsilon\} [D^s(\tau_\upsilon)]^* \geq \frac{C_{\omega,\upsilon}}{\ln(\upsilon + x)}$$

$$\times \mathbf{M}\left\{\sum_{s=0}^{\kappa-1} D^s(\tau_\upsilon) E(\tau_{\upsilon+\kappa-s-1}) E^*(\tau_{\upsilon+\kappa-s-1}) [D^s(\tau_\upsilon)]^* | F^\upsilon\right\},$$

where $C_{w,\upsilon} = \min(C_w, \lambda_{R_v})$ is the least eigenvalue of the matrix R_v. Since the pair of matrices $\{(D(\tau_\upsilon), E(\tau_\upsilon)\}$ is controllable (vide the lemma 6.P.7 and the equality of $\lim_{v\to\infty} | E(\tau_{\upsilon+\kappa}) - E(\tau_\upsilon) | = 0$), there exists a positive random quantity CDE such that

$$\mathbf{M}\left\{\sum_{s=0}^{\kappa-1} D^s(\tau_\upsilon) E(\tau_{\upsilon+\kappa-s-1}) E^*(\tau_{\upsilon+\kappa-s-1}) [D^s(\tau_\upsilon)]^* | F^\upsilon\right\} \geq C_{DE} I_\kappa$$

for sufficiently large υ. Hence for the same υ the following inequality holds with probability 1

$$\mathbf{M}\{z'_{\upsilon+\kappa}[z'_{\upsilon+\kappa}]^* | F^\upsilon\} \geq \tilde{C} I_\kappa / \ln \upsilon$$

where $\tilde{C} > 0$ is a positive random quantity. Since from the condition (5) of the theorem, the vector $\tilde{z}_t$ appears to be a part of the vector (P.57), then the following inequality holds for the same, *i.e*,

$$\mathbf{M}\{\tilde{z}'_{\upsilon+\kappa}(\tilde{z}'_{\upsilon+\kappa})^* | F^\upsilon\} \geq \tilde{C} I_\kappa / \ln \upsilon \tag{P.70}$$

where $\tilde{C} > 0$ is a positive random quantity. Because of the condition (5) of the theorem we have :

$$\mathbf{M}\{\Phi'_{\upsilon+\kappa}(\Phi'_{\upsilon+\kappa})^* | F^\upsilon\} = \sum_j Q^{(j)} \mathbf{M}\{z'_{\upsilon+\kappa}(z'_{\upsilon+\kappa})^* | F^\upsilon\} Q^{(j)}$$

$$\geq \frac{\tilde{C}}{\ln \upsilon} \sum_j Q^{(j)} [Q^{(j)}]^* \geq C_\Phi [\ln \upsilon]^{-1} I.$$

It is same as (P.66). Comparing the vector z''_t with the matrix Φ''_t, the validity of the inequality (P.65) is established. The lemma 6.P.8 is proved.

We complete the proof of the theorem. By virtue of (P.51) and (P.65), we have

$$\sum_{t=\varkappa}^{\infty} \mathbf{M}\{\gamma_{t-\kappa} |\Phi_t'(\tau_{t-\kappa}-\tau)|^2 F^{t-\kappa}\} < \infty. \tag{P.71}$$

The following relationship results from the inequality (P.66) :

$$\mathbf{M}\{\gamma_{t-\kappa} |\Phi_t'(\tau_{t-\kappa}-\tau)|^2 | F^{t-\kappa}\}$$

$$= \gamma_{t-\kappa}(\tau_{t-\kappa}-\tau)^* \mathbf{M}\{(\Phi_t')^* \Phi_t' | F^{t-\kappa}\}(\tau_{t-\kappa}-\tau)$$

$$\geq \gamma_{t-\kappa} \frac{C_\Phi}{\ln(t-\kappa)} |\tau_{t-\kappa}-\tau|^2 \geq \frac{C_\Phi |\tau_{t-\kappa}-\tau|^2}{(t-\kappa)\ln(t-\kappa)}.$$

Here C_Φ is a positive random quantity. Taking (P.71) into account, we arrive at the inequality

$$\sum_{t=\kappa}^{\infty} \frac{|\tau_{t-\kappa}-\tau|^2}{(t-\kappa)\ln(t-\kappa)} < \infty. \tag{P.72}$$

Since $\sum_{t=1}^{\infty}(t \ln t)^{-1} = \infty$ then from (P.72) follows the existence of a subsequence $\{t_k\}$, such that $\lim_{k\to\infty} \tau_{t_k} = \tau$. From the relations (P.40) and (P.46) we have $\lim_{t\to\infty} \gamma_{t-1}\rho_t = 0$.

Hence because of (P.41) and (P.42) the limit (2.27) exists with probability 1. The assertion of the theorem 6.2.4, then follows from the lemma 6.2.3.

6.P.9 *Proof of the Theorem 6.2.2*

Let $\theta = \theta(\tau)$ be the set of coefficients of optimal regulator (2.69). Then from (2.72) we get

$$|\theta_{t+1+k}-\theta|^2 \leq \theta_t - \theta|^2 - 2\gamma_t \phi_t^*(\tau_t-\tau)(y_{t+k}-y_{(t+k)*})$$

$$+\gamma_t^2 |\phi_t|^2 | y_{t+k}-y_{(t+k)*}|^2. \tag{P.73}$$

Considering (2.70) and (2.69), we get

$$y_{t+k}-y_{(t+k)*} = \phi_t^*(\theta_t-\theta) + F(\nabla, \tau)v_{t+k},$$

$$\mathbf{M}\{(y_{t+k}-y_{(t+k)*}) | v^t\} = \phi_t^*(\theta_t-\theta), \tag{P.74}$$

$$\mathbf{M}\{|y_{t+k}-y_{(t+k)*}|^2 | v^t\} = \sigma_v^2 \sum_{j=0}^{k-1} |F_j(\tau)|^2 = J_*.$$

Carrying out a conditional mean in (P.73) and considering (2.74), we get

$$\mathbf{M}\{|\theta_{t+1+k}-\theta|^2 | v^t\} \leq |\theta_t-\theta|^2 - \gamma_t |\phi_t^*(\theta_t-\theta)|^2 + \gamma_t^2 J_* |\phi_t|^2.$$

Because of (2.76)

$$\gamma_t^2 |\phi_t|^2 \leq \gamma_t\gamma_{t-1} |\phi_t|^2 \leq \gamma_t\gamma_{t-1}(\gamma_t^{-1}-\gamma_{t-1}^{-1}) = \gamma_t - \gamma_{t-1}.$$

Consequently,

$$\sum_{t=0}^{\infty} \gamma_t^2 J_* |\phi_t|^2 \leq J_* \gamma_0 = J_*.$$

In conformity with the theorem 2.A.1, the limit below exists

$$\lim_{t\to\infty} |\theta_t - \theta|^2 = \Gamma_*,\ M\Gamma_* < \infty, \tag{P.75}$$

and the inequality below holds

$$\sum_{t\to\infty}^{\infty} M\gamma_t |\phi_t^*(\theta_t - \theta)|^2 < \infty. \tag{P.76}$$

Hence as in (P.40), we have

$$\lim_{t\to\infty} \gamma_t \sum_{s=1}^{t} |\phi_s^*(\theta_s - \theta)|^2 = 0. \tag{P.77}$$

We shall show that the control strategy examined above is realizable, i.e., the inequality (2.5) holds. From the equation of (2.63), because of the stability of the polynomial $d(\lambda, \tau)b_+(\lambda, \tau)$ and (2.74), we can very easily find out the estimate

$$\overline{\lim_{t\to\infty}} t^{-1} \sum_{s=1}^{t} |u_s|^2 \leq C_4\left[1 + \overline{\lim_{t\to\infty}} t^{-1} \sum_{s=1}^{t} |y_s|^2\right]. \tag{P.78}$$

Here $C_4 > 0$ is a finite random quantity. From (P.77), (2.76) and (P.78) we have the inequality below for an arbitrary $\varepsilon > 0$ and constants C_j

$$\overline{\lim_{t\to\infty}} t^{-1} \sum_{s=1}^{t} |\phi_s^*(\theta_s - \theta)|^2 \leq \varepsilon \overline{\lim_{t\to\infty}} t^{-1} \sum_{s=1}^{t} |\phi_s|^2$$

$$\leq \varepsilon C_1\left(\overline{\lim_{t\to\infty}} t^{-1} \sum_{s=1}^{t} |y_s|^2 + \overline{\lim_{t\to\infty}} t^{-1} \sum_{s=1}^{t} |u_s|^2\right)$$

$$\leq \varepsilon C_2\left(\overline{\lim_{t\to\infty}} t^{-1} \sum_{s=1}^{t} |y_s|^2 + 1\right).$$

Taking into account the above inequality, we find out from (P.74) along with (2.75) (with constants C_3 and C_5 and as small as possible ε)

$$\overline{\lim_{t\to\infty}} t^{-1} \sum_{s=1}^{t} |y_s|^2 \leq C_* + \varepsilon C_5 \overline{\lim_{t\to\infty}} t^{-1} \sum_{s=1}^{t} |y_s|^2. \tag{P.79}$$

The inequality $\lim_{t\to\infty} t^{-1} \sum_{s=1}^{t} |y_s|^2 < \infty$ results from (P.79). This alongwith (P.78) results in fulfilment of the inequality (2.5), i.e., establishment of the realizability of the control strategy examined. Because of (2.76), we get, from (2.5) $\underline{\lim}_{t\to\infty} t\gamma_t > 0$.

When it is taken into account in (P.77), we arrive at the equality below

$$\overline{\lim_{t\to\infty}} t^{-1} \sum_{s=1}^{t} |\phi_s^*(\theta_s - \theta)|^2 = 0. \tag{P.80}$$

We introduce the inequality of (2.72) from the first equality of (P.74) taking into account (P.80). This means that the strategy examined is adaptive and hence the proof.

6.P.10 *Proof of the Theorem 6.2.3*

The proof is similar to the proof of the theorem 6.2.2. Consideration of the frequency condition (2.82) is done in the same manner as in the proof of the theorem 6.2.1. Since the theorem 6.2.3 is known (vide (213, 214) and also (177, theorem 5.1.1)), hence we confine our attention to these directions only.

6.P.11 *Proof of the Lemma 6.2.6*

The control system of (2.1) and (2.13) is written in the standard form of (P.30) for $\tau_t \equiv \tau$, $\eta_t \equiv 0$ with respect to the vector (P.29). Since the matrix $D(\tau)$ is stable, then the inequalities of (P.7) hold for it with positive numbers C and ρ, ($\rho < 1$), which may be chosen to be independent of $\mathcal{T}$(because of the compactness of the set $\mathcal{T}$). Next we find out from (P.30)

$$z_t = D^{t-t_0} z_{t_0} + \sum_{s=0}^{t-t_0-1} D^s B \begin{Vmatrix} c(\nabla, \tau) v_{t-s-1} \\ w_{t-s-1} \end{Vmatrix}, \tag{P.81}$$

where t_0 is an arbitrary instant of time. Taking into account (P.7), we get from (P.81) by obvious transformation (for $t \geq t_0$)

$$|z_t|^2 \leq C_1[\rho^{2(t-t_0)} |z_{t_0}|^2$$

$$+ \sum_{s=1}^{t-t_0} \sum_{s'=1}^{t-t_0} \rho^{s+s'} \sqrt{\sum_{j=0}^{n} |v_{t-s-j-1}|^2 + |w_{t-s-1}|^2}$$

$$\times \sqrt{\sum_{j'=0}^{n} |v_{t-s'-j'-1}|^2 + |w_{t-s'-1}|^2} \leq C_2[\rho^{2(t-t_0)} |z_{t_0}|^2$$

$$+ \sum_{s=1}^{t-t_0} \sum_{s'=1}^{t-t_0} \rho^{s+s'} \left(\sum_{j=0}^{n} |v_{t-s-j-1}|^2 + |w_{t-s-1}|^2 \right)$$

$$\leq C_3[\rho^{2(t-t_0)} |z_{t_0}|^2 + \sum_{s=1}^{t-t_0} \rho^s (\sum_{s=1}^{t-t_0} \rho^s \left(\sum_{j=0}^{n} |v_{t-s-j-1}|^2 + |w_{t-s-1}|^2 \right), \tag{P.82}$$

where C_j are constants. Taking into account :

$$\sum_{s=1}^{t-t_0} \rho^s |v_{t-s-j-1}|^2 = \sum_{s=t_0+1}^{t} \rho^{t-s} |v_{s-j-1}|^2,$$

$$\sum_{s=t_0+1}^{t} \sum_{s'=t_0+1}^{s} \rho^{s-s'} |v_{s'-j-1}|^2 = \sum_{s'=t_0+1}^{t} \sum_{s=s'}^{t} \rho^{s-s'} |v_{s'-j}|^2 =$$

$$= \sum_{s'=t_0+1}^{t} |v_{s'-j-1}|^2 \sum_{t'=0}^{t-s'} \rho^{t'} \leq \frac{1}{1-\rho} \sum_{s=t_0+1}^{t} |v_{s-j-1}|^2,$$

we can find out from (P.82)

$$\sum_{s=t_0+1}^{t} |z_s|^2 \leq C_4[|z_{t_0}|^2 + \sum_{s=t_0+1}^{t} \left(\sum_{j=0}^{n} |v_{s-j-1}|^2 + |w_{s-j}|^2 \right).$$

Hence from above (by virtue of (2.94)) we get

$$\tilde{\upsilon} = \sup_{t>0} t^{-1} \sum_{s=t_0+1}^{t} |z_s|^2 \leq C_5\left[|z_{t_0}|^2 / t_0 + n\mu\right]. \tag{P.83}$$

Here C_4 and C_5 are constants. Assuming that the inequality below holds at the instant t_0,

$$\mathbf{M}\left(|z_{t_0}|^2 / t_0\right) < \infty, \tag{P.84}$$

the summability of the random quantity $\bar{\nu}$ is verified (because of (P.83)) and (alongwith it (for a choice of $t_0 = 1$)) of the random quantity in (2.95).

6.P.12 *Proof of the Theorem 6.2.4*

The consistency of the estimates $\{\tau_t\}$ is shown in a manner similar to that in the theorem 6.2.1. The computations during the intervals of switching on of the regulator (2.25) and also those during the intervals of switching on of the regulator (2.87) are to be done directly using the relations (2.1) and (2.87) in a manner such that the equation (P.58) during these intervals possesses constant coefficients, and the matrix $D(\tau_t) = D(\tau)$ is stable, as because the regulator (2.87) is stabilizing. Hence we shall proceed from the consistency of the estimates $\{\tau_t\}$.

The inequality of (2.90) holds when the regulator of (2.25) is on. Let t_0 be the random instant of time for switching over to the regulator (2.87). Then from (P.83) we have

$$\frac{1}{t}\sum_{s=1}^{t} |z_s|^2 = \frac{t_0}{t}\sum_{s=1}^{t_0} |z_s|^2 + \frac{1}{t}\sum_{s=t_0+1}^{t} |z_s|^2 \le \frac{t}{t_0} C_z$$

$$+ C_6\left[\frac{t_0 |z_{t_0}|^2}{t\ \ t_0} + \mu\right] \le C_z + C_6[C_z + \mu] \qquad \text{(P.85)}$$

for any t so long as the regulator (2.87) is on. The switching over to the regulator (2.25) takes place only when the inequality of (2.90) is satisfied. Thus, the random quantity (2.89) is summable.

If the constant C_z is chosen to be sufficiently large, then each 'on period' of the regulator (2.87) is finite with probability 1. In fact, it follows from the lemma 6.2.1 that for an infinite period of operation of the regulator (2.25), the following inequality is satisfied

$$\overline{\lim_{t\to\infty}}\, t^{-1} |z_t|^2 \le 2(q+1)\left(\overline{\lim_{t\to\infty}}\, t^{-1}\sum_{s=1}^{t} |y_s|^2 + \overline{\lim_{t\to\infty}}\, t^{-1}\sum_{s=1}^{t} |\mu_s|^2\right)$$

$$= 2(q+1)\left(\overline{\lim_{t\to\infty}}\, t^{-1}\sum_{s=1}^{t} \mathbf{M}|y_s|^2 + \overline{\lim_{t\to\infty}}\, t^{-1}\sum_{s=1}^{t} \mathbf{M}|u_s|^2\right)$$

$$= 2(q+1)J_c < C_z.$$

Here the quality J_c is as given in the footnote in page 509. Moreover, (2.27) is satisfied and the inclusion of $\tau_t \in \mathcal{T}'$ will be also satisfied from some finite instant. (Since the set $\mathcal{T}'$ is open.) Consequently, the conditions (2.90) and in conformity with the rule for switching, the changeover to the regulator (2.25) must be realized. Thus the finiteness of each interval of starting the regulator (2.87) is established. Because of the consistency of the estimates $\{\tau_t\}$ from any $\varepsilon > 0$, there exists a random, but finite (with a probability 1) instant t_* such that $|\tau_t - \tau| < \varepsilon$ for $t > t_*$. For such t the inequality (2.90) is fulfilled as a corollary of the method of formulation of the feedback (2.25) (for $\tau_t \equiv \tau$ the regulator of (2.25) is stabilizing and optimal). Hence, the changeover to the regulator (2.87) from the instant t_* for realization of the corresponding control process is not possible. By this the finiteness of the general operation period of the stabilizing regulator is established.

We write the feedback of the control system examined in the form

$$\alpha(\nabla, \tau)u_t = \beta(\nabla, \tau)y_t + \bar{w}_t, \qquad \text{(P.86)}$$

where

$$\tilde{w}_t = w_t + \theta_t\{[\alpha(\nabla,\tau) - \alpha(\nabla,\tau_t)]y_t + [\beta(\nabla,\tau_t) - \beta(\nabla,\tau)]y_t\}$$

$$+(1-\theta_t)\{[\alpha(\nabla,\tau) - \alpha^c(\nabla)]u_t + [\beta^c(\nabla) - \beta(\nabla,\tau)]y_t\}, \quad \text{(P.87)}$$

$\theta_t = 1$ when the optimal regulator (2.25) is on, and $\theta_t = 0$ when the stabilizing regulator (2.87) is on. We get from here

$$|\tilde{w}_t|^2 \le 2\{|w_t|^2 + |z_t|^2[|l(\tau_t) - l(\tau)|^2 + (1-\theta_t)\tilde{C}]\}.$$

Here $l(\tau)$ and $l(\tau_t)$ denote the aggregate of the coefficients of respective polynomials $\alpha(\lambda,\tau)$, $\beta(\lambda\,\tau)$ and $\alpha(\lambda,\tau_t)$ $\beta(\lambda,\tau_t)$. $\tilde{C} > 0$ is a deterministic constant. Since $\tau_t \in \mathcal{T}$ and $\mathcal{T}$ is a compact set, then the random quantity $S_t = |l(\tau_t) - 1(\tau)|^2 + \tilde{C}\ (1-\theta_t) \le C_1$, where C_1 is a deterministic constant. In addition, because of the continuous dependence of the coefficients of the polynomials $\alpha(\lambda,\tau)$ and $\beta(\lambda,\tau)$ on τ, and of the fulfilment of the inequalities (2.27) (with a probability 1) and since $\lim_{t\to\infty}\theta_t = 1$ we have for the random quantities s_t, $\lim_{t\to\infty} s_t = 0$ (with a probability 1). Thus, $|w_t|^2 \le 2\ (|w_t|^2 + s_t\,|z_t|^2)$, where the random quantities s_t are bounded by the deterministic constant and $\lim_{t\to\infty} s_t = 0$. Because of the summability of the random quantities (2.89), the equality $\lim_{t\to\infty} t^{-1}\sum_{s=1}^{t}\mathbf{M}\,|\tilde{w}_p|^2 = 0$ is established for the random quantities $\tilde{w}_t$, as had been done in case of the proof of the lemma 6.2.4. Because of the lemma 6.2.2, the feedback in (P.86) is optimal in $\mathbf{U}^p$ and this completes the proof of the theorem.

6.P.13 *Proof of the Lemma 6.2.7*

Because of (2.95), the unbinding η_{t+1} (vide (2.98)) can be expressed as

$$\eta_{t+1} = \Phi_t^*(\tau - \tau_t) + c(\nabla,\tau)v_{t+1}. \quad \text{(P.88)}$$

Making use of (P.88), (2.92) and (2.98), we have from (2.97)

$$|\tau_{t+1} - \tau|^2 \le |\tau_t - \tau|^2 + 2\theta_t\frac{[\Phi_t(\tau_t - \tau)]^*\eta_{t+1}}{|\Phi_t\Phi_t^*|}$$

$$+\theta_t\frac{|\eta_{t+1}|^2}{|\Phi_t\Phi_t^*|} \le |\tau_t - \tau|^2 - \theta_t\frac{|\eta_{t+1}|^2}{|\Phi_t\Phi_t^*|} + 2\theta_t C_3\frac{|\eta_{t+1}|^2}{|\Phi_t\Phi_t^*|}$$

$$\le |\tau_t - \tau|^2 - \theta_t\frac{|\eta_{t+1}|}{|\Phi_t\Phi_t^*|}\left(2C_v + \varepsilon\sqrt{|\Phi_t\Phi_t^*|} - 2C_v\right)$$

$$\le |\tau_t - \tau|^2 - \varepsilon\theta_t\frac{|\eta_{t+1}|}{\sqrt{|\Phi_t\Phi_t^*|}} \le |\tau_t - \tau|^2 - \varepsilon^2\theta_t.$$

The inequality of (2.99) as well as the finite convergence of the algorithm ((2.99) and (2.98)) result from above.

6.P.14 ***Proof of the Theorem 6.2.5***
The auxiliary assertion will be proved first.

Lemma 6.P.9 *Let the condition* $\theta_t \equiv 0$ *(i.e.,* $\hat{\tau}_t = \tilde{\tau}_t$*) be satisfied for the control system of (2.1), (2.93) and (2.97) under the conditions of the theorem 6.2.5 and in the interval* $[t', t'']$. *Then there exist deterministic positive constants C and* ρ, $(\rho < 1)$ *such that for sufficiently small* $\varepsilon > 0$ *and for all* $t \in (t', t'')$, *the inequality below is satisfied for the vector (2.91):*

$$|z_t|^2 \le C[\rho^{t-t'}|z_{t'}|^2 + 1]. \tag{P.89}$$

Proof The equation (2.1) may be written in the form

$$a(\nabla, \tilde{\tau}_{t-1})y_t = b(\nabla, \tilde{\tau}_{t-1})u_t + \eta_t \tag{P.90}$$

where the quantity η_t has the expression as in (2.95) and it satisfies the inequality $|\eta_t| \le 2C_v + \varepsilon\sqrt{\tilde{\Phi}_t\Phi_t^*}$ (vide (2.98)) in the interval (t', t''). The control system of (2.1) and (2.95) is written in the standard form (P.30) in respect of the vector (2.91) (for $\tau_t \equiv \tilde{\tau}_{t'}$) with the vector-function

$$f_t = \mathrm{col}(\eta_{t+1}, w_{t+1}). \tag{P.91}$$

Since the matrix $D(\tau)$ is stable for $\tau \in \mathcal{T}$ then there exists a positive matrix $H = H(\tau)$ that satisfies the inequality,

$$D^*(\tau)H(\tau)D(\tau) \le \rho_0^2 H(\tau). \tag{P.92}$$

Here ρ_0^2 is a constant $(\rho_0^2 < 1)$ which can be chosen (by virtue of the compactness of $\mathcal{T}$) independent of τ. Due to the same compactness of $\mathcal{T}$, there exist constants $\lambda_{\mathcal{T}}$ and $\Delta_{\mathcal{T}}$ such that

$$0 < \lambda_{\mathcal{T}} I \le H(\tau) \le \Lambda_{\mathcal{T}} I, \tau \in \mathcal{T}. \tag{P.93}$$

Here I is the unity matrix of proper dimension. The Liapunov function $V(z, \tau) = [z^*H(\tau)z]^{1/2}$ will be then introduced. Because of (P.91) and (P.92) we have on the system trajectory (P.30) (for $t \in (t', t'')$):

$$V(z_{t+1}, \tilde{\tau}_{t'}) \le \sqrt{(Dz_t + Ef_{t+1})^*H(Dz_t + Ef_{t+1})}$$

$$\le \sqrt{z_t^*D^*HDz_t} + \sqrt{|E^*HE|(|\eta_{t+1}|^2 + |w_{t+1}|^2}$$

$$\le \rho_0 V(z_t, \tilde{\tau}_{t'}) + C_2(|\eta_{t+1}| + |w_{t+1}|).$$

Here, $C_2 = \sup_{\tau \in \mathcal{T}} |E^*(\tau)H(\tau)|^{1/2}$ and

we have considered that $D(\tilde{\tau}_t) = D(\tilde{\tau}_{t'})$ and $E(\tilde{\tau}_t) = E(\tilde{\tau}_{t'})$ during the interval (t', t'').

The following can be found out after the estimate η_{t+1} for $t \in (t', t'')$ is considered:

$$V(z_{t+1}, \tilde{\tau}_{t'}) \le \rho_0 V(z_t, \tilde{\tau}_{t'}) + \varepsilon C_2\sqrt{|\Phi_t\Phi_t^*|} + C_3$$

$$\le \rho_0 V(z_t, \tilde{\tau}_{t'}) + \varepsilon C_4|z_t| + C_3 \le \rho_0 V(z_t, \tilde{\tau}_{t'})$$

$$+ \frac{\varepsilon}{\sqrt{\lambda_{\mathcal{T}}}}V(z_t, \tilde{\tau}_{t'}) + C_3 = \left(\rho + \frac{\varepsilon C_4}{\sqrt{\lambda_{\mathcal{T}}}}\right)V(z_t, \tilde{\tau}_{t'}) + C_3. \tag{P.94}$$

Here, C_2, C_3 and C_4 are deterministic constants (here we have taken $|w_t| \le C_\omega < \infty$). Choosing ε to be small so that

$$\rho = \rho_0 + \varepsilon C_4 \lambda_T^{-1/2} < 1, \tag{P.95}$$

we find out from (P.94)

$$V(z_t, \tilde{\tau}_{t'}) \le \rho^{t-t'} V(Z_t', \tilde{\tau}_{t'}) + C_3/(1-\rho).$$

Hence using (P.93), the inequality (P.89) is obtained.

From the lemma 6.P.9, it follows that the regulator of (2.93) with finitely converging algorithm of parameter tuning ((2.97) and (2.98)) is the stabilizing for the control object (2.1). In addition to that, for sufficiently small $\varepsilon > 0$, there exists for this regulator a deterministic constant $C > 0$, such that the inequality below is satisfied :

$$\sup_t (|y_t| + |u_t|) < C. \tag{P.96}$$

Again C does not depend on the choice of the realizing disturbance and test signal (but it can depend on the choice of the control system initial conditions). In reality, for sufficiently small ε the inequality of (P.95) is satisfied, and so in each of the intervals of constancy of the parameter τ_t, the inequality (P.89) is fulfilled. Since by virtue of (2.99), the number of such intervals do not exceed the values $\varepsilon^2 |\tau - \tau_0|^{-2}$, then for a bounded set of initial values, the vectors z_t will be found out in the bounded set for all t, irrespective of the realizations of the noise and test signal. The remaining discussions are practically a repetition of the proof of the theorem 6.2.4 and its reproduction here is not essential.

6.P.15 *Proof of the Theorem 6.3.1*

The control signal u_t may be presented in the form

$$u_t = \bar{u}_t + x_t w_t, \tag{P.97}$$

where by virtue of (3.10) and (3.11) the control signals $\bar{u}_{pt+j}$ are determined, by the relations :

$$\bar{u}_{pt+j} = [1 - \alpha(\nabla, \tau_t)]\bar{u}_{pt+j} + \beta(\nabla, \tau_t) y_{pt+j},$$

$$j = 0, \ldots, p-1. \tag{P.98}$$

The formula (P.60) of section 5.P can be written in the form below, when the formula (5.6) of section 5.5 is taken into account :

$$y_{pt+k+j} = f_j(y_{pt+k-n}^{pt+k-1}, u_{pt-m}^{pt-1})$$

$$+ \sum_{l=0}^{j} \theta_l \bar{u}_{pt+j-l} + \theta_j x_{pt} \omega_{pt} + \bar{f}_j(v_{pt+k}^{pt+k+j}). \tag{P.99}$$

It results from (P.98) and (P.99) that under the conditions of the theorem 6.3.1, the inequalities (5.16) and (5.17) of section 5.5 are fulfilled for the values of $\bar{u}_{pt+j}, j = 0, \ldots, p-1$. The conditions of the theorem 5.5.1 appeared to be fulfilled by the same considerations and by the strong consistency of the estimates $\{\tau_t\}$. Considering this and following the proof of the lemma (6.2.5) (vide section 6.P.7), the realizability of the control strategy as in the theorem 6.3.1 can be easily shown. The relationship of (3.10) can be written in the form

$$\alpha(\nabla, \tau)u_t = \beta(\nabla, \tau)y_t + e_t, \tag{P.100}$$

where $e_t = \tilde{w}_t + [\alpha(\nabla, \tau) - \alpha(\nabla, \tau_s)]\, y_t + [\beta(\nabla, \tau_s) - \beta(\nabla, \tau)]\, u_t$.

Because of the consistency of the estimates $\{\tau_t\}$ and the continuous dependence of the polynomials $\alpha(\lambda, \tau)$ and $\beta(\lambda, \tau)$ on τ, and also the consideration of the realizability of the control strategy to be examined, we arrive at $\lim_{t \to \infty} e_t = 0$.

This signifies the fulfilment of the inequality (3.9) for the functional (3.7), *i.e.*, the control objective is satisfied. The theorem 6.3.1 is proved.

COMMENTS

To Chapter 1

§ 1.1 Control theory is written in greater details in [42, 44, 45, 59, 60, 74, 82, 110, 112, 124, 144, 165, 167, 169, 185].

§ 1.2, 1.3 The information presented here are sufficiently standard ones. The statement follows the work [177].

§ 1.4 The statement is from [173, 175].

To Chapter 2

§ 2.1 Dynamic programming [37] has been studied by many workers. Here, the standard form is presented.

§ 2.2 Stochastic control is reported in a large number of works and the interest in its application is constantly increasing [30, 42, 109, 114, 116, 120, 123, 126, 138, 139, 144, 145, 153, 167, 168]. The optimal control problem is not clearly formulated in this case. To be precise, the class of control where optimal control is to be located, is not indicated. Dependence of the least value of the functional on class of admissible control is discussed, for example, in [168], where references to corresponding literature can be obtained. The contents of this paragraph are close to that of the corresponding section in [168].

§ 2.3, 2.4, 2.5 Stochastic version of the dynamic programming is discussed in details in [30] (vide also [116]). Bayesian formulations related to finding the recursive relations for a *posteriori* densities, direct the beginning in the works [157, 158].

§ 2.A The standard results of the probability theory can be studied in more complete form from the works [80, 54, 105, 189]. The statement has been taken from [170, 175].

To Chapter 3

§ 3.1 In recent years the method of Liapunov functions is widely applied to the synthesis of stabilizing feedbacks [93, 77, 115, 179, 180, 182, 217, 220, 230, 19-23, 47]. Similar results, as in the theorem 3.1.3, are given in [93]. Results, close to those in theorem 1.3.4, are given in [182].

§ 3.2 The problem of analytical formulation of the regulators was studied mainly for continuous systems [101, 102, 111]. The system of solving equations of Lur'e was studied for the first time in [106] and for their connection with the Riccati equation (vide [48], page 58-60, [99]). The theorem 3.2.1 is due to S.G. Semenov and is published for the first time. The theorem 3.2.2 is apparently known (vide, for example, [165]).

§ 3.3 The trasfer function method applied to the problem of linear optimization is published, apparently, for the first time.

§ 3.4 The material in sections 3.4.2 and 3.4.3 is mostly the systematic processing of [147] and corresponding section in [177]. The results of the section 3.4.4 are generalised ones [177, 188, 173, 175] and are substantially presented in [35]. The lemma 3.4.2 and the theorems 3.4.5 and 3.4.6 are the restatement and insignificant generalisation of the corresponding results of [34] obtained for the controllability case. The transfer function method itself is a variation of the spectral method of optimal feedback synthesis [26, 100, 69]. The results in section 3.4.5 are due to S.G. Semenov and published for the first time.

§ 3.5 The minimax control for special classes of objects is given in [199, 200]. The problem is presented and studied from a different angle in [73, 84-86, 97, 98]. The unpublished resutls of A.E. Barbarinov and O.N. Granichin, are restated and supplemented in the section.

§ 3.A The proof of the frequency theorems and application of these theorems to different stabilization and optimization problems may be found in [48, 29, 197, 137, 228]. It is shown in [197] that the problem of matrix factorisation of polynomials may lead to the analogous problem of scalar polynomials. The factorisation problems of F.R.F. are discussed in details in [137] (Also vide [270, 201, 99]). Equivalent transformations of the polynomial functions are studied in [46]. The statement in section 3.A.4 is from [175].

To Chapter 4

§ 4.1 The formal determination of adaptive systems for a special class of control systems was first formulated by V.A.Yakubovich, in [193, 194] and later it was gradually generalised and improved in [195, 196, 198, 155, 156, 170, 173, 177]. The statement has been taken from [175].

§ 4.2 The results of [150] are restated and supplemented. A closely similar approach to control problems is given in [94, 95, 96].

§ 4.3 The self-tuning method was primarily developed for continuous systems [127-129, 125, 154, 78, 43, 56-58, 191, 192, 31, 32, 177, 181] and as a rule without additive disturbances. A few discrete self-tuning problems are studied in [182]. The method of 'frozen' Liapunov functions used in the section 4.3.3 have been taken from [3, 4, 1].

The method of recursive objective inequalities has developed as an independent topic in the theory of adaptive control and a large number of works is devoted to this method, namely, [1, 3, 4, 7, 15-18, 24, 40, 38, 39, 50-53, 90, 119, 121, 122, 107, 136, 149, 151, 152, 155, 160-164, 170, 173, 177, 179, 195-198]. The state of this topic is described with sufficient completeness of [177]. Hence, in the section 4.3.4 we have only mentioned this method so as to establish its relation with discrete self-tuning.

To Chapter 5

§ 5.1 A large number of works is devoted to the problem of identification and estimation. We shall mention only [233, 236, 148, 63-65, 68, 83, 103, 146, 160, 190, 203, 204, 36, 215, 221, 223, 224, 231-234, 226, 227]. Studies based on the Kalman-Bucy filter starting from the fundamental works as in [62, 218, 219] are still being intensively pursued [5, 206, 216, 41, 104, 120, 144].

The Kalman-Bucy filtration theory is the only development of the Weiner-Kolmogorov filtration theory for stationary processes [79, 237, 89, 54]. The approach explained in section 5.1.1 is the generalisation of the corresponding sections in [175]. The minimax version of the Kalman-Bucy filter is based on the work [76]. Closely similar results (for continuous time systems) are given in [97]. The lemma 5.1.3 is from [10].

§ 5.2 The problem of a parameter drift in the form presented (and also in a more generalised form) has been discussed in [70-72, 67]. The theorem 5.2.1 has been taken from [175]. The results of the section 5.2.2 are due to S.G. Semenov and V.N.Fomin. These are published for the first time (vide also [9]).

§ 5.3 The idea behind the use of forecasts for obtaining estimates of unknown parameters is well-known [203, 205, 207]. It is presented in [225, 226, 227] in a clearer and sequential form. The contents of the sections 5.3.2 and 5.3.3 happen to be the restatement of a few sections of these works.

§ 5.4 The method of converting the identification problem to a functional minimization one (for determining the mean square value of the forecast) has been discussed in works [226, 227, 235 and 203]. There are assertions in these works, which are close to the lemma 5.4.1. The theorem 5.4.1 is published for the first time.

§ 5.5 A version of the stochastic approximation method, namely, algorithm 'with noise at the input' has been proposed in [232]. Here, identifications are ensured by feeding a specially measureable noise at the input of the system. This algorithm was generalised to a feedback control system in [231]. But in all these works a *priori* stability of the control object and other fairly restraining conditions on the test signal and appropriate controls, are to be assumed. This may be satisfied only for special forms of feedback and sufficiently accurate control object parameters. Under these circumstances the possibility of suboptimal control has been studied, as because the test signal has a positive dispersion. A significant improvement on the test signal method has been obtained in [2]. Here the estimate consistency is established without any presupposition of the control object stability and there are no limits on statistical properties of the noise in the object, except the assumptions on foundedness of the second moments and uncorrelation with test signal. The extension of the method to the case of dependence of the control object paramter on a test signal has been obtained in [12]. The statement in section 5.5 has been taken from [49].

To Chapter 6

§ 6.1 The concept of dual control as introduced by A.A. Fel'dbaum, [166, 167] appears to be useful for interpretation by Bayesian formulations applied to parametric uncertainty of a control object. It has served as a powerful stimulus to the formation of adaptive systems [30, 148, 168]. Bayesian formulations are degenerated from stochastic parameter approximations and Markovian states widely used in various estimation and filtration problems [158, 159, 30, 104, 37].

The following statement made as early as 1962 [209] will be produced in connection with the above discussion : 'It appears that the systematic approach to adaptive control problem leads to a meta-problem which is not an adaptive control problem'. The interpretation of the Bayesian formulations as adaptive techniques becomes interesting only when the asymptotic properties of the estimates (which are independent of the choice of a *priori* distributions) so obtained are established. In such cases their stochastic optimal property may, however, appear to be redundant. As a corollary, the excessive computational complexity of the Bayesian formulations becomes unjustified.

The idea of adaptive control has been widely accepted by the control experts and presently the flow of work is becoming difficult to be followed. We shall point out only the work which has a bearing on the text of this book [25, 140-142, 145, 202, 186]. Kul'bak-Leibler information number is widely used in mathematical statistics [55]. The theorem 6.1.2 is the generalisation of the similar work (for observable-noise-free states of the system) carried out in [61]. The computer simulation of the example in section 6.1.4 was realised by S.V.Eliseev.

§ 6.2 As known [222], the estimates from LSM appear to be mixed when the noise is correlated. In the presence of the correlated disturbance, the extended LSM is used. In this method, side by side with the estimates of the unknown parameters, the past values of the augmented process are also estimated.

The stochastic approximation method was evolved within the structures of the mathematical statistics for determination of the root of the regression equation in terms of observations for realizations of stochastic gradient regression function (Robbins-Monro method, 1951) or of the random quantity generating a regression function (method of Kiefer-Wolfovitz, 1952). In 60's Ya.Z. Tsypkin [186] was interested in the fact that many recursive learning and adaptation algorithms including the algorithms of the potential function method [14], may be examined as an stochastic approximation method. Since then the stochastic approximation method continues to draw the attention of the specialists to the area of theoretical and technical cybernetics. The general methods for investigation of consistency of the estimates, obtained by the recursive methods, have been well-studied in [91, 92, 87, 88, 6, 27, 28, 108, 131, 133-135, 170, 177, 183, 188, 221-224]. The estimation procedures are themselves widely used as learning and adaptation algorithms [8, 66, 33, 35, 53, 75, 171, 172, 174, 178, 186, 187, 202, 203, 210, 215, 229, 235, 236, 238, 239].

The estimation schemes analogous to that used in the section 6.2.1. were studied in many works (vide, for example, [214, 222, 234]). However, the majority of the cases, the identification problem in them is soved for 'nontunable' stable feedback or under similar assumptions. The direct approach to the adaptive optimization problem has been studied in [235]. The algorithm for feedback coefficient tuning many itself include the procedure for estimation of unknown control object parameters (but not those parameters on which noise may depend). It has been suggested in [235] to choose the instrumental variable method for the latter case. Special variants of this method have been studied in [183, 184].

The adaptive control problem with the random functional type (2.6) of chapter 6 has been studied in [177]. A more precise study of the optimal problem for the functionals below has been made in chpater 5 of [177], such as :

$$J[U^{\infty}(\cdot), \tau, v^{\infty}] = \overline{\lim_{t \to \infty}} \ \frac{1}{t} \ \sum_{s=1}^{t} \mathbf{M}\{ | y_s - y_{s_*} |^2 \ \ | y^{s-k} | \}. \tag{K.1}$$

Here k is the control delay.

$$J[U^{\infty}(\cdot), \tau, v^{\infty}] = \overline{\lim_{t \to \infty}} \ \frac{1}{t} \ \sum_{s=1}^{t} \mathbf{M}\{ q(y_s, u_{s-1}) \ \ | y^{s-1}, u^{s-1} \}. \tag{K.2}$$

The dissipation problem with the control objective below has also been studied,

$$J[U^{\infty}(\cdot), \tau, v^{\infty}] = \overline{\lim_{t \to \infty}} \ \frac{1}{t} \ \sum_{s=1}^{t} \mathbf{M}\{ (| y_s |^2 + | u_{s-1} |^2) \ \ | y^{s-1}, u^{s-1} \} < \infty. \tag{K.3}$$

The functional (K.1) differs very little from the functional (2.62) presented in chapter 6 and the proof of the theorem 6.2.3 is a less significant complication of the corresponding proof in [177]. In [177], identification approach has been adopted for problems with functionals as in (K.2) and (K.3). The consistency of the estimates of the regulator coefficients has been provided with the help of the method of 'Shake up' when the movement of the object discontinues at any instant and resumes from the new initial conditions. Side by side with the method of 'Shake up' another method was suggested (for the same objectives) in [177], which was based on switching of stabilizing and optimizing regulators (vide above sections 6.2 (c) and (d)).

There are unfortunate misprints in the expressions for the formulae (K.2) and (K.3) as given in [177]. The operation $\frac{1}{t}\sum_{s=1}^{t}$ has been dropped in time averaging of the expressions. Although these prints can be detected and eliminated during analysis of the proof given in [177], its presence, side by side with loconicism of a few proofs in [177] (which is stipulated by the excess of the permissible volume of the book), undoubtedly complicates the understanding of the results of chapter 5 in [177]. These difficulties were kindly pointed out by V.G.Sragovich and the remarks put forth by him have been considered in the present work while writing section 6.2.

The statement of Section 6.2 follows the works in [5, 11, 13, 176].

§ 6.3 is based on [49].

REFERENCES

1. *Agafonov, S.A.*, Global Behaviour of the Linear Minimal-phase System with Bounded Noise during Adaptive Control. Abstract No. 4140-81, August 20, 1981, VINITI, p. 18 (in Russian).
2. *Agafonov, S.A.*, Stochastic Approximation Algorithm with Noise at the Input in an Adaptive Linear Control Problem. Abstract No. 5682-81, December 15, 1981, VINITI, p. 28 (in Russian).
3. *Agafonov, S.A.* and *Barabanov, A.E.*, Adaptive Stabilization and Linear System Tracking with Bounded Noise. Abstract No. 2841-80. July 7, 1980, VINITI, p. 11 (in Russian).
4. *Agafonov, S.A.* and *Barabanov, A.E.*, Adaptive Stabilization and Suboptimal Tracking of Linear Systems. Proceedings of the Second All Union Seminar on 'Optimization of Dynamic Systems'. Minsk, 1980, pp. 113-114 (in Russian).
5. *Agafonov, S.A.*, *Barabanov, A.E.* and *Fomin, V.N.*, Adaptive Filtration of the Random Processes : in the book – *Problems of Cybernetics*. Real life adaptive control problems. Moscow, 1982, pp. 4-31 (in Russian).
6. *Agafonov, S.A.*, *Krasulina, T.P.* and *Fomin, V.N.*, Stochastic Approximation Method in Linear Dynamic System Identification. Vestnik Leningrad University, 1981, No.1, pp. 5-10 (in Russian).
7. *Agafonov, S.A.*, *Semenov, S.G.* and *Fomin, V.N.*, Adaptive Limited Optimal Control Regulators in the Discrete Stochastic Control Problems. Proceedings of the Fifth All Union Conference on statistical methods in control processes. Moscow, 1981, pp. 124-126 (in Russian).
8. *Agafonov, S.A.*, *Semenov, S.G.* and *Fomin, V.N.*, Adaptive Limited Optimal Control of Linear Discrete Systems in the Presence of Additive Correlated Noise. Abstract No. 4828-83. August 30, 1983. VINITI, p. 26 (in Russian).
9. *Agafonov, S.A.*, *Semenov, S.G.* and *Fomin, V.N.*, Continuity of the Solutions of Nonstationary Riccati Equations in the Small. Abstract No. 4695-83, August 26, 1983, VINITI, p. 12 (in Russian).
10. *Agafonov, S.A.* and *Fomin, V.N.*, Crudeness of the Least Square Methods as Regards Non-stationary Observation-noise. Vestnik Leningrad University, 1982, No. 1, pp. 5-11 (in Russian).
11. *Agafonov, S.A.* and *Fomin, V.N.*, Adaptive Limited Optimal Control of Discrete Linear Systems under White Noise. Abstract No. 3500-80, August 7, 1980 VINITI, p. 29 (in Russian).
12. *Agafonov, S.A.* and *Fomin, V.N.*, Identification of Control Objects using Test Signals, Abstract No. 4226-82, August 3, 1982, VINITI, p. 22, (in Russian).
13. *Agafonov, S.A.* and *Fomin, V.N.*, Adaptive Limiting Optimal Control of Stochastic Systems under Correlated Noise. Automation and Remove control, 1982, No. 5, pp. 117-126 (in Russian).
14. *Aizerman, M.A.*, *Braverman, A.M.* and *Rosonoyer, L.I.*, Potential Function Method in the Theory of Learning Machines. Moscow, 1970, p. 384 (in Russian).
15. *Aksenov, G.S.* Adaptive Control Synthesis for Problems on Linear Nonstationary System Dissipativity : in the book – *Regulator Synthesis in Some Adaptive Control Problems* (Ed. Fomin, V.N.), Abstract No. 1441-77, April 15, 1977, VINITI, pp. 5-10 (in Russian).
16. *Aksenov, G.S.*, *Gusev, S.V.*, *Timofeev, A.V.*, *Fomin, V.N.* and *Yakubovitch, V.A.*, Adaptation in Mechanical systems : in the book - *Annotation of the Lectures in the Fourth All Union Conference on Theoretical and Applied Mechanics*, Kiev, 1976, p. 7 (in Russian).
17. *Aksenov, G.S.*, *Pavlov, V.A.* and *Fomin, V.N.*, Adaptive regulator synthesis in problems on stability of multivariable linear discrete control systems : in the book - *Transactions of the Ninth All Union School-seminar on Adaptive Systems*. Alma-Ata, 1979, pp. 7-11 (in Russian).
18. *Aksenov, G.S.* and *Fomin, V.N.*, On Adaptive Discrete Regulator Synthesis in Problems on Dissipativity of Multivariable Linear Systems. Abstract No. 3961-79, November 22, 1979, VINITI, p. 20 (in Russian).
19. *Aksenov, G.S.* and *Fomin, V.N.*, Liapunov Function Method in Problems on Adaptive Regulator Synthesis : in the book – *Problems in Cybernetics. Adaptive Control Systems*, Moscow, 1979, pp. 69-93 (in Russian).
20. *Aksenov, G.S.* and *Fomin, V.N.*, Design of Adaptive Control with the Aid of Degenerate Liapunov Functions. Vestnik Leningrad University, 1979, No. 19, pp. 5-8 (in Russian).
21. *Aksenov, G.S.* and *Fomin, V.N.*, Design of Adaptive Controlless using the Method of Liapunov Functions. Automation and Remote Control, 1982, No. 6, pp. 126-137 (in Russian).
22. *Aksenov, G.S.* and *Fomin, V.N.*, Liapunov Function in the Stabilizing Regulator Design Problem : in the book - *Adaptation and Learning in Control Systems and Solutions*. Novosivirsk, 1982, pp. 27-32. (in Russian).

23. *Aksenov, G.S.* and *Fomin, V.N.*, On Extension of Applicability of Liapunov Function Method to Adaptive Control Design Problem : in the book – *Computational Technique and Problems in Cybernetics*. Leningrad, 1982, No. 18, pp. 19-30 (in Russian).
24. *Aksenov, G.S. Fomin, V.N.* and *Khryaschev, S.M.*, On Linear Adaptive Control Systems : in the book - *Computational Methods*. Leningrad, Part 1, - 1973, No. 8, pp. 95-116: Part 2 - 1974, No.9, pp. 73 - 104; Part 3 - 1976, No. 10, pp. 164-175 (in Russian).
25. *Aleksandroskii, N.M., Egorov, S.V.* and *Kuzin, R.E.*, Adaptive Control Systems for Automatic Complex Technological Processes. Moscow, 1973, p. 272 (in Russian).
26. *Aliev, F.A., Larin V.B., Naumenko, K.I.* and *Sunthsev, V.N.*, Linear Time Invariant Control System Optimization, Kiev, 1978. p. 327 (in Russian).
27. *Al'ber, Ya. I* and *Shil'man, S.V.*, Non Asymptotic Estimates of the Convergence Rate of Stochastic Iterative Algorithms. Automation and Remote Control, 1981, No. 1, pp. 41-52 (in Russian).
28. *Al'ber, Ya. I* and *Shil'man, S.V.*, A Unique Approach to the Minimization Problem for Smooth and Nonsmooth Functions. Proceeding of the Academy of Sciences, USSR. Technical Cybernetics, 1982, No. 1, pp. 26-33 (in Russian).
29. *Andreev, V.A.* and *Shepeljavyi, A.I.*, Discrete Optimal Control System Design for Quadratic Functional Minimization Problems. Elektronische informations verarbeitung und Kybernetik, 1972, 8, No. 8/9, pp. 549-568 (in German).
30. *Aoki, M.*, Optimization of Stochastic Systems. Academic Press, New York, 1967.
31. *Afanas'ev, V.N.* and *Furasov, V.D.*, Design of Self-tuning Controllers in the Case of Incomplete Information about the State of the Object. Automation and Remote Control, 1976, No. 8, pp. 87-95 (in Russian).
32. *Alimov, A.A., Sydzykov, D.J.* and *Tokhtabayev, G.M.*, Nonsearching Self-tuning Identification Systems. Automation and Remote Control, 1973, No. 2, pp. 184-188 (in Russian).
33. *Barabanov, A.E.*, Adaptive Suboptimal Linear Control System with Stationary (Deterministic) Noise : in the book – *Problems in Cybernetics. Adaptive Control Systems*. Moscow, 1979, pp. 107-116 (in Russian).
34. *Barabanov, A.E.*, Optimal Linear Control Systems with Stationary Noise and Quadratic Performance Criterion. Abstract No. 3478-79, October 7, 1979, VINITI, p. 22 (in Russian).
35. *Barabanov, A.E.*, Least Square Methods in Adaptive Optimal Control Problems. Abstract No. 2842-80, July 7, 1980, VINITI, p. 23 (in Russian).
36. *Barabanov, A.E.*, Strong Convergence of the Method of Least Squares. Automation and Remote Control, 1983, No 10, pp. 119-127 (in Russian).
37. *Bellman, R.* and *Kalaba, R.*, Dynamic Programming and Modern Automatic Control. Academic Press, New York, 1965.
38. *Bondarko, V.A., Likhtarnikov, A.L.* and *Fradkov, A.L.*, Design of an Adaptive System for Stabilization of a Linear Distributed Parameter System. Automation and Remote Control, 1979, No. 12, pp. 95-103 (in Russian).
39. *Bondarko, V.A.* and *Yakubovich, V.A.*, Design of an Adaptive Control System using a Reference Model without Forced Stoppage of the Control System : in the book - *Adaptation and Learning in Control Systems and Solutions*, Novosibirsk, 1980, pp. 45-43 (in Russian).
40. *Bondarko, V.A.* and *Yakubovich, V.A.*, Recursive Objective Inequalities Method in Adaptive System Theory. Results and Problems : in the book - *Problems in Cybernetics. Problems and Adaptive Control Methods*, Moscow, 1981, pp. 19-39 (in Russian).
41. *Brammer, K.* and *Ziffling, G.*, Kalman-Bucy Filter. Moscow, 1982, p. 200 (in Russian).
42. *Bryson, A.E. Jr.*, and *Ho, Y.C.*, Applied Optimal Control. Blaisdell Publishing Co., Waltham, MA, 1969.
43. *Brusin, V.A.* Design of a Non Searching Self-tuning System using the Absolute Stability Theory Method. Automation and Remote Control, 1978, No. 7, pp. 61-67 (in Russian).
44. *Voronov, A.A.*, Fundamentals of Automatic Control Part III. Optimal, Multivariable and Adaptive Systems. Leningrad, 1970, p. 326 (in Russian).
45. *Voronov, A.A.*, Stability, Controllability and Observability. Moscow, 1979, p. 335 (in Russian).
46. *Gantmacher, F.R.*, The Theory of Matrices, 2 Vol. Chelsea Publishing Company, New York, 1959.
47. *Gelig, A.H.*, Dynamics of Impulse Systems and Neuro Networks. Leningrad, 1982, p. 192 (in Russian).
48. *Gelig, A.H., Leonov, G.A.* and *Yakubovich, V.A.*, Stability of Nonlinear Systems with Non Unique Equilibrium State. Moscow, 1978, p. 400 (in Russian).

49. *Granichin, O.N.* and *Fomin, V.N.*, Adaptive Control Design using Test Signals in the Feedback Channel. Abstract No. 1339-84, March 7, 1984, VINITI, p. 19 (in Russian).
50. *Gusev, S.V.*, Adaptive Control of Systems with Parameters Changing in an Unknown Manner : in the book - *Design of Some Adaptive Control Regulators.* (E.V.N. Fomin), Abstract No. 1441-77, April 15, 1977, VINITI, pp. 10-17 (in Russian).
51. *Gusev, S.V.*, A Criterion for Comparison of Terminal Convergent Algorithms for Solutions of Inequality Systems. Abstract No. 3781-78, July 21, 1978, VINITI, p. 15 (in Russian).
52. *Gusev, S.V.* and *Yakubovich, V.A.*, Adaptive Control Algorithm for a Manipulator. Automation and Remote Control, 1980, No. 9, pp. 101-111 (in Russian).
53. *Derevitskii, D.P.* and *Fradkov, A.L.* Applied Discrete Adaptive Control System Theory. Moscow, 1981, p. 216 (in Russian).
54. *Doob, J.L.*, Stochastic Proces. John Wiley and Sons, New York, 1953.
55. *Zacks, S.*, Theory of Statistical Inferences. Moscow, 1976, p. 621 (in Russian).
56. *Zemlyakov, S.D.* and *Rutkovskii, V. Yu.*, Generalised Adaptation Algorithms for a Class of Nonsearching Self-tuning Systems with a Model. Automation and Remote Control, 1967, No. 6, pp. 88-94 (in Russian).
57. *Zemlyakov, S.D.* and *Rutkovskii, V. Yu.*, Synthesis of Coordinate Parameter Control Systems Based on Nonsearching Selftuning Systems with a Reference Model. Transactions of the Academy of Sciences, USSR. Technical Cybernetics, 1973, pp. 168-178 (in Russian).
58. *Zemlyakov, S.D.* and *Rutkovskii, V.Yu.*, Conditions for Functioning of Multivariable Self-tuning Control Systems with a Reference Model under Continuous Presence of Parametric Noise. Lectures of the Academy of Sciences, USSR, 1978, Vol 241, No. 2, pp. 301-304 (in Russian).
59. *Zubov, V.I.*, Lecuture Notes on Control Theory. Moscow, 1975, p. 495 (in Russian).
60. *Zubov, V.I.*, Dynamics of Control Systems. Moscow, 1982 p. 285 (in Russian).
61. *Kazarinov, Yu. F.* and *Karelin, V.V.*, Adaptive Optimal Strategies in Controllable Markov Processes. Abstract No. 843-81, February 2, 1981, VINITI, p. 23 (in Russian).
62. *Kalman, R.E.* and *Bucy, R.S.*, New Results in Linear Filtering and Prediction Theory. Journal of Basic Engineering, March 1961, pp. 95-108.
63. *Kaminskas, V.*, Dynamic System Identification through Discrete Observations. Vilnius, 1982, p. 244 (in Russian).
64. *Katkovnik, V. Ya.*, Linear Estimates and Stochastic Optimization Problems. Moscow, 1976, p. 488 (in Russian).
65. *Katkovnik, V. Ya.*, and *Kul'chitskii, O.Yu.*, Linear Dynamic System Identification with Random Noise under Incomplete Observation of Vector States : in the book - *Cybernetics and Computational Techniques. Discrete Systems.* Leningrad, 1975, Vol. 28, pp. 18-22 (in Russian).
66. *Katkovnik, V. Ya.* and *Kul'chitskii, O. Yu.*, Applicability of Stochastic Approximation Methods for Adaptive Stabilization of Discrete Linear Dynamic System. Automation and Remote Control, 1976, No. 9, pp. 113-123 (in Russian).
67. *Katkovnik, V. Ya.* and *Kul'chitsku, O. Yu.*, and *Kheisin, V.E.*, Approximation of Solutions to Strongly Nonstationary Stochastic External Problems in Continuous Time. Automation and Remote Control, 1982, No.11, pp. 73-80 (in Russian).
68. *Katkovnik, V. Ya.* and *Pervozvanskii A.A.*, Extremum Search Methods for Multivariable Extremum Control Problems : in the book - *Adaptive Automatic Control Systems*, Moscow, 1972, pp. 17-42 (in Russian).
69. *Kathovnik, V.Ya.* and *Poluektov, R.A.*, Multivariable Discrete Control Systems. Moscow, 1966, p. 418 (in Russian).
70. *Katkovnik, V. Ya.* and *Kheisin, V.E.*, Iterative Algorithms of Optimization for Extremum Drift Tracking. Automation and Computational techniques, 1976, No.6, pp. 34-40 (in Russian).
71. *Katkovnik, V. Ya.* and *Kheisin, V.E.*, Dynamic Stochastic Approximation of Polynomial Drifts. Automation and Remote Control, 1979, No.5, pp. 89-98 (in Russian).
72. *Katkovnik, V. Ya.* and *Kheisin, V.E.*, Dynamic Adaptation Algorithms using Universal Residual Functions : in the book - *Problems in Cybernetics. Problems and Methods of Adaptive Control.* Moscow, 1981, pp. 39-51 (in Russian).
73. *Kats, I. Ya.* and *Kurzhanskii, A.V.*, Minimax Multistep Filtering in Statistically Uncertain Situations. Automation and Remote control, 1978, No. 11, pp. 7-9 (in Russian).
74. *Kwakernaak, H.* and *Sivan, R.*, Linear Optimal Control systems. Wiley Interscience, New York, 1972.

75. *Kel'mans, G.K.*, *Poznyak, A.S.* and *Chernitser, A.V.*, 'Local' Optimization Algorithms in Asymptotic Control of Nonlinear Dynamic Systems. Automation and Remote Control, 1977, No. 11, pp. 73-88 (in Russian).
76. *Kirichenko, N.F.* and *Nakonechnii, A.G.*, Minimax Approach to Linear Dynamic System State Estimation. Cybernetis, 1977, No. 4, pp. 52-55 (in Russian).
77. *Klimentov, S.I.* and *Prokopov, V.I.*, Design of an Asymptotic State Algorithm for Adaptive Systems with Reference Model using Direct Liapunov Method. Automation and Remote Control, 1977, No. 10, pp 97-104 (in Russian).
78. *Kozlov, Yu. M.* and *Yusupov, R.M.*, Non Searching Self-tuning Systems. Moscow, 1969, p. 455 (in Russian).
79. *Kolmogorov, A.N.*, Interpolation and Extrapolation of Random Sequences. Transactions of the Academy of Sciences, USSR Mathematics, 1941, No. 5, pp. 3-14 (in Russian).
80. *Kolmogorov, A.N.* Fundamentals of Probability Theory. Moscow, 1974, p. 119 (in Russian).
81. *Cramer, G.* and *Leadbetter, M.*, Stationary and Related Stochastic processes. Wiley, New York, 1967.
82. *Krasovskii A.A.*, Automatic Flight Control Systems and their Analytical Design. Moscow, 1973, p. 560 (in Russian).
83. *Krasovskii A.A.*, Optimum Algorithms in Identification Problems with an Adaptive Model. Automation and Remote Control, 1976, No. 12, pp. 75-82 (in Russian).
84. *Krasovskii N.N.*, Motion Control Theory, Moscow, 1968, p. 476 (in Russian).
85. *Krasovskii N.N.*, Control Problem with Incomplete Information. Transactions of the Academy of Sciences, U.S.S.R. Technical Cybernetics, 1976, No. 2, pp. 3-7 (in Russian).
86. *Krasovskii N.N.*, On Control with Incomplete Information. Applied Mathematics and Mechanics, 1976, Vol. 40, No.2, pp. 197-206 (in Russian).
87. *Krasulina, T.P.*, Application of Stochastic Approximation Algorithms to Automatic Control Problems. Automation and Remote control 1969, No. 5, pp. 104-107 (in Russian).
88. *Krasulina, T.P.*, Stochastic Approximation. Automation and Remote Control 1980, No. 12, pp. 72-75 (in Russian).
89. *Krein, M.G.*, On Basic Approximation Problem of Theory of Stationary Random Process Extrapolation and Filtering Lectures of the Academy of Sciences, U.S.S.R., 1954, Vol. 94, pp. 13-16 (in Russian).
90. *Kulinich, A.S.* and *Penev, G.D.*, Multi-variable Adaptive Control Algorithms. Vladivostok 1978, p. 21 (in Russian).
91. *Kul'chitskii, O. Yu.*, Sufficient Conditions for Convergence of Stochastic Approximation Algorithms for Random Processes with Continuous Time. Cybernetics, 1979, No.6, pp. 114-126 (in Russian).
92. *Kul'chitskii, O. Yu.*, Stochastic Approximation Algorithms Operating in the Adaptation Loop of a Discrete Stochastic Linear Dynamic System. Automation and Remote Control, Part I, 1983, No. 9, p. 102-118; Part II, 1984, No.3, pp. 104-113 (in Russian).
93. *Kuntsevitch, V.M.* and *Lychak, M.M.*, Automatic Control System Design using Liapunov Functions, Moscow, 1977, p. 400 (in Russian).
94. *Kuntsevitch, V.M.*, Optimal Control of Discrete Dynamic Systems with Unknown Nonstationary Parameters. Automation and Remote Control 1980, No. 2, pp. 79-88 (in Russian).
95. *Kuntsevitch, V.M.* and *Lychak, M.M.*, Optimal and Adaptive Control of Dynamic Objects Under Conditions of Uncertainty. Automation and Remote Control 1979, No. 1, pp. 79-83 (in Russian).
96. *Kuntsevitch, V.M.* and *Lychak, M.M.*, External Control Game Systems. Automation, 1982, No. 6, pp. 74-86 (in Russian).
97. *Kurzhanszkii, A.B.*, Control and Observation under Conditions of Uncertainity. Moscow, 1977, p. 392 (in Russian).
98. *Kurzhanszkii, A.B.*, Dynamic Problems of Acceptance of Solutions under Conditions of Uncertainty in the book - *Recent State of the Theory of Investigation of Operations*. Moscow, 1979, pp. 197-235 (in Russian).
99. *Larin, V.B.*, Techniques of Solution of Algebraic Riccati Equations. Transactions of the Academy of Sciences, U.S.S.R. Technical Cybernetics, 1983, No. 2, pp. 186-199 (in Russian).
100. *Larin, V.B.*, *Naumenko, K.I.* and *Synthsev, V.N.*, Spectral Methods of Linear Feedback System Design. Kiev, 1971, pp. 138 (in Russian).
101. *Letov, A.M.*, Analytical Design of Regulators. Automation and Remote Control, 1960, No. 6, pp. 5-14 (in Russian).
102. *Letov, A.M.*, Flight Dynamics and Control. Moscow, 1969, p. 360 (in Russian).

103. *Lee, R.*, Optimal Estimates, Determination of Characteristic and Control. Moscow, 1966, p. 176 (in Russian).
104. *Liptser, R. Sh.* and *Shiryaev, A.N.*, Statistics of Random Processes. Moscow, 1974, p. 696 (in Russian).
105. *Loeve, M.*, Probabaility Theory. Van Nostrand Reinhold, 1960.
106. *Lure', A.I.*, Some Nonlinear Automatic Control Problems. Moscow, 1951, p. 216 (in Russian).
107. *Malyshev, V.A.* and *Timofeev, A.V.*, Manipulator Dynamics and Adaptive Control Automation and Remote Control, 1981, No. 8, pp. 90-88 (in Russian).
108. *Medvedev, G.A.*, Adaptive Estimation based on Realisations of Stochastic Processes and Fields. Automation and Remote control, 1978, No. 10, pp. 87-94 (in Russian).
109. *Medvedev, G.A.*, and *Tarasenko, V.P.*, Probability Methods for Extremal Systems. Moscow, 1967, p. 456 (in Russian).
110. *Merriam III, C.W.*, Optimization Theory and the Design of Feedback Control Systems. Mcgraw-Hill Co., 1964.
111. *Moiseev, N.N.*, Infinite Time Period Optimal Control Theory. Computational Mathematics and Mathematical Physics, 1974, Vol. 14, No. 4, pp. 851-861 (in Russian).
112. *Neimark, Yu. I.*, Dynamic Systems and Controllable Processes. Moscow, 1978,pp. 333 (in Russian).
113. *Neimark, Yu. I* and *Fufaev, N.A.*, Dynamics of Unbarenamed Systems. Moscow, 1967, p. 519 (in Russian).
114. *Nemirovskii, A.S.* and *Yudin, D.V.*, Complexity and Effectivity of the Optimization Methods. Moscow, 1979, p. 383 (in Russian).
115. *Nosov, V.R.*, Investigation of the Stability and the Transient Quality in Adaptive Systems with a Reference Model : in the book - *International Konferenz Nichtlinear Schwingungen.* Band II, No.2, Berlin, 1977, pp. 173-182 (in German).
116. *Aström, M.*, Optimization of Stochastic Systems, Academic Press, New York, 1970.
117. *Pavlov, V.A.* and *Fomin, V.N.*, Design of Optimal Regulators by Spectral Methods : in the book - *Regulator Design in Some Adaptive Control Problems* (Ed.) Fomin, V.N. Abstract No. 1441-77, April 15, 1977, VINITI, pp. 44-51 (in Russian).
118. *Pavlov, V.A.* and *Fomin, V.N.*, Sensitivity of Optimal Control Systems to Small Changes in Regulator Coefficients. Vestnik Leningrad University, 1978, No.1, pp. 67-71 (in Russian).
119. *Pavlov, V.A.*, *Fomin, V.N.*, and *Khryaschev, S.M.*, Adaptive Control of Linear Difference Equation under Uncontrollable Gaussian Noise. Abstract No. 2027-76, May 28, 1976, VINITI, p. 16 (in Russian).
120. *Paraev. Yu. I.*, Introduction to statistical dynamics of Controlled Processes and Filtering. Moscow, 1976, p. 184 (in Russian).
121. *Penev, G.D.*, A Few Problems on Adaptive Control Design for Dynamic Objects : in the book - *Computational Methods, Series 9*, Leningrad, 1974, pp. 105-115 (in Russian).
122. *Penev, G.D.*, Parametric Identification Algorithms for Completely Controllable and Observable Linear Systems. Abstract No. 2082-79, April 17, 1979, VINITI, p. 25 (in Russian).
123. *Pervozvanskii, A.A.*, Adoption of Markov Chain to Computation of Set Error in External Controllers. Transactions of the Academy of Sciences, USSR. Energy and Automation, 1960, No. 3, pp. 64-72 (in Russian).
124. *Pervozvanskii, A.A.* and *Gaitsgori, V.G.*, Decomposition, Aggregate and Approximation of Optimization. Moscow, 1979, p. 344 (in Russian).
125. *Petrov, A.I.*, An Optimal Self-tuning Algorithm Problem. Transactions of the Academy of Sciences, USSR. Technical Cybernetics. 1971, No. 1, pp. 206-217 (in Russian).
126. *Petrov, A.I.*, Statistical Design of Model Reference Adaptive Terminal Control Systems. Lectures of the Academy of Sciences, USSR, 1978, Vol. 242, No. 2, pp. 298-301 (in Russian).
127. *Petrov, B.N.*, *Rutkovskii, V. Yu.* and *Zemlyakov, S.D.*, Adaptive Coordinate-parameter Control of Nonstationary Objects. Moscow, 1980, p. 244 (in Russian).
128. *Petrov, B.N.*, *Rutkovskii, V. Yu.*, *Zemlyakov, S.D.*, *Krutova, I.N.* and *Yadykin, I.B.*, On the Theory of the Non Searching Selftuning Systems, Transactions of the Academy of Sciences, USSR. Technical Cybernetics, 1976, No.2. p. 154-162; 1976, No. 3, pp. 142-154 (in Russian).
129. *Petrov, B.N.*, *Rutkovskii, V.Yu.*, *Krutova, I.N.* and *Zemlyakov, S.D.*, Principles of Design and Planning of Self-tuning Systems. Moscow, 1972, p. 260 (in Russian).
130. *Petrov, Yu. P.*, Variational Methods for Optimal Control Theory. Leningrad, 1977, p. 280 (in Russian).
131. *Poznyak, A.S.*, Convergence of Stochastic Approximation Algorithms in Parameter Identification of Dynamic Plants. Automation and Remote Control 1979, No. 8, pp. 186-190 (in Russian).

132. *Polyak, B.T.*, Introduction to Optimization. Moscow, 1983, p. 384 (in Russian).
133. *Polyak, B.T.* and *Tsypkin, Ya. Z.*, Pseudogradient Algorithms for Adaptation and Bearing. Automation and Remote Control, 1973, No. 3, pp 45-69 (in Russian).
134. *Polyak, B.T.* and *Tsypkin, Ya. Z.*, Stable Control under Incomplete Information : in the book - *Problems in Cybernetics. Adaptive Control Systems*. Moscow, 1977, pp. 6-15 (in Russian).
135. *Polyak, B.T.* and *Tsypkin, Ya. Z.*, Adaptive Estimation Algorithms (Convergence, Optimality Stability). Automation and Remote Control, 1979, No. 3, pp. 71-84 (in Russian).
136. *Ponomarenko, V.I.* and *Yakubovich, V.A.*, Recursive Objective Inequality Method in Problems Sub-optimal Adaptive Control of Dynamic Objects : in the book - *Problems in Cybernetics. Adaptive Control Systems*. Moscow, 1977, pp. 16-28 (in Russian).
137. *Popov, V.M.*, Hyperstability of Automatic Systems. Moscow, 1970, p. 189 (in Russian).
138. *Pugachev, V.S.*, Statistical Methods in Technical Cybernetics, Moscow, 1971, p. 189 (in Russian).
139. *Pugachev, V.S.*, *Kazakov, I.E.* and *Evlanov, P.G.*, Fundamentals of Statistical Theory of Automatic Systems. Moscow, 1974, p. 400 (in Russian).
140. *Raibman, N.S.* and *Chadeev, V.M.*, Adaptive Models in Control Systems. Moscow, 1966, p. 159 (in Russian).
141. *Raibman, N.S.* and *Chadeev, V.M.*, Production Process Modelling, Moscow, 1975, p. 374 (in Russian).
142. *Rastryagin, L.A.*, Adaptation in Complex Systems. Riga, 1981, p. 376 (in Russian).
143. *Rastryagin, L.A.*, External Control Systems. Moscow, 1974, p. 630 (in Russian).
144. *Roitenberg, Ya. N.*, Automatic Control. Moscow, 1978, p. 552 (in Russian).
145. *Saridis, J.*, Self-organising Stochastic Control Systems. Moscow, 1980, p. 400 (in Russian).
146. *Sage, A.P.* and *Melsa, J.L.*, Systems Identification. Academic Press, New York, 1971.
147. *Semenov, S.G.* and *Fomin, V.N.*, On linearity of the Optimal Control of Linear Discrete Objects having Stationary Noise. Vestnik Leningrad University, 1981, No. 19, pp. 59-65 (in Russian).
148. Trends and Progress in System Identification (Ed.) *Eykhoff, P.*, Pargamon, Oxford, 1981.
149. *Sokolov, B.M.* and *Fomin, V.N.*, Adaptive Control Design for the Problem of Stabilization of the Temperature of Synthetic Rubber Polymerization. Vestnik Leningrad University, 1983, No. 1, pp. 59-63 (in Russian).
150. *Sokolov, V.F.*, Recursive Procedure for Adaptive Optimal Control with Minimax Approach : in the book - *Regulated Sets and Operator Equations. Syktyvkar*, 1982, pp. 63-70 (in Russian).
151. *Sokolov, V.F.* and *Fomin, V.N.*, Adaptive control of Linear Non-minimal Phase Systems. Abstract No. 1864-76, May 17, 1976, VINITI, p. 16 (in Russian).
152. *Sokolov, V.F.* and *Fomin, V.N.*, Adaptive Stabilization of Linear Systems using Variable Strucutre Regulators. Abstract No. 1985-77, May 12, VINITI, p. 19.
153. *Solodovnikov, V.V.*, Introduction to Statistical Dynamics of Automatic Control Systems. Moscow, Leningrad, 1952, p. 367 (in Russian).
154. *Solodovnikov, V.V.* and *Shramko, L.S.*, Design and Planning of Analytic Self-tuning Systems with Reference Models. Moscow. 1972, p. 270 (in Russian).
155. *Sragovich, V.G.*, Adaptive Control, Moscow, 1981, p. 384 (in Russian).
156. *Sragovich, V.G.*, Adaptive System Theory. Moscow, 1976, p. 319 (in Russian).
157. *Stratonovich, R.L.*, Conditional Markov Processes – Probability Theory and its Application, 1960, Vol. 5, Series 2, pp. 172-193 (in Russian).
158. *Stratonovich, R.L.*, Conditional Markov Processes and their Application to Optimal Control Theory. Moscow, 1966, p. 319 (in Russian).
159. *Tertychnii, V. Yu.* and *Fomin, V.N.*, An Adaptive Control Algorithm not Requiring Computation of Higher Derivatives. Abstract No. 3504-80, August 7, 1980, VINITI, p. 14 (in Russian).
160. *Timofeev, A.V.*, Recursive Terminal Convergent Algorithms for Adaptive Identification of Discrete Dynamic Systems. The transactions of the Academy of Sciences, U.S.S.R., Technical Cybernetics, 1973, No. 5, pp. 206-213 (in Russian).
161. *Timofeev, A.V.*, Principles and Algorithms for Design of an Adaptive Control for Robots : in the book - *Robotechnics*, Leningrad, 1977, pp. 35-43 (in Russian).
162. *Timofeev, A.V.*, Robots and Artificial Intelligence. Moscow, 1978, p. 191 (in Russian).
163. *Timofeev, A.V.*, Design of Adaptive Control Systems for a Programmed Motion. Leningrad, 1980, p. 85 (in Russian).
164. *Timofeev, A.V.*, Parametric Optimization of Programmed Motion and Adaptive Terminal Control. Lectures of the Academy of Sciences, U.S.S.R., 1981, Vol.256, No.2, pp. 310-313 (in Russian).
165. *Wonham, W.M.*, Linear Multivaribale Control : a Geometric Approach. Springer, New York, 1979.

166. *Fel'dbaum, A.A.*, Dual Control Problems in the book - *Dynamic System Optimization Methods*. Moscow, 1972, pp. 89-108 (in Russian).
167. *Fel'dbaum, A.A.*, Fundamentals of Optimal Control Systems, Moscow, 1966, p. 623 (in Russian).
168. -------------------, Filtering and Stochastic Control of Dynamic Systems. (Ed.) *Leondes, C.T.*, Moscow, 1980, p. 407 (in Russian).
169. *Fleming, U.* and *Raechelle R.*, Optimal Control of Deterministic and Stochastic System. Moscow, 1978, p. 316 (in Russian).
170. *Fomin, V.N.*, Mathematical Theory of Learning Pattern Recognition Systems. Leningrad, 1976, p. 235 (in Russian).
171. *Fomin, V.N.*, Recursive Estimation of Adaptive Control Synthesis. Abstract No. 3868-76, October 22, 1976, VINITI, p. 20, (in Russian).
172. *Fomin, V.N.*, On Discrete Control System Stabilization. Abstract No. 2506-78, July 19, 1978, VINITI, p. 25 (in Russian).
173. *Fomin, V.N.*, Adaptive Controller Synthesis for Linear Discrete Systems. Abstract No. 2202-79, July 19, 1979. VINITI, p. 65 (in Russian).
174. *Fomin, V.N.*, Adaptive Limited Optimal Control Design for Linear Stochastic Systems : in the book - *Questions of Cybernetics. Problems and Methods of Adaptive Control*. Moscow, 1981, pp. 62-65 (in Russian).
175. *Fomin, V.N.*, Recursive Estimation and Adaptive Filtering. Moscow, 1984, p. 278 (in Russian).
176. *Fomin, V.N.*, Adaptive Limited Optimal Control of Linear Discrete Stochastic Systems. Part 1, Abstract No. 4909-83, August 31, 1983, VINITI, p. 32; part 2, No. 2101-84, April 6, VINITI, p. 51 (in Russian).
177. *Fomin, V.N.*, *Fradkov, A.L.* and *Yakubovich, V.A.*, Adative Control of Dynamic Plants. Moscow, 1981, p. 448.
178. *Fomin, V.N.* and *Khryaschev, S.M.*, An Adaptive Control Problems for a Linear System Subjected to Random Noises. Automation and Remote Control. 1976, No. 10, pp. 109-117 (in Russian).
179. *Fradkov, A.L.*, Adaptive System Design for Linear Dynamic System Stabilization. Automation and Remote Control, 1974, No. 12, pp. 96-103 (in Russian).
180. *Fradkov, A.L.*, Quadratic Liapunov Functions in the Problem of Adaptive Stabilization of a Dynamic Plant. Siberian Mathematical Journal, 1976, No. 2, pp. 436-445 (in Russian).
181. *Fradkov, A.L.*, Dynamic Gradient Scheme and its Application to Adaptive Control Problems. Automation and Remote Control, 1979, No. 9, pp. 90-101 (in Russian).
182. *Furasov, V.D.*, Stability and Stabilization of Discrete Systems. Moscow, 1982, p. 192 (in Russian).
183. *Khryaschev, S.M.*, Estimation of the Parameters of the Linear Systems in the Presence of Correlated Noise. Abstract No. 3368-77, August 28, 1977, VINITI, p. 17 (in Russian).
184. *Khryaschev, S.M.*, Consistency of Estimates of the Matrix of Coefficients of a Linear System with Correlated Noise. Automation and Remote Control. 1982, No. 8, pp. 68-75 (in Russian).
185. *Tsypkin, Ya. Z.*, Theory of Sampled Data Systems. Moscow, 1963, p. 968 (in Russian).
186. *Tsypkin, Ya. Z.*, Self Adaptive and Learning Systems. Moscow, 1968, p. 399 (in Russian).
187. *Tsypkin, Ya. Z.*, Adaptive Optimization Algorithms when there is a *priori* Uncertainty. Automation and Remote Control, 1979, No. 6, pp. 94-108 (in Russian).
188. *Shil'man, S.V.*, Iterative Methods of Adaptive Filter Design with Finite Memory. Transactions of the Academy of Sciences, U.S.S.R. Technical Cybernetics. 1982, No. 4, pp. 3-10, (in Russian).
189. *Shiryaev, A.N.*, Probability, Moscow, 1980, p. 574 (in Russian).
190. *Eykhoff, P.*, System Identification. John Wiley and Sons, London, 1974.
191. *Yadykin, I.B.*, Stochastic Stability of a Class of Model Reference Non-seeking Adaptive Control Systems. Automation and Remote Control. 1981, No. 3, pp. 56-67 (in Russian).
192. *Yadykin, I.B.*, A Property of Adaptability of a Regulator in Adaptive Control Systems. Lectures of the Academy of Sciences, 1981, Vol. 259, No.2, pp. 310-313 (in Russian).
193. *Yakubovich, V.A.*, Theory of Adaptive Systems. Lectures of the Academy of Sciences, USSR, 1968, Vol. 182, No. 3, pp. 518-521 (in Russian).
194. *Yakubovich, V.A.*, A Problem of Self-learning Towards Desirable Behaviour. Automation and Remote Control. 1969, No. 8, pp. 119-139 (in Russian).
195. *Yakubovich, V.A.*, Finitely Convergent Algorithms for Solution of Accounting Systems of Inequality and their Application to Adaptive System Problems. Lectures of the Academy of Sciences, U.S.S.R., 1969, Vol. 189, No. 3, pp. 495-498 (in Russian).

196. *Yakubovich, V.A.*, A Method of Design of an Adaptive Control of a Linear Dynamic Plant under the Conditions of Large Uncertainty : in the book - *Questions of Cybernetics. Adaptive Systems*, Moscow, 1974, pp. 46-61 (in Russian).
197. *Yakubovich, V.A.*, Frequency Theorem in Control Theory, Siberian Mathematics Journal, 1973, Vol. 14, No. 2, pp. 384-420 (in Russian).
198. *Yakubovich, V.A.*, Adaptive Suboptimal Control of Linear Dynamic Systems in the Presence of Control Delay. Cybernetics. 1976, No. 1, pp. 26-43 (in Russian).
199. *Yakubovich, V.A.*, A Solution to a Problem of Optimal Discrete Linear Control System. Automation and Remote Control. 1975, No. 9, pp. 73-79 (in Russian).
200. *Yakubovich, V.A.*, Optimal Control of a Linear Discrete Systems in the Presence of Non-measurable Disturbances. Automation and Remote Control, 1977, No. 4, pp. 49-54 (in Russian).
201. *Anderson, B.D.O.*, *Hitz, K.L.* and *Diem, N.D.*, Recursive Algorithm for Spectral Factorization. IEEE Transactions on Circuits and Systems, 1974, Vol.CS-6, No.6, pp. 742-750.
202. *Aström, K.J.*, *Borrisson, U*, *Ljung, L.* and *Wittenmark, B.*, Theory and Applications of Self-tuning Regulators. Automatica, 1977, Vol. 13, No.5, pp. 457-476.
203. *Aström, K.J.* and *Söderesröm,, T.*, Uniqueness of the Maximum Likelihood Estimates of the Parameters of an ARMA Model. IEEE Transactions and Automatic Control, 1974, Vol. AC-19, No.6, pp. 769-773.
204. *Bucy, R.S.* and *Joseph, P.D.*, Filtering for Stochastic Processes with Application to Guidance. New York, London, 1968, p. 195.
205. *Caines P.E.*, Prediction error identification methods for stationary stochastic processes. IEEE Trans. Aut. Contr., 1976, Vol. AC-21, N 4, pp.500-506.
206. *Deutsch R.*, Estimation Theory. New York, 1965, p. 269.
207. *Dugard L.*, *Landau I.D.* Recursive Output Error Identification Algorithms : Theory and Evaluation- –Automatica, 1980, Vol. 16, N 5, pp. 443–462.
208. *Egardt B.*, Stability Analysis of Discrete-time Adaptive Control Schemes–IEEE Trans. Aut. Contr., 1980, Vol. AC-25, N 4, pp. 710-716.
209. *Florentin J.J.*, Optimal, Probing, Adaptive Control of a Simple Bayesian System – J. Electr. Contr., 1962, N 11, pp. 165–177.
210. *Fuchs J.*, Discrete Adaptive Control : A Sufficient Condition for Stability and Applications. – IEEE Trans. Aut. Contr., 1980, Vol. AC-25, N 3, pp. 940–946.
211. *Goodwin G.C.*, *Ramage P.J.*, *Caines P.E.*, Discrete-time Multivariable Adaptive Control. – IEEE Trans. Aut. Contr;l 1980, Vol. AC-25, N 3, pp. 449–456.
212. *Goodwin G.C.*, *Ramadge P.J.*, *Caines P.E.*, A Globally Convergent Adaptive Predictor. – Automatica, 1981, Vol. 17, N 1, pp. 135–140.
213. *Goodwin G.C.*, *Sin K.S.*, Stochastic Adaptive Control. The general delay-coloured noise case. – Techn. Report N 7904. The Univ. of Newcastle, New South Wales, Australia. March, 1979.
214. *Goodwin G.C.*, *Sin K.S.*, *Saluja K.K.*, Stochastic Adaptive Control and Prediction : The General Delay-coloured Noise Case. – IEEE Trans. Aut. Contr., 1980, Vol. AC-25, N 5, pp. 946-950.
215. *Gustavsson I.*, *Ljung L.*, *Söderström T.*, Identification of Processes in Closed Loop Identifiability and Accuracy Aspects. – Automatica, 1977, Vol. 13, N 1, pp. 59–75.
216. *Jazwinski A.N.*, Stochastic Processes and Filtering Theory. New York, London, 1970.
217. *Kalman R.E.*, Liapunov Functions for the Problem of Lur'e in Automatic Control. – Proc. Nat. Acad. Sci. USA, 1963, Vol. 49, pp. 201–205.
218. *Kalman R.E.*, New Approach to Linear Filtering and Prediction Problems.–J. Basic Eng. Amer. Soc. Math. and Eng., ser. D, 1960, Vol. 82, N 1, pp. 35–45.
219. *Kalman R.E.*, *Bucy R.S.*, New Result in Linear Filtering and Prediction Theory.–J. Basic Eng. Amer. Soc. Math. and Eng., Ser. D, 1961, Vol. 83, N 1, pp. 95–108.
220. *Lindorff D.P.*, *Carrol R.L.*, Survey of Adaptive Control using Liapunov Design. – Int. J. Contr., 1973, Vol. 18, N 5, pp. 897–914.
221. *Ljung L.*, Consistency of the Least-squares Identification Method. – IEEE Trans. Aut. Contr. 1976, Vol. AC-21, N 5, pp. 779-881.
222. *Ljung L.*, Analysis of Recursive Stochastic Algorithms. – IEEE Trans. Aut. Contr., 1977, Vol. AC-22, pp. 551–575.

223. *Ljung L.*, On Positive Real Transfer Functions and the Convergence of Some Recursive Schemes. – IEEE Trans. Aut. Contr., 1977, Vol. AC-22, N 4, pp. 539–551.

224. *Ljung L.*, Convergence Analysis of Parametric Identification Methods. – IEEE Trans. Aut. Contr., 1978, Vol. AC-23, N 5, pp. 770–783.

225. *Ljung L.*, The ODE Approach to the Analysis of Adaptive Control Systems – Possibilities and Limitations. – Report, LiTH-ISY-I-0371. Dept. of Electr. Eng., Linköping University, Sweden, 1980. p. 9.

226. *Ljung L.*, System Identification. – Internal Report, LiTH-ISY-I-0457, Dept. of Electr. Eng., Linköping University, Sweden, 1981. p. 18.

227. *Ljung L.*, and *Söderström T.*, Theory and Practice of Recursive Identification. MIT Press. Cambridge, 1983. p. 529.

228. *Monopoli R.V.*, The Kalman – Yakubovich Lemma in Adaptive Control System Design. – IEEE Trans. Aut. Contr., 1973, Vol. AC-18, N 5, pp. 527–529.

229. *Narendra K.S. and Valavani L.S.*, Direct and Indirect Adaptive Control. – Automatica, 1979, Vol. 15, N 6, pp. 653–664.

230. *Parks P.C.*, Lyapunov Redesign of Model Reference Adaptive Control Systems. – IEEE Aut. Contr., 1966, Vol. AC-11, N 3, pp. 362–367.

231. *Saridis G.N.*, *Lobbia R.N.*, Parameter Identification and Control of Linear Discrete-time Systems. – IEEE Trans. Aut. Contr., 1972, Vol. AC-17, N 1 p. 52–60.

232. *Saridis G.N.*, *Stein G.*, A New Algorithm for Linear System Identification. – IEEE Trans. Aut. Contr., 1968, Vol. AC-13, p. 592–594.

233. *Söderström T.*, Convergence Properties of the Generalized Least Squares Identification Method. – Automatica, 1974, Vol. 10, N 6, pp. 617–626.

234. *Solo V.*, The Convergence of AML. – IEEE Trans. Aut. Contr., 1979, Vol. AC-24, N 6, pp. 958–962.

235. *Trulsson E.*, *Ljung L.*, Direct Minimization Methods for Adaptive Control on Non-minimal Phase System. – Internal Report, LiTH-ISY-I-0469, Dept of Electr. Eng., Linköping University, Sweden, 1981. p. 24.

236. *Tsypkin Ya. Z.*, *Aved'jan E.D.*, *Gulinskii O.V.*, On Convergence of the Recursive Identification Algorithm.–IEEE Trans. Aut. Contr., Vol. AC-26, 1981, N 5, p. 1009–1017.

237. *Wiener N.*, Extrapolation, Interpolation and Smoothing of Stationary Time Series. Cambridge, 1949, p. 243.

238. *Willems J.C.*, Least Square Stationary Optimal Control and the Algebraic Riccati Equation. – IEEE Trans. Aut. Contr., 1971, Vol. AC-26, N 6, pp.621–634.

239. *Wittenmark B.*, Stochastic Adaptive Control Methods : A Survey. – Int. J. Contr., 1975, Vol. 21, N 5, pp. 705–730.

240. *Youla D. S.*, On the Factorization of Rational Matrices. – IRE Trans. 1961, IT-7 N 3, pp. 172–189.

OPERATORS AND NOTATIONAL CONVENTIONS

$\boldsymbol{R} = \boldsymbol{R}^1$	field of real numbers
$\boldsymbol{R}^n$	Eucledean n-dimensional space
$\boldsymbol{X}, \boldsymbol{Y}, \boldsymbol{U}$ etc.	subsets of Eucledean spaces
$\boldsymbol{A}^T$	transpose of the matrix A
$\boldsymbol{A}^*$	complex-conjugate of transpose of the matrix A
A^{-1}	inverse of the matrix A
$\lvert x \rvert = (x^* x)^{1/2}$	Eucledean norm of the vector x
$\lVert A, B \rVert, \left\lVert \begin{matrix} A \\ B \end{matrix} \right\rVert$	aggregate matrices with the constituents A,B (with conformed dimensions)
$\lvert \lambda \rvert$	absolute value of the complex number λ
$\Pi(\lambda)$	matrix function of the complex argument λ
$\Pi^\nabla(\lambda)$	$\Pi^{-1}(\lambda^{-1})$
$\boldsymbol{P}\, B$	probability of the event B
$\boldsymbol{M}\, \zeta$	mathematical expectation (or mean) of the random variable ζ
$M\{\zeta \mid \mathrm{A}\}$	conditional expectation (for σ-algebra $\boldsymbol{A}$) of the random variable ζ
$M\{\zeta \mid x, y, \ldots\}$	the same but A produced by the random variables x,y,...
$\oint d\lambda$	contour integral along to the unit circumference ($\lvert \lambda \rvert = 1$) of complex plane, $\oint \frac{d\lambda}{\lambda} = 2\pi i, \quad i = \sqrt{-1}$
U^a	a set of admissible strategies
U^r	a set of realizable strategies

$x_s^t = (x_s, x_{s+1}, \ldots, x_t), \quad t \geq S$

$U_s^t(\cdot) = (U_s(\cdot), U_{s+1}(\cdot), \ldots, U_t(\cdot)), \quad t \geq S$

$U^\infty(\cdot) = (U_0(\cdot) \ldots,)$	control strategy (for infinite control time)
$J\left[U_s^t(\cdot)\right], J[U^\infty(\cdot)]$	objective (cost) functionals
$q(x, u)$	quadratic positive form of vector variables x,u
$\overline{\lim}$	Supper limit
Sp A	the sum of the diagonal elements of the matrix A
$\operatorname{argumin}_x$ f(x)	value of x that minimizes f(x)

$A{:}R^n \to R^n \;\&\; A > 0 \Leftrightarrow x^T A x > 0 \; (\forall x \in R^n, x \neq 0)$

∇	one step backward operation (backward shift), $\nabla y_t = y_{t-1}$, etc.

I_n	$n \times n$ unit matrix $I_n x = x, \forall x \in R^n$,
$D_1 = \{\lambda : \lvert\lambda\rvert \leq 1\}$	the unit disk (circle)
$P(\cdot) \sim N(a_1 R) \Leftrightarrow$	p(.) is Gaussian density of probability (with the mean a and covariancy R)
$x \in X$	the vector x belongs to the set X
$x \overline{\in} X$	the vectore x does not belong to the set X
ARMA =	Auto Regressive Moving Average
LSM =	Least Square Method
RLS =	Recursive Least Square
F.R.F. =	Fractionally Rational Function (or matrix function with such elements)
GTO =	Generalized Tunable Object
G.C.D. =	Greatest Common Divisor
deg =	degree (index) of a polynomial
def =	determinant of matrix
w.r.p. =	with respect to

SUBJECT INDEX

K

L

M

N

O

P

Q

R

S